18th International Conference "Aviation and Cosmonautics" (AviaSpace-2019)

IOP Conference Series: Materials Science and Engineering
Volume 868

Moscow, Russia
18 - 22 November 2019

Editors:

Dean Vucinic
Anna Zykova
Peter Ogbuna Offor

ISBN: 978-1-7138-3174-7
ISSN: 1757-8981

Printed from e-media with permission by:

Curran Associates, Inc.
57 Morehouse Lane
Red Hook, NY 12571

Some format issues inherent in the e-media version may also appear in this print version.

Institute of Physics
Dirac House, Temple Back
Bristol BS1 6BE UK

Phone: 44 1 17 929 7481
Fax: 44 1 17 920 0979

techtracking@iop.org

Additional copies of this publication are available from:

Curran Associates, Inc.
57 Morehouse Lane
Red Hook, NY 12571 USA
Phone: 845-758-0400
Fax: 845-758-2633
Email: curran@proceedings.com
Web: www.proceedings.com

TABLE OF CONTENTS

IOP Conf. Series: Materials Science and Engineering **868** (2020) 011001 doi:10.1088/1757-899X/868/1/011001

Preface
18th International Conference "Aviation and Cosmonautics" (AviaSpace-2019)

The 18th International Conference "Aviation and Cosmonautics" (AviaSpace-2019) was held by Moscow Aviation Institute (National Research University) on 18-22 November, 2019 in Moscow, Russia. The Program Committee was presented by scientists from Nanyang Technological University (Singapore), University of San Diego and Massachusetts Institute of Technology (USA), Polytechnic University of Milan (Italy) and Central Aerohydrodynamic Institute (Russia).

The conference was a great assembly of specialists, scientists and experts from enterprises and organizations of the aerospace and other hi-tech industries, researchers, Ph.D. students from higher educational institutes who gathered to share their research findings and wealth of experiences in the following scientific sessions: (a) Artificial intelligence for the aerospace industry; (b) Aviation and cosmonautics; (b) Design of morphing surfaces: structural and technological issues; (c) Multi objective optimization of the dynamic behavior in robotics; (d) Modern challenges of aircraft development and some ways of their realization; (e) Multi-objective optimization approach covering two flight conditions for S-Duct inlets design; (f) Up-to-date status of International Nuclear Power and Propulsion System flagship (INPPS) and future high power space transportation; (g) Design of a rotor supported by aerodynamic bearings suitable for achieving high circumferential speed; and (h) Assessment and demonstration of the fire testing compliance for Civil Propulsion System.

AviaSpace-2019 offered opportunity for interaction of engineers, scientists, academicians and young researchers, which enabled the participants to discuss and exchange ideas on development of aviation and space industry as one of the world priorities and to find solutions to current problems and challenges facing the aviation and space industries and experience and knowledge exchange between the scientific and industrial enterprises and universities of the world.

The Organizing Committee sincerely appreciate the different bodies and individuals who worked hard to ensure the spectacular success of the important event. All manuscripts were analyzed according to the acceptable criteria as novelty of the presented work, reasonable and detailed description of the experimental procedure, understandable interpretation of the presented data, significant and evident proves to verify the proposed hypothesis, the absence of plagiarism, adequacy of references, the quality of language by eminent specialists in different scientific subject areas in Nigeria, China, Russia, USA, France, Croatia, South Africa, Italy, Singapore, Egypt, United Kingdom, Ukraine, India, Spain, Brazil, Japan, Malaysia, Netherlands, Belgium, and Poland. The Editor is thankful to all the reviewers, whose devotedness boosted the high standard and quality of the research papers.

AviaSpace-2019 was supported by the Russian Foundation for Basic Research (project 19-08-20073).

Organizing Committee:

Prof. Yury Ravikovich,
Moscow Aviation Institute, Russia

Editors:
Prof. Dean Vučinić, Vrije University, Brussels, Belgium
Dr. Anna Zykova, Institute of Strength Physics and Materials Science of Siberian Branch of Russian Academy of Sciences, Russia
Dr. Peter Ogbuna Offor, University of Nigeria, Nsukka, Nigeria

IOP Conf. Series: Materials Science and Engineering **868** (2020) 011002 doi:10.1088/1757-899X/868/1/011002

Peer review statement

All papers published in this volume of *IOP Conference Series: Materials Science and Engineering* have been peer reviewed through processes administered by the proceedings Editors. Reviews were conducted by expert referees to the professional and scientific standards expected of a proceedings journal published by IOP Publishing.

Optimization of series-parallel connection of PV array to mitigate negative influence of partial shading conditions

A Kurmanbay*, K Baktybekov, K Sakhanov, A Syzdykov and A Mukhamediyev

Research and Development Department, *Kazakhstan Garysh Sapary,* 89 Turan Street, Astana 010000, Republic of Kazakhstan

*Email: a.kurmanbay.sh@gmail.com

Abstract. Rapid development in the photovoltaic systems demanding more and more efficient solutions, not only in micro-fabrication techniques but in energy extraction methods as well. In recent decades large number of Maximum Power Point Tracking algorithms with different complexity accompanied by the capability to efficiently locate global maximum under partial shading were developed. However, a deeper look at the problem reveals that to further enhance the performance of the PV system under non-uniform irradiance there is a need for hardware intervention to the interconnect, or in other words adjustments to the panel configuration. In this paper, we propose a simple technique for optimizing the electrical scheme of the solar panel (connection of individual cells) by using a simple algorithm which disperses shadow across all structure. The method allows to mitigate partial shading effect on solar panels of any size and shape, therefore rise efficiency of energy extraction. Thus static optimization of the PV panel configuration disperses partial shading across all structure and ensures more synchronous functioning of the PV cells. The resulting structure not only capable of extracting more available power but, straighten ups power curve to loosen requirements for MPPT computations.

1. Introduction

Ever-increasing energy demand of the world needs to be fulfilled with renewable sources to save valuable resources of the planet. Solar energy is one of the means of achieving it, which posses distinctive advantages over competing alternatives: low maintenance cost, long lifetime, simplicity of manufacturing as well as the absence of moving parts. However, several challenges are preventing PV systems to be the energy of the future. One of the crucial factor which diminish output power of PV array is partial shading [1-6]. Dust collected on panel, bird droppings, passing clouds, and other standing objects create a barrier through which sunlight cannot pass. It was observed that under thus circumstances the performance of the PV system diminishes considerably and harvesting power becomes very difficult. Furthermore, shaded cells result in the creation of hot spots that deteriorate panel and reduces its lifetime. PV panel manufacturers often use bypass diodes to combat this issue. However, a completely efficient solution that minimizes the effect of partial shading is yet to come. Photovoltaic plants for industrial use are located in places where environmental conditions are ideal for harvesting sunlight. Nonetheless, panels for private use could be placed in areas where partial shading is unavoidable. Therefore, the issue of non-uniform solar irradiance is one of the major concerns in solar industry. One of the trick to pass around the destructive effect of partial shading is to optimize the electrical interconnect of the PV array. In this paper, we propose a method to construct a partial shading resilient series-parallel configuration of PV array.

1.1. PV system decomposition

The theoretical limit of extracting energy from the PV cell is primarily set by material property (bandgap of semiconductor) [7]. However, in the PV system, not all that energy could be used due to non-ideality in power extraction techniques. The efficiency of the solar panel is further reduced under partial shading conditions. Orthoganilzation of PV system into distinct modules, reveal several optimization options to boost performance under non-uniform solar irradiance [2]:

1. Development of MPPT algorithms able to locate the global maximum.
2. Modification of the PV array electrical scheme.
3. Alteration in the PV system architecture.
4. Reconfiguration of converter topology.

In the past decades, MPPT received more and more attention, and numerous algorithms with different complexity able to track global maximum under partial shading conditions were developed [8-10]. Recently, in literature, a rising interest in the optimization of the electrical interconnect of the PV array could be observed. The most common methods of schematic reconfiguration are presented below:

1. Static optimization of the PV array scheme – Serial Parallel (SP), Series (S), Total Cross Tied (TCT) configurations have different resilience toward partial shading [11-13]. Depending on the geometry of the shade, their performance varies. However, such configurations have little to no resistance to partial shading. Later, studies show that Sudoku [14-15] or Magic square formation [16] hold higher resilience to common patterns of shade.

2. Dynamic optimization of the PV array scheme – is the method of extracting maximum power based on the reconfiguration of the electrical scheme using switches [17-19]. Such a method allows to adaptively alter the configuration of the PV array based on the solar irradiance, therefore, to reduce mismatch effects actively. Among listed hardware solutions, the dynamic method offers a higher energy yield but requires complex hardware intervention and advanced control. Thus modifications would result in higher unit price as well as a reduced lifetime of the device due to a reduction in reliability.

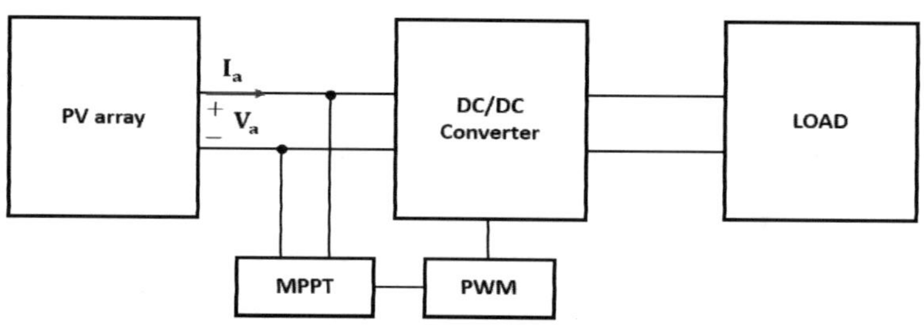

Figure 1. In this case simply justify the caption so that it is as the same width as the graphic.

In this paper, we propose a method of reconfiguring the electrical interconnect of the SP PV array (figure 1). By the offered technique, an array of any size and shape could be restructured to make it more resilient to partial shading (by static optimization). The rearranged structure relaxes requirements for the MPPT algorithm as it smooths the power curve preventing generation of numerous local maximums. Other modules of the PV system will not be touched. However, the derived optimization procedure for interconnect is independent of other modules, and resulting PV arrays could be integrated into any PV system.

IOP Conf. Series: Materials Science and Engineering **868** (2020) 012001 doi:10.1088/1757-899X/868/1/012001

1.2. Model of a PV cell

The starting point of our analysis is a basic building block of the PV system - PV cell. Several mathematical models for understanding the electrical behavior of PV cells exists. In this paper, we focused on the widely accepted PV model as shown in figure 2. More complex models [20] considering recombination current and correction factors to shunt resistance, as well as a model constructed by artificial neural network [21] were also studied and could be used for the simulation with minor changes. However, for the optimization of the electrical interconnect, it would be more important to see possible outcomes of the partial shading to the PV system and pick algorithm accordingly rather than putting efforts on determining the most accurate model. Therefore, the focus was kept on a simpler model. The performance of the PV cell was simulated on MATLAB and results of single diode solutions could be seen from figure 3.

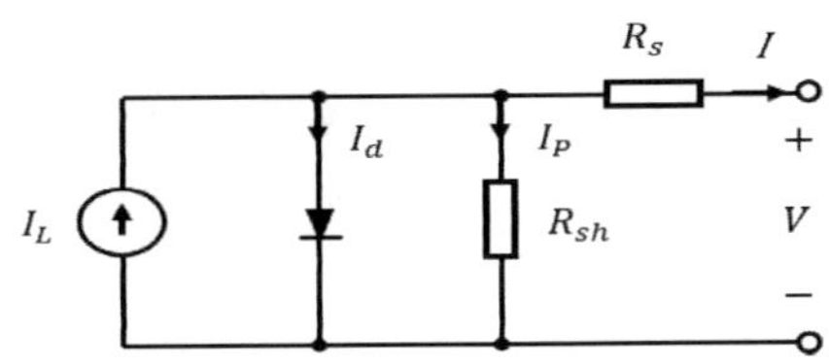

Figure 2. PV cell electrical scheme.

$$I = I_L - I_d - I_P \tag{1}$$

where I – output current of the cell (A); I_L – current generated by solar irradiance (A); I_d – current passing through the diode (A); I_P – current dissipated in shunt resistor (A).

Rewriting equation (1) would give:

$$I = I_L + I_0\left[e^{\frac{q(V+IR_S)}{nkT}} - 1\right] - \frac{V+IR_s}{R_{sh}} \tag{2}$$

where $q = 1.602{\cdot}10^{-19}C$ (C) (electronic charge); $k = 1.380{\cdot}10^{-23}$ (J/K) (Boltzmann constant); n – is ideality factor; I_o – is diode reverse saturation current (A); R_{sh} – shunt resistance (Ω); R_s – series resistance (Ω).

Current generated by the solar irradiance was taken as shown in [22].

$$I_L = \frac{G}{G_{ref}}[\,I_{L,ref} + \mu_{I,sc}\left(T_c - T_{c,ref}\right)] \tag{3}$$

where G – solar irradiance (W/m^2); T - temperature (°C); $\mu_{I,sc}$ – temperature coefficient of short circuit current (A).

If the reference parameters of the solar panel are not known, the simulation could be performed by parameter estimation methods based on the data from the PV manufacturers (models constructed empirically) [20-24]. However, it is not always possible to do so as not all manufacturers provide the necessary information. Therefore, single diode computation was performed on imaginary PV cell table 1. Simulation results for various temperature and irradiance ranges are presented in figure 3.

Table 1. Single diode reference parameters at standard reference conditions (SRC: T = 25 °C and G = 1000 W/m^2).

Parameters	Values (explanation)
$I_{L,ref}$	7.53 A
$I_{0,ref}$	82.689 nA
$E_{g,ref}$	1.121 eV *(Band gap of semiconductor)*
$R_{s,ref}$	0.094 Ω
$R_{sh,ref}$	15.72 Ω
$\mu_{I,sc}$	3 mA/°C

From a single diode numerical computations, the influence of environmental factors on the PV cell could be observed. This dependence changes over time due to the degradation of materials, therefore, altering the property. However, data is key in the further analysis as operation range under different temperatures and solar irradiance are very important information of efficient array structure. From figure 3 it could be already observed how non-uniform temperature or irradiance might disrupt synchronous functioning of the PV array. Furthermore, from Figure 3c significance of bypass diodes becomes obvious, as shaded cells will block all current in the string [25].

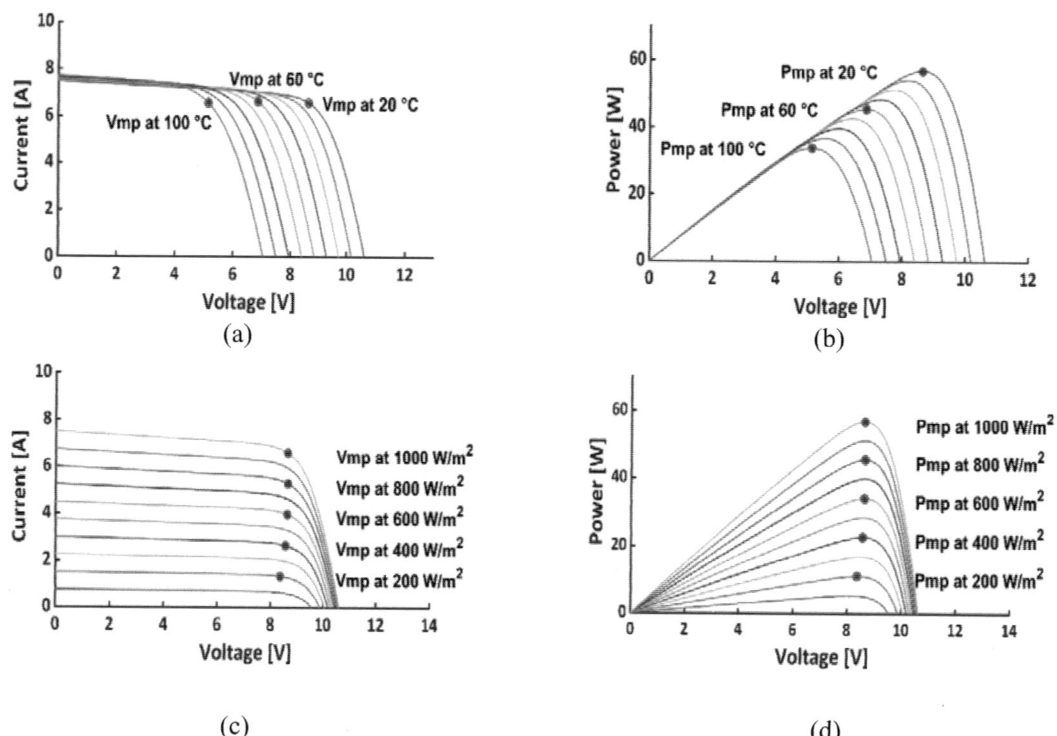

Figure 3. I-V and P-V characteristics of PV cell at various temperature and irradiance ranges: (a) I-V under different temperatures (1000 W/m^2); (b) P-V under different temperatures (1000 W/m^2); (c) I-V under different solar irradiance (20 °C); (d) P-V under different solar irradiance (20 °C).

1.3. PV array

In the previous section, electrical response of the PV cell to the changes in irradiance as well as to the temperature was presented. As a next step, the electrical performance of the PV array at a series-parallel (SP) configuration was simulated. Computation was conducted on (20×20) PV arrangement (figure 4). In other words, 20 elements connected in series were connected in parallel to the other 20 modules of the same size. Solar irradiance was inputted separately to each cell. In principle performance of an array of any dimension could be represented by the following procedure.

IOP Conf. Series: Materials Science and Engineering **868** (2020) 012001 doi:10.1088/1757-899X/868/1/012001

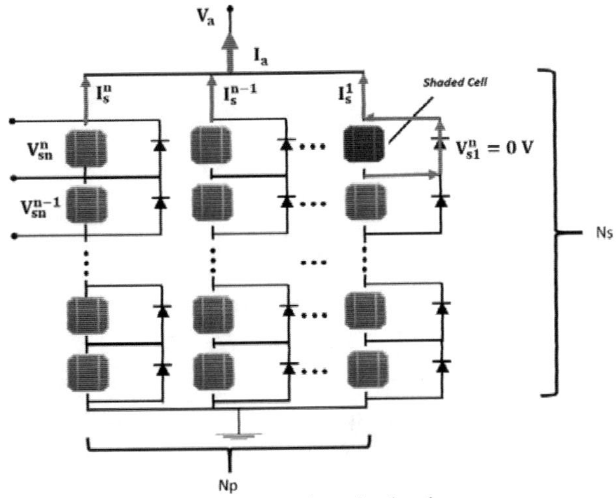

Figure 4. SP electrical scheme.

Sum of voltage drop across the cells connected in series is equal to the output voltage (Kirchhoff's voltage law):

$$V_a = \sum_{i=1}^{n} V_s^i \tag{4}$$

where V_a – output voltage of the PV array (V); V^i – voltage drop across the functioning cell (V).
According to the Kirchhoff's current law, current at the output is sum of all currents generated in each series connection (module):

$$I_a = \sum_{i=1}^{n} I_s^i \tag{5}$$

where I_a – output current of the PV array (A); I_i – current passing through the cells connected in series (A).

Using the data from the PV cell model and applying it to the scheme in Figure 4 several test scenarios of partial shading were simulated figure 5. The results are presented in figures 6-8. Depending on the shape of the shade, P-V curves change dramatically.

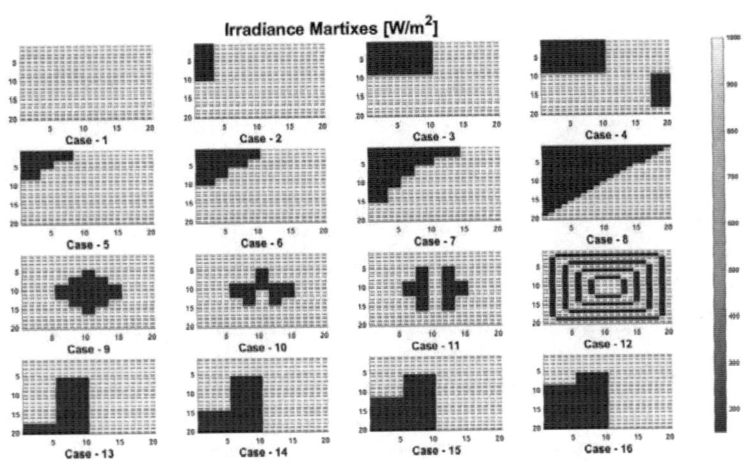

Figure 5. Example irradiances inputed to the PV array.

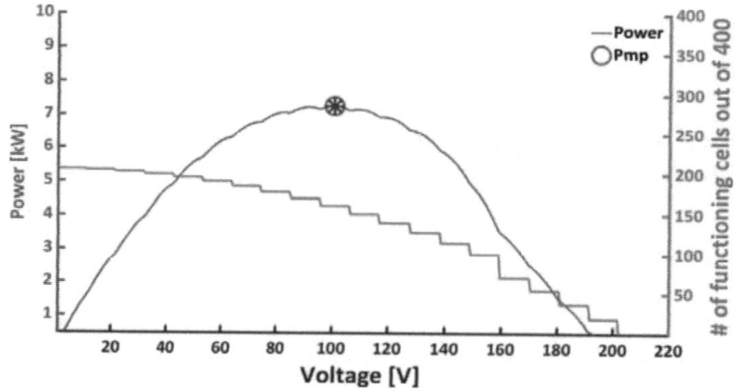

Figure 6. P-V curve along with number of working cells for case 8.

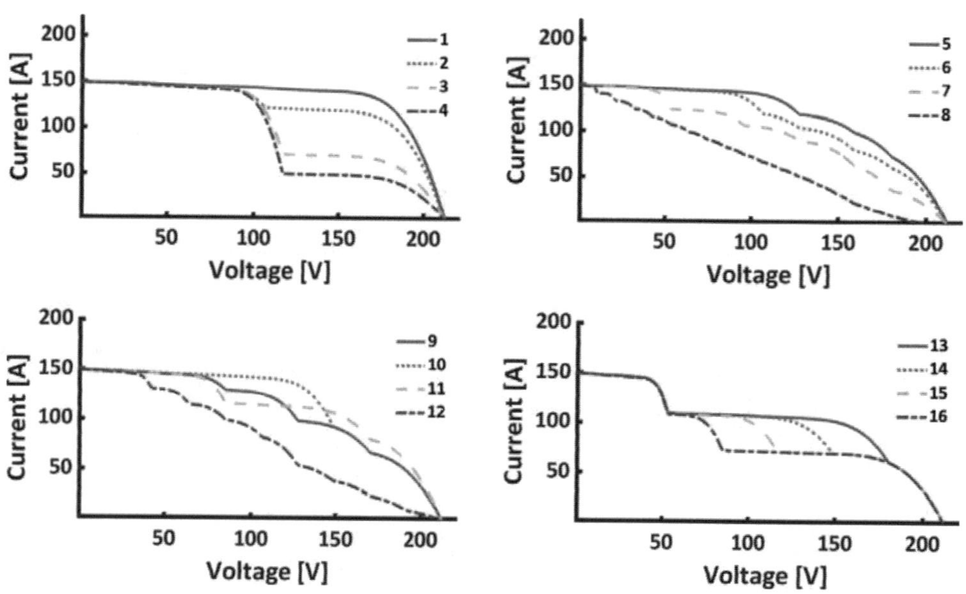

Figure 7. Estimated I– V curve for the irradiance cases shown in figure 5 before rearrangements.

Simulation Assumptions:

1. Voltage drop across the functioning cells connected in series are same [26].
2. Temperature across the PV array is uniform. (Temperature of each cell could be specified as well.

However, to keep it simpler, it was decided to keep uniform temperature (20 °C) distribution across the panel.)

Partial shading of the PV panel regions prevents to harvest all the available power. Such an event is taking place because of shaded modules (strings of cells connected in series) that go out from the operation range earlier. Thus mismatched functioning could be seen from figure 6. The SP configuration is not able to extract all the power produced by the PV array.

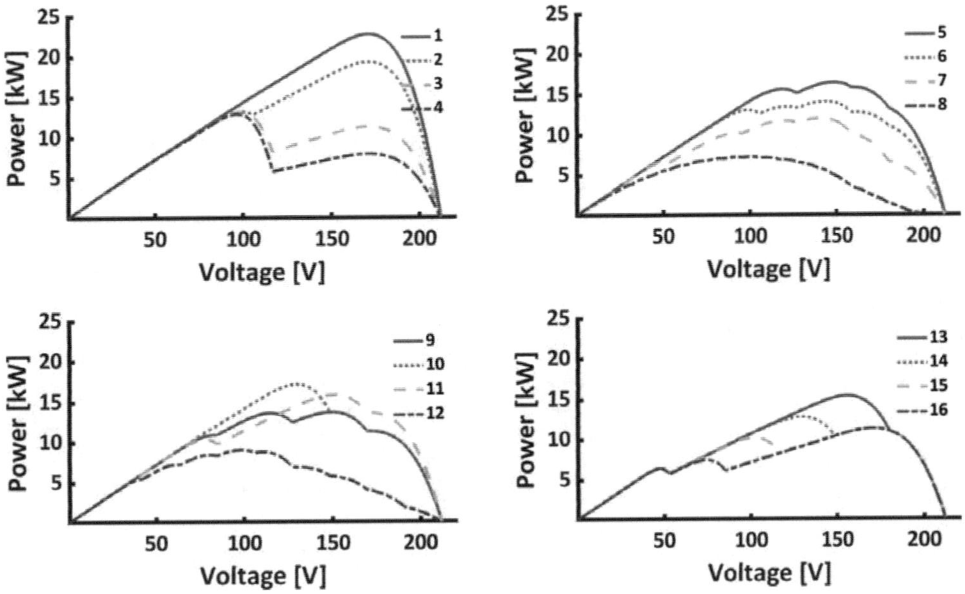

Figure 8. Estimated P–V curve for the irradiance cases shown in figure 5 before rearrangements.

2. Optimization and simulations

2.1. Problem formulation

The problem was simplified for simulation so that each PV cell represents an element in a matrix. A procedure that repositions elements (permanently) to minimize mismatch effects, needs to be derived (to prevent the early reaching of open-circuit voltage due to shaded cells). In the SP configuration, to mitigate the negative effect of shaded cells, the array needs to be dispersed in a way that each module (a string of series-connected cells) contain an equal number of shaded elements. To achieve mismatch reduction cells were relocated to the furthest distance relative to original nearest neighbors. In other words, the distance between original neighbors is the controlling parameter, which needs to be maximized for all elements in a uniform manner. This would ensure spread of all elements across the panel surface and any hypothetical shaded region will include elements from all the matrix (maximum diversity). Bluntly searching for a whole state space would require an analysis of $(n \times n)!$ combinations which is not feasible and will require computational time more than the age of the universe [27]. The combinatorics problem formulated in this paper although with some differences resembles the maximum diversity problem. State of the art algorithms for solving it includes greedy randomized adaptive search procedure (GRASP) [28], as well as the implementation of memetic and tabu search algorithm [29], etc. Any of the mentioned heuristics methods cannot be directly applied as nearly all problems need some sort of customization of the algorithm [30]. We attempted to find close to the optimal solution by setting manageable constraints and designing a very simple hybrid evolutionary algorithm (EA) with the ability of local search to rearrange shade across all circuitry. Such a class of EA often called memetic algorithms [30]. The choice fell on the memetic algorithm due to its straightforward structure, yet powerful local search ability.

For the simplicity of simulation and to not deviate from previous sections all the computations were performed on $n \times n$ ($n = 20$) matrix. Number of elements in the matrix is $N = n^2 = 400$. For each element i (from 1 to N) there are set of neighbors located (d) distance away.

$$A_i^d = \{a_{i,j}\}, \qquad 1 \le d \le (n-1), \qquad 1 \le i \le N, \qquad n, N \in \mathbf{Z} \tag{6}$$

where d (Chebychev distance) could be found from standard (Euclidean) coordinates by the equation (7) and j is size of the neighbor set.

$$d(x, y) := max_i(|x_i - y_i|). \tag{7}$$

For instance, $A_i^1 = \{a_{i,1}^S, a_{i,2}^N, a_{i,3}^E, a_{i,4}^W, a_{i,5}^{SE}, a_{i,6}^{SW}, a_{i,7}^{NE}, a_{i,8}^{NW}\}$ – (figure 9a) is set of nearest neighbors or elements located one Chebychev distance away in the original matrix. Chebychev space was used for the convenience, while other coordinate systems are suitable as well. If the shape of the PV cell is not square (*hexagonal,triangular ... etc.*) or if there is a non PV element in the panel, sets would change but optimization algorithm should still be able to solve it.

For each element expected distance between its nearest neighbors is described by equation (8)

$$\mu_i = \sum_{j=1}^{8} (\tfrac{1}{8}) d_{i,j} \tag{8}$$

While mean distance between original neighbors for whole matrix is:

$$\mu = \frac{1}{N} \sum_{i=1}^{N} \sum_{j=1}^{8} (\tfrac{1}{8}) d_{i,j} = \frac{1}{N} \sum_{i=1}^{N} \mu_i \tag{9}$$

Our goal is to maximize the equation (9) keeping uniformity of equation (8). Condition for uniform spread of matrix elements is dictated by spread of distance (standard deviation) which needs to be minimized:

$$\sigma = \sqrt{\frac{\sum_{i=1}^{N} (\mu_i - \mu)^2}{N}} \tag{10}$$

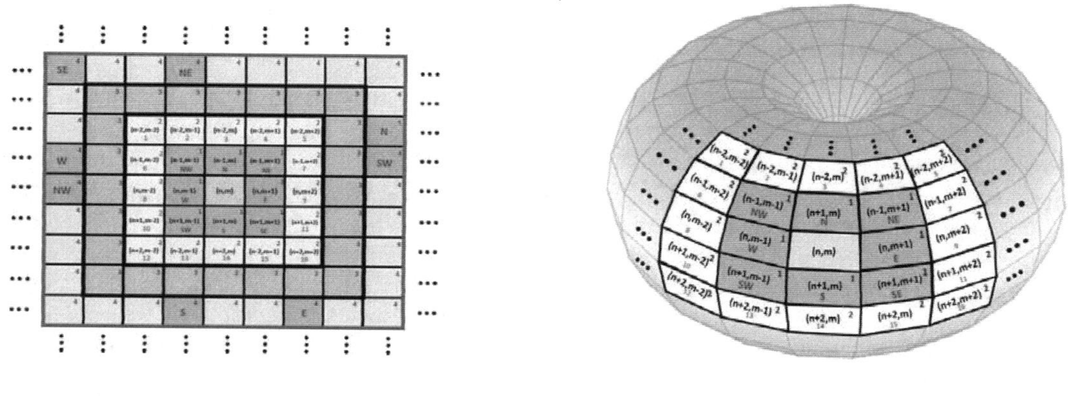

(a) (b)

Figure 9: Visualization of set operation and coordinate system which was used: (a) visual explanation of Equation 15 (for $d1 = 4$); (b) simulation was done on toroid.

Furthermore, additional condition for uniformity is minimum distance below which no neighbor should be located:

$$\{B_i^{(1,2\ldots m)} \mid b \notin A_i^1\}, \quad 1 \le i \le N, \quad m, N \in \mathbf{Z} \tag{11}$$

In other words, to keep uniform dispersion of elements from each other we set a minimum distance m and make sure that no neighbors are located closer than the value set. In order for search to be successful more elements need to satisfy the condition in equation (11) and ensure decrease in spread as well as increase in mean distance equation (12).

$$\sigma' \le \sigma \qquad \mu' \ge \mu \tag{12}$$

Here σ' and μ' is solution offered at each iteration of the algorithm. To avoid boundary conditions of non equal number of neighbors computation were performed on toroid (figure 9) and results were transformed back to original coordinates.

2.2. Perturbation schedule

The problem formulation Equation (8-9) resembles maximum diversity problem. However, the difference and the major challenge which arises is each time element is relocated it has an effect on all the matrix, reshaping the whole structure creating an immense amount of possible combinations. To tackle this issue we attempted to narrow down the search by segmenting perturbation schedule. Sets described equation (6) mimics orbits of each element and different repelling strategies are used in each stage of perturbation (table 2).

Table 2. Algorithm structure.

Algorithm
1: Generate Initial solution *X*
2: Generate a neighbor solution around *X* with perturbation schedule. Perturbation rules:
Stage 1:
Perturb according to Equation (13-15).
Stage 2:
Perturb according to Equation (16-18).
3: Evaluate: If the neighbor solution is better - update *X*.
4: Return to step 2 until termination criteria or maximum number of iterations is reached.
5: End

Stage 1: In this stage, all the matrix elements repeal nearest neighbors in a certain range to the further distance. In other words, all the closest elements in the original matrix which are currently in the range of m are randomly redistributed through ought the matrix ($\forall i \in \{1,2,\dots N\}$) equation (13-15) is performed randomly selecting i till it is not performed for all elements). Using such a perturbation rule can reshape the solution significantly in a single iteration. The result is accepted only if all conditions set in equation (11-12) are satisfied therefore ensuring the formation of desired final topology. Structures formed in this stage are subject to change and as better solutions are found and more elements are located further than mdistance away the formations are less susceptible to change and the perturbation schedule is changed to the 2nd stage.

$$B_i^{(1,2\dots m)} = \{b_1, b_2, \dots b_j\} \quad b_j \in A_i^1 \tag{13}$$

$$B_i^{(d_1)} = \{b'_1, b'_2, \dots b'_j\} \quad b'_j \notin A_i^1 \tag{14}$$

$$B_i^{(1,2\dots m)} = \{B_i^{(1,2\dots m)} \backslash \{b_j\}\}, \quad B_i^{(d_1)} = \{B_i^{(d_1)} \backslash \{b'_j\}\},$$
$$append\ (\{b_j\} \text{ to } B_i^{(d_1)}), \ (\{b'_j\} \text{ to } B_i^{(1,2\dots m)}) \tag{15}$$

Stage 2: As long as no neighbors are left within m distance away search space is narrowed. Perturbation is continued only by selecting elements randomly and shifting nearest neighbors further away. Similar to previous stage solutions are accepted IFF requirements set in equation (11-12) are satisfied. As perturbation rule is applied to a single element and its neighbors at a time, each iteration takes significantly less time in this stage. Furthermore, the matrix cannot be reshaped significantly in single step anymore as search space is significantly narrowed.

$$B_i^{(m+1)} = \{b_1, b_2, \dots b_j\} \quad b_j \in A_i^1 \tag{16}$$

$$B_i^{(d_2)} = \{b'_1, b'_2, \dots b'_j\} \quad b'_j \notin A_i^1 \tag{17}$$

$$B_i^{(m+1)} = \{B_i^{(m+1)} \backslash \{b_j\}\}, \quad B_i^{(d_2)} = \{B_i^{(d_2)} \backslash \{b'_j\}\},$$
$$append\ (\{b_j\} \text{ to } B_i^{(d_2)}), \ (\{b'_j\} \text{ to } B_i^{(m+1)}) \tag{18}$$

The execution of the algorithm was done by the flow given in figure 10. d_1 and d_2 in equation (13-18) were chosen as {5, 6, 7} and { $m+2$, $m+3$} where $m \leq \frac{n}{2} - 3$. m is incremented as soon as all nearest neighbors are shifted to m distance away. Thus however doesn't necessarily fix values. We experimented with several d_1 and d_2 values and selected the ones which had desirable convergence speed and could produce enough diversity of solutions. With our setup of n = 20 matrices, the best result received in toroid was shifting all the neighbors to at least 5 Chebychev distance away and achieving $\mu = 7.64$, $\sigma = 1.48$. The result of the simulation could be seen from figure 11. The final structure after rearrangements resembles Sudoku arrangement as the probability of neighbors being in the same row, column, and diagonal is minimized. Thus any shaded region would contain elements throughout the whole matrix, due to dispersion of neighbors in a uniform manner.

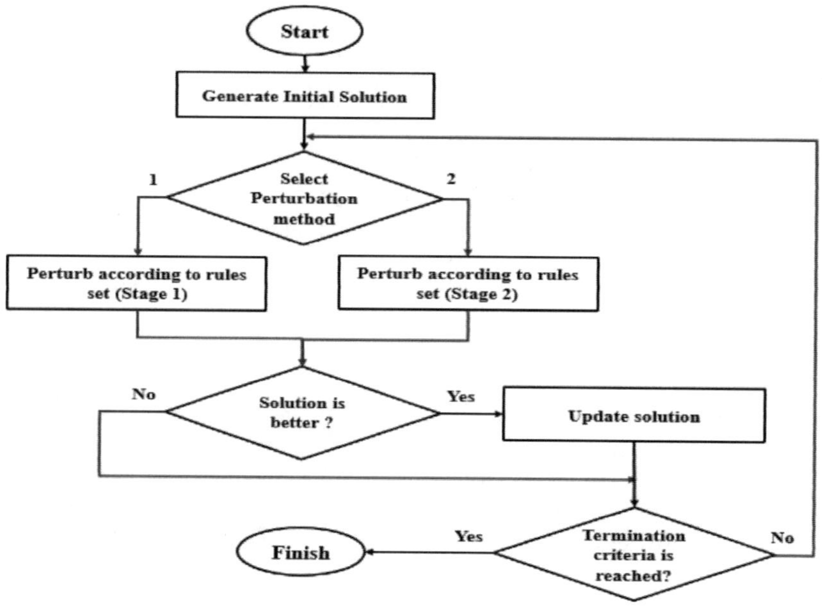

Figure 10. Computation is executed according to the flow shown above.

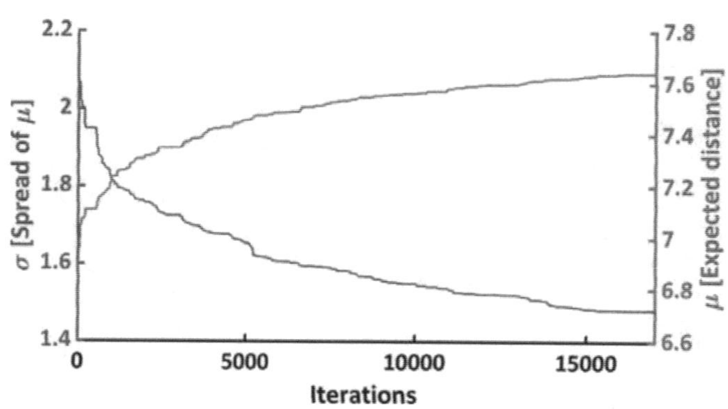

Figure 11. Simulations results.

IOP Conf. Series: Materials Science and Engineering **868** (2020) 012001 doi:10.1088/1757-899X/868/1/012001

3. Results and discussion

For different test case scenarios shown in figure 5, after rearrangements of PV cells confident increase in maximum power was detected. In real operating conditions it is impossible to predict the shape and sizeof the shadow. Therefore, different shadow patterns were experimented. It has to be noted that it is physically impossible to disperse some of the shadow patterns uniformly across the circuitry due to size limitations. Nevertheless, such a pattern also had to be tested to see the efficiency of the method. The results of simulations are shown in figures 12-13. In all cases, more synchronous functioning of the PV array was observed, and I-V, P-V curves had fewer peaks compared to the regular SP interconnect. Numerical results are shown in table 3.

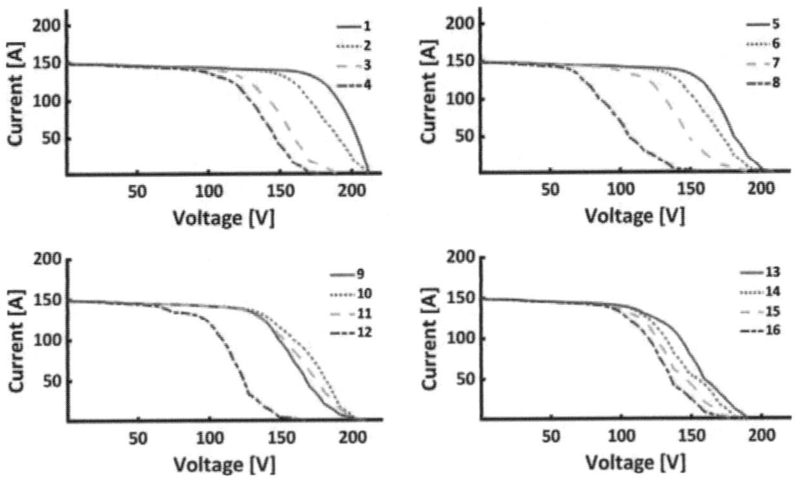

Figure 12. Estimated I–V curve for the irradiance cases shown in figure 5 after rearrangements.

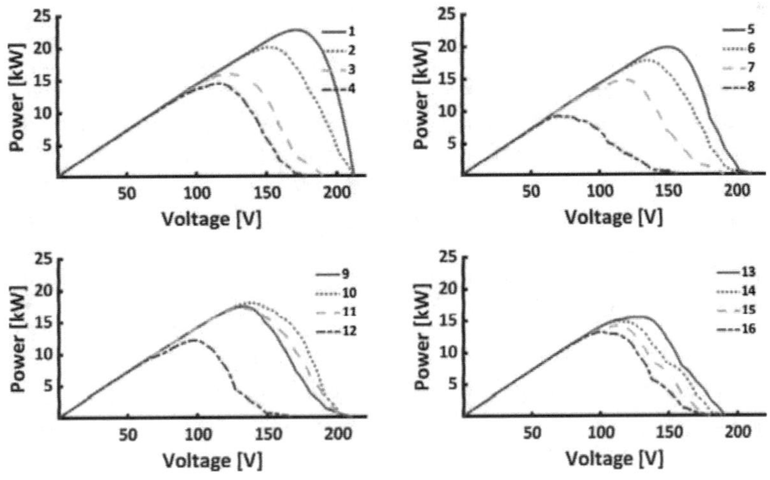

Figure 13. Estimated P – V curve for the irradiance cases shown in figure 5 after rearrangements.

Table 3. Maximum Power results for irradiance cases shown in Figure 5 after rearrangements.

Case	Original [kW]	After rearrangement [kW]
1.	22.74	22.74
2.	19.33	20.57
3.	13.21	16.01
4.	12.95	14.49
5.	16.48	19.82
6.	14.11	17.77
7.	12.05	14.80
8.	7.27	9.24
9.	13.75	17.45
10.	17.13	18.04
11.	15.85	17.23
12.	9.08	12.17
13.	15.42	15.46
14.	12.83	14.96
15.	11.37	14.14
16.	11.37	13.12

Adjusting positions of PV cells come with a substantial cost of complex wiring patterns. The issue could be partially compensated by the introduction of central connection unit circuitry to configure all the cells respectively. However, the same challenge arises for all the dynamic electrical configuration optimization techniques with an additional reduction in reliability [25] due to the installation of switches. Therefore, selecting an optimization technique should be done according to the priorities (cost, efficiency, fabrication complexity, etc.) set for the specific application.

4. Conclusion

In this paper optimization of an SP electrical configuration of PV cells was discussed. The developed method could be applied to a PV array of any shape and dimensions as there is no need for symmetry or size restrictions. Furthermore, if there are non-PV elements present in the panel method can take it into account by specifying initial boundary conditions. Although it is not an exact solution, it was observed that our heuristic approximation for simple shadow shapes increases power yield up to 20%, and up to 10 - 15% for more complex patterns. Numerical computations from a basic building block of PV array until the simulation of a whole panel were performed on MATLAB. Interconnect optimization takes place according to the algorithm which disperses shadow across all circuitry to mitigate the effect of electrical mismatch. The results show a promising increase in power harvesting and more synchronous functioning of PV cells under partial shading. Furthermore, the resulting power curve was simpler, therefore making it easier for the MPPT algorithm to find a global maximum.

References

[1] Patel H and Agarwal V 2008 Matlab-based modeling to study the effects of partial shading on PV array characteristics. *IEEE Transact. on Energy Convers.* **23(1)** doi: 10.1109/TEC.2007.914308
[2] Bidram A, Davoudi A D, and Balog R S 2012 Control and circuit techniques to mitigate partial shading effects in photovoltaic arrays. *IEEE Journal of Photovoltaics* **2(4)** doi: 10.1109/JPHOTOV.2012.2202879
[3] Karatepe, E Syafaruddin and Hiyama T 2010 Simple and high-efficiency photovoltaic system under

non- uniform operating conditions. *IET Renewable Power Generation* **4(4)** 354 doi:10.1049/iet-rpg.2009.0150

[4] Teo J C, Rodney H G, Vigna V R V, Mok H and Tan C 2018 Impact of partial shading on the p-v characteristics and the maximum power of a photovoltaic string. *Energies* **11** 1860 doi: 10.3390/en11071860

[5] Young-Hyok, J Jung D Kim J_G Kim J H Lee T W and Won C Y 2011 A real maximum power point tracking method for mismatching compensation in PV array under partially shaded conditions. *IEEE Transactions on Power Electronics* **26(4)** doi: 10.1109/TPEL.2010.2089537

[6] Maki A and Valkealahti S 2012 Power losses in long string and parallel-connected short strings of series- connected silicon-based photovoltaic modules due to partial shading conditions. *IEEE Transactions on Energy Conversion* **27(1)** doi: 10.1109/TEC.2011.2175928

[7] Markvart T 2016 From steam engine to solar cells: can thermodynamics guide the development of future generations of photovoltaics? *WIREs Energy and Environment* doi: 10.1002/wene.204

[8] Mao M Duan Q Zhang L Chen H, Hu B and Duan P 2017 Maximum power point tracking for cascaded PV -converter modules using two-stage particle swarm optimization. *Nature Scientific Reports* **7(1)** DOI: 10.1038/s41598-017-08009-7

[9] Seyedmahmoudian M Mekhilef S Rahmani R Yusof R and Shojaei A A 2014 Maximum powerpoint tracking of partial shaded photovoltaic array using an evolutionary algorithm: A particle swarm optimization technique. *J. of Renewable and Sustainable Energy* **6(2)** doi: 10.1063/1.4868025

[10] Pavlovic T and Ban Z 2013 An improvement of incremental conductance MPPT algorithm for PV systems based on the Nelder–Mead optimization. *IEEE* 6

[11] Kumar A Pachauri R K and Chauhan Y K 2016 Experimental analysis of SP TCT PV array configurations under partial shading conditions. *IEEE International Conference on Power Electronics, Intelligent Control and Energy Systems* (Delhi, India) doi: 10.1109/ICPEICES.2016.7853403

[12] El-Dein M Z S, Kazerani M and Salama M M A 2013 An optimal total cross tied interconnection for reducing mismatch losses in photovoltaic arrays. *IEEE Transactions on Sustainable Energy* **4(1)** doi: 10.1109/TSTE.2012.2202325

[13] Tian H, Mancilla-David F, Ellis K, Jenkins P and Muljadi E 2012 *A Detailed Performance Model for Photovoltaic Systems* (Alliance for Sustainable Energy, LLC)

[14] Rao P S, Dinesh P Ilango G S and Nagamani C 2015 Optimal SU-DO-KU based interconnection scheme for increased power output from PV array under partial shading conditions. *Frontiers in Energy* **9** 199 doi: 10.1007/s11708-015-0350-1

[15] Malathy S and Ramaprabha R 2020 Shade Resilient Total Cross Tied Configurations to Enhance Energy Yield of Photovoltaic Array Under Partial Shaded Conditions. *Lecture Notes on Data Engineering and Communications Technologies* vol 35 (Springer, Cham) doi: 10.1007/978-3-030-32150-5_13

[16] Reddy G, Reddy T and Kumar M 2018 Shadow dispersion of PV array under variable irradiance for superior power generation by magic square configuration. *International Journal of Engineering & Technology* **7(1)** 172 doi: 10.14419/ijet.v7i1.8.11543

[17] Velasco-Quesada G, Guinjoan-Gispert F, Piqué-López R, Román-Lumbreras M and Conesa-Roca A 2009 Electrical PV array reconfiguration strategy for energy extraction improvement in grid-connected PV systems. *IEEE Transactions on Industrial Electronics* **56(11)** doi: 10.1109/TIE.2009.2024664

[18] Nguyen D and Lehman B 2008 An adaptive solar photovoltaic array using model-based reconfiguration algorithm. *IEEE Transactions on Industrial Electronics* **55(7)** doi: 10.1109/TIE.2008.924169

[19] Chao K H, Lai P L and Liao B J 2016 The optimal configuration of photovoltaic module arrays based on adaptive switching controls. *Energy Conversion and Management* **100(2015)** 157 doi: 10.1016/j.enconman.2015.04.080
[20] Sauer K J, Roessler T and Hansen C W 2015 Modeling the irradiance and temperature dependence of photovoltaic modules in PVsyst. *IEEE Journal of Photovoltaics* **5(1)** 152 doi: 10.1109/JPHOTOV.2014.2364133
[21] Karatepe E, Boztepe M and Colak M 2005 Neural network based solar cell model. *Energy Conversion and Management* **47** 1159
[22] Deceased J A and Beckman W A 2013 *Solar Engineering of Thermal Processes, Fourth Edition* (John Wiley & Sons, Inc.) DOI: 10.1002/9781118671603
[23] King D, Boyson W and Kratochvill J 2004 *Photovoltaic array performance model* (Sandia National Laboratories)
[24] Soto W D, Klein S and Beckman W 2005 Improvement and validation of a model for photovoltaic array performance. *Solar Energy* **80** 78–88 doi: 10.1016/j.solener.2005.06.010
[25] Lashway C 1988 Photovoltaic system testing techniques and results. *IEEE Transactions on Energy Conversion* **3(3)** doi: 10.1109/60.8058
[26] Masters G M 2004 *Renewable and Efficient Electric Power Systems* (John Wiley & Sons, Inc., Hoboken, New Jersey)
[27] Norvig P 2009 Solving every sudoku puzzle, available at: https://norvig.com/sudoku.html
[28] Prokopyev O A, Kong N, and Martinez-Torres D L 2009 The equitable dispersion problem. *European Journal of Operational Research* **197** 59 doi: 10.1016/j.ejor.2008.06.005
[29] Wanga Y, Wub Q, and Glover F, 2017 Effective metaheuristic algorithms for the minimum differential dispersion problem. *European Journal of Operational Research* **258(3)** 829 doi: 10.1016/j.ejor.2016.10.035
[30] Konaka A, Coitb D and Smith A E 2006 Multi-objective optimization using genetic algorithms: Atutorial. *Reliability Eng. and System Safety* **91(2006)** 992 doi: 10.1016/j.ress.2005.11.018

Stressed state "boundary layer" in a round plate of variable thickness according to refined theory

V V Firsanov[1], Q H Doan[1*] and T T Bui[2]

[1]Department 906, *Moscow Aviation Institute* (National Research University), 4 Volokolamsk Highway, Moscow 125993, Russian Federation
[2]Bureau of In-survice Training Administration, *The People's Police University*, 36 Nguyen Huu Tho Street, District 7, Ho Chi Minh 72915, Vietnam

[*]E-mail: dqhieu57@gmail.com

Abstract. Based on the refined theory, the edge stress state of an isotropic round plate of variable thickness under the influence of local load was considered. In constructing the mathematical model of the plate, three-dimensional equations of the theory of elasticity and the variation Lagrange principle were used. Displacements were represented in the form of polynomials along with the coordinate normal to the middle surface, which was two degrees higher than the classical theory of the Kirchhoff – Love type. The resolving system of equations includes eleven ordinary differential equations with variable coefficients. The solution of the formulated boundary-value problem was carried out by finite difference methods and matrix sweeps. The deformations and tangential stresses of the plate were determined from the corresponding geometric and physical equations of the elasticity theory. This article has been focused on identifying the stress state of the boundary layer type near rigidly and elastically fixed edges by the round plate where the destruction of thin-walled structural elements in machinery, including aviation and space technology, takes place.

1. Introduction

Analysis or solution of plates and shells means the determination of displacement, moment, stress, etc. at various points in space of it. This analysis is very important to ensure the stability of the plate to resist the design load. A number of research works were performed on the solution of plate problems by various classical, numerical and engineering methods in the past years. Recently, there has been increasing interest in constructing refined theories of plates and shells. This is due to the need of study on the stress-deformed state (SDS) when designing elements of modern structures, including aircraft and rocketry products. When constructing an approximate theory of plates that is free of Kirchhoff – Love hypotheses, the method of direct asymptotic integration of differential equations of the three-dimensional theory of elasticity became widespread in [1]. The problem of an additional "boundary layer" type in relation to the classical theory of SDS near the clamped edge of rectangular and round plates was solved in a variation formulation by the Vlasov– Kantorovich method using a specially constructed polynomial approximating function. Another approach [2] was based on the representation of displacements in polynomial series along with the coordinate normal to the median plane and relates to the number of terms in these expansions in tangential and transverse directions.

The methods for calculating SDS based on analytical models including structures such as plates and shells with significantly varying stiffness characteristics that are more effective than numerical methods are still relevant [3-5]. It should also be noted that the works [6-9], in which other methods of refined calculation of plates and shells are considered. Rajasekaran Sundaramoorthy and Wilson Antony [10] studied the determination of exact buckling loads and vibration frequencies of variable thickness isotropic plates using well known finite difference technique. The plates are subjected to uni, biaxial compression and shear loadings with various combinations of boundary conditions. In most of the cases, researchers carried out their analysis using the refined theory and then compared their results with other available known solutions to check validity of their analysis.

Therefore, this article aims to build a solution to calculate the SDS of the round plates of isotropic material with a thickness symmetrically changing in the radial direction. Here, we present a refined theory for determining SDS with the same accuracy in all areas of the plate, including the narrow boundary zones near the rigidly fixed edge. Comparison of the results obtained by the refined theory with the data of the classical theory showed that in the edge zone of the plate, the normal tangential stresses are substantially refined and the transverse normal stresses, which are neglected in the classical theory, are of the same order as the maximum values of the main bending stress.

2. Materials and methods

Let us consider the elastic equilibrium of a round isotropic plate with a thickness symmetric with respect to the median plane, loaded with a transverse local load $q(r,\varphi)$, referred to a cylindrical coordinate system (r,φ,z) (figure 1). Denoted by a and b are the external and internal radii of the plate, and by $2h(r)$ its variable thickness. The outer and inner edges of the plate $r=a$, $r=b$ are rigidly pinched.

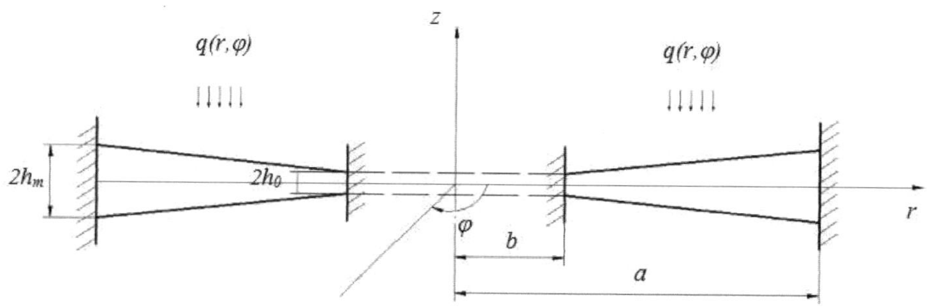

Figure 1. A round plate of variable thickness.

In accordance to [2], the desired elastic displacements U_1, U_2, U_3 are represented in the form

$$U_1(r,\varphi,z)=u_0(r,\varphi)+u_1(r,\varphi)z+u_2(r,\varphi)\frac{z^2}{2!}+u_3(r,\varphi)\frac{z^3}{3!},$$
$$U_2(r,\varphi,z)=v_0(r,\varphi)+v_1(r,\varphi)z+v_2(r,\varphi)\frac{z^2}{2!}+v_3(r,\varphi)\frac{z^3}{3!},\qquad(1)$$
$$U_3(r,\varphi,z)=w_0(r,\varphi)+w_1(r,\varphi)z+w_2(r,\varphi)\frac{z^2}{2!}.$$

where the indices 1, 2, 3 correspond to the axes x, y and z. Decomposition (1) corresponds to a two-degree increase in polynomials approximating the desired displacements.

The geometric equations of the three-dimensional elasticity theory have the form

$$\varepsilon_r = \frac{\partial U_1}{\partial r},\ \varepsilon_\varphi = \frac{1}{r}(\frac{\partial U_2}{\partial \varphi} + U_1),\ \varepsilon_z = \frac{\partial U_3}{\partial z},$$
$$\gamma_{r\varphi} = r\frac{\partial}{\partial r}\frac{U_2}{r} + \frac{1}{r}\frac{\partial U_1}{\partial \varphi},\ \gamma_{rz} = \frac{\partial U_3}{\partial r} + \frac{\partial U_1}{\partial z},\ \gamma_{\varphi z} = \frac{1}{r}\frac{\partial U_3}{\partial \varphi} + \frac{\partial U_2}{\partial z}. \tag{2}$$

Putting formulas (1) in (2), we find

$$\varepsilon_r = \sum_{i=0}^{3}\frac{\partial u_i}{\partial r}\frac{z^i}{i!},\ \varepsilon_\varphi = \frac{1}{r}(\sum_{i=0}^{3}\frac{\partial v_i}{\partial \varphi}\frac{z^i}{i!} + \sum_{i=0}^{3} u_i\frac{z^i}{i!}),\ \varepsilon_z = w_1 + w_2 z,$$
$$\gamma_{r\varphi} = \sum_{i=0}^{3}\left(\frac{1}{r}\frac{\partial u_i}{\partial \varphi} + \frac{\partial v_i}{\partial r} - \frac{1}{r}v_i\right)\frac{z^i}{i!}, \tag{3}$$
$$\gamma_{rz} = \left(u_1 + \frac{\partial w_0}{\partial r}\right) + \left(u_2 + \frac{\partial w_1}{\partial r}\right)z + \left(u_3 + \frac{\partial w_2}{\partial r}\right)\frac{z^2}{2!},$$
$$\gamma_{\varphi z} = \left(v_1 + \frac{1}{r}\frac{\partial w_0}{\partial \varphi}\right) + \left(v_2 + \frac{1}{r}\frac{\partial w_1}{\partial \varphi}\right)z + \left(v_3 + \frac{1}{r}\frac{\partial w_2}{\partial \varphi}\right)\frac{z^2}{2!}.$$

The physical equations of the three-dimensional elasticity theory for the plate are written as following

$$\sigma_r = A_{11}\varepsilon_r + A_{12}\varepsilon_\varphi + A_{13}\varepsilon_z,\ \sigma_\varphi = A_{21}\varepsilon_r + A_{22}\varepsilon_\varphi + A_{23}\varepsilon_z,$$
$$\tau_{r\varphi} = A_{44}\gamma_{r\varphi},\ \sigma_z = A_{31}\varepsilon_r + A_{32}\varepsilon_\varphi + A_{33}\varepsilon_z,\ \tau_{rz} = A_{55}\gamma_{rz},\ \tau_{\varphi z} = A_{66}\gamma_{\varphi z}, \tag{4}$$

where the coefficients A_{ij} $(i = \overline{1,6}, j = \overline{1,6})$ are the elastic constants of the isotropic material of the plate.

We find differential equilibrium equations and natural boundary conditions for plates on the basis of the Lagrange variation principle [2]

$$\iiint\left(\sigma_r\delta\varepsilon_r + \sigma_\varphi\delta\varepsilon_\varphi + \sigma_z\delta\varepsilon_z + \tau_{r\varphi}\delta\gamma_{r\varphi} + \tau_{rz}\delta\gamma_{rz} + \tau_{\varphi z}\delta\gamma_{\varphi z}\right)drd\varphi dz -$$
$$-\iint q(r,\varphi)\delta\left[w_0(r,\varphi) + w_1(r,\varphi)h + w_2(r,\varphi)\frac{h^2}{2}\right]drd\varphi = 0. \tag{5}$$

First we represent the external load, strain (3) and stress (4) in the form of trigonometric series along the circumferential coordinate. Substituting the obtained decompositions into equality (5), we obtain the system of equations of equilibrium in displacements:

$$\left(K_1^{u00} + K_1^{u01}\frac{d}{dr} + K_1^{u011}\frac{d^2}{dr^2} - K_1^{u022}\cdot m^2\right)U_{0m}(r) +$$
$$+\left(K_1^{u20} + K_1^{u21}\frac{d}{dr} + K_1^{u211}\frac{d^2}{dr^2} - K_1^{u222}\cdot m^2\right)U_{2m}(r) -$$
$$-m(K_1^{v02} + K_1^{v012}\frac{d}{dr})V_{0m}(r) - m(K_1^{v22} + K_1^{v212}\frac{d}{dr})V_{2m}(r) +$$
$$+(K_1^{w10} + K_1^{w11}\frac{d}{dr})W_{1m}(r) = 0,$$

$$\left(K_2^{u10} + K_2^{u11}\frac{d}{dr} + K_2^{u111}\frac{d^2}{dr^2} - K_2^{u122}\cdot m^2\right)U_{1m}(r) +$$
$$+\left(K_2^{u30} + K_2^{u31}\frac{d}{dr} + K_2^{u311}\frac{d^2}{dr^2} - K_2^{u322}\cdot m^2\right)U_{3m}(r) -$$

$$-m(K_2^{v12}+K_2^{v112}\frac{d}{dr})V_{1m}(r)-m(K_2^{v32}+K_2^{v312}\frac{d}{dr})V_{3m}(r)+$$

$$+(K_2^{w20}+K_2^{w21}\frac{d}{dr})W_{2m}(r)+K_2^{w01}\frac{d}{dr}W_{0m}(r)=0,$$

$$\left(K_3^{u00}+K_3^{u01}\frac{d}{dr}+K_3^{u011}\frac{d^2}{dr^2}-K_3^{u022}\cdot m^2\right)U_{0m}(r)+$$

$$+\left(K_3^{u20}+K_3^{u21}\frac{d}{dr}+K_3^{u211}\frac{d^2}{dr^2}-K_3^{u222}\cdot m^2\right)U_{2m}(r)-$$

$$-m(K_3^{v02}+K_3^{v012}\frac{d}{dr})V_{0m}(r)-m(K_3^{v22}+K_3^{v212}\frac{d}{dr})V_{2m}(r)+$$

$$+(K_3^{w10}+K_3^{w11}\frac{d}{dr})W_{1m}(r)=0,$$

$$\left(K_4^{u10}+K_4^{u11}\frac{d}{dr}+K_4^{u111}\frac{d^2}{dr^2}-K_4^{u122}\cdot m^2\right)U_{1m}(r)+$$

$$+\left(K_4^{u30}+K_4^{u31}\frac{d}{dr}+K_4^{u311}\frac{d^2}{dr^2}-K_4^{u322}\cdot m^2\right)U_{3m}(r)-$$

$$-m(K_4^{v12}+K_4^{v112}\frac{d}{dr})V_{1m}(r)-m(K_4^{v32}+K_4^{v312}\frac{d}{dr})V_{3m}(r)+$$

$$+(K_4^{w20}+K_4^{w21}\frac{d}{dr})W_{2m}(r)+K_4^{w01}\frac{d}{dr}W_{0m}(r)=0,$$

$$m\left(K_5^{u02}+K_5^{u012}\frac{d}{dr}\right)U_{0m}(r)+m\left(K_5^{u22}+K_5^{u212}\frac{d}{dr}\right)U_{2m}(r)+$$

$$+(K_5^{v011}\frac{d^2}{dr^2}-m^2K_5^{v022})V_{0m}(r)+(K_5^{v211}\frac{d^2}{dr^2}-m^2K_5^{v222})V_{2m}(r) \qquad (6)$$

$$+m\cdot K_5^{w12}W_{1m}(r)=0,$$

$$m\left(K_6^{u12}+K_6^{u112}\frac{d}{dr}\right)U_{1m}(r)+m\left(K_6^{u32}+K_6^{u312}\frac{d}{dr}\right)U_{3m}(r)+$$

$$+(K_6^{v111}\frac{d^2}{dr^2}-m^2K_6^{v122}+K_6^{v10})V_{1m}(r)+(K_6^{v311}\frac{d^2}{dr^2}-m^2K_6^{v322}+K_6^{v30})V_{3m}(r)+$$

$$+m\cdot K_6^{w02}W_{0m}(r)+m\cdot K_6^{w22}W_{2m}(r)=0,$$

$$m\left(K_7^{u02}+K_7^{u012}\frac{d}{dr}\right)U_{0m}(r)+m\left(K_7^{u22}+K_7^{u212}\frac{d}{dr}\right)U_{2m}(r)+$$

$$+(K_7^{v011}\frac{d^2}{dr^2}-m^2K_7^{v022})V_{0m}(r)+(K_7^{v211}\frac{d^2}{dr^2}-m^2K_7^{v222}+K_7^{v20})V_{2m}(r)+$$

$$+m\cdot K_7^{w12}W_{1m}(r)=0,$$

$$m\left(K_8^{u12}+K_8^{u112}\frac{d}{dr}\right)U_{1m}(r)+m\left(K_8^{u32}+K_8^{u312}\frac{d}{dr}\right)U_{3m}(r)+$$

$$+(K_8^{v111}\frac{d^2}{dr^2}-m^2K_8^{v122}+K_8^{v10})V_{1m}(r)+(K_8^{v311}\frac{d^2}{dr^2}-m^2K_8^{v322}+K_8^{v30})V_{3m}(r)+$$

$$+m\cdot K_8^{w02}W_{0m}(r)+m\cdot K_8^{w22}W_{2m}(r)=0,$$

$$K_9^{u11}\frac{d}{dr}U_{1m}(r)+K_9^{u31}\frac{d}{dr}U_{3m}(r)-m\cdot K_9^{v12}V_{1m}(r)-m\cdot K_9^{v32}V_{3m}(r)+$$

$$+(K_9^{w011}\frac{d^2}{dr^2}-m^2K_9^{w022})W_{0m}(r)+(K_9^{w211}\frac{d^2}{dr^2}-m^2K_9^{w222})W_{2m}(r)+$$

$$+K_9^{Q330}Q33_m(r)=0,$$

$$(K_{10}^{u01}\frac{d}{dr}+K_{10}^{u00})U_{0m}(r)+(K_{10}^{u21}\frac{d}{dr}+K_{10}^{u20})U_{2m}(r)-m\cdot K_{10}^{v02}V_{0m}(r)-$$

$$-m\cdot K_{10}^{v22}V_{2m}(r)+(K_{10}^{w111}\frac{d^2}{dr^2}-m^2K_{10}^{w122}+K_{10}^{w10})W_{1m}(r)=0,$$

$$(K_{11}^{u11}\frac{d}{dr}+K_{11}^{u10})U_{1m}(r)+(K_{11}^{u31}\frac{d}{dr}+K_{11}^{u30})U_{3m}(r)-m\cdot K_{11}^{v12}V_{1m}(r)-$$

$$-m\cdot K_{11}^{v32}V_{3m}(r)+(K_{11}^{w011}\frac{d^2}{dr^2}-m^2K_{11}^{w022})W_{0m}(r)+$$

$$+(K_{11}^{w211}\frac{d^2}{dr^2}-m^2K_{11}^{w222}+K_{11}^{w20})W_{2m}(r)+K_{11}^{Q330}Q33_m(r)=0.$$

On rigidly clamped edges, the boundary conditions have the form

$$r=a:\quad U_{im}=V_{im}=W_{jm}=0,\ i=0,1,2,3,\ j=0,1,2,$$

$$r=b:\quad U_{im}=V_{im}=W_{jm}=0,\ i=1,2,3,\ j=0,1,2.$$

In equations (6), the coefficients K with alphabetic and numerical indices are variable coefficients depending on the geometric parameters and elastic constants of the plate material. To solve system (6), we use the finite-difference method [11-12]. Approximating the first and second order derivatives by the central differences of the second order of accuracy [13-14], we obtain a system of linear algebraic equations, the eleventh order matrix of which is solved by the matrix sweep method using a computer program Maple v.18.

The deformations and tangential stresses of the plate are found from the geometric and physical relations of the theory of elasticity. Transverse stresses are obtained by direct integration of the equilibrium equations of the three-dimensional elasticity theory

$$\tau_{rz}=-\int_{-h}^{z}\left(\frac{\partial\sigma_r}{\partial r}+\frac{1}{r}\frac{\partial\tau_{r\varphi}}{\partial\varphi}\right)dz,\quad \tau_{\varphi z}=-\int_{-h}^{z}\left(\frac{1}{r}\frac{\partial\sigma_\varphi}{\partial\varphi}+\frac{\partial\tau_{r\varphi}}{\partial r}\right)dz,$$

$$\sigma_z=-\int_{-h}^{z}\left(\frac{\partial\tau_{rz}}{\partial r}+\frac{1}{r}\frac{\partial\tau_{\varphi z}}{\partial\varphi}-\frac{1}{r}\sigma_\varphi\right)dz.$$

3. Results and discussion

Consider a steel round plate, rigidly clamped at two edges $r=a$, $r=b$ and under the influence of local load

$$q(\mathrm{r},\varphi)=\begin{cases}0, & b\le r<r_1\\ Q_0\sin(\varphi), & r_1\le r\le r_2,\\ 0, & r_2<r\le a\end{cases}\qquad \text{where } Q_0=const,$$

$r_1=(a+19b)/20$, $r_2=(19a+b)/20$. Poisson's ratio $\mu=0.3$, elastic modulus $E=2\cdot10^{11}\,Pa$. Plate dimensions: outer and inner radii $a=0.5(m)$, $b=0.15(m)$, $h_m=0.05(m)$, $h_0=0.015(m)$; linearly varying thickness is determined by the ratio $h(r)=h_m - tg(\alpha)\cdot r$, where $tg(\alpha)=(h_m - h_0)/(a-b)$. The results of calculating the SDS of a round plate are presented in figures 2-5. Here the abbreviations "Class" and "Non-class" correspond to the results of calculations according to classical and refined theories. In figure 2 shows the variation in the additional stresses of the "boundary layer" in thickness according to refined theories at the rigidly clamped edge of the plate.

An analysis of the results shows that when moving away from the edge, the stresses obtained from the refined and classical theories practically coincide, which confirms the reliability of the refined theories; maximum mismatch of the calculation results takes place (figure 3) in determining normal stresses σ_r and is about 10%.

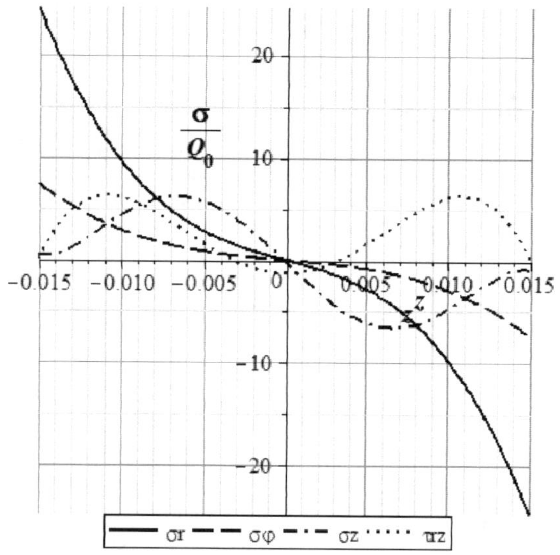

Figure 2. Change of stresses $\sigma_r, \sigma_\varphi, \sigma_z, \tau_{rz}$ along the thickness at the edge $r=b$.

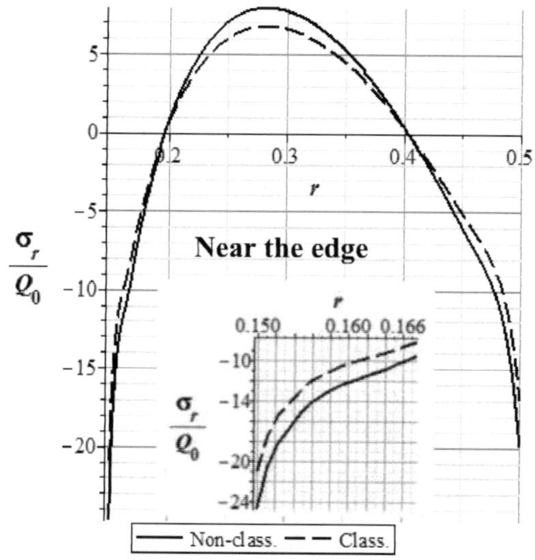

Figure 3. Change σ_r in radius.

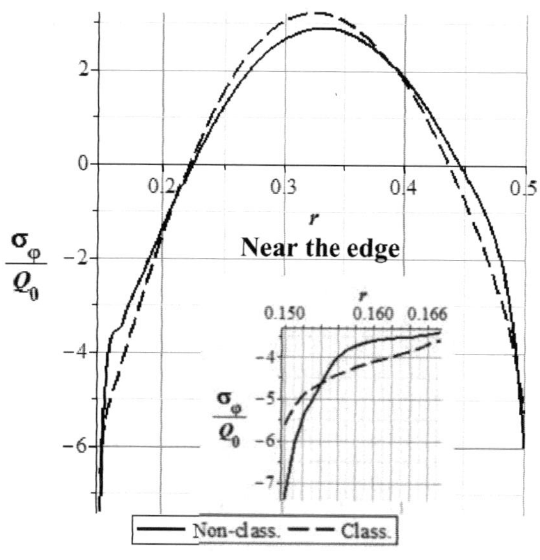

Figure 4. Change σ_φ in radius.

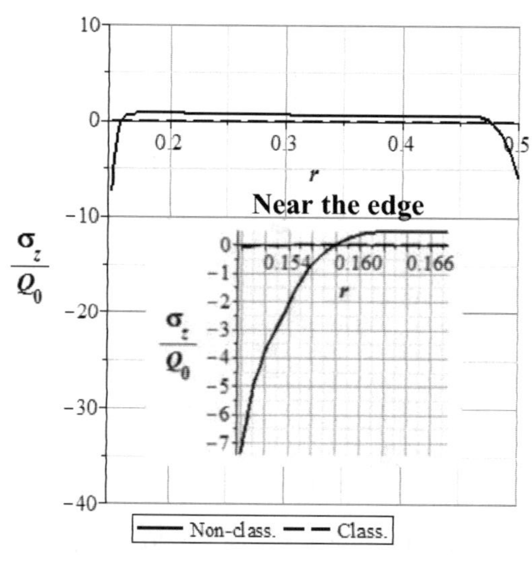

Figure 5. Change σ_z in radius.

It can be noted that the stresses in the edge zone are substantially refined: normal tangential stresses σ_r – by 21.4% (figure 3) and σ_φ – by 30.9% (figure 4). In addition, the normal transverse stresses σ_z, negligible in the classical theory, according to the revised theory, account for about 28% of the maximum bending stresses σ_r (figures 2, 3, 5).

Compared with the classical theory, the refined theory of round plates considers quickly decaying additional boundary SDSs of the "boundary layer" type that make a significant contribution to the total SDS of the plate. When moving away from the edge of the plate - a zone of the distorted stressed state – the results obtained according to the refined and classical theories almost coincide, thus confirming the validity of the proposed theory.

It can be noted that the additional stresses of the border layer type determined in rectangular plates by other methods [1] (asymptotic integration of differential equations of three-dimensional problem of the elasticity theory) nearly coincide in damping magnitude and nature with the results of this work. For cylindrical shell by this method [2] of solving three-dimensional elasticity theory problems, obtained the same result. Using the modified semi-inverse Saint-Venant method, an approximate solution of the spatial problems of the theory of elasticity, in the work [3] also obtained such results for thin shells. In the future, the method presented in this paper can be used to solve SDS problems for spheres, cones, and other shapes. In the next research process, experiments will be conducted to verified results of this research.

4. Conclusion

Based on the refined theory, a boundary-value problem is constructed for a circular plate of variable thickness under the influence of local load. The results of calculations of the stress state of the plate according to the classical and refined theories are presented and their comparison with each other is given. It was established that in the study of SDS in the marginal zones of the plate, the proposed mathematical model should be used, since the maximum stresses are substantially refined. It is shown that the transverse normal stresses turn out to be of the same order with the maximum values of the main bending stress. This result is important because it allows to reliably assess the strength and crack resistance of the power hulls of aircraft.

References

[1] Firsanov Val V 2016 Study of stress-deformed state of rectangular plates based on nonclassical theory. *J. of Machinery, Manufacture and Reliability* **45(6)** 515 DOI: 10.3103/S1052618816060078

[2] Firsanov V V 2018 The stressed state of the "boundary layer" type cylindrical shells investigated according to a nonclassical theory. *J. of Machinery, Manufacture and Reliability* **47(3)** 241 DOI: 10.3103/S1052618818030068

[3] Zverayaev E 2016 Constructive theory of thin elastic shells. *Keldysh Institute Preprints* **33** 25 DOI:10.20948/prepr-2016-33

[4] Demidov A, Popov V 2019 Stress State in a Finite Cylinder with Outer Ring-Shaped Crack at Non-stationary Torsion. *Proceedings of the Second International Conference on Theoretical, Applied and Experimental Mechanics* (Springer, Cham) vol. 8 pp 215-221 DOI: 10.1007/978-3-030-21894-2_41

[5] Groh R M J and Weaver P M 2016 Deleterious localized stress fields: the effects of boundaries and stiffness tailoring in anisotropic laminated plates. *Proceedings of The Royal Society A* **472(2194)** 22 DOI:10.1098/rspa.2016.0391

[6] Groh R M J, Weaver P M 2015 On displacement-based and mixed-variational equivalent single layer theories for modelling highly heterogeneous laminated beams. *Int. J. of Solids and Structures* **54** 147 DOI:10.1016/j.ijsolstr.2015.01.020

[7] Benselama K, Noureddine E, Bedia E, Tounsi A 2015 Buckling analysis in hybrid cross-ply composite laminates on elastic foundation using the two variable refined plate theory.

Structural Engineering and Mechanics **55** 47 DOI: 10.12989/sem.2015.55.1.047

[8] Zhou S S, Gao X L 2013 Solutions of half-space and half-plane contact problems based on surface elasticity. *Zeitschrift für angewandte Mathematik und Physik ZAMP* **64(1)** DOI: 10.1007/s00033-012-0205-0

[9] White S C, Weaver P M 2016 Towards imperfection insensitive buckling response of shell structures-shells with plate-like post-buckled responses. *Aeronautical Journal - New Series* **120** 233 DOI:10.1017/aer.2015.14

[10] Rajasekaran S, Wilson A 2013 Buckling and vibration of rectangular plates of variable thickness with different end conditions by finite difference technique. *Structural Engineering and Mechanics* **46** DOI: 10.12989/sem.2013.46.2.269

[11] Roknuzzaman Md, Hossain Md, Haque Md. Rashedul, Ahmed Dr 2015 Analysis of Rectangular Plate with Opening by Finite Difference Method. *American Journal of Civil Engineering and Architecture* **3** 165 DOI: 10.12691/ajcea-3-5-3

[12] Bilbao S 2008 A family of conservative finite difference schemes for the dynamical von Karman plate equations. *Numerical Methods for Partial Differential Equations* **24(1)** 193 DOI: 10.1002/num.20260

[13] Ergün A, Kumbasar N 2011 A New Approach of Improved Finite Difference Scheme on Plate Bending Analysis. *Scientific Research and Essays* **6(1)** 6 DOI: 10.5897/SRE09.388

[14] Pisacic K, Horvat M, Botak Z 2019 Finite difference solution of plate bending using Wolfram Mathematica. *Tehnički glasnik* **13** 241 DOI: 10.31803/tg-20190328111708

IOP Conf. Series: Materials Science and Engineering **868** (2020) 012003 doi:10.1088/1757-899X/868/1/012003

Combined surface hardening of parts for friction pairs of gas turbine engines

D S Moeni Tabatabai, A G Boytsov*, V V Kuritsyna and S A Kazantsev

Institute of Aerospace High Technology and Production, *Moscow Aviation Institute* (National Research University), 4 Volokolamskoye shosse, Moscow 125993, Russia

*E-mail: tpdla@mail.ru

Abstract. The paper presents an analysis of the methods of surface hardening of parts. The authors experimentally worked out and determined the technology of combined hardening of parts of friction pairs of hydraulic pneumatic units, which allows increasing the efficiency and reliability of the units. In order to increase the life and reliability of precision friction pairs, the production feasibility of combining spark alloying and diamond smoothing has been substantiated. Technological and software for the formation of regular micro reliefs, surface reinforcement with combined processing are proposed.

1. Introduction

Statistics show that the largest number of malfunctions of fuel and hydraulic system units is associated with a malfunction of precision pairs and seal elements. Moreover, most failures, including failure of hydraulic units, occur due to the malfunctioning of control and distribution devices, as well as plunger, piston and plate pairs, which perform the functions of displacing or power elements of pumps and hydraulic motors.

The most common cause of increased friction, which causes jamming and failure of parts of slide control devices, is the setting of rubbing surfaces and fretting corrosion, which is a corrosion-abrasive process of destruction of the mating metal surfaces of parts subject to vibration.

The wear of the spool pair, caused most often by the delayed replacement of consumables (filters, and the hydraulic fluid itself) in the hydraulic dispersion mechanism, causes a decrease in pressure on the working body, which entails a decrease in the machine's working capacity, and a decrease in efficiency. Spontaneous movements or intermittent operation of the actuating mechanism of the servo hydraulic drive are caused by an increase in friction in the switchgear. Damage to the pumping pump assembly and the destruction of hydraulic motors are often the result of jamming of the plunger, plate or piston rotor pairs. In this regard, the analysis of the operating conditions and the establishment of the reasons for the failure of the precision pairs deserve special attention when developing measures to improve the reliability of hydraulic units.

Improving the reliability and service life of hydropneumatic units can be carried out by combined hardening, which will improve both the mechanical properties and the corrosion resistance of the working surfaces. To select the most suitable combined hardening method, it is necessary to analyze the operating conditions of hydropneumatic units, to determine the requirements for the quality of the surface layer.

2. Requirements for the quality of the surface layer

The main requirements for spool pairs are high stability of low friction forces and good tightness, i.e. the presence of minimal, not increasing during operation, above the permissible limit of leakage of the working fluid through the gaps between the parts.

In order to ensure minimum gaps in the pumping unit and at the same time to prevent jamming of parts and large wear of rubbing surfaces, stringent requirements are imposed on the geometric dimensions and cleanliness of the working surfaces of the parts of the plunger pumps. Plungers are made from cemented steels of tool and high-speed steels with hardness of working surfaces HRA>80. Plunger blocks are made of alloy cast iron or special bronzes [1].

Clearance space the plunger into the cylinder is carried out with a permissible value between 0.015-0.02 mm, which is provided by pair selecting the plunger and the cylinder or lapping them. The surface roughness of the plungers is usually performed within the range of Ra = 0.1...0.32 microns. One of the main technical conditions is the requirement that the working surfaces of the plunger and cylinder are cylindricity; their roundness deviation and taper should not exceed 0.002 and 0.005 mm, respectively.

3. The choice of combined hardening method

For surface hardening, methods differing in nature of the hardening effect are used: surface plastic deformation (SPD), chemical heat treatment, highly concentrated energy flows (laser, electron beam, ion), coating of materials with high performance properties, etc. [2-4].

For various reasons, such as, for example, an increase in roughness, a decrease in the fatigue strength of adhesion of a coating to a substrate, the scope of many effective hardening processes is significantly limited. In such cases, by combining with another method of hardening (reducing roughness, increasing fatigue strength, decreasing porosity, and increasing adhesion strength), it is possible to achieve the required state of the surface layer and provide a high level of operational properties, expand technological capabilities [2, 5, 6].

Such combinations are called combined hardening [7]. By combining surface hardening methods, they also solve the problems of increasing the oil absorption of friction surfaces, hardening the substrate for the applied coating, and increasing accuracy. Combined hardening methods should be used for the most critical parts when traditional technology does not provide the required level of performance [8, 9].

The variety of hardening methods determines the breadth of possibilities for combining them. However, not all combinations are of practical value. In many cases, the technical indicators of the combined methods are similar. Some methods are incompatible, and their combination can lead to a decrease in the quality of the surface layer.

Based on the required characteristics of the surfaces of parts of friction pairs of hydropneumatic units, it is necessary, first of all, to provide resistance to fretting corrosion, and also to provide high antifriction properties. Electric spark doping makes it possible to significantly improve (by 2 … 10 times in comparison with the source material) such exploitation characteristics of parts as abrasion resistance, fretting resistance, hardness, surface strength, temperature constancy. Therefore, the first step in combined hardening should be the selection of spark alloying. Reinforcing the surfaces of parts of friction pairs of hydropneumatic units with a bronze alloying electrode makes it possible to obtain alloyed sections providing antifriction, and wear resistance is determined by the main material.

At the second stage of combined hardening, the surface is diamond-smoothed, resulting in increased surface hardness.

4. Electrospark alloy surface reinforcement

With electrospark alloying, the formation of a hardened layer is achieved by transferring the material of the alloying electrode (anode) to the hardened product (cathode) during electrical impulse discharges [10, 11]. Particles of the alloyed electrode materials melted by an electric spark discharge and the parts are mixed and deposited on the hardened surface, forming a layer of alloy saturated with elements and quenched to high hardness (more than 82 HRA up to 90 HRA). The microhardness of the surface of the parts is reached by the value of 16000 ... 20000 MPa at the depth of the hardened layer up to 400 microns.

Hard alloys (VK6, T15K6, etc.), carbides, nitrides and borides of transition metals are used as alloying electrodes (TiC, WC, TiN, TiB_2, NbC, CrC, ZrB_2 , etc.), metals (Ti, Cr, Co, Ni, Al, W, Mo, Re, Ta, Hf , etc.), as well as alloys based on them.

The required value of the interelectrode gap is supported by the movement of the electromechanical tracking system (U). In order to increase the flatness of the hardened layer and uniform erosion of the alloying electrode, it is given rotation from an electric motor with a frequency of 400 ... 4000 rpm. The alloying electrode with a diameter of 0.5 ... 2.0 mm is installed in a collet chuck.

The alloying process scheme is shown in figure 1. Alloy electrode (1) is an anode, and hardened product (2), installed on the desktop (3), the cathode. The parameters of electric spark discharges are selected so that predominant erosion of the alloying electrode occurs. The necessary power, frequency and duration of the discharges are provided by a pulse generator, which supplies rectangular voltage pulses to the doping electrode.

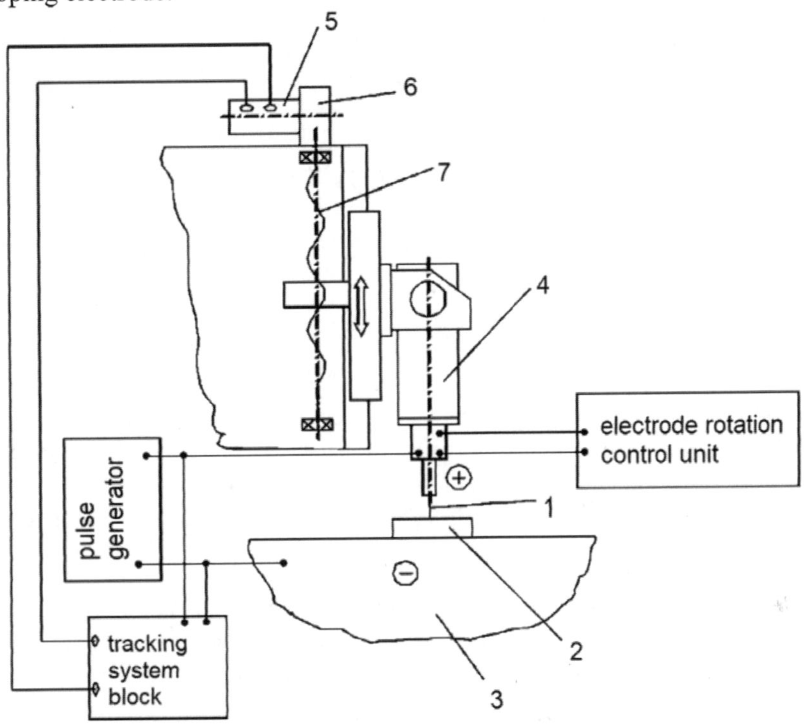

Figure 1. Electrospark alloying process scheme: (1) alloy electrode; (2) hardenable product; (3) platten; (4) application head; (5) servomotor of tracking system; (6) gearbox; (7) feed-screw.

The required value of the interelectrode gap is supported by an electromechanical servo system, by moving the servo head (4) in the vertical direction by means of an actuator motor (5), a worm gearbox (6) and a lead screw (7). The engine (5) is controlled by a servo unit.

The performance of alloying is determined by the speed of movement of the alloying electrode relative to the hardened surface. The equipment for Electrical Discharge Machining (EDM) “ELFA-731” provides the movement of the alloying electrode along the coordinates X, Y, Z (figure 2).

“ELFA-731” machine is equipped with a CNC system “Fanuc- 3M”. This CNC device is focused on manual input of control information in a dialogue mode. Due to this, it is possible to prepare a control program directly at the machine, based on the drawing data.

Surface reinforcement consists in hardening local areas evenly spaced on the surface. Such processing can provide the necessary level of operational properties with a significant increase in productivity. In addition, surface reinforcement allows you to combine the properties of the base material and the material of the hardened sections.

IOP Conf. Series: Materials Science and Engineering **868** (2020) 012003 doi:10.1088/1757-899X/868/1/012003

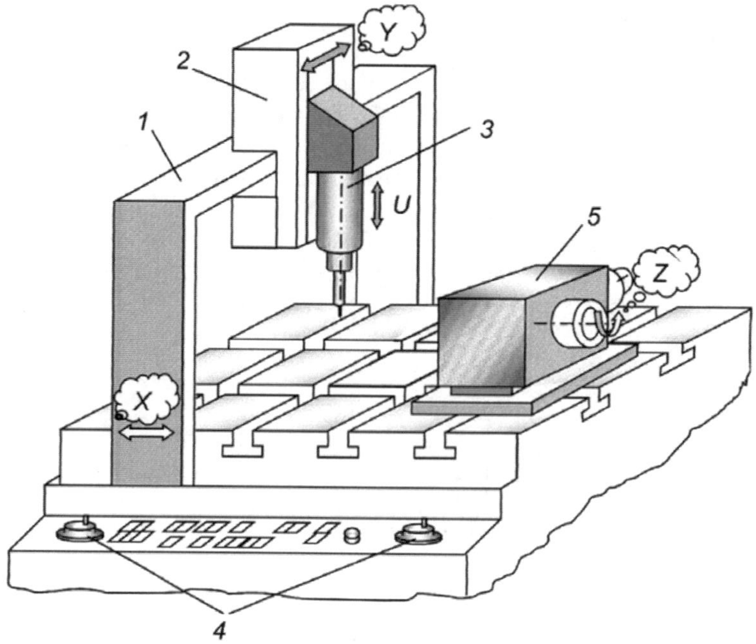

Figure 2. Working zone of electric-discharge milling system: (1) gantry; (2) support; (3) head; (4) pulse generators; (5) bath.

The geometric shape of the hardened sections can be very different. It is important that it ensures the uniformity (regularity) of the distribution of plots on the surface. Some possible site shapes are shown on the figure 3. The choice of form depends on the capabilities of the used CNC equipment.

Other things being equal, the operational properties of reinforced surfaces depend on the ratio of hardened (S_H) and not hardened (S_0) areas (*bp*), as well as relief density (D_R), that is, the number of cells per surface unit (per sq.cm).

When surface reinforcing on CNC equipment, for example, the "ELFA-731" machine, it is advisable to apply cells in a hexagonal or square shape (figure 3a and 3b). In these cases, the specified parameters are calculated by the formulas (1) for hexagonal cells, and (2) for square cells:

$$bp_6 = \frac{S_H}{S_0} = \frac{a_6^2}{(a_6 - \frac{d_e}{\sqrt{3}})^2} - 1, \ D_{R6} = \frac{200}{3\sqrt{3}a_6^2}, \quad (1)$$

$$bp_4 = \frac{a_4^2}{(a_4 - d_e)^2} - 1, \ D_{R4} = \frac{100}{a_4^2}, \quad (2)$$

where d_e – diameter of alloying electrode; a – side of a hexagon or square (cell parameter); indexes (4) and (6) refer to a square or hexagon, respectively.

It is also advisable to apply a relief in the form of circular sections having an electrode diameter by periodically stopping the alloying electrode for periods of time during which the area under the electrode is alloyed. It is desirable to arrange the hardened round sections in the corners of the square (figure 3c).

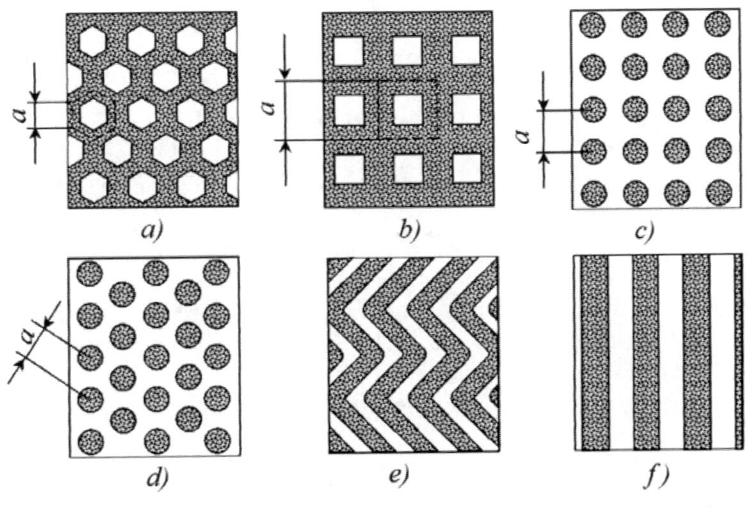

Figure 3. Forms of hardened areas used in surface reinforcement.

The relief density indicators are calculated using the following formulas:

$$bp_4 = \frac{\pi\, d_e^2}{a_4^2 - \pi\, d_e^2}, \quad D_{R\ 4} = \frac{100}{a_4^2}, \tag{3}$$

or ensure their tight packing (figure 3 d), then the calculation of the relief density indices is performed according to the formulas:

$$bp_6 = \frac{\pi\, d_e^2}{\frac{\sqrt{3}}{2} a_6^2 - \pi\, d_e^2}, \quad D_{R\ 6} = \frac{200}{9\sqrt{3} a_6^2}. \tag{4}$$

The advantage of such a reinforcement technology is the possibility of obtaining a greater relief density at the same values *bp* [12].

Surface reinforcement can be combined with unbroken spark alloying. For example, first a layer is applied with an alloying electrode of wear-resistant material, and then a relief of antifriction material is formed [13]. It is also obvious that it is possible to cross-apply reliefs shifted relative to each other by a certain amount, with the selection of the sizes of square or rectangular cells and the magnitude of the shift, providing hardening of the entire surface, and it will consist of sections hardened with various materials. For example, first a grid of cells is alloyed with a wear-resistant material, and then, with a shift, a grid of cells of anti-friction material is applied. The possibilities here are very wide. Figure 4 shows a photograph of a part with a surface treated by this technology.

Figure 4. BK6-M coating, on top bronze.

Since grooves or grooves are formed during such processing, if there are requirements for the tightness of the mating parts, it is recommended to apply the relief in the form of a closed mesh with

IOP Conf. Series: Materials Science and Engineering **868** (2020) 012003 doi:10.1088/1757-899X/868/1/012003

hexagonal or other cells. On the contrary, in order to improve the flow of oil in the contact surfaces, it is advisable to apply reliefs of the open type (figures 3e and 3f).

5. Diamond smoothing as a means of forming a regular microrelief

The next stage of combined hardening is diamond smoothing. By combining surface reinforcement and diamond smoothing, it is possible to obtain reinforced regular microreliefs, i.e. surfaces with protrusions made of hardened material [14, 15].

Due to the greater hardness of the hardened sections, the depth of the smoothing agent on the initial (not hardened) material is greater than on the hardened one. Respectively, the plastic displacement of the material is also large. As a result, after ironing, a regular microrelief is formed, the protrusions of which are made of a material of higher hardness (up to 90 HRA) hardened by electrospark alloying, than other sections (about 75 HRA) (figure 5).

The maximum possible depth of the microrelief formed after smoothing (R_{max}) can be determined from the dependence:

$$R_{\max} = h_0 - h_H + \Delta h = h_0 (1 - \frac{H_0}{H_H}) + \Delta h \,, \tag{5}$$

where h_0 and h_H – the smoothing depths when smoothing the initial and hardened areas, respectively; Δh – the difference between the levels of the original and hardened electrodischarge hardening of surfaces; H_0 and H_H – hardness of the initial and hardened sections, respectively.

The microrelief depth increases with a decrease in the smoothing radius R_{Sf}:

$$R_{\max} = R'_{\max} \frac{1}{R_{Sf}} \,, \tag{6}$$

where R'_{max} – microrelief depth when machining with a radius smoother R_{Sf} = 1 mm.

Due to the mutual influence of hardened areas, with their close proximity, the depth of the regular microrelief decreases with a decrease in the length of the initial sections, and its determination must be carried out experimentally.

Figure 6 shows photographs of parts treated with surface reinforcement.

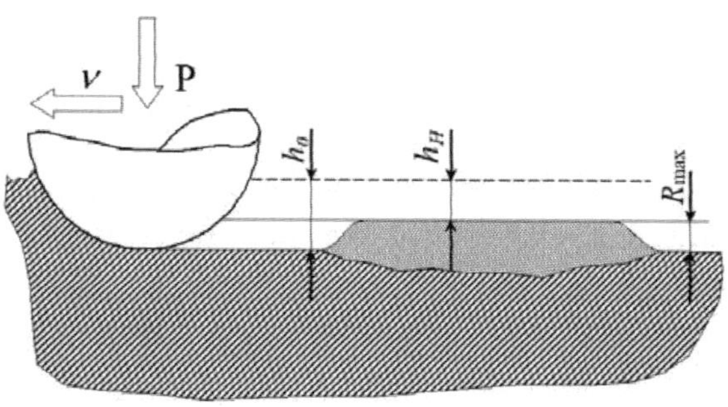

Figure 5. The scheme of formation of the hardened microrelief.

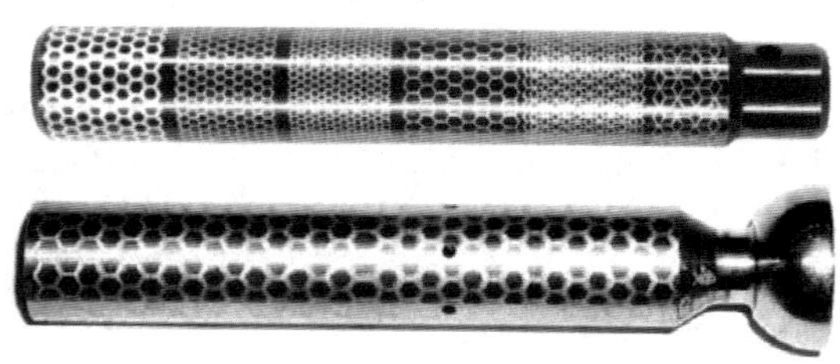

Figure 6. Parts hardened by electrodischarge hardening and smoothing.

Figures 7 and 8 show a technological scheme and an algorithm for a flexible control program, which is advisable to use for applying regular microreliefs in the form of a set of hexagonal cells with side a (a_z) on flat surfaces bounded by a rectangle with dimensions of sides (B) and (L), or surfaces of revolution with a diameter (D) and length (B).

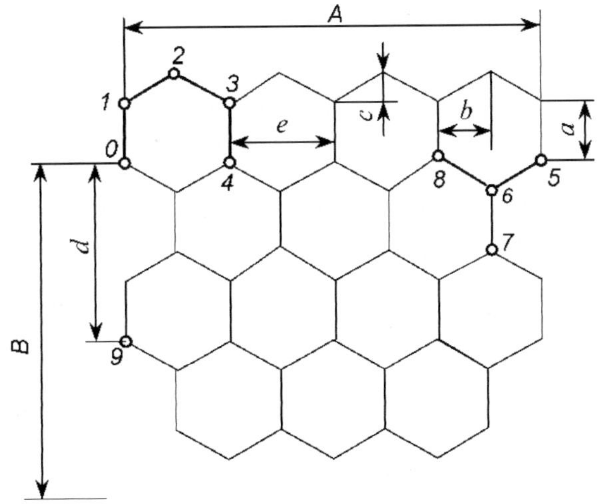

Figure 7. Operational design of a hexagonal microrelief formation program.

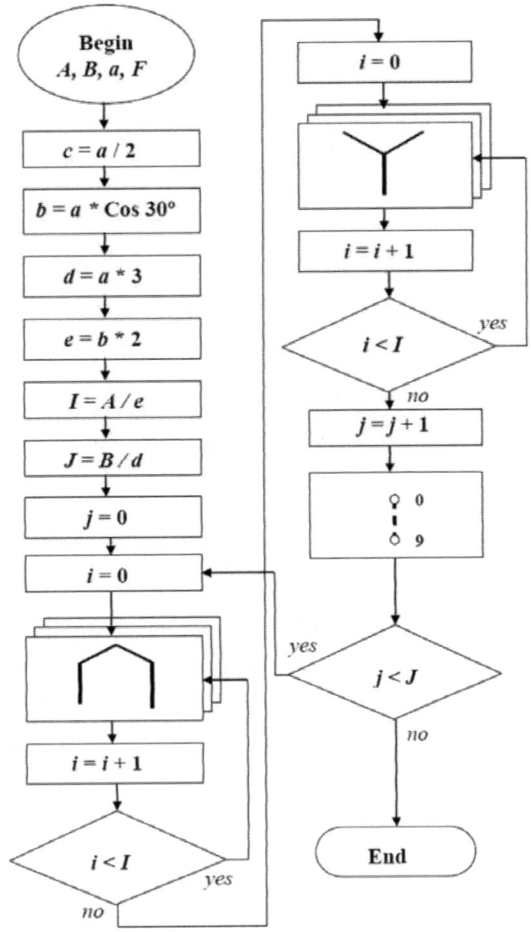

Figure 8. The algorithm of the control program for the formation of the hexagonal relief.

The method of forming regular microreliefs proposed in the work by combined processing based on electroerosive hardening followed by diamond smoothing is expediently used to harden the surfaces of friction pairs of hydropneumatic units. By means of electroerosive hardening, sections regularly located on the surface are hardened, and after diamond smoothing, due to differences in hardness of the hardened and initial sections, protrusions are formed from the material hardened by electroerosive hardening.

The obtained analytical expressions allow us to calculate the height of the microrelief. The developed method of combined hardening allows to improve the lubrication conditions of the surfaces of friction pairs, to optimally combine the properties of the base and alloyed materials.

An express assessment of the wear resistance of surfaces treated with the combined hardening technology was carried out under the conditions of reciprocating movements with the counterbody of the test samples and witness samples. The methods of experimental rapid assessment and data processing are of a special nature and allow you to perform a comparative analysis in terms of wear resistance, bearing capacity, fatigue resistance and other tribotechnical indicators [16]. Tests of samples with a surface-reinforced layer of combined hardening under various loading conditions showed a decrease in the friction coefficient up to ten times, a multiple increase in wear resistance and resistance to fretting corrosion damage.

An assessment of the production feasibility and industrial potential of research and development in the field of hardening and restoration of parts for aircraft gas turbine engines and assemblies [17-19] shows, that electric spark alloying and combined methods based on it are an effective tool to ensure the service properties of a wide range of aircraft products.

6. Conclusion

A new technology of combined hardening is proposed, based on the combination of electroerosive hardening and diamond smoothing with the creation of the effect of surface reinforcement and the creation of special regular microreliefs. Diamond smoothing is the most suitable method of surface plastic deformation for processing surfaces hardened by electric spark alloying. This is due to the high hardness of the diamond and the specific conditions of its contact interaction with the alloyed layer, which achieves plastic deformation of the hardest deposited materials. A combination of surface reinforcement and diamond smoothing can produce reinforced regular microreliefs, i.e. surfaces with protrusions made of hardened material. The advantages of these developments in comparison with the known ones are: locality of hardening, high adhesion strength of the coating to the substrate, the ability to actively control the composition, properties and microgeometry of the surface layer.

The proposed technology of combined surface hardening, based on a combination of electrospark alloying processes in the form of local reinforcement and diamond smoothing, was applied on the details of the rods of plunger pairs of hydraulic units. The proposed technological solutions, software methods and means of implementing the process made it possible to meet the requirements for accuracy and micro geometry of the working surface with a significant increase in the life of a friction pair under conditions of hydrodynamic lubrication. Completed investigation of the operational properties of parts of pneumatic units showed the high efficiency of combined hardening to ensure the service characteristics and service life of machines, devices and assemblies.

Thus, combined hardening with the formation of regular microreliefs makes it possible to increase the life of pneumatic hydraulic units by an order of magnitude without making any design changes to the geometry of the part.

References

[1] Eliseev Yu S, Boytsov A G, Krymov V V, Hvorostukhin L.A 2003 Production technology of aircraft gas turbine engines (Moscow: Mashinostroyeniye) p 512

[2] Boitsov A G, Kovalev A P and Novikov A S 2007 *Mechanical and Physicochemical Processing in Production of Aircraft Engines* (Moscow: The Bauman Moscow State Technical University) p 584

[3] Morgunov Yu A, Panov D V, Saushkin B P and Saushkin S B *Advanced Technologies in Machine*

Engineering Industry: Physical-Chemical Methods and Technologies (Moscow: Forum) p 928

[4] Kochergin S A, Morgunov Yu A and Saushkin B P 2016 Surface manufacturing under pulse fiber laser. *Proc CIRP* **42** 470 https://doi.org/10.1016/j.procir.2016.02.234

[5] Isachenko V A, Astakhov Yu P and Saushkin B P 2016 Aerospace engineering technology: problems and prospects. *Eng. Technol.* 1:10

[6] Yeliseyev Yu S, Saushkin B P 2010 *EDM of Products of Aerospace Equipment* (Moscow: Bauman MSTU Publishing Houses) p 437

[7] GOST 18295-72 *Surface Working, Terms and Definitions* (Moscow: Izdatel'stvo Standartov)

[8] Boitsov A G, Mashkov V N, Smolentsev V A and Khvorostukhin L A 1991 *Surface Hardening of the Parts by Combined Methods* (Moscow: Mashinostroenie) p 144

[9] Gruzdev A A, Morgunov Y A, Saushkin B P 2019 Tools for electrical-discharge hole drilling *Lect. N. Mech. Eng.* 1467 doi.org/10.1007/978-3-319-95630-5_156

[10] Verkhoturov A D, Podchernyaeva I A, Ivanov V I and Konevtsov L A 2010 On the problem of creating a new scientific school in the field of electric erosion machining: electrode material science. *Surf. Eng. Appl. Elect.* **46(5)** 523 doi: 10.3103/S1068375510050194

[11] Verhoturov A D, Vlasenko V D and Konevtsov L A 2018 Investigation of the adhesive strength of antifriction coatings depending on the energy parameters of electro-spark alloying. *J. Frict. Wear+* **39(3)** 232 doi: 10.3103/S1068366618030145

[12] Shneyder Yu G 2001 Regularization of surface microgeometry by plastic deformation method. Metalloobrabotka **3(3)** 31 [in Russian]

[13] Khvorostukhin L A, Shishkin S V, Kovalev A P and Ishmakov R A 1988 *Increase of the Bearing Capacity of Machine Parts by Surface Hardening* (Moscow: Mashinostroenie) p 144

[14] Khvorostukhin L A, Metelkin V V, Safronev V V 1990 Patent SU 1555113 (07 April 1990)

[15] Shishkin S V 2019 Application of finish-strengthening process with diamond smoothing for bearing capacity increase of joints with tightness. *Bulletin of Bryansk State Technical University* **6(79)** 4 doi: 10.30987/article_5d10851eda9a19.64369355

[16] Khopin P N 2018 Test analysis of friction couples with solid lubricant coatings under ground–space conditions and prediction of tribological characteristics. *J. Frict. Wear+* **39(2)** 137 doi: 10.3103/S1068366618020071

[17] Grachev M V 2019 Determination of parameters of the surface layer according to indentation diagram for titanium alloy TI-6AL-2.5MO-1.5CR-0.2SI-0.5FE. *International Transaction Journal of Engineering, Management, & Applied Sciences & Technologies* **10 (10)** 10A10G doi: 10.14456/ITJEMAST.2019.128.

[18] Denisov L V, Boitsov A G and Siluyanova M V 2018 Surface hardening in hydraulic cylinders for airplane engines. *Russian Engineering Research* **38(12)** 1080 doi: 10.3103/S1068798X18120237

[19] Boitsov A G, Denisov L V and Siluyanova M V 2018 Assuring performance characteristics of gas-turbine engine parts and assemblies using electrospark doping *Vestnik of Samara University. Aerospace and Mechanical Engineering* **17(2)** 58 doi: 10.18287/2541-7533-2018-17-2-58-67

IOP Conf. Series: Materials Science and Engineering **868** (2020) 012004 doi:10.1088/1757-899X/868/1/012004

Automation of scientific and technical expertise procedures in the methodology of technological audit of aircraft manufacturing

V V Kuritsyna[1], D N Kuritsyn, S M Shahrivar and S A Kazantsev

Institute of Aerospace High Technology and Production, *Moscow Aviation Institute* (National Research University), 4 Volokolamskoye shosse, Moscow 125993, Russia

[1]E-mail: kuritzynav@mail.ru

Abstract. The work is devoted to issues of conceptual, informational and software problems of expert assessment of the current state, problems and development potentials of technological and production systems. The purpose of the research is to develop methods and tools for variative forecasting of operational characteristics of aircraft manufacturing facilities and automation of audit procedures for complex technological systems. The originality of this method is the use of the principles of technological inheritance of quality parameters of parts in the process of processing under the influence of technological energy and their relationship to performance characteristics. The advantage is the ability to define process controllers in the image of progressive technological operations of forming and modification of precision surfaces in order to ensure a guaranteed increase in operational efficiency by their flexible integration into production systems. A hierarchical model of the technological system implemented in the MatLAB Simulink visual modeling environment is proposed. The operation of the process controller is shown by the example of a diamond smoothing operation aimed at increasing the service life of precision friction pairs. The developed methods and tools are a tool for forming an evidence base for evaluating the technological potential, reliability, stability, and repeatability of technological processes in the total quality management system.

1. Introduction

The tasks of production quality management and ensuring the competitiveness of high-tech products are solved in a multidimensional space of objects, technologies, management tools. The field of aircraft construction, creation of transport equipment of new generation is the most difficult from the point of view of technological support [1].

The need to coordinate long-term strategic planning [2] and implementation in practice of specific mechanisms for the introduction and support of innovative technologies is noted in the state economic and political strategies of countries that determine the technical, economic and socio-political balance of the world [3, 4]. The scientific and technological knowledge and skills underlying expert forecasting become a strategic resource for the development of industrial sectors, national companies and incorporated corporations [5]. At any of these levels, it is extremely important to have information and be able to form ideas about the main directions of scientific and technological development in the medium and long term [6].

At the stages of the life cycle of complex science-intensive products of special engineering, including development, production, operation, it is often necessary to make decisions in conditions of insufficient information, unclear algorithms for solving the problem, in conditions of risk [7]. The problem of adequate information support, accuracy and reliability of decisions in the scientific and technological sphere is now considered as one of the key.

Aircraft engineering products are complex technical systems, each component of which, in turn, represents a multi-dimensional production technological system for forming specific geometric shapes, physical and mechanical properties of parts under the influence, operational properties of components, reliability and performance of the product as a whole [8]. A particular difficulty in predicting and ensuring the required quality is caused by the huge number of influencing technological factors of technological operations and the inability to account for the synergistic effect of their influence through private analytical models.

At the same time, the accumulated experience of research, description and modeling of advanced technological methods for increasing the resource of high-loaded parts, such as surface-plastic deformation with a diamond tool, application of functional coatings, special chemical and thermal treatment, remains without analysis of the technological history of their formation, their joint influence, and mutual technological inheritance. Tools for describing the phenomena and effects of technological inheritance in production audit procedures are of considerable scientific and practical interest.

For the first time, the concept of technological inheritance in the field of mechanical engineering technology was proposed by Dr. A M Dal'sky, and advanced by scientists of the scientific school of the Bauman Moscow State Technical University [9-11]. In the field of aircraft engine production technology, the theory of technological inheritance was developed by the scientific school of Dr. L A Hvorostukhin of the Moscow Aircraft Technology University named by K E Tsiolkovsky [12] and his followers of the Moscow Aviation Institute (National Research University) [13].

Research and software solutions in the field of innovation management [14-16] offer mainly financial and economic models of investments and strategies, but do not take into account the technological risks of introducing new processing methods at the level of physical models. In the publishing space, the concept of financial, economic, and law audit of manufacturing and industries has long existed, while the concept of technological audit, proposed for the first time by Dr. A G Boitsov [1, 8], it is only now gaining strength, bringing to the forefront of analysis the mutual connection of physical phenomena and processes in the field of key technologies for forming and modifying surfaces.

Of particular importance in the field of mathematical representation of complex analytical networks, analysis of hierarchies, and decision-making are the works of Dr. S L Saati [17, 18]. This organizational approach and modern software tools for simulating MatLAB Simulink allow us to solve the problems of developing new methods and tools for technological audit procedures for aviation production.

The purpose of the research presented in this paper is to develop methods and software for variative forecasting of operational characteristics of objects manufactured in the environment of complex technological systems of aircraft manufacturing industries. The developed methods and tools are a tool for forming an evidence base for evaluating the technological potential, reliability, stability, and repeatability of technological processes in the universal quality management system. One of the main features of the analytical system being developed is the ability to search for and identify process controllers in order to ensure guaranteed operational efficiency by flexibly integrating advanced technological operations with precision surface modifications into production systems.

2. Technological audit decision extent

The analysis of areas of professional activity of objects, technologies and means of creation and operation of high-tech products, such as high-tech transport and space systems, allowed us to identify three vectors of space, reflecting the functional and support systems (figure 1): *the object vector* – the structure of an object segment transport systems, involving directly transport equipment, maintenance systems functioning in its structure management system (transport systems), it infrastructure deployment and maintenance; *the technological vector* – the structure of instrumental subsystems of complex objects maintenance at the stages of the life cycle, including technologies, methods and means of design, manufacture, operation of equipment and objects of transport systems; *the control vector* –

the structure of providing subsystems including methods and means of ensuring reliability, safety of objects and systems, management of innovative development, knowledge engineering, tools of economic and state regulation.

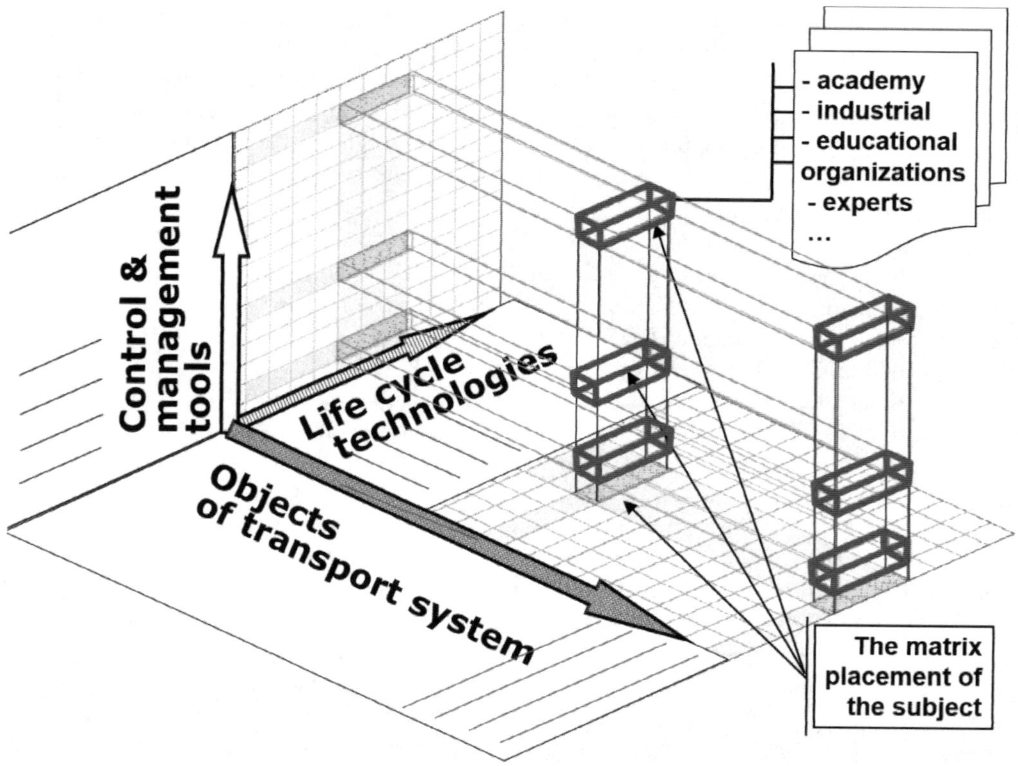

Figure 1. The multidimensional structure of the entity-relationship model for industry segment information of the cluster.

The composition of information for each vector is a hierarchical structure that can be detailed to the preferred level of analysis. It should be noted that if the object vector has individual specificity for each of the industry segments by means of transport, then, on the contrary, the other two vectors – technological and management tools – have similar composition and structure for any segment by means of transport and can be taken as typical.

When constructing the coordinate space "segment objects – life cycle technologies – management and regulation tools", a clear system of cells that meet the competencies of the cluster subject is determined.

Complex technical and technological systems have a very large number of parameters that reflect the design and functional perfection of the product (such as operating temperatures, pressures, speeds, as well as operational reliability, resource, durability, wear resistance, and so on). The creation of a product is determined by numerous indicators that reflect the perfection of manufacturing technologies, economic aspects of production, exploitation [19, 20]. Despite the numerous factors of the assessment tasks and under risk conditions, it is necessary to make scientifically-based decisions on which depend not only tactical moments, but also strategy of development of the enterprises, branch.

Nevertheless, in such conditions it is necessary to make scientifically grounded decisions on which depend not only tactical moments, but also strategy of development of the enterprises, branch.

The problem of technological support of operational properties of machine parts, and, consequently, and indicators of reliability of the product is solved at the present time due to the appropriate choice of processing conditions [1, 8]. Current trends in this direction are the study of the influence of the processing conditions on the formation of the surface layer and based on operational properties of details of the status parameters of this layer, as well as in the calculation of the numerical values of these parameters and to ensure appropriate conditions of technological process of manufacturing of details [8, 10, 11, 20]. The surface quality of machine parts is determined by the combination of roughness and undulation characteristics, physical and chemical properties and microstructure of the surface layer. This variety of parameters reflects the complexity of the concept of "surface layer". These parameters currently form the basis for selecting a specific scheme of characteristics that are regulated in the design and technological documentation.

The urgency of solving the problem of ensuring the quality of technological processes with a sufficient probability of obtaining the required output parameters sets the task of developing a methodology for the integration of software and information systems of electronic support of production processes.

3. Methodological base

The development of the methodological base for modeling and evaluating advanced technologies is also dictated by the need to create and subsequent certification of quality assurance systems for products and its elements. The lack of a scientific approach in this case, as a rule, leads to irrational, energy-intensive methods of technological supply of parameters, requiring a large investment of time and labor, to increase the cost of manufactured parts without a proper increase in their reliability.

Technological audit serves as the basis for the development of design and technological measures and related documentation for implementation and implementation in production [2, 19]. Expert methods should be considered as one of the possible approaches to a comprehensive study of complex problems in which the final solution is not explicit. The expert method for assessing the effectiveness of technological solutions in the production of complex equipment is rationally used in cases where it is difficult to apply methods of objectively determining the values of complex integrated quality indicators due to their significant number, many relationships and regulators.

The methodology of technological inheritance is used to describe the mechanism of transferring product properties during the technological route of production. The concept of heredity, which originally arose in the field of analysis of the transfer of properties in biological organisms, is now beginning to enter such areas of knowledge as cybernetics, object-oriented programming, and system analysis. The development of methods of mathematical, structural-parametric, and functional analysis in relation to the production processes of high-tech engineering technology ensures the validity of this term. Technological inheritance is the phenomenon of transferring properties from previous manufacturing operations to the next; these properties can be both beneficial and negative. The scope of the term covers the stages of production and operation and is consistent with the ideology of product information support at all stages of the life cycle (CALS).

As a tool for a comprehensive description of complex phenomena arising in the process of forming precision parts and modifying their working surfaces, the most optimal use of matrix computing software MatLab (Math Works, Inc.) and the optional Simulink visual simulation extension package for modeling dynamic systems. The implementation of the principles of visual programming allows you to design models of complex structure and hierarchy. The researching of technological systems subject to random perturbations is carried out using statistical modeling and testing methods, experimental design techniques, statistical data treatment [21, 22].

4. Modeling the structure of the technological process

A technological system is a set of functionally interconnected elements of an enterprise's production structure, technological equipment and performers performing technological processes of product manufacturing under regulated conditions in accordance with the requirements of regulatory and technological documentation. According to the concept of an electronic description of technical and technological systems, in the course of their life cycle, their models should be presented in a computer environment in the form of a hierarchy of information models that make up a single whole and have a subordination, where each subsequent model is more detailed than the previous one and contains additional information.

A technological system, considered in time with discrete states of technological operations and transitions, is a tightly connected information network of a hereditary nature. The carrier of hereditary information is the actual surface of the part with a variety of parameters describing its condition.

The structure of the technological process can be represented in the form of a graph, which is based on an oriented elementary link that characterizes the conditions for a separate operation of the technological process while ensuring the considered parameter. A specific property of the workpiece, expressed by the input characteristic of the link ($X_{v(p-1)}$), changes during the technological operation in accordance with the transmission coefficient of the hereditary relationship (K) for this operation and is characterized by a new value (X_{vp}): $X_{vp} = f\left(K_{v(p-1)p}, X_{v(p-1)}\right)$. These changes are determined by the action of a set of technological factors $\{t_1, t_2, \ldots\}$ for each operation of the technological process (figure 2).

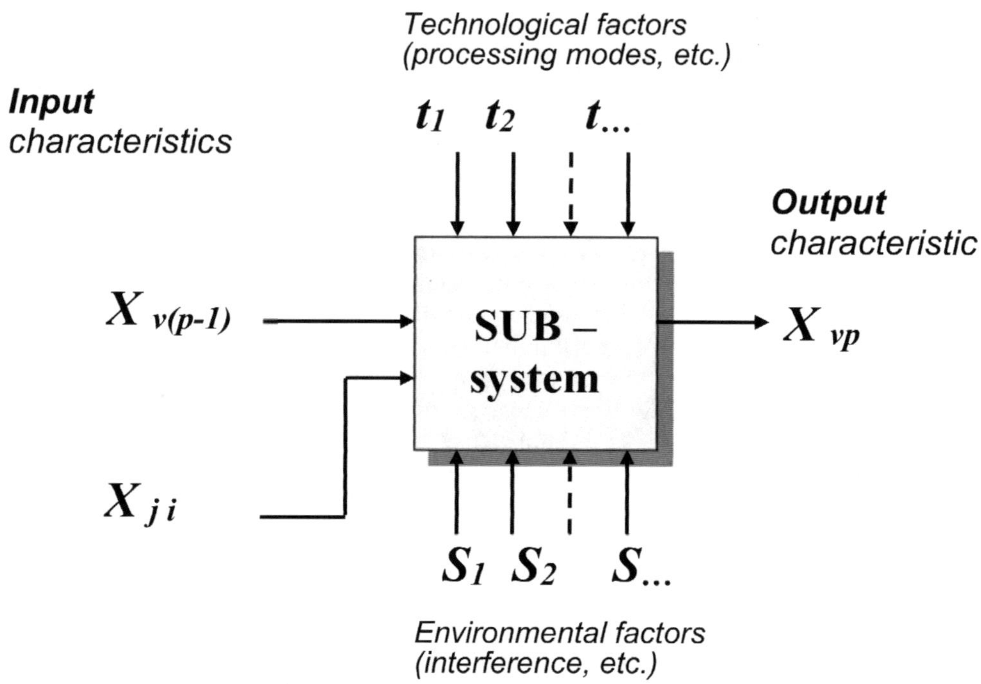

Figure 2. Concept model of a link of a technological system (operation): p – the observable operation number; v – the observable parameter number; j – some number of some influencing parameter (if there is a transverse connection); i – some number from previous operations.

From the point of view of system analysis, the technological operation is presented in the form of a “black box”, the input of which contains the following information signals: 1) the value of the investigated surface quality parameter before processing in this operation; 2) technological parameters (processing modes, tool characteristics, etc.); 3) random environmental factors that may affect the output characteristic (interference). At the output of the SUB-system, we obtain the analyzed output characteristic, which is the value of the investigated surface quality parameter after processing in this operation.

With a complex manifestation of technological inheritance, correlations of heterogeneous parameters of the surface layer are traced, such as, for example, the effect of residual stresses on the geometry of the product. The so-called transverse hereditary bonds are formed. In the general case, the technological process is penetrated by hereditary bonds of longitudinal and transverse species (figure 3) [2, 13].

A comprehensive assessment of the manifestation of technological inheritance can be represented by the subsystem of the integral characteristics of technological inheritance (figure 4), where the transmission coefficients of hereditary relationships (*K*) are presented as weighting factors for the quality indicators of the part (*X*) formed during the operations of the technological process:

$$X_{vp} = \frac{1}{K_{v(p-1)p}} \cdot X_{v(p-1)} + \frac{1}{K_{vip}} \cdot X_{vi} + \frac{1}{K_{vipj}} \cdot X_{ji} \quad . \tag{1}$$

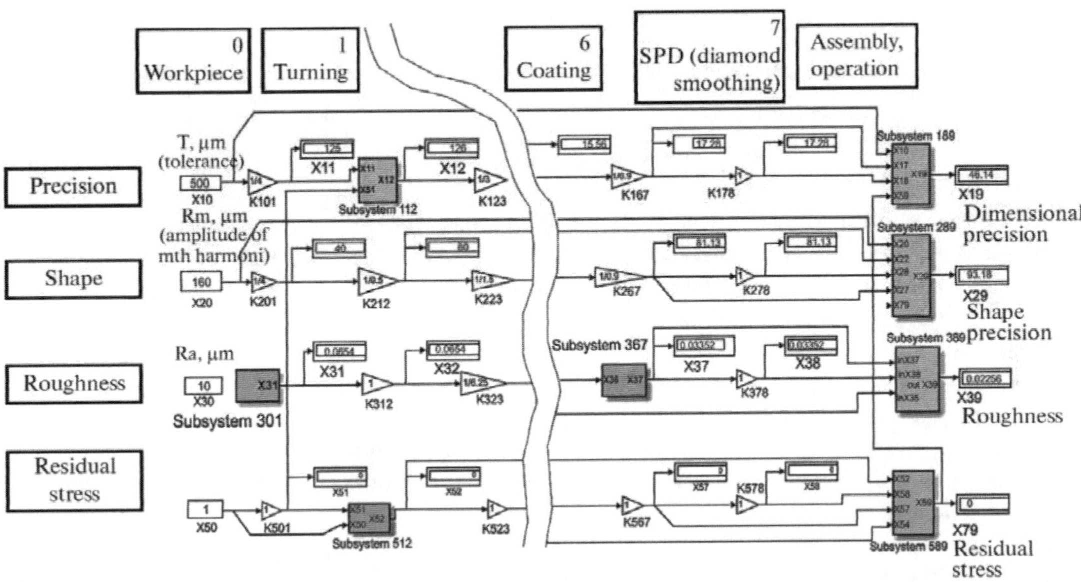

Figure 3. Implementation of a simulation model of technological inheritance in MatLab Simulink.

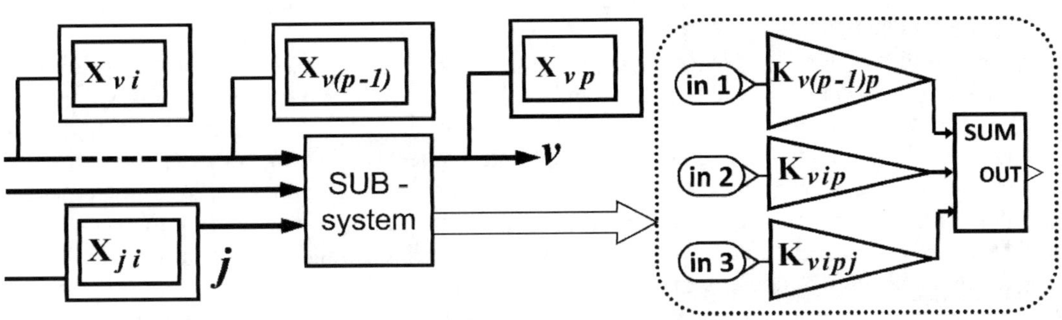

Figure 4. Subsystem of complex hereditary relations of the generalized form.

The mathematical model of the general form of technological inheritance is very multidimensional:

$$X_{vp} = \sum_{j=1}^{s} \left[\sum_{i=0}^{p-1} \left(\frac{1}{K_{vipj}} X_{ji} \right) \right], \tag{2}$$

where X_{vp} – observable of *v*-th detail parameter quality after *p*-th process operation; *v* – the number of the observable parameter of the quality of the part (workpiece); *p* – the number of the observable technological operation; *i* – the number in the sequence of technological operations from the beginning of the process to the operation under consideration ($i = 0, 1, \dots, p$); *j* – current number from the number taken into account ($j = 1, 2, \dots, s$) basic geometrical and physico-mechanical parameters.

When forming each of the links of the complex model of technological inheritance, it is possible to use both one method and different ways of describing transfer mechanisms. The coefficients of transmission of hereditary connections (K) have one of the following types:

- linear heredity coefficients (*K=const*);
- analytical dependencies, functional relationships with initial parameters, technological modes of operation, and environmental factors $K = f\left(X_{ij}, \mathbf{t}_{ij}, \mathbf{S}_{ij}\right)$;
- regression models of various degrees indicating weight estimates of the influence of initial characteristics, technological factors, and their combined influence $K = f\left(X_{ij}, \mathbf{t}_{ij}, \mathbf{S}_{ij}, \beta_{XtS}\right)$
- stochastic dependencies, statistical relationships with the characteristics of mathematical expectation, variance, etc.

The source of information about the type and particular analytical expressions of the coefficient of transmission of hereditary connections are determined by one of the following methods:

- experimental research in the subject area of material processing technology with statistical processing of experimental data;
- from published sources of research results, dissertations in the subject area;
- predicting the expected dependencies from the acceptable range for both standard situations and the risk zone.

When studying the process of property formation, a technological operation should be selected as the structural unit of the process. It is necessary to use special models that characterize the contribution of individual mechanisms in the formation of the value of a particular parameter of the quality of the part. When building individual links of the technological system, it is possible to consider them both locally, without connection with other operations of the technological process with the interruption of external links, and comprehensively, in conjunction with other elements of the system.

The model of technological inheritance of a real technological process is a subset of the general model, which includes only significant relationships of a hereditary nature. A description of the process of technological inheritance using a system of equations facilitates the task of quantifying properties. The number of variables included in its full description at the stage of analysis determines the scale of the object.

The simulation model of the technological system in the environment of MatLab Simulink is built using the folding mechanism of the structure, and each transmission link of the hereditary connection, in turn, can be a subsystem determined on the basis of scientific research, statistical analysis, technological audit and expert evaluations. The simulated object (technological system) is represented graphically by its functional parametric block diagram, which includes blocks of system elements and communication between them. In terms of Simulink, such a block diagram is referred to as the S-model. Thus, figure 3 presents a simulation model of a technological system for ensuring the quality of the surface layer of parts using the example of manufacturing a hydraulic cylinder body of a precision pneumatic-hydraulic system, implemented using the MatLab Simulink visual simulation system using drag-and-drop technology. The functional blocks of the elements of the simulated system can, in turn, be nested subsystems with their organization, forming hierarchical structures. Subsystems, the structure of which during modeling can be either visually hidden or expanded if necessary, decomposition and detailed analysis, in the terminology of Simulink are referred to as SUB systems (SUB system).

An example of a model of a subsystem of a technological operation for diamond smoothing is presented on the figure 5. Diamond smoothing refers to methods of surface plastic deformation processing and plays a special role in shaping the quality and performance properties of parts [12, 23].

The use of diamond smoothing at the final stages of the technological process allows you to multiply the bearing capacity of the surface and the resource of friction units, by increasing the surface hardness, reducing roughness, and distributing residual mechanical stresses in the surface layer [24]. The result of the technological operation depends on such modes as kinematic scheme, feed, processing speed, tool clamping forces, material and geometry of the tool, etc.

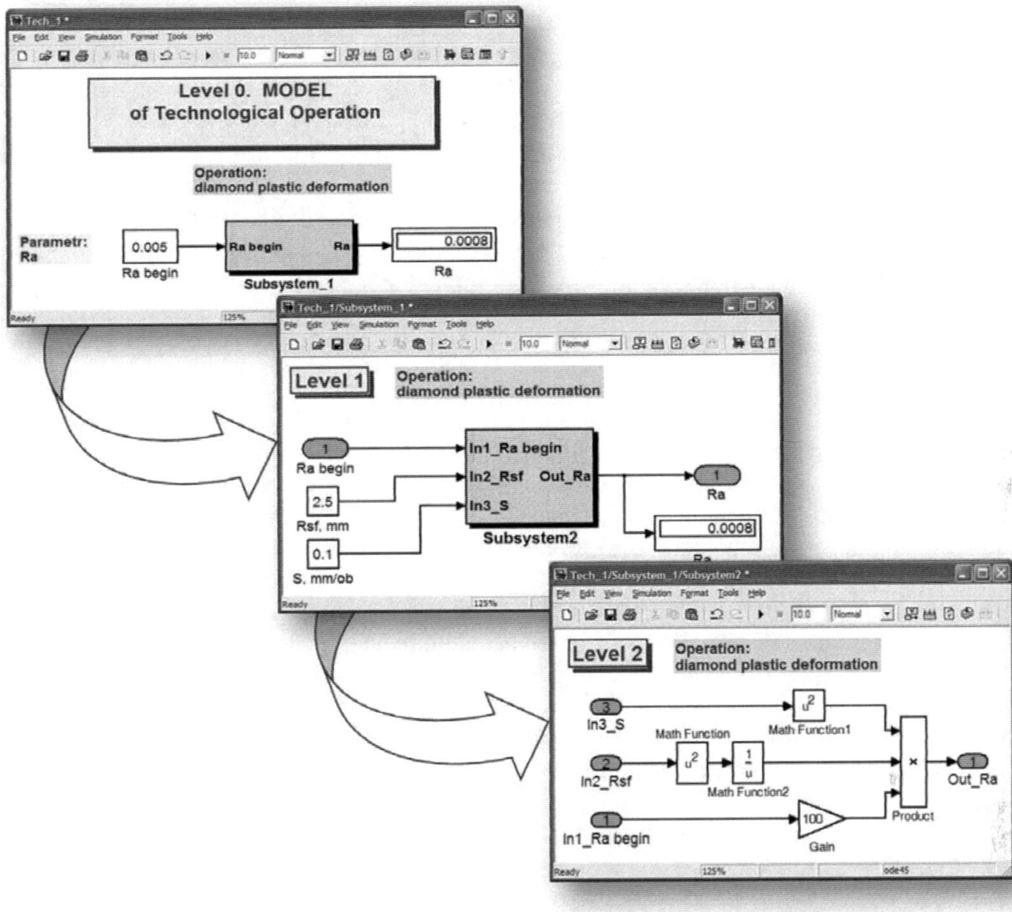

Figure 5. Hierarchical modeling in the refinement of technological systems.

The principle of visual modeling of a link in a technological system is shown by the example of the formation of surface roughness (R_a) when smoothed with a diamond indenter. The longitudinal and transverse roughness, the degree and depth of hardening of the machined surface depend on the initial roughness ($R_{a\,previous}$) of the part, the pressing force (P_y) and the radius of the sphere (R_{sph}) of the diamond indenter. Reducing the initial roughness before processing reduces the value R_a after processing. The value of the speed during processing slightly affects the value R_a. The feed has the greatest effect on the resulting surface quality, on the one hand, in the case of a small feed, each surface point undergoes deformation several times, on the other hand, a small feed reduces the processing productivity, an increase of 1.1÷1.2 times leads to an increase in roughness by 1.5÷2.0 times. After processing with optimal conditions, there is a relation, was the basis of the model of the transmission link of the technological system:

$$R_a = K \cdot R_{a\ previous},\ \text{where}\ K = f\left(\mathbf{t},\ R_{a\ previous}\right),\ \text{thus}\ R_a = \frac{110 \cdot s^2}{R_{sph}^2} \cdot R_{a\ previous} \quad (3)$$

where $\mathbf{t}$ – vector process parameters, such as feed per revolution (s), tool radius (R_{sph}).

The tools of the simulation environment allow the seamless integration of S models into the overall system, using the principle of folding the structure. In this case, it is possible to embed subsystems of individual links of a technological system without affecting its general structure, or with minimal transformations of complex longitudinal and transverse connections.

The MatLAB Sumulink s-model is implemented as an example of the method of technological audit of production of precision pneumatic hydraulic system parts. The dimension of the model is $(9 \times 9 \times 9)$ (9 positions of the observed technological operations; 9 positions of the observed parameters of the surface layer; from 3 to 9 dynamically varying technological parameters and conditions of individual operations). To study the effect of processing conditions, the following limits of variation were selected: Feed per revolution s = 0.02-2 mm/rev, $R_{a\ previous}$ = 0.64-8 μm, R_{sph} = 1-4 mm, P_y = 120-300 N.

Connecting to the system of simulation of the random effect (the signal source as a random number generator of the normal distribution law) allowed us to confirm the technological stability of the operation.

The effectiveness of embedding the diamond smoothing operation in the technological process of obtaining precision holes of pneumatic cylinders as a regulator of technological inheritance of such parameters as hardness, roughness, and residual stresses was shown by subsequent production testing. Full-scale products have shown that diamond smoothing is the most suitable method of surface-plastic deformation for processing surfaces of hard metals and coatings [23, 25]. This is due to the high hardness of the diamond and the specific conditions of its contact interaction with the doped layer, which results in plastic deformation of the hardest deposited materials.

The synergistic effect of the combined effect of coating and diamond smoothing operations provides broad prospects for combined processing in order to increase the resource and reliability of critical tribotechnical units of aircraft. The new technology of combined hardening is based on a combination of electroerosive hardening and diamond smoothing with the creation of the effect of surface reinforcement and the creation of special regular microreliefs.

Progressive methods of exposure to concentrated energy flows of various nature, such as surface plastic deformation [25], electrospark alloying [26], combined methods of hardening and shaping [23, 26] are considered as powerful control links in the technological chain of ensuring operational properties of critical parts and assemblies aviation technology.

5. Technological expertise in making production decisions

In practice, the implementation of technological systems for shaping and modifying surfaces is a multivariate creative task under fuzzy conditions and a lack of information [13]. Alternative technological methods have both advantages and disadvantages. Therefore, it is not always obvious which method is optimal for use in a particular technological process. Making a decision in this case is an action on a multitude of alternatives, which results in a subset or a single selected option. In the conditions of a limited resource of time and means for making technological decisions, an effective tool is the use of automated systems for supporting the adoption of technical decisions based on multivariate and multicriteria optimization algorithms, mathematical statistics, and processing of expert opinions.

The evaluation criterion when choosing a solution is a system parameter that reflects a weighted set of features by which a single solution can be distinguished from many alternative ones. An array of assessment criteria is a subset of the full array of parameters of a technical (technological) system selected according to the purpose of the assessment. The classification of criteria is similar to the classification of parameters of technical systems [2] and includes groups of indicators reflecting the degree of perfection of the product design, its operability, production and operational manufacturability;

the degree of perfection of the production system; commercial, environmental, social and other indicators.

Integral assessment of technological alternatives by criteria includes the following steps: establishing for each option the degree of satisfaction (fulfillment, quality) of requirements for each criterion (F_{jm}); calculation of the relative quality indicator of the compared options for each criterion $f'_{jm} = F_{jm} \cdot q'_j$ of the criterion m, taking into account the significance (importance) of the criterion (q'_j); calculation of a complex (integral) quality indicator of each option as the sum of local estimates taking into account the weight contribution of the significance of the criteria: $Q_m = \sum_j f'_{jm}$; generalization of opinions received from each expert, taking into account qualification factors and the significance of the members of the expert group. The conclusion of the technological expertise is formed with the calculation and indication of quality according to the parameters of consistency, accuracy and reliability.

Based on the results of calculating the eigenvectors of the priority matrix of influence on the failure-free operation of precision pneumatic hydro units, the weight of the influence of technological operations was shown (in normalized estimates):

- honing and superfinishing of cylindrical surfaces – 0.3;
- surface hardening with a diamond tool – 0.25;
- application of special wear-resistant coatings – 0.25;
- other operations (summary) – 0.2;

and weight of influence of surface layer parameters:

- geometric accuracy – 0.29;
- roughness – 0.26;
- surface hardness – 0.25;
- other parameters (summary) – 0.21.

The infological data model when implementing information processes for processing expert assessments in multivariate and multicriteria problems represents a multidimensional matrix structure. The software implementation of the automated system organically lays on the combination of the methods of relational (tabular) data representation in Excel and information representation of hierarchical networks in MatLAB [13].

The advantages of the software implementation of the developed automated system of technological expertise include: automation of statistical calculations; prompt response to a change in production conditions, objects and subjects of analysis; the ability to forecast development options and the consequences and risks of the adoption of various innovative decisions and projects; modular structure, flexibility and the possibility of modification with the aim of further expanding the range of tasks; visualization and interactive dialogue support [13, 19]. Modern information technologies provide an opportunity for modeling processes, operational analysis, preparation and presentation of results for subsequent decision-making.

6. Conclusion

As a result of research, a methodology has been developed and tools for variative forecasting of operational characteristics of aircraft manufacturing facilities and automation of audit procedures for complex technological systems have been proposed.

The novelty of this method is the use of the principles of technological inheritance of quality parameters of parts in the process of processing under the influence of technological energy and their relationship to performance characteristics.

A hierarchical model of the technological system implemented in the MatLAB Simulink visual modeling environment is proposed. Technological regulators are defined in the image of progressive technological operations of forming and modifications of precision surfaces aimed at increasing the resource of precision friction units. Establishing patterns of technological inheritance can improve the reliability of technological systems by using special technologies as a regulatory link in the technological chain.

A model of a technological system built using an integrated technique is a dynamically updated structure and allows us to analyze the effectiveness of technological support both in individual operations and in the technological process as a whole. When constructing the individual links of the technological system, it seems possible to consider them both locally, without communication with other operations of the technological process with interruption of external connections, and comprehensively in conjunction with other elements of the system.

The integration of one of the fastest matrix mathematical systems - MatLab - with the Simulink simulation package opens up new possibilities for using the most advanced mathematical methods to solve the problems of dynamic and situational modeling of complex technological systems. MatLab Simulink has the ability to organize simulation models of dynamic systems, analyze their functioning and evaluate the output characteristics of processes. The modular construction principle and intuitive interface provide dynamic model management, structure editing and the addition of functional modules to the model.

The development of automated means of monitoring and control of the system of parameters of the state of the surface layer during the technological process, using the laws of technological inheritance, allows you to dynamically combine the experimental results of the study of methods of technological support of operational properties and computer-aided design of technological processes, which makes it possible to analyze many options for technical solutions to increase reliability results coding, forecasting and management.

Tools for processing expert assessments in multidimensional and multi-criteria tasks can improve the efficiency and effectiveness of technological management of industrial enterprises.

References

[1] Eliseev Yu S, Boytsov A G, Krymov V V, Hvorostukhin L A 2003 *Production technology of aircraft gas turbine engines* (Moscow: Mashinostroyeniye) p 512

[2] Siluyanova M V, Kuritsyna V V and Boitsov A G 2017 *Models and Methods of Technological Audit of Hi-Tech Industries* (Moscow: Mosk. Aviats. Inst.) p 160

[3] Morgunov Yu A, Panov D V, Saushkin B P and Saushkin S B 2014 *Advanced Technologies in Machine Engineering Industry: Physical-Chemical Methods and Technologies* (Moscow: Forum) p 928

[4] Morgunov Y A, Saushkin B P and Shandrov B V 2016 Development of conceptual equipment of engineering technology. *Sprav. Inzh. Zh. Prilozh.* **4(229)** 3 doi: 10.14489/hb.2016.04.pp.003-007

[5] Isachenko V A, Astakhov Yu P and Saushkin B P 2016 Aerospace engineering technology: problems and prospects. *Journal "Tekhnologiya Mashinostroeniya"* **1** 10 [in Russian]

[6] Morgunov Y A, Saushkin B P and Shandrov B V 2019 Knowledge intensity if engineering production and its elements. *Science Intensive Technologies in Mechanical Engineering* **6(96)** 37 doi: 10.30987/article_5ce675a22352c1.74868398

[7] Kraev V M, Siluyanova M V and Tikhonov A I 2019 Assessment and improvement of rationality methods of modern aircraft engines design and technological solution. *Journal of Physics: Conference Series* **1353(1)** 012033 doi: 10.1088/1742-6596/1353/1/012033

[8] Boitsov A G, Kovalev A P and Novikov A S 2007 *Mechanical and Physicochemical Processing in Production of Aircraft Engines* (Moscow: The Bauman Moscow State Technical University) p 584

[9] Dal'sky A M, Bazrov B M and Vasilyev A S 2000 *Technological heredity in machine-building production* (Moscow: Mosk. Aviats. Inst.) p 364

[10] Bazrov B M, Rodionova N A 2019 Reconciling Quality Requirements over the Product Life Cycle. *Russ. Engin. Res.* **39** 407 https://doi.org/10.3103/S1068798X19050058

[11] Vasilyev A S, Yamnukov A S, Matveev I A, Yamnikova O A 2019 Influence of hereditary technological errors of production of a basic pipe on parameters of the assembled propulsion

section. *Chernye Metally* **1** 67

[12] Khvorostukhin L A, Shishkin S V, Kovalev A P and Ishmakov R A 1988 *Increase of the Bearing Capacity of Machine Parts by Surface Hardening* (Moscow: Mashinostroenie) p 144

[13] Siluyanova M V and Kuritsyna V V 2018 Automated management in aerospace production. *Russ. Eng. Res.* **38(3)** 201 doi: 10.3103/S1068798X18030085

[14] Pausits A 2019 Innovation Audit: Measuring Innovation Management Capabilities. In: Carayannis E. (eds) *Encyclopedia of Creativity, Invention, Innovation and Entrepreneurship.* (Springer, New York) https://doi.org/10.1007/978-1-4614-6616-1

[15] Bross U, Inzelt A, Reib T 1998 Principles of Technology Audit. In: Bio-Technology Audit in Hungary. Technology, Innovation and Policy (Series of the Fraunhofer Institute for Systems and Innovation Research) *Physica, Heidelberg* vol **7** https://doi.org/10.1007/978-3-642-52472-1_4

[16] Grigor'ev S N, Dolgov V A, Krasnov A V and Andreev N S 2015 A method of technologic audit of technical re-equipment projects in aircraft production enterprises. *Russ. Aeronaut.* **58** 244 https://doi.org/10.3103/S106879981502018X

[17] Saaty T L and Vargas L G 2012 The possibility of group choice: pairwise comparisons and merging functions. *Soc Choice Welf* **38** 481 https://doi.org/10.1007/s00355-011-0541-6

[18] Saaty T L 2016 The Analytic Hierarchy and Analytic Network Processes for the Measurement of Intangible Criteria and for Decision-Making. In: Greco S, Ehrgott M, Figueira J (eds) *Multiple Criteria Decision Analysis. International Series in Operations Research & Management Science,* (Springer, New York) **233** https://doi.org/10.1007/978-1-4939-3094-4_10

[19] Siluyanova M V, Kuritsyna V V and Iosifov P A 2016 *Strategies, Methods, and Models of Technological Development of Aerospace Industry* (Moscow: Mosk. Aviats. Inst.) p 160

[20] Kondakov A I and Vasiljev A S 2018 Evaluation of the types of multiproduct manufacturing of machine components and some aspects of their design. MATEC *Web Conf.* **224** 01121 doi: 10.1051/matecconf/201822401121

[21] Schenk H Jr 1979 *Theory of engineering experimentation. 3rd Revised edition* (Washington New York London: Hemisphere McGraw-Hill) p 320

[22] Orlov A I 2020 Organizational and economic modeling in the organization of production in the epoch of digital economy. MATEC *Web of Conferences* **311** 02001 https://doi.org/10.1051/matecconf/202031102001

[23] Boitsov A G, Mashkov V N, Smolentsev V A and Khvorostukhin L A 1991 *Surface Hardening of the Parts by Combined Methods* (Moscow: Mashinostroenie) p 144

[24] Shishkin S V 2019 Application of finish-strengthening process with diamond smoothing for bearing capacity increase of joints with tightness. *Bulletin of Bryansk State Technical University* **6(79)** 4 doi: 10.30987/article_5d10851eda9a19.64369355

[25] Kuritsyna V V, Siluyanova M V and Kuritsyn DN. et al. 2020 Local Surface Deformation of Precision Airplane-Engine Components. *Russ. Engin. Res.* **40** 266 doi: 10.3103/S1068798X20030144

[26] Denisov L V, Boitsov A G and Siluyanova M V 2018 Surface hardening in hydraulic cylinders for airplane engines. *Russ. Eng. Res.* **38(12)** 1080 doi: 10.3103/S1068798X18120237

IOP Conf. Series: Materials Science and Engineering **868** (2020) 012005 doi:10.1088/1757-899X/868/1/012005

Experimental determination of the influence of the radome on the characteristics of the radio-technical system

N Kh Gyulmagomedov

JSC "MIC" NPO Mashinostroeniya", 33 Gagarina Street, Moscow District, Reutov 143966, Russian Federation

E-mail: 940@vpk.npomash.ru

Abstract. This article proposes a methodology and considers the results of experimental studies of the influence of the antenna radome on the characteristics of the radio system installed under it. To study the influence, a technique was developed and tested to determine the level of signal attenuation in the radome wall during thermal heating. A detailed description of the methodology for measuring the level of signal attenuation in the radome wall is given. For research, a special stand has been created that allows measurements of the level of signal attenuation in the radome wall when it is heated. Measurements were made of the level of signal attenuation in the radome wall during heating in accordance with a given temperature dependence on exposure time. Based on the analysis of the results obtained, it was found that an increase in the level of signal attenuation can lead to a deterioration in the radio characteristics of the microwave system «antenna- radome» and the radio system as a whole.

1. Introduction

The dielectric materials intended for the manufacture of radome for aircraft should have special properties, which is associated with the need for the radome to work under conditions of heating with an oncoming aerodynamic flow (figure 1).

The radome must withstand particularly high thermal and mechanical loads as part of high-speed aircraft. The radome performs two main functions: it must provide effective thermomechanical protection of the aircraft equipment at high temperatures and pressure created by the oncoming unsteady air flow in flight, and on the other hand, it must maintain the specified radio characteristics of the radome, providing high-precision measurements of the kinematic characteristics of the target by the airborne radar, in particular angular measurements, in various modes and flight conditions of the aircraft.

Operating experience and numerous theoretical and experimental studies have shown that radomes made of even the best dielectric materials with high accuracy noticeably change the radiation characteristics of the antennas they protect [1].

Radomes of large elongation, usually installed on high-speed aircraft, noticeably worsen (radio transparency decreases, angular measurement errors increase) if the radome is unevenly heated by the oncoming aerodynamic flow [2]. The magnitude and laws of temperature changes on the side surface of the radome as a result of aerodynamic heating depend on a number of physical factors: speed and altitude, thermophysical properties of the material, compartment design, configuration of the radome, etc.

IOP Conf. Series: Materials Science and Engineering **868** (2020) 012005 doi:10.1088/1757-899X/868/1/012005

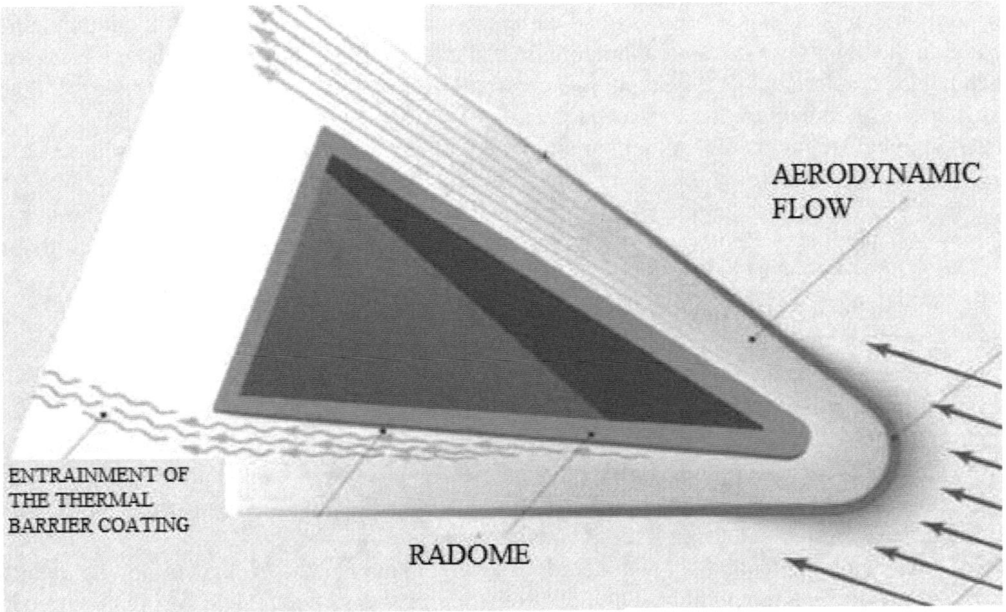

Figure 1. Radome heating with oncoming aerodynamic flow.

Depending on the design of the radome and operating conditions, a significant temperature gradient arises: along the generatrix of the radome (figure 2). Even for a constant speed of high-speed flight, the surface temperature along the generatrix of the pointed radome will be different. During thermal heating, the temperature appears to be maximum at the surface near the tip of the pointed radome and minimum at its base.

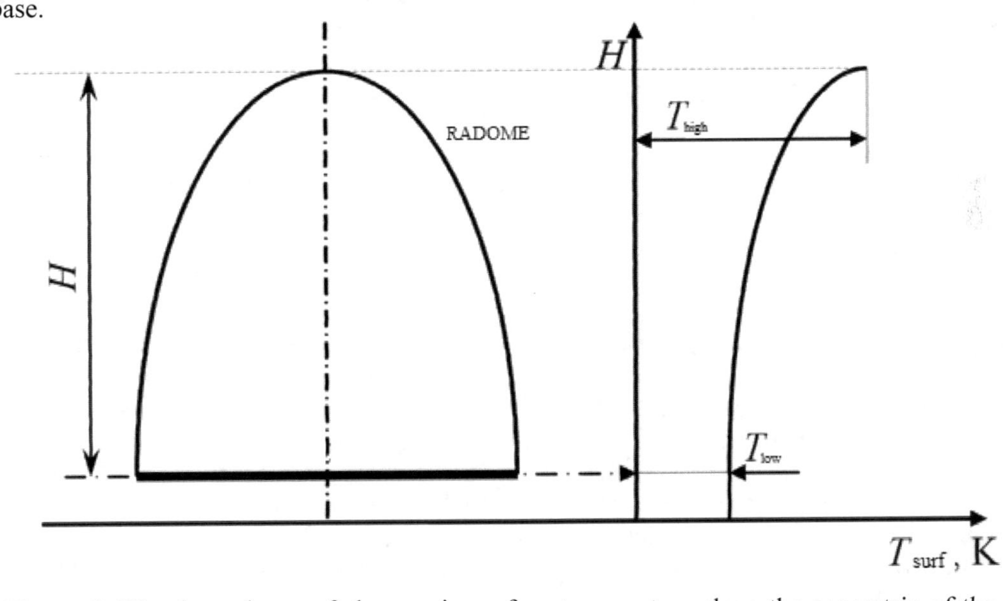

Figure 2. The dependence of changes in surface temperature along the generatrix of the pointed radome during thermal heating.

In [3], a numerical simulation of aerodynamic flow around a high-speed flight of an aircraft for different angles of attack is presented. An analysis of the simulation results shows that when the angle of attack changes, pressure and temperature are redistributed on the radome surface.

Thermoradiotechnical tests, which must be carried out both in the cold state of the radome, after heat heating and in the process of its thermal heating, in order to establish the dependences of changes in the radio technical characteristics of the radome on external influences, are of great importance for ensuring high radio technical characteristics of the radome.

Currently, measurements of radio characteristics under normal conditions based on anechoic chambers are not difficult. However, there is a need to create techniques for predicting dynamic changes in the parameters of radomes, which correlate well with experimental results during thermal heating of the radome. External influences (especially with regard to temperature effects) can significantly change the parameters of dielectrics.

The main methods for measuring dielectric parameters during thermal heating are [4]:

- resonance method;
- waveguide method;
- optical method;
- free space method, etc.

Each of the above methods for measuring physical parameters has its own characteristics. So it should be noted that optical methods solve the problem of measuring the physical parameters of dielectrics in the millimeter frequency range using interferometers and open resonators.

It should be emphasized that all waveguide and resonance methods, except for the restrictions determined by the limiting temperatures, do not allow measurements to be made during dynamic heating of the dielectric, as well as when its state changes. As shown, the waveguide and resonator methods are suitable only for statistical investigation of the temperature dependences $\varepsilon(T)$ and $tg\ \delta(T)$. However, they are the main methods for measuring physical parameters in the decimeter and centimeter frequency ranges.

The greatest successes in measuring the parameters of dielectric materials with a change in their physical state associated with high-temperature heating were achieved in [5]. In this paper, the physical and technical properties and technological features of obtaining a number of materials of radiolucent ceramic radome are considered. The main frequency dependences of the materials of dielectric parameters under the influence of thermal heating are revealed.

The dynamic measurements of the radio characteristics of aircraft radome have been the subject of several works. So in [6] a method for measuring the radio technical characteristics of a radome is presented.

To implement this method of measuring the radio characteristics of radome, a special stand has been created that ensures the following actions. The bottom line is the sequence of heating and measurements. A radome with an antenna system mounted on a slewing ring is first heated by infrared emitters to a certain temperature. After heating, the slewing ring is rotated 180° and carry out the measurement of radio characteristics.

Another method for measuring the radio technical characteristics of radome is the method proposed in [7]. The essence of the method lies in the fact that the temperature field on the outer surface of the radome is created by focusing on the surface using parabolic reflectors of radiation from individual linear infrared emitters, which are located along the focal lines of the reflectors. Emitters with reflectors are located along and around the axis of the radome at such a distance that its ratio to the radius of the radome is more than ten. A receiving antenna is installed inside the radome, and a transmitting antenna on the outside from the nose of the radome, or vice versa.

The mentioned methods, in fact, cannot always give accurate information about the dynamic dependence of the change in radio technical characteristics, taking into account the peculiarities of uneven heating of the entire surface of the radome. Another reason for using these methods is related to their technical implementation. In the first case, in the conditions of measurements it is impossible to talk about the continuous and accurate determination of the radio technical characteristics during heating, since the test object has time to cool down during the dilution of the heaters, which is a necessary procedure for this method. In the second case, zonal heating does not always ensure heating of the radome surface as in real aerodynamic heating.

To conduct accurate dynamic reference to the gradient of temperature measurement on the surface of the radome, a method is proposed for measuring the radio technical characteristics of the radome during heating.

The purpose of these studies is to test experimental studies of the influence of the antenna fairing on the characteristics of the radio system installed under it.

The novelty of this work, in a similar context, is experimental radio technical research under conditions as close as possible to aerodynamic heating. Conducting experimental radio engineering research is complex and expensive in terms of technical implementation. Therefore, most similar studies have theoretical results [8-10].

2. The method of measuring the level of attenuation of the signal in the wall of the radome during heating

This method allows measurements of the radio technical characteristics of the radome during heating, as close as possible to real heating during flight. A general view of the installation for carrying out measurements is presented in Figure 3. Heating is carried out using infrared emitters (heaters) [11]. With this heating method, infrared heaters are usually located at a close distance from the test object. Therefore, it is necessary to take into account their presence when conducting measurements of radio technical characteristics. The functional diagram of the installation with the main blocks is shown in figure 3.

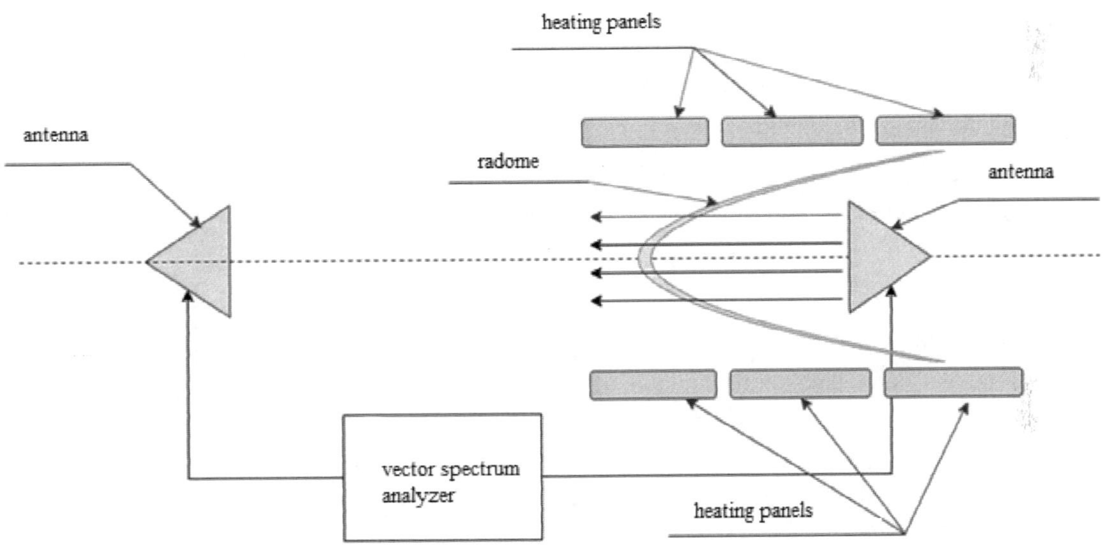

Figure 3. Functional diagram of the installation.

For heating, heating panels with quartz tube lamps with bent ends are used. Reflected polished aluminum panel screens are water cooled. Lamps with the aim of cooling a quartz bulb in forced mode are blown with air. For this, a collector for blowing lamps is installed on each heating panel. The heating panels are attached to the sliding frame. The panels are arranged in three rows of eight panels in a row, forming, when the frame is closed, three rings around the radome (figure 4). The radome is heated in 6 independent control zones (bottom, top, side, bow, middle and aft) according to the appropriate temperature regime.

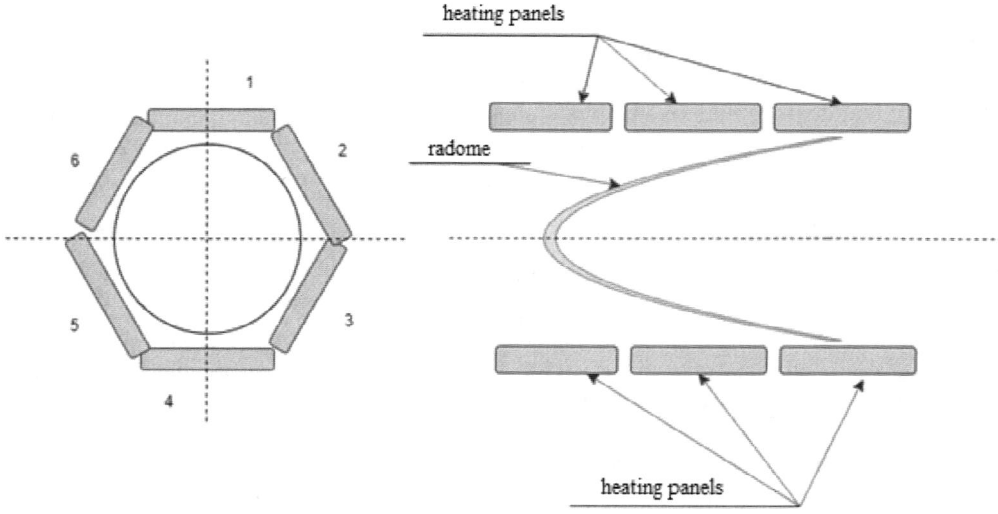

Figure 4. Location of heating panels around the radome.

To measure temperature and control the heating mode, thermocouples are installed on the radome opposite each heating panel (figure 5). This arrangement of heating elements and measuring thermocouples allows you to precisely control the heating course and temperature in all sections of the radome. The experiment accepted the condition that thermocouples do not affect the measurement results, since the thickness of thermocouples is much less than the wavelength λ.

Measurement, processing and recording of the readings of all temperature sensors and control of the heaters is carried out by the heating control and management system.

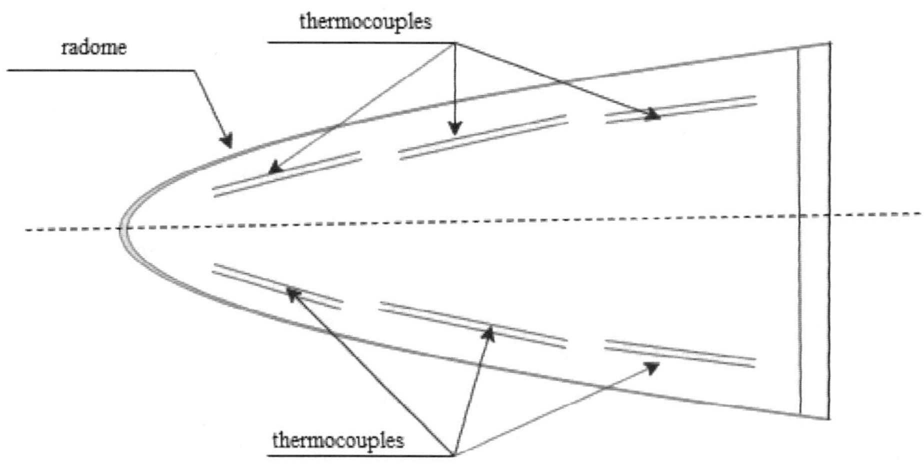

Figure 5. Location of thermocouples on the radome.

According to the methodology being created, the test object is a prototype of the radome, which will be subjected to thermal heating.

Dynamics of temperature change is chosen arbitrarily. In this case, during testing, the temperature gradually increases on the surface of the radome to T_{max} (° C). A typical temperature dependence on exposure time is shown in figure 6.

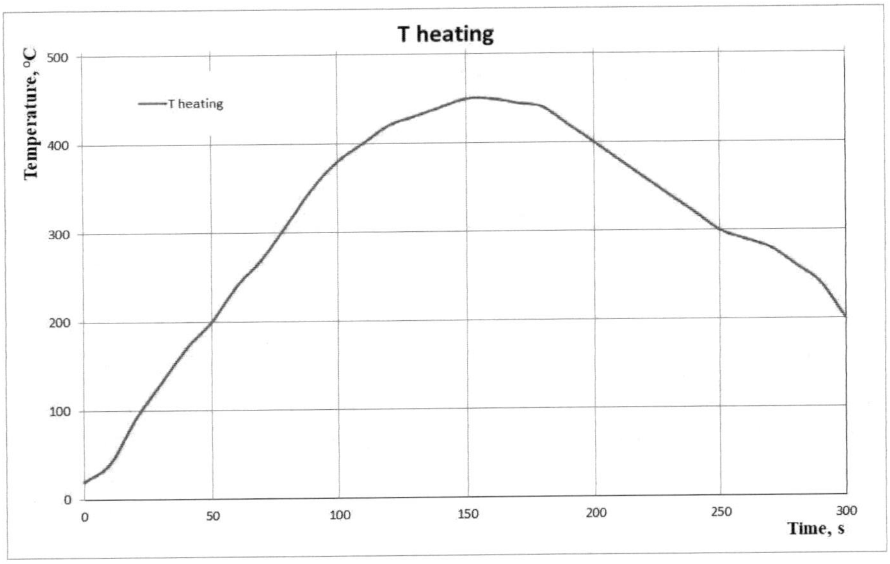

Figure 6. Temperature change during heating.

3. Measurement of the signal attenuation level in the radome wall during heating

The main purpose of the tests is to measure the level of attenuation of the signal in the radome during its aerodynamic heating.

Signal attenuation in the radome under normal climatic conditions is compensated and reduced to zero dB.

Measurements of the radio transparency coefficient are relative. Therefore, the tests are carried out sequentially in 3 stages in the following sequence:

- measurement of the attenuation level of the signal without radome (calibration);
- measurement of the level of attenuation of the signal in the radome under normal conditions;
- measuring the level of attenuation of the signal in the radome during heating according to the dynamics of temperature changes on the surface of the radome as in Figure 6.

The measurements are performed by the free space method [12], when the change in the signal power level after passing through the radome wall is controlled. The measurement results were recorded using a vector network analyzer.

4. Results and discussion

According to the results of the tests, the dependence of the signal attenuation level in the radome during heating was obtained, according to the dynamics of temperature changes on the radome surface. The dependence is shown in figure 7.

According to the results of the analysis, it was found that in the maximum heating temperature the highest value of the signal attenuation level in the radome wall was noted. Analyzing the obtained experimental results, a conclusion follows about the influence of thermal heating of the radome on the characteristics of the radio engineering system installed under the radome.

An increase in temperature led to a change in the level of signal attenuation in the radome wall. A probable reason for the increase in the attenuation level is a change in the dielectric parameters of the material [13]. Thus, a change in the dielectric constant of the material of the radome wall leads to a shift in the working frequency of the radome, and an increase in the level of attenuation of the signal indicates

IOP Conf. Series: Materials Science and Engineering **868** (2020) 012005 doi:10.1088/1757-899X/868/1/012005

an increase in active losses in the radome, which are directly proportional to the tangent of the angle of dielectric losses.

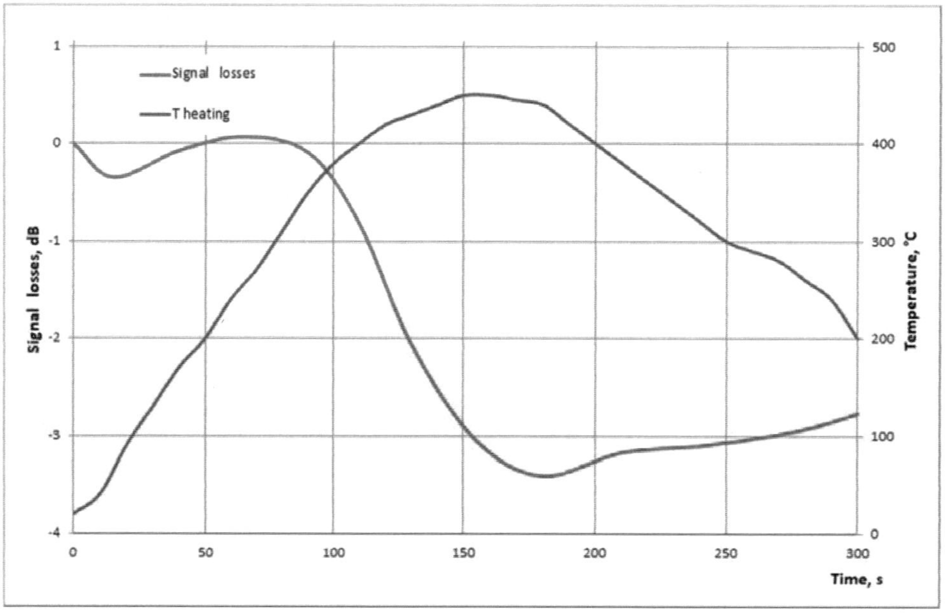

Figure 7. The dependence of the level of attenuation of the signal in the radome during heating.

It should also be noted that the process of heating the radome led to irreversible changes inside the wall of the radome that the radio technical characteristics measured after cooling do not coincide with the characteristics before heating. A possible reason for such changes is a partial or complete burnout of the binder in the material of the radome wall. Burnout of the binder may have led to a decrease in the dielectric loss tangent, which led to a decrease in the signal attenuation level at the final heating stage.

The methods proposed in the international literature cannot always give accurate information about the dynamic dependence of changes in radio technical characteristics, taking into account the peculiarities of uneven heating of the entire surface of the fairing due to the complexity of their technical implementation. The main reason is the complicated technical implementation of continuous and accurate reproduction of aerodynamic heating.

5. Conclusion

The above quantitative estimates for calculating the temperature changes in the radio characteristics of a radome undergoing thermal heating show a significant temperature deviation of the properties of the used dielectrics and, as a result, often unacceptable deterioration of the radio characteristics of the microwave system "antenna- radome" and the radio system as a whole. All of the above fully applies to antenna windows (radomes) and dielectric sealing inserts of any configuration and size, operating in powerful microwave paths.

The study of dielectrics in various temperature conditions allows us to proceed with the prediction of operational parameters and characteristics of radomes made of these materials.

References

[1] Kaplun V A 1974 *Microwave Radomes (Calculation and Design)* (Moscow: Sovetskoye radio) p 240
[2] Voronin M Ya, Karmanov I N, Koshelev A V, Noskov M F, Lesnykh I V, Mitsenko I D, Ushakov O K 2008 *Physics of Aircraft Microwave Dielectric Heating and Their Protection: Monographs* (Novosibirsk: SSUGT) p 156

[3] Grishin N V, Prikhod'ko A A 2016 Numerical simulation of supersonic flow around a Zenit-2SLB launch vehicle. *Theoretical mechanics* **4** 24 [in Russian]

[4] Vorob'yev Ye A, Mikhaylov V F, Kharitonov A A 1977 *Microwave dielectrics at high temperatures* (Moscow: Sovetskoye radio) p 208

[5] Yegorov V N, Kashchenko M V, Masalov V L, Tokareva Ye Yu 2013 Installation for measuring the dielectric parameters of materials when heated to 1800 °C. *Higher Istituions News: Physics* **56(8/2)** 347 [in Russian]

[6] Dergachov A A, Slivko S A, Bobrov A V, Shekhtman M A, Shilo V K, Patent RU 2525844 (23.01.2013)

[7] Raylyan V S, Rusin M Yu, Fokin V S, Shadrin A P, Krylov V P, Patent RU 2626406 (18.08.2016)

[8] Haris M Y, Fernandez-Lopez A, Mohd Aris K D, Muzafar Z, Zuhudi N Z M, Laila D, Nurhafiza K 2019 Preliminary analysis of hybrid laminate composite for aircraft radome application. *IOP Conf. Series: Materials Science and Engineering* **697** 012032 doi: 10.1088/1757-899X/697/1/012032

[9] Raveendranath U N, Vandana S, Sandhya S, Jha R M 2015 Temperature-dependent Electromagnetic Performance Predictions of a hypersonic streamlined radome. *Progress In Electromagnetics Research* **154** 65 doi: 10.1109/TAP.2009.2026595

[10] Zhou L, Liu Z, Tang L, Pei Y 2017 Design and characterization for a high-temperature dual-band radome wall structure for airborne applications. *Materials and Design* **114** 264 http://dx.doi.org/10.1016/j.matdes.2016.11.067

[11] Baranov A N, Belozerov L G, Il'in Yu S, Kut'inov V F 1974 *Static strength tests of supersonic aircraft* (Moscow: Mashinostroyeniye) p 344

[12] Gurtovnik I G, Sokolov V I, Trofimov N N, Shalgunov S I 2003 *Radio-products from fiberglass* (Moscow: Mir) p 368

[13] Prigoda B A, Kokun'ko V S 1970 *Aircraft Antennas Radomes* (Moscow: Mashinostroyeniye) p 288

IOP Conf. Series: Materials Science and Engineering **868** (2020) 012006 doi:10.1088/1757-899X/868/1/012006

Functional hazards assessment of an integrated flight control system validation using model-based design

A Savelev[1,2], N Lituev[2] and E Olidaev[2]

[1]Faculty of Robotic and Intelligent Systems, *Moscow Aviation Institute,* 4 Volokolamscoe Schosse, Moscow 125993, Russian Federation
[2]Automatic Flight Control Systems Department, *Irkut Corporation,* 68 Leningradsky Avenue, Moscow 125315, Russian Federation

[1]E-mail: Artem.Savelev@irkut.com

Abstract. Functional Hazard Assessment is the beginning of the work on the system safety assessment and it directly affects the design, safety and final cost of the project. Errors in the Functional Hazard Assessment can lead to insufficient system safety or unreasonably expensive development. The main purpose of the work is to introduce the Functional Hazard Assessment validation process which minimizes the risk of incorrect development of MC-21 aircraft Flight Control System. To achieve this goal, Model-Based Design methods are used as validation methods. The time benefit in detecting and isolating errors in algorithms and software with the help of the Model-Based Design is shown. Integrated Flight Control System which includes fly-by-wire control system and automatic flight control system of the Irkut's airliner MC-21 safety requirements validation process using means of Model-Based Design is considered. As a result of the work, Integrated Flight Control System models were adapted for assessing functional hazards, and a user interface was developed for the flight crew to participate in the simulation of the failures.

1. Introduction

Nowadays, due to the MC-21 project approaching certification phase more tasks connected with safety documentation release are conducted. The Functional Hazard Assessment (FHA) is one of the main documents, which must be carried out before all safety analysis [1].

The FHA is used to assess the impact of the aircraft function failures on the aircraft and its occupants considering existing or assumed mitigation means and derive safety requirements for the aircraft design [2]. The purpose of the FHA is to clearly define each failure condition along with justifying their classification by criticality [3].

Previously, the safety assessment process in Irkut Corporation, consists in the developing, the updating and the verification of safety requirements at the same time as the MC-21 design as required by SAE ARP 4761. From the aircraft level the requirements descend to the Flight Control System (FCS) and Automatic Flight Control System (AFCS) level ("top-down" process) and are supplemented with derivatives according newly identified end effects (i.e. newly functional hazards) from Failure Mode and Effect Analysis ("down-up" process) of each component of Integrated Flight Control System that includes FCS and AFCS.

The main problem of carrying out the FHA is the lack of regulatory procedures and accepted processes for validating the classification of the requirements, according to the criticality of each failure condition. The relevance of this problem is connected with the fact that the validation materials (including the validation of safety requirements that are the result of the FHA) should be prepared during certification [4]. Validation documents (procedures and results) of the FHA thus should be prepared by the request of a certification authority.

Depending on the expected criticality of a functional failure safety requirements are formed, using the Failure Tree Analysis (FTA) and budgeting. These safety requirements include assignable Development Assurance Levels for functions (FDAL) and items (IDAL), which are responsible for these functions. Also, the required minimum allowable probabilities of occurrence of failures of system elements are included in the safety requirements.

Requirements for independence are based on the rigidity of DAL and directly affect the final cost of the project, the necessary time and labor.

An incorrect assumption of the fail severity thus requires making adjustments in the design at the later project stages (bench / flight tests):

– The product satisfies safety requirements, but inadequately expensive; there is a risk of reduced demand in favour of more affordable competitors;

– The product does not satisfy the safety requirements; redesign is required, which entails a delay in the production, certification and delivery of the developed system.

In order to exclude the possibility of incorrect determination of the criticality of a failure, it becomes necessary to validate each failure condition. One of the validation methods is mathematical modeling. If the mathematical models reflect the behaviour of the systems under consideration and their interactions with the sufficient accuracy, these models can be used to determine the effect of failure on the stability and controllability of the aircraft. It is also possible to assess the failure situation by the flight crew with the help of pilot control simulators. Simulation with a pilot in the loop also allows to evaluate the available methods for recognizing failure and neutralizing actions. To obtain adequate results, models must be developed correctly. Model-based design allows to do this.

Model-based design (MBD) at present is becoming more widespread in a development: in the Aeroengine Control System [5], Intelligent Dynamic Systems [6], Modular Inspection Equipment Design for Modular Structured Mechatronic Products [7]. MBD allows identifying requirements violations and finding software errors in the early stages of the developed product.

The article proposes the use of a new method from a scientific and engineering point of view in the FHA requirements validation - the MBD approach. Integrated Flight Control System (IFCS) of the Irkut's airliner MC-21 safety requirements validation process is expected to be improved and accelerated by the usage of the MBD. For this, the following tasks need to be solved:

- Confirmation of the applicability of the MDB for the FHA validation;
- Development of a software for a modeling bench;
- Setting up work with real-time simulation equipment using pilot control simulators;
- Determination of quantitative criteria for failure conditions.

2. Methods and materials

To use the results obtained with the help of the MBD, the models of the IFCS and associated equipment must be reliable, that is, similar to real equipment on board the aircraft. The correctness of validation results must be proven. To prove this, the necessity and sufficiency of detailing IFCS models and associated equipment should be demonstrated. The equipment used as pilot controls should be identical to the real one so that the operator (engineer or pilot) can perform similar actions that are performed on board the aircraft. The actions of the pilot (if the pilot is in the aircraft control loop) must comply with the Flight Crew Operations Manual (FCOM) and take into account the problems of the human-machine interface, such as the time to recognise and parry the emerging failure situations.

2.1. Applicability of model-based design to FHA validation

The model-based design approach consists of designing models using elementary blocks and using these models as a basis for the development of embedded software. This means working with models as opposed to working with code (e.g. debugging, designing the architecture, reducing the complexity, testing) and generating the code from them. This approach aims to reduce the costs associated with testing. Using a model-based design approach makes it easy to create tests associated with a model at the early milestones. Powerful tools which are widespread nowadays allow generating tests automatically, thereby making it easy to test the development of your embedded software early and robustly. Model-Based Design is transforming the way engineers and scientists work by moving design tasks from the lab and field to the desktop. When software and hardware implementation requirements are included, such as fixed-point and timing behavior, you can automatically generate code for embedded deployment and create test benches for system verification, saving time and avoiding the introduction of manually coded errors [8].

The applicability of the MBD approach to the validation of the criticality of IFCS FHA is achieved taking into account the detailed study of the necessary components - internal and interface.

The algorithms that are used in the model are identical to those that are verified in flight tests of the MC-21 aircraft with the participation of EASA [9]. This allows engineers to get the most relevant data from the simulation results. Detailed actuators models of the integrated control system are also used: a set of electro-hydraulic actuators of ailerons, elevators, rudder, spoilers and speed brakes, flaps and slats actuation systems and electromechanical stabiliser actuator. These models also take into account reconfiguration of the system in case of failures of active actuators on the surface. Models of complex avionics systems, chassis and engine operation provide a sufficient level for validation of IFCS FHA. The model of the AFCS mode control panel (AFCP) is also included. It takes into account time delays in generating and transmitting signals, and furthermore is fully in line with the real AFCP logic of switching and indication of modes on AFCP indicators. Various weather conditions could be set thanks to complexity of the model and flexibility of simulation preferences. This ensures that the FHA is carried out for the worst conditions. The development of models at this level of complexity is a necessity for ensuring the validation of IFCS requirements.

There are some sufficient conditions that should be met to use the MBD approach, such as application of the methods for managing the configuration of the model (including tracking changes in models, integration of FBW and AFCS models). Moreover, the convergence of models, current software (used at the semi-natural test benches) and flight test results with current software and real actuators are also required. Convergence analysis showed that the implemented models are adequate for researching problems connected with the behaviour of FBW and AFCS in normal and failed flight conditions. Correct configuration management of the model ensures the use of current versions of algorithms and software.

2.2. Functional diagram of the mini test bench

MATLAB Simulink is used as the modeling environment, as the most progressively developing software package today for solving such problems. Mathematical models of an IFCS, interacting systems, and aircraft flight dynamics were prepared using the standard libraries built into MATLAB Simulink.

Interaction with visualisation and indication tools (Flight Ind, Flight Gear) and fault input interface (Fault Input Panel) is performed via UDP (Universal Datagram Protocol). This protocol allows for the exchange of information between components of the test bench without affecting the speed of modelling. Furthermore, as UDP is a network protocol, separate machines in a local network could be used for different human-machine interface implementation or even for simulating different systems. This allows improving the test bench with the help of distributed calculations.

AFCP is developed using the Qt cross-platform software libraries and connected to MATLAB Simulink in the same way using UDP to increase user convenience and information processing speed. The IFCS Fault Input Panel is similarly configured interface, which includes the ability to disable one or more actuators on the surface, hydraulic and electrical system failures, as well as other related FHA failures.

Interaction with physical pilot control simulators (Sidestick Unit, Throttle Quadrant Assembly, Pedals Control Unit) is carried out using the appropriate drivers and FlightSim library for MATLAB Simulink.

The resulting complex test bench was configured for modelling both in real time and in the "accelerated" time modes for solving various types of problems.

Real-time simulation allows you to conduct a flight assessment of the criticality of the situation at the employee's workplace, since the available simulators of the pilot controls and visualisation bodies ensure that the employee has a correct understanding of the system's behaviour and the impact on safety and flight in general. At the same time, in the evaluation process, it is possible to quickly change the logic or algorithms of the developed system component by adding, disabling or switching blocks from the MATLAB Simulink library. There is no need for time and financial costs thus that occur during the similar validation on test benches with real flight computers and a real aircraft, requiring a "flashing" of the software when making any changes to it.

Simulation in "accelerated" time allows performing validation in cases where an assessment of a given situation is necessary with a large number of initial conditions. This problem can be solved when assessing the consequences at different stages of flight, under different weather conditions (wind, wind shear), abnormal configurations that exacerbate the impact on safety (failure of various hydraulic systems or electrical systems, engines). If the necessary test vectors are correctly prepared and the simulation is launched in a cycle, the engineer or pilot does not need to conduct simulation of each case step-by-step, he only has to monitor the progress of the simulation, performing other tasks in parallel, and analyse the results in the end of the simulation. A special checklist for assessing the flight situation and the impact on safety is used in modelling in "accelerated" time. This approach has also found application in validating the functional requirements for FBW and AFCS.

Besides traditional testing, that involves simulating different flight scenarios, complex modelling system allows performing a kind of laboratory tests, such as frequency analysis. This kind of research is needed to prove aeroelasticity stability, one of the most important characteristics of an airplane. Traditionally these tests are carried out on a real aircraft using laboratory equipment, so there is no possibility to perform tests with a full range of initial conditions (due to the costs of prototypes destruction). Performing frequency tests on the complex model allows receiving more credible results than analytical (mathematical) method results that takes into consideration only simple linear models. At the same time all ranges of frequencies and amplitudes could be tested without danger of destruction. As it was said before, MBD approach gives engineers tools to tune the system to satisfy frequency criteria in their workplaces.

As it can be seen from the description above, the developed modelling system in terms of simulating controls and software is currently widely used. As a result of, the means of integrating various components of the "mini test bench" became available not only to the programmer, but also to the control system engineer. The development of such kind of system thus does not require the involvement of third-party companies and additional financial investments. An assessment of safety and flight situation is adequate and can be used to make adjustments to the relevant sections of the FHA.

The functional diagram of the test bench is shown in figure 1.

2.3. Comparing results with criteria

At the end of the simulation, it is possible to assess the consequences of failure. The criteria available in regulatory documents [1] are qualitative:

- Catastrophic (CAT): risks which would result in multiple fatalities, usually with the loss of the aircraft.
- Hazardous (HAZ): risks which would reduce the capability of the aircraft or the ability of the crew to cope with:
  - A large reduction in safety margins or functional capabilities,
  - flight crew cannot perform their tasks accurately or completely due to physical distress or excessive workload,

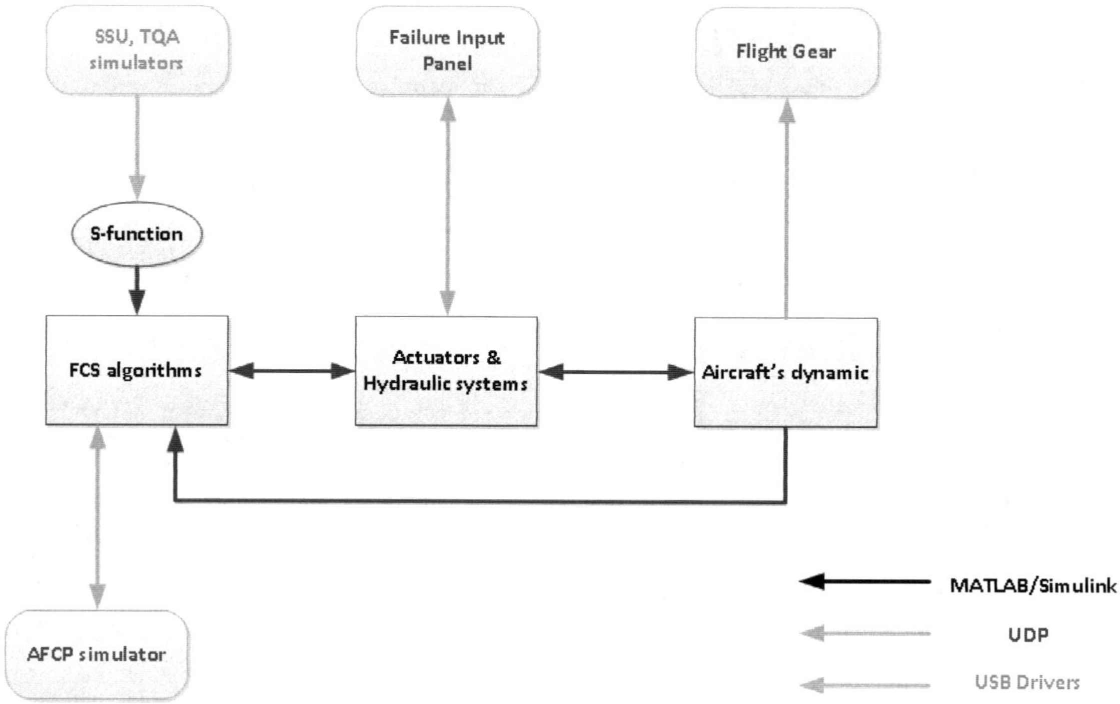

Figure 1. The functional diagram of the mini test bench.

- serious or fatal injury to a relatively small number of the occupants other than the flight crew.

– Major (MAJ): risks which would reduce the capability of the aircraft or the ability of the crew to cope with:

- a significant reduction in safety margins or functional capabilities,
- a significant increase in crew workload or in conditions impairing crew efficiency,
- discomfort to the flight crew,
- physical distress to passengers or cabin crew, possibly including injuries.

– Minor (MIN): risks which would not significantly reduce aircraft safety, and which involve crew actions that are well within their capabilities. Minor risks may include, for example, a slight reduction in safety margin or functional capabilities, a slight increase in crew workload or some physical discomfort to passengers or cabin crew.

– No Safety Effect (NSE): risks which would not impact the safety of the aircraft.

Qualitative criteria do not allow accurate analysis of the modeling results. For accurate analysis, quantitative criteria based on qualitative classification have been developed. These quantitative criteria take into account stability and controllability, as well as structural strength of the aircraft.

Comparing each characteristic obtained during the modeling with the criteria, the correct criticality of the failure condition is determined. The proposed failure validation method allows us to determine at the functional level the actual criticality of each violation.

3. Results and discussion

Regulatory documentation does not govern the validation of functional hazard assessments. The authors suggest validation of failure situations for fear of the possible misjudged criticality.

Regulatory documentation [1-4] does not provide clear quantitative criteria for determining the degree of criticality of a hazard. Therefore, some failure conditions in the IFCS FHA can be assessed inadequately. Therefore, based on the qualitative criteria, quantitative criteria have been developed that take into account stability, controllability, and also aircraft strength.

The authors analyzed the experience of a model-based design of aircraft systems [5] and other technical systems [6, 7]. According to the analysis of the available tools, the MATLAB was chosen as a main IDE.

As a result, a mini test bench was developed, quantitative criteria for FHA, as well as methods for using the stand both in real and accelerated time.

The IFCS MC-21 FHA has become the result of the work. The results of FHAs validation have allowed accelerating the work without using common test benches with real equipment. MBD has reduced the financial costs of FHA carrying out.

The software and hardware complex that has been implemented is generally universal. The experience gained in the construction and debugging mini test bench will significantly reduce the time for developing new projects in the future due to the gained experience and well-established processes when working with such models.

Based on the validation results, the criticality of functional failures identified during the previously performed FHA was clarified. In accordance with the safety process [1, 3], these refinements will be included in the design process.

The obtained result proves that the MBD can be used in the validation of FHA at aircraft and FCS levels of abstraction.

Currently, this approach, in addition to early detection of design errors, can also reduce the time spent on large stands.

Further work in this direction should be aimed at improving the quality of the model in terms of avionics (developing more detailed system models), as well as the introduction of MBD at the aircraft level. To spread the model-based design approach to the Irkut Corporation, significant cooperation of all interested departments is required. The tasks of storing, processing and detailing large volumes of information should be solved. Many related tasks, such as configuration management, model validation (convergence analysis), will also arise and need to be addressed.

For the successful functioning of MBD, the methods and standards of Irkut Corporation must be developed and approved, which will allow using these materials in the certification of new types of aircrafts. This is the perspective ways for future projects in this area.

4. Conclusion

MBD is a real and effective way in the validation of FHA, which is confirmed by the validation of IFCS MC-21 FHA. A number of necessary tasks were solved to use of MBD in the validation of IFCS FHA - integration of software and hardware complex modelling system based on MATLAB Simulink, an analysis of model convergence with existing software and equipment on test benches and aircraft was carried out, a model configuration management process was developed.

The implemented mini test bench reliably fulfils tasks both in real-time modelling, the main purpose of which is the possibility of direct flight assessment of the situation, and in "accelerated" time for stream-based solution of similar problems under various conditions.

Work with the model is carried out both with the participation of pilots for accurate assessments of the flight situation and the impact on safety, and is available to the engineer of the automatic control systems department. The developed criteria for assessing the impact on safety make it possible to analyse the simulation results according to the characteristics of aerobatic and navigation information (such as roll angle, lateral and normal overload, pitch angle, angle of attack, glideslope angle and many others).

The results FHA validation with the use of MBD has become the IFCS MC-21 FHA, which will be submitted for certification of the aircraft.

Currently, the safety assessment process in Irkut is improved compared to the requirements of SAE ARP 4761. This process includes the developing, validation, updating and verification of safety requirements.

Acknowledgments
The authors are grateful to the Irkut Corporation for the publication of this article and support for using modern methods of designing flight control systems.

References

[1] European Aviation Safety Agency 2007 *Certification Specifications for Large Aeroplanes CS-25 Amendment 3*

[2] Bieber P, Metge S, Morel M, Ple J 2019 Aircraft Safety Model Development and Integration in a Risk Observatory. *EASN International Conference on Innovation in European Aeronautics Research* (ONERA, Université de Toulouse) https://hal.archives-ouvertes.fr/hal-02078264

[3] SAE ARP 4761 1996 *Guidelines and Methods for Conducting the Safety Assessment Process on Civil Airborne Systems and Equipment* (SAE International) p 311 https://doi.org/10.4271/ARP4761

[4] SAE ARP 4754 2010 *Guidelines for Development of Civil Aircraft and Systems* (SAE International) p 115 https://doi.org/10.4271/ARP4754A

[5] Zhanga D, Luab J, Wanga L and Lia J 2015 Research of Model-based Aeroengine Control System Design Structure and Workflow. *Proc. Eng.* **99** 788 https://doi.org/10.1016/j.proeng.2014.12.603

[6] Oestersötebier F, Abrishamchian F, Lankeit C, Just V and Trächtler A 2016 Approach for an Integrated Model-based Design of Intelligent Dynamic Systems Using Solution and System Knowledge. *Proc. Technology* **26** 436 https://doi.org/10.1016/j.protcy.2016.08.056

[7] Lukei M, Hassan B, Dumitrescu R, Sigges T and Derksen V 2016 Modular Inspection Equipment Design for Modular Structured Mechatronic Products – Model Based Systems Engineering Approach for an Integrative Product and Production System Development. *Proc. Technology* **26** 455 https://doi.org/10.1016/j.protcy.2016.08.058

[8] Why Use Model-Based Design? available at: https://www.mathworks.com/solutions/model-based-design.html

[9] *EASA flight test experts completed the second session of certification flights*, available at: http://eng.irkut.com/press-centre/news/2392/

Dynamic engine charge simulation for unmanned aerial vehicles

Yu A Grishin[1,*] and V N Bakulin[2,3]

[1]Department of Piston Engines, *Bauman State Technical University*, 5 Second Baumanskaya Street, Moscow 107005, Russian Federation
[2]Department of Aircraft Design and Sertification, *Moscow Aviation Institute* (National Research University), 4, Volokolamskoye Avenue, Moscow 125993, Russian Federation
[3]Laboratory of Non-Classical Models of Mechanics of Composite Materials and Structures, *Institute of Applied Mechanics*, Russian Academy of Sciences, 7 Leningradskaya Avenue, Moscow 125040, Russian Federation

*E-mail: grishin.bmstu@mail.ru

Abstract. Rational use of wave processes of significant intensity in the intake pipes of piston engines allows providing dynamic charge, which leads to a noticeable increase in engine power, reduction of fuel consumption. This is especially true for engines used on unmanned aerial vehicles. For numerical calculation of the dynamic charge, a modification of the method of decay of an arbitrary discontinuity has been developed, based on the application of special gas-dynamic functions of the unsteady flow, which does not use a simplifying linearization and thus provides increased accuracy of the results required for wave processes of significant intensity. Corresponding calculations of the engine cylinder charge filling at dynamic charge have shown the possibility of the power increase by 24%, which corresponds well with the experimental data.

1. Introduction

The periodic behavior of the working processes in the piston engine cylinders is the cause of vibrational dynamic phenomena in their gas-air paths. This phenomenon can be used for significant improvement of the main parameters of the engines, like power and fuel efficiency [1-3]. In this case, the inlet manifolds are adjusted for dynamic charge, i.e. the length and section of the individual manifolds for each of the engine cylinders are chosen. At the beginning of the inlet into the cylinder in front of the valve, a vacuum is formed, and decompression wave is transmitted to the open end of the inlet manifold, where it is reflected by a compression wave. Selection of the geometry of the individual manifolds allows ensuring that the compression wave arrives to the cylinder at the most favorable moment before closing the valve. This leads to a significant increase of the volumetric coefficient η_v and, consequently, the power by 10-20% [4-6] and improves engine fuel efficiency by 5-10% [2, 7, 8].

Engine power increase together with simultaneous increase of the volumetric coefficient is especially important for the piston engines installed on unmanned aerial vehicles (UAVs) of various applications, as it will provide increase of flight characteristics, increase of a flight time. At the same time, there is a tendency to use diesel engines fueled by kerosene as the most economical ones for the mid-range UAVs.

Numerical methods of gas dynamics simulation tasks are now widely used for calculations of unsteady gas-dynamic processes [9-11]. The numerical method of the decay of an arbitrary discontinuity (DAD) – method of Godunov is the most effective for calculation of the dynamic charge in extended inlet manifolds in a one-dimensional simulation [1, 12, 13]. The ideal gas flow (p = ρRT) in coordinates "length-time", x - t is described by the system of basic equations of gas dynamics [1, 11, 14]:

$$\begin{cases} \oint(\rho dx - \rho u dt) = 0; \\ \oint\left[\rho u dx - \left(p + \rho u^2\right)dt\right] = -\iint \frac{\lambda \rho u^2}{2D} dxdt; \\ \oint\left[\rho\left(e + \frac{u^2}{2}\right)dx - \rho u\left(e + \frac{u^2}{2} + \frac{p}{\rho}\right)dt\right] = -\iint \frac{4\alpha\left(T - T_c\right)}{D} dxdt. \end{cases} \quad (1)$$

where, ρ, u, p, T, e – density, speed, pressure, temperature, specific internal energy (e=c_vT); D and T_C – hydraulic diameter and local pipe wall temperature; λ, α – friction and heat transfer coefficients. Numerical implementation of the method after dividing the calculation area into segments (cells) Δx most often uses simplified dependencies (linearization) [12-14] for speeds $u_{i+1/2}$ and pressure $p_{i+0.5}$, with them the flow crosses the border i+0.5 between the cells i and i+1 (figure 1):

$$p_{i+0.5} - p_i + a_i\rho_i(u_{i+0.5} - u_i), \quad p_{i+1} - p_{i+0.5} - a_{i+1}\rho_{i+1}(u_{i+1} - u_{i+0.5}).$$

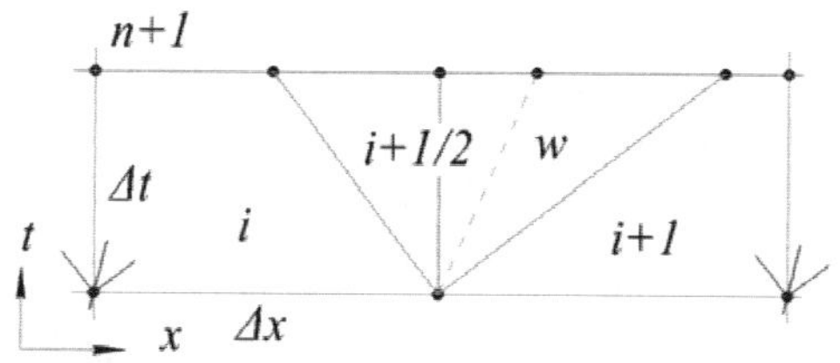

Figure 1. Decay of an arbitrary discontinuity (DAD) on the boundary between the cells of the numerical calculation.

The presented relations are obtained from finite-difference recording of the conditions of preservation of Riemann's invariants at transitions through the fronts of elementary waves, respectively r+ - backwards and r^- - straightforward with an equal speed $u_{i+0.5} = u_w$ and pressure $p_{i+0.5} = p_w$ on either side of the contact surface that separates the zones i+0.5 and w behind the wave fronts after the DAD. After the determination $u_{i+0.5}$, $p_{i+0.5}$ and the corresponding density $\rho_{i+0.5}$ we calculate mass, pulse and total enthalpy fluxes across the boundary i+0.5, which allows us to move to the final stage of the calculation scheme in accordance with the system (1). Unfortunately, the use of linearized relations in the DAD method gives satisfactory calculation results only for wave processes with insignificant amplitudes [11, 12, 14]. At dynamic charging in front of the valve, the wave amplitudes can reach 0.5 Bar, and such algorithm does not provide the necessary accuracy of the results even with a significant reduction of the calculation step Δt. For modeling of such processes it is recommended to use the classical scheme of the decay of discontinuity method, taking into account the shock fronts [11, 12, 14]. However, the algorithm of such a scheme is heterogeneous, logically complex and requires iterations at each calculation step.

2. Description of the research

The paper presents a new modification of the DAD method, based on the usage of simple Riemann waves without simplifying linearization for both - compression front and the depression front. A Ferri [15] showed that the replacement in the calculations of shock waves by simple waves of Riemann within the amplitudes of compression 0.5 Bar was justified and provided the same accuracy of the results. In this paper, it was proposed to use specially developed non-stationary gas-dynamic functions (NSGDF) from the Mach number $M = u/a$ to express the differences at the fronts of these waves [1, 16].

To cross the front of the backward wave, we have: $r^+ = \frac{2}{k-1}a + u = \frac{2}{k-1}a''^*$, where a''^* - sound speed if the stream is slowed down by the wave ($u = 0$). Obviously, the speed a''^* is unambiguously related to the value of Riemann's invariant r^+, as well as the corresponding temperature T''^*, density ρ''^* and pressure p''^*. Then, for the difference in sound speed, temperature, density and pressure at the front of the wave going against the flow, it is possible to set the following equations:

$$a_{i+0.5} = a_i \frac{a''^*}{a''^*}\frac{\alpha''(M_{i+0.5})}{\alpha''(M_i)} = a_i \frac{\alpha''(M_{i+0.5})}{\alpha''(M_i)}, \quad T_{i+0.5} = T_i \frac{T''^*}{T''^*}\frac{\tau''(M_{i+0.5})}{\tau''(M_i)} = T_i \frac{\tau''(M_{i+0.5})}{\tau''(M_i)},$$

$$\rho_{i+0.5} = \rho_i \frac{\rho''^*}{\rho''^*}\frac{\varepsilon''(M_{i+0.5})}{\varepsilon''(M_i)} = \rho_i \frac{\varepsilon''(M_{i+0.5})}{\varepsilon''(M_i)}, \quad p_{i+0.5} = p_i \frac{p''^*}{p''^*}\frac{\pi''(M_{i+0.5})}{\pi''(M_i)} = p_i \frac{\pi''(M_{i+0.5})}{\pi''(M_i)},$$

where, the NSGDF is present as follows:

$$\alpha''(M) = \left(1 + \frac{k-1}{2}M\right)^{-1}, \ \tau''(M) = [\alpha''(M)]^2, \ \varepsilon''(M) = [\alpha''(M)]^{\frac{2}{k-1}}, \ \pi''(M) = [\alpha''(M)]^{\frac{2k}{k-1}}.$$

Obviously, to cross the front of the forward wave in the flow, $r^- = \frac{2}{k-1}a - u = \frac{2}{k-1}a'^*$, where a'^* - sound speed when the wave is slowing down. The corresponding variations in sound speed, temperature, density and pressure at the front of this wave can be expressed as:

$$a_w = a_{i+1} \frac{a'^*}{a'^*}\frac{\alpha'(M_w)}{\alpha'(M_{i+1})} = a_{i+1} \frac{\alpha'(M_w)}{\alpha'(M_{i+1})}, \quad T_w = T_{i+1} \frac{T'^*}{T'^*}\frac{\tau'(M_w)}{\tau'(M_{i+1})} = T_{i+1} \frac{\tau'(M_w)}{\tau'(M_{i+1})},$$

$$\rho_w = \rho_{i+1} \frac{\rho'^*}{\rho'^*}\frac{\varepsilon'(M_w)}{\varepsilon'(M_{i+1})} = \rho_{i+1} \frac{\varepsilon'(M_w)}{\varepsilon'(M_{i+1})}, \quad p_w = p_{i+1} \frac{p'^*}{p'^*}\frac{\pi'(M_w)}{\pi'(M_{i+1})} = p_{i+1} \frac{\pi'(M_w)}{\pi'(M_{i+1})},$$

where, the NSGDF $\alpha'(M) = \left(1 - \frac{k-1}{2}M\right)^{-1}$, $\tau'(M) = [\alpha'(M)]^2$, $\varepsilon'(M) = [\alpha'(M)]^{\frac{2}{k-1}}$ and $\pi'(M) = [\alpha'(M)]^{\frac{2k}{k-1}}$.

Apparently, on the contact surface $p_{i+0.5} = p_w$, and for pressure drops between the wave configuration zones we have:

$$\frac{p_i}{p_{i+1}} = \frac{p_i}{p_{i+0.5}} \cdot \frac{p_{i+0.5}}{p_w} \cdot \frac{p_w}{p_{i+1}} = \frac{\pi''(M_i)\pi'(M_w)}{\pi''(M_{i+0.5})\pi'(M_{i+1})} = \left[\frac{\alpha''(M_i)\alpha'(M_w)}{\alpha''(M_{i+0.5})\alpha'(M_{i+1})}\right]^{\frac{2k}{k-1}}; \quad (2)$$

speed parity $u_{i+0.5} = M_{i+0.5}a_{i+0.5} = u_w = M_w a_w$ lets us to record the following:

$$M_{i+0.5}a_i \frac{\alpha''(M_{i+0.5})}{\alpha''(M_i)} = M_w a_{i+1} \frac{\alpha'(M_w)}{\alpha'(M_{i+1})}. \quad (3)$$

Using the formulas $F = \frac{\alpha'(M_{i+1})}{\alpha''(M_i)}$; $X = F\left(\frac{p_i}{p_{i+1}}\right)^{\frac{k-1}{2k}}$ and $Y = F\frac{a_i}{a_{i+1}}$, containing the initial parameters p_i, p_{i+1}, a_i, a_{i+1}, M_i, M_{i+1}, by revealing the record $\alpha'(M)$, $\alpha''(M)$ from M and performing transformations of the system of equations (2) and (3), it is possible to obtain simple expressions of the Mach numbers M:

$$M_{i+0.5} = \frac{2}{k-1}\frac{X-1}{Y+1}, \quad (4)$$

$$M_w = \frac{Y}{X} M_{i+0.5}. \quad (5)$$

To identify flows over the border i+0.5 it only takes $M_{i+0.5}$. NSGDF calculates the corresponding $p_{i+0.5}$, $\rho_{i+0.5}$, $u_{i+0.5}$, then flow rate, impulse and full enthalpy.

Undoubted advantages of the use of NSGDF are shown in the calculations at the boundaries of the calculation areas with a one-dimensional non-stationary flow, which ends in some of the local resistances (LR) considered as concentrated boundary conditions (BC) [1]. Calculation methods for the stationary flow of compressible gas through such elements are based on gas-dynamic functions [17] from the M-number or the speed limit $\lambda = u/a_{cr}$, where a_{cr} – critical speed of sound. When developing models of interaction of non-stationary flow with such BC, it is necessary to use the notion of DAD. Here, NSGDFs are used in combination with known gas-dynamic functions of the stationary flow since every calculated step of the DAD uses the notion of a quasi-stationary flow between the waves that passed and reflected from BC [3, 15, 18].

Let us imagine the calculation model of interaction of a non-stationary flow with the BC "inlet valve" in the adjusted inlet manifolds at dynamic engine charge. The DAD scheme for the design step is shown in figure 2. The parameters known before the DAD are p_m and M_m in the last cell of the calculation area, the current value of the valve slot passage area F_v and the pressure in the cylinder p_c. The index $m + 0.5$ indicates the parameters in the manifold behind the edge of the reflected wave required to calculate the mass flows, pulse and total enthalpy across the conditional.

The pressure difference between the DAD zones can be recorded in the form of stationary and non-stationary functions:

$$\frac{p_m}{p_v} = \frac{p_m}{p_{m+0.5}} \cdot \frac{p_{m+0.5}}{p_v} = \frac{\pi''(M_m)\pi(M_{m+0.5})}{\pi''(M_{m+0.5})\pi(M_v)}. \quad (6)$$

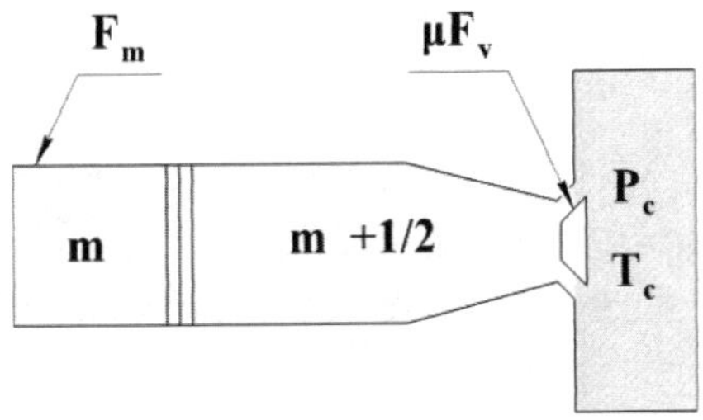

Figure 2. Scheme for calculating the DAD when flowing into the cylinder through the valve.

For quasi-stationary flow between zones $m + 0.5$ and out of the valve slot v the rightly known expenditure ratio [17]:

$$F_m p^*_{m+0.5}(T^*_{m+0.5})^{-1}q(M_{m+0.5}) = \mu F_v p^*_v (T^*_v)^{-1} q(M_v) \tag{7}$$

Where $\pi(M) = \left(1+\frac{k-1}{2}M^2\right)^{-\frac{k}{k-1}}$ $q(M) = M\left[\frac{2}{k+1}\left(1+\frac{k-1}{2}M^2\right)\right]^{-\frac{k+1}{2(k-1)}}$ – stationary gas-dynamic functions [17]. To consider the unevenness of the flow in the section F_v the valve slot has a flow coefficient μ, which is pre-determined by purging the valve. Air flow through the valve in the engine cylinder head is not accompanied by heat exchange and is assumed to be adiabatic. Then for stationary braking temperature we have $T^*_{m+0.5} = T^*_v$ in addition, it is known from the experiments that the total pressure losses in the short channels with restricted flow section, which is the inlet nozzle in the valve head, are negligibly small, i.e. $p^*_{m+0.5} = p^*_v$. As a result, from (7) regarding the designation $\mu F_v F_m^{-1} = f$ we have:

$$q(M_{m+0.5}) = fq(M_v)\,. \tag{8}$$

Flow into the cylinder can be pre-critical or critical, with flow locked in the valve slot. In pre-critical flow, the pressure at the valve slot exit is obviously p_v equals the pressure in the cylinder p_c. In critical mode p_v is determined by the condition $M_v = 1$.

At the time of $M_v = 1$ from (8) you get the ratio, which for this f allows you to determine the value $M_{m+0.5,cr}$ in the area of m+0.5 in critical mode:

$$q(M_{m+0.5,cr}) = f\,. \tag{9}$$

Now, by setting up M_v= 1 and $M_{m+0.5,cr}$ into (6), you can find the maximum difference $\left(p_m p_c^{-1}\right)_{cr}$. If, in the calculated step $p_m p_c^{-1} \geq \left(p_m p_c^{-1}\right)_{cr}$, then we have a critical flow mode with locking at the valve slot outlet, and the number required to calculate the DAD number $M_{m+0.5,cr}$ thus defined. At the same time $p_m p_c^{-1} = \left(p_m p_c^{-1}\right)_{cr}$ and obviously, the pressure at the valve slot exit of the p_v will be greater than the p_c. If that assessment turns out to be $p_m p_c^{-1} < \left(p_m p_c^{-1}\right)_{cr}$, the regime is pre-critical. The pressure at the outlet of the valve slot is equal to the pressure in the cylinder, and in equation (6) instead of p_v it is necessary to substitute the pressure p_c known at this design step. In this mode, the system of equations (6) and (8) with two unknowns $M_{m+0.5}$ and M_v, is solved by iteration. Number $M_{m+0.5}$ necessary to calculate the wave process in the manifold and the flow into the cylinder. To describe the process in the cylinder, the technique described in [19] is used.

3. Results of calculations

In accordance with the above described, the wave process simulation for dynamic charge was performed for diesel engine 1Ch8.5/11 (One cylinder four tact diesel engine, Bore 85 mm, Stroke 110 mm), which can serve as a prototype for the development of the aviation diesel engine for UAV. General design parameters of this engine are: speed n = 1500 *rpm* (revolutions per minute), volumetric coefficient η_v = 0.85 in the absence of dynamic charge have been used as the input data. In work [8] classical experimental results on research of dynamic charge of this motor are presented. Figure 3 shows obtained in the calculation and published experiment [8], showing dependence of μ_v on the length of the inlet manifold L. The calculated results are well consistent with the experimental data in both qualitative and quantitative terms. Here is shown the wave character of the change of η_v along the length of the tube, caused by the number of periods of free oscillations, with full phase coordination with the experiment. In the maximum area for dynamic boost, the η_v calculation error is no more than 1.5%. It is also shown that dynamic boost can provide an increase in η_v above 1.05. Compared to a non-configured system, the volumetric coefficient, and therefore the engine power, which is proportional to η_v, increases by 24%.

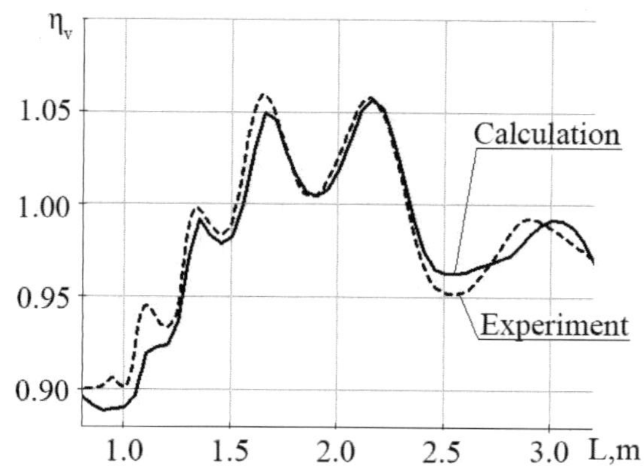

Figure 3. Dependence of the volumetric coefficient η_v on the length L of the adjusted pipe by calculations results and experiment description in [8].

Note that when using a propeller unit with a propeller screw, extended inlet pipes can be placed along the drone fuselage with a forward flow inlet opening, that will provide an additional 5-8% increase in η_v and capacity due to the high-speed head.

4. Conclusion

The paper presents the results of numerical simulation of the dynamic charge for piston engine, which can be used as a prototype engine for UAV. It has been shown that such charge can significantly increase the engine power. It is important to note that the simulation was carried out based on a new modification of the numerical method of DAD, which uses special non-stationary gas-dynamic functions from the Mach number, expressing the differences in parameters at the fronts of simple Riemann waves. Here we obtain a simple homogeneous computational algorithm that does not require iterations at the computational step, caused by the extraction of the shock front, as it takes place in the classic Godunov scheme for the wave processes of significant amplitude. In addition, the use of Mach numbers allows us to conveniently express the calculated relations at the boundaries, in particular, at the valve and the inlet to the manifold.

Comparison of the simulation results carried out by means of a new modification of the DAD method with experimental data of well-known authors, testifies the adequacy of the proposed method. It can be recommended for solving practical problems of gas dynamics with wave processes of significant intensity. In particular, when designing the inlet manifolds of piston engines with adjustment of the dynamic charge with considerable increase in power.

References

[1] Grishin Yu A, Zenkin V A and Khmelev R N 2017 Boundary conditions for numerical calculation of gas exchange in piston engines. *Journal of Engineering Physics and Termophysics* **90** (4) 965 doi: 10.1007/s10891-017-1644-4
[2] Grishin Yu A and Sonkin V I 2013 Intake Systems. *Internal Combustion engines, Mechanical Engineering, Encyclopedia* ed. A A Aleksandrov and N A Ivashchenko (Moscow: Machine Building) vol IV-14 pp 461-474 [in Russian]
[3] Emmons H W 1958 *Fundamentals of Gas Dynamics* (Princeton, New Jersey: Princeton University Press) p 704
[4] Stehlig J, Taylor J, Dingelstadt R and Gurney D 2012 Variable-length air intake module for turbocharged engines. *MTZ Worldw* **73** 34 https://doi.org/10.1365/s38313-012-0130-9
[5] Weber O, Wild S and Kiener T 2002 Highly-variable intake systems *MTZ Worldw* **63** 7 https://doi.org/10.1007/BF03227532
[6] Vihert M M and Grudskii Yu G 1982 *Design of Inlet Systems for High-speed Diesel Engines* (Moscow: Machine Building Publishing) p 151
[7] Weaving J H 1990 *Internal Combustion Engineering: Science & Technology* (London and New York: Elsevier Science Publishers LTD) p 873
[8] Semenov B N and Komov A P 1972 *Acoustic Supercharge as a Tool for Reducing Specific Effective Fuel Consumption* (Moscow: NIIINFORM-TiAZhMASh) [in Russian]
[9] Grishin Yu A and Bakulin V N 2015 New calculation schemes based on the large-particle method for modeling gas_dynamic problems. *Dokady Physics* **60(12)** 555 doi: 10.1134/S1028335815120058
[10] Toro E F 1997 *Riemann Solvers and Numerical Methods for Fluid Dynamics. A Practical Introduction* (Berlin: Springer) p 724
[11] Godunov S K, Zabrodin A V, Ivanov M Y, Krayko A N and Prokopov G P 1976 *Numerical Methods of Multi-Dimensional Gas Dynamic Problems* (Moscow: Science) p 400 [in Russian]
[12] Pirumov U G and Rosliakov G S 1987 *Numerical Methods of Gas Dynamics* (Moscow: Hight School) p 232 [in Russian]
[13] Gogrichiani G V and Shipilin A V 1986 *Pneumatic System Transients* (Moscow: Machine Building) p 160
[14] Rozhdestvenskii B L and Yanenko N N 1978 *Quasi-Linear Equation Systems* (Moscow: Science) p 688 [in Russian]
[15] Ferri A 1961 *Fundamental Data Obtained from Shock-tube Experiments* (Oxford – London – New York-Paris: Pergamon Press) p 442
[16] Enikeev R D, Nozdrin G A and Chernousov A A 2017 The model and the methods for numerical simulation of wave action of real working fluids in pipelines. *Proc. Engineering* **176** 461 https://doi.org/10.1016/j.proeng.2017.02.345
[17] Abramovich G N 1991 *Applied Gas Dynamics* (Moscow: Science) p 600 [in Russian]

[18] Liepmann H V and Roshko A 1957 *Elements of Gasdynamics* (New York: John Wiley & Sons, Inc., London: Chapman & Hall, Limited) p 506

[19] Barchenko F B and Bakulin V N 2017 Calculation of the termal loading of the cylinder-piston group of the automobile engine. *Journal of Engineering Physics and Termophysics* **90** 657 https://doi.org/10.1007/s10891-017-1613-y

IOP Conf. Series: Materials Science and Engineering **868** (2020) 012008 doi:10.1088/1757-899X/868/1/012008

Dynamic modification of the G Rasch's logistic model for the tasks of examination the engineers' knowledge

A Kirillin

Technology of testing and operation Department, *Moscow Aviation Institute (National Research University)*, 4 Volokolamskoe shosse, Moscow 125993, Russian Federation

E-mail: kirillinav@mati.ru

Abstract. Problem of efficient and safe operation of complex technical systems is always one of the most important in its life cycle. Analysis of the failures and emergencies statistics shows a trend in the growth of the human factor and a decrease of failures for technical reasons. Unfortunately, it is not possible to completely exclude human from the control system due to adherence to safety standards and industry requirements. However, issues of evaluating the professional suitability of personnel operating complex equipment, as well as issues of organizing and planning the training process, are becoming relevant. Based on a comparison of the statistical model of Georg Rasch, used to assess the preparedness of students and the degree of complexity of control tasks, with the dynamic universal model of reliability growth in the development process, used in various practical applications, a methodology for the analysis and planning of training effectiveness is developed. The algorithm of operation and application of the described methodology is illustrated by a number of examples. The dynamic modification of the presented Rasch's statistical model allows predicting the degree of preparedness of homogeneous groups of students, thereby making it possible to plan the number of personnel training cycles for admission to work with complex and potentially dangerous technical systems.

1. Introduction

The reliability and safe operation of complex technical systems cannot be evaluated from the point of only the technical part. Since in this case it is assumed that the performance characteristics of the operating engineering personnel are accepted as optimal and there is no risk of error. Obviously, such a situation cannot arise in actual operation. For complex technical systems, in which human is an integral part of the control loop, there are two options for responding to a possible violation of the operating mode: the first one is automatic, when signs of violation appear; the second – duplicate manual control carried out by an engineer in case of violation of the operating mode according to light, sound or other indications. At the same time, a decrease in the inertia of the engineer-operator reaction, as well as the accuracy of his actions, significantly affects the negative consequences of the possible failures. Therefore, there is a need to develop various methodological methods for evaluation and ensuring the reliability of the ergonomic component of the system which is an engineer-operator.

Currently there are many approaches to evaluate reliability of human, as part of a technical system. Moreover most of them are the basis of the HRA methods (Human Reliability Analysis) and are well described by E Hollnagel [1] and H Liao [2-4]. In engineering practice the applicability of various HRA methods is considered. For example, Zhang Li [5] considers approaches to evaluate the human cognitive reliability (HCR) as an addition to the standard THERP procedure (The Technique for Human Error-Rate Prediction) for Qinshan NPP staff. Engineers of University in Brno (Czech Republic) applied the

Monte Carlo method to calculate the most important indicator of the human reliability working in the control loop of a complex technical system – the Mean Time Between Failures (MTBF) [6].

Analysis of the publications showed that HRA methods and models evaluate reliability either from the position of a probabilistic analysis of errors or from the position of cognitive reliability. At the same time, an engineer is considered as a specialist having 100% knowledge about the facility being operated, the rules and methods of working with it. In practice this situation is far from reality. Therefore, the author sets the task of developing a methodology for evaluation of the level of competence and knowledge of engineer-operator, which in combination with the HRA methods application would provide more complete information about the reliability of human as an element of a complex technical system.

2. Methodology and research methods

2.1. G Rasch's logistic model

The simplest and most common method of assessing the success is test control estimating the answers to control questions in a nominal scale (right - wrong) and an integral assessment of a group of students – an evaluation of the probability of success by frequency $R = \frac{M}{N}$, where M – the number of right answers, $N = k \cdot n$ – selection size, n – the number of students in the group, k – the number of control questions [7].

This probability, which characterizes the assimilation degree of the taught material, depends both on the initial preparedness of the students and on the content of the taught discipline and the professional skill of the teacher. The level of complexity of the material should be correlated with the degree of preparedness of students and the perception of this material [8-14]. A statistical model that allows to evaluate the effectiveness of training, taking into account both the degree of preparedness of students and the degree of complexity of the taught material, is the logistic model of Georg Rasch [15]:

$$P(\theta - \beta) = \frac{\exp\{(\theta - \beta)\}}{1 + \exp\{(\theta - \beta)\}} \qquad (1)$$

where β – level of difficulty of control task, θ – level of preparedness of students.

Logistic probability distribution $P(\theta - \beta)$ is symmetrical with zero expected value (at $\theta = \beta$, $P(\theta - \beta) = 0.5$), dispersion $\pi^2/3$ and $\beta^2 = 4.2$.

The logistic distribution in the middle part is close to normal. However, it has more "heavy tails", i.e. the probabilities of both small and large values of a random variable $(\theta - \beta)$ are greater than that of the normal law of probability distribution ($\beta^2 > 3$ for a normal distribution and for a logistic distribution it is more peaked).

The logistic distribution is simpler than normal because the calculation of the function values of this distribution does not require the use of special tables or programs.

To carry out the calculations using the logistic distribution of Rasch, the transformation $\ln\frac{p}{q} = \theta - \beta$ or $\ln\frac{q}{p} = \beta - \theta$ was introduced, where $q = 1 - p$.

When organizing an active experiment it is possible to evaluate the preparedness degree of a group of students with a fixed complexity degree as well as the complexity degree with a fixed group composition [16-18].

The Rasch's model is statistical, i.e. it allows a one-time evaluation of the training effectiveness.

2.2. Analysis of G Rasch's model from the perspective of the general reliability theory

When preparing operators for working with complex equipment, the reliability and operational safety of the equipment depend on the human factor reliability. Therefore, the requirements for the preparedness of the human operator are extremely high. In this case one-time training may not be enough. The question of planning the entire training process arises, i.e. firstly the question of the number of training cycles.

This problem can be solved using well-known models of reliability growth. These models are widely used in variable areas as development models, for example, in substantiating the number of test cycles during experimental testing of aircraft [19]. The task of evaluating the dynamics of training effectiveness close to the task of the dynamics reliability assessment of the product, depending on the number of test cycles. After each cycle the product is refined in order to eliminate defects identified during testing, increasing its reliability.

Indeed, the control of students' knowledge can be considered as their test, and the identified undeveloped fragments, as learning defects that require more detailed study.

3. Results and discussion

3.1. Dynamic modification of the G Rasch's logistic model

One of the models of reliability growth is the logistic model [20]

$$R(n)=\frac{R_0}{R_0+(1-R_0)\exp(-\Omega n)} \qquad (2)$$

At the number of cycles $n=0$, $R=R_0$; at $n\to\infty$, $R\to 1$.

It is easy to see that model (2) is a special case of model (1), when $R_0=0.5$, $n=1$, $\Omega=\theta-\beta$ and only positive values Ω are considered leading to increased reliability R. Also in the model (2), the parameter $\Omega=const$, but in the model (1) $\Omega=\theta-\beta$ – random variable, the indirect measurement of which is based on an estimate of the probability P by the frequency.

Model (2) can be written in the equivalent discrete recurrent form

$$R_i=\frac{R_{i-1}\exp\{\Omega\}}{R_{i-1}\exp\{\Omega\}+(1-R_{i-1})},$$

whence after simple transformations the following is obtained

$$\ln\frac{R_i}{q_i}=\ln\frac{R_{i-1}}{q_{i-1}}+\Omega,$$

and finally passing to the designation of model (1), the prediction equation for 1 step is obtained:

$$\ln\frac{p_i}{q_i}=\ln\frac{p_{i-1}}{q_{i-1}}+(\theta_{i-1}-\beta_{i-1}) \qquad (3)$$

where $(\theta_{i-1}-\beta_{i-1})$ – the estimates obtained by processing information on previous cycles up to $(i-1)$.

It should be noted that there is the simplest linear model of the dependence of the growth rate Ω_i on the number of cycles is used:

$$\Omega_i=(\theta-\beta)n_i.$$

It is rational to use the arithmetic mean estimate as estimate of a random variable $(\theta_{i-1}-\beta_{i-1})$

$$\Omega_{i-1}=\frac{1}{n-1}\sum_{j=1}^{n-1}\ln\frac{p_j}{q_j}.$$

Equation (3) can be expanded to predict more than one step forward, for example:

$$\ln\frac{p_{i+1}}{q_{i+1}}=\ln\frac{p_{i-1}}{q_{i-1}}+(\theta_{i-1}-\beta_{i-1})\cdot 2$$

etc.

3.2. Examples of using dynamic modification of the G Rasch's model

Example 1. The set value of P_{set}=0.99, i.e. $\ln\frac{0.99}{0.01}=4.595$. Then the value $\ln\frac{p_1}{q_1}$, obtained after the first familiarization control of knowledge is:

$$\ln\frac{p_1}{q_1}=\ln\frac{0.6}{0.4}=0.405=\theta_1-\beta_1, \text{ as in (3) } \ln\frac{p_0}{q_0}=\ln\frac{0.5}{0.5}=0.$$

Then, after the second training cycle the following values are expected to be reached: $\ln\frac{p_2}{q_2}=0.405\cdot 2$, $\frac{p_2}{q_2}=2.25$, whence $P_2=0.692$.

With such reliability growth rate, for reaching the value P_{set}=0.99 it will be required $4.595-0.405=0.405\cdot n$ training cycles, whence $n\approx 10$.

Example 2. There is the preparedness of the group of students is higher, $p_1=0.8$. Then $\ln\frac{0.8}{0.2}=1.386$, and for reaching the value P_{set}=0.99 it will be already required $n\approx 3$ training cycles.

Example 3. The set value P_{set} is very high, and equal to 0.999999, i.e. $\ln\frac{0.999999}{0.000001}=13.8$. At the level of preparedness $p_1=0.8$, in this case $n=\frac{13.8-1.386}{1.386}\approx 9$ training cycles will be required.

In the practical application of the proposed approach there is the question of the decisions reliability is raised. The probability in the model (1) is a monotonically increasing function of the value $(\theta-\beta)$. Thus, the lower confidence limit of the evaluation of this probability by frequency corresponds to the lower confidence limit of an indirect measurement $(\theta-\beta)$. The result is a simple practical technique for taking errors into account when replacing the true probability value P with its estimate $\hat{P}=\frac{M}{N}$, shown below with a specific example.

Example 4. Assume that as a result of familiarization test of 10 students on 10 control questions ($N=n\cdot k=10\cdot 10=100$), there were received 80 correct answers, i.e. $\hat{P}=\frac{80}{100}=0.8$.

Using the normal approximation of the binomial distribution, the lower 90% confidence limit is obtained at $\gamma=0.9$:

$$P_L=\hat{P}-u_{1-\gamma}\sqrt{\frac{\hat{P}(1-\hat{P})}{n}}=0.8-1.28\sqrt{\frac{0.8\cdot 0.2}{100}}\approx 0.75,$$

where $u_{1-\gamma}$ – quantile of standard normal distribution.

Further calculation is carried out analogously to example 3, i.e. $\ln\frac{0.75}{0.25}=\theta_1-\beta_1=1.099$, and the upper number of cycles is $n_H=\frac{13.8-1.099}{1.099}\approx 11$.

Herewith, the predicted value after the second training cycle will be $\ln\frac{p_2}{q_2}=1.099\cdot 2$, whence $p_2=0.9$.

Suppose further, that during testing on the second cycle, 90 correct answers were obtained, i.e. $\hat{P}=\frac{90}{100}=0.9$.

The lower confidence limit at $\gamma = 0.9$ will be:

$$P_L = 0.9 - 1.28\sqrt{\frac{0.9 \cdot 0.1}{100}} \approx 0.862,$$

whence $\theta_2 - \beta_2 = \ln\frac{0.862}{0.138} - \ln\frac{0.75}{0.25} = 1.832 - 1.099 = 0.733$.

The average growth rate will be $\frac{1.099 - 0.733}{2} = 0.916$ and the predicted number of cycles for reaching $P_L = 0.999999$ will be $n_H = \frac{13.8 - 1.832}{0.916} \approx 13$, i.e. estimation errors $\hat{P}$ already significantly affect the number of training cycles.

To reduce the required number of training cycles it is necessary to increase its effectiveness.

Example 5. Suppose that under the remaining conditions of Example 4, the number of correct answers after the second training cycle increased to 95. Then $\hat{P} = 0.95$, the lower confidence limit will be:

$$P_L = 0.95 - 1.28\sqrt{\frac{0.95 \cdot 0.05}{100}} \approx 0.922$$

whence $\theta_2 - \beta_2 = \ln\frac{0.922}{0.078} - \ln\frac{0.75}{0.25} = 2.47 - 1.099 = 1.371$.

The average growth rate will be $\frac{1.099 - 1.371}{2} = 1.235$, and $n_H = \frac{13.8 - 2.47}{1.235} \approx 9$ is required, i.e. an increase in the training effectiveness by 5% leads to a decrease in the training volume by more than 30%.

3.3. Discussion

The dynamic modification of the G. Rasch's logistic model can also be used in problems of comparing different groups of operators from the standpoint of preparedness. In contrast to traditional models of 2-way-ANOVA, which is the main mathematical model for solving such problems, the Rasch's model allows not only to assess the influence of complexity factors of control tasks and the preparedness of engineers, but also to calculate the probability of successful training in both discrete and continuous scales.

The dynamic modification of Rasch's model also allows forecasting and planning the number of training cycles, which is not possible in case using ANOVA models.

In contrast to the models of ANOVA, the presented mathematical models, in the opinion of the author of the report, look much simpler.

4. Conclusion

However, the above methods are very sensitive to the composition of students groups, as described in detail in [18], therefore the author does not recommend their use for dynamically recruited groups, the composition of which is constantly changing. Results of the analysis of these groups have very uncertain character. At the same time, the proposed technique showed itself very well in planning the training cycles for a group during a fixed period of time, for example, a semester. It can be recommended for use as an alternative to G Ebbinghaus's approach for planning the training cycles to ensure the validity of the results. Thus, the developed technique allows to plan and manage a multi-stage learning process, which ensures its effectiveness.

References

[1] Hollnagel E 1996 Reliability Analysis and Operator Modelling. *Reliability Engineering and System Safety* **52** 327 doi: 10.1016/0951-8320(95)00143-3

[2] Liao H, Forester J, Dang V N, Bye A, Chang Y H, Lois E 2019 Assessment of HRA method predictions against operating crew performance: Part I: Study background, design and methodology. *Reliab. Eng. and System Safety* **191 106509** doi: 10.1016/j.ress.2019.106509

[3] Liao H, Forester J, Dang V N, Bye A, Chang Y H, Lois E 2019 Assessment of HRA method predictions against operating crew performance: Part II: Overall simulator data, HRA method predictions and intra-method comparison. *Reliability Engineering and System Safety* **191 106510** doi: 10.1016/j.ress.2019.106510

[4] Liao H, Forester J, Dang V N, Bye A, Chang Y H, Lois E 2019 Assessment of HRA method predictions against operating crew performance: Part III: Conclusions and achievements. *Reliability Engineering and System Safety* **191 106511** doi: 10.1016/j.ress.2019.106511

[5] Zang L, He X, Dai L, Huang X 2007 The Simulatior experimental study on the Operator Reliability of Qinhan nuclear power plant. *Reliability Engineering and System Safety* **92** 252 doi: 10.1016/j.ress.2005.12.005

[6] Jirgl M, Havlikova M, Bradac Z, Stastny L 2016 Monte Carlo Reliability Analysis of System with Human Operator. *IFAC-PaperOnLine* **49(25)** 272 doi: 10.1016/j.ifacol.2016.12.046

[7] Avanesov V S 1980 The problem of psychological tests. *Soviet Education* **22(6)** 6

[8] Donihue M, Mattos J, Theoharides C, Tiffany C 2006 Meeting the Standards: An Analysis of Educational Assessment Test Scores in Maine. *Goldfarb Center Working Paper* 2006-0003

[9] Haertel E 2018 Tests, Test Scores, and Constructs. *Educational Psychologist* **53** 14 doi: 10.1080/00461520.2018.1476868

[10] Thissen D, Wainer H 2002 Test Scoring. *Journal of Educational Measurement* **39(3)** 265 doi:10.1111/j.1745-3984.2002.tb01178.x

[11] Engelhardt L, Goldhammer F 2019 Validating Test Score Interpretations Using Time Information. *Frontiers in Psychology* **10** 1131 doi:10.3389/fpsyg.2019.01131

[12] Kleinberg J, Raghu M 2015 Team Performance with Test Scores. *Proceedings of the Sixteenth ACM Conference on Economics and Computation* - EC'15 (Portland, OR, USA) doi:10.1145/2764468.2764496

[13] Rasch G E 1993 *Probabilistic Models for Some Intelligence and Attainment Tests* (Chicago, IL, USA)

[14] Peterson J K 2016 Logistics Models. *Cognitive Science and Technology* **257** doi:10.1007/978-981-287-874-8_13

[15] Hand D J, van der Linden W J, Hambleton R K 1998 Handbook of Modern Item Response Theory. *Biometrics* **54(4)** 1680 doi:10.2307/2533698

[16] Aleksandrovskaya L N, Kirillin A V, Shumskaya L P 2014 Mathematical models in assessing the effectiveness of continuing education. *Quality. Innovations. Education* **8(111)** 30 [in Russian]

[17] Aleksandrovskaya L N, Kirillin A V, Iosifov P A, Mitrofanova I P 2015 A model for evaluating the effectiveness of retraining of specialists in multilevel studying. *Quality. Innovations. Education* **1(116)** 3 [in Russian]

[18] Aleksandrovskaya L N, Kirillin A V, Borisova E V 2018 Statistical analysis of homogeneity of students' preparedness. *Quality. Innovations. Education* **6(157)** 5 [in Russian]

[19] Aleksandrovskaya L N, Kruglov V I, Kuznetcov A G, Kuznetcov V A, Kutin A A, Sholom A M 2003 *Theoretical basis of the test and experimental development of complex technical systems* (Moscow: Logos)

[20] Tsoularis A, Wallace J 2002 Analysis of logistic growth models. *Mathematical Biosciences* **179(1)** 21 doi:10.1016/s0025-5564(02)00096-2

IOP Conf. Series: Materials Science and Engineering **868** (2020) 012009 doi:10.1088/1757-899X/868/1/012009

The application of radar absorbing materials to reduce interference emissions from instruments and devices of spacecraft electrical systems

P A Zhukov* and V Yu Kirillov

Department of Theoretical Electrical Engineering, *Moscow Aviation Institute* (National Research University), 4 Volokolamskoe shosse, Moscow 125993, Russia

*E-mail: kaf309@mai.ru

Abstract. The study of interference emissions, the model of the spacecraft module in the case when the radio-absorbing material TPMV-1C is placed in the internal space of the module model is carried out. The interference emissions are created by an on-board instrument simulator located in the internal space of the spacecraft module model. A noticeable decrease in the interference emission of the spacecraft module model is demonstrated with the use of radar absorbing material. The expediency of using the TPMV-1C radar absorbing material to reduce the interference emissions of on-board instruments and devices of electrical systems of spacecraft is shown.

1. Introduction

To ensure electromagnetic compatibility of instruments and devices of the electrical systems of spacecraft, it is necessary to reduce the interference emissions of airborne instruments and devices [1]. The interference emissions in the form of a high-frequency electromagnetic field propagates within the design of the spacecraft and creates a difficultly predicted resulting electromagnetic environment. The resulting electromagnetic environment is formed due to the reflection of the waves of the electromagnetic field from the bodies of on-board instruments and devices, structural elements of the spacecraft. Within the design of the spacecraft, an electromagnetic field is formed with high intensity in local spatial areas. An electromagnetic field with high intensity acts on-board electronic devices, worsens the quality of their functioning and creates conditions for failures and irreversible violations in their work [2-4]. One of the common ways to reduce interference emissions is the use of electromagnetic shields [5]. Electromagnetic shields have inhomogeneities and apertures and therefore cannot always provide an effective reduction in interference emissions [6-8]. To reduce the interference emissions and, accordingly, to increase the efficiency of shielding, in cases where the capabilities of the electromagnetic shield are fully used, it is possible to use radar absorbing materials (RAM) [9]. Radar absorbing materials have the property of decreasing the energy of the electromagnetic field within their structure. Radar absorbing materials are widely used to create anechoic rooms for testing the electromagnetic compatibility of technical equipment. At present, radar-absorbing materials studied in solving scientific problems [10-14], but these studies not intended to use radar-absorbing materials on board spacecraft to reduce interference emissions. For spacecraft, radar-absorbing materials that are resistant to low and high temperatures in outer space should be used. Radar-absorbing materials placed in the module of the spacecraft must save their electrophysical properties in the temperature range from –150 to +150°C.

IOP Conf. Series: Materials Science and Engineering **868** (2020) 012009 doi:10.1088/1757-899X/868/1/012009

At the enterprise, "Radiostream" developed radio-absorbing material TPMV-1C. This material is made of mineral fibers coated with activated carbon particles. A special material manufacturing technology allows it to save its electrophysical properties at low and high outer space temperatures. Previous experimental studies by the authors showed that the TPMV-1C radar absorbing material reduces the electric field strength [9]. The purpose of this work is to prove that the TPMV-1C radar absorbing material can be effectively used to reduce interference emissions from on-board electronic instruments and devices of spacecraft.

2. Materials and methods

To achieve this goal, in the course of the study, the electric field strength of interference emissions of the spacecraft module model was measured. The electric field strength was measured for two cases. In the first case, there were no elements of the radio-absorbing material in the internal space of the module model. In the second case, the elements of the TPMV-1C radio-absorbing material were placed in the internal space of the spacecraft module model. The elements of the radar absorbing material were made in the form of plates with dimensions 300x200x20 mm. During the study, three elements of the radar absorbing material were attached to the internal sides of walls of the module model. The interference emissions of the spacecraft module model were emitted by the on-board device simulator located in the module model. The on-board device simulator is a cubic-shaped electrically conductive body with a rib size of 100 mm, connected with a coaxial cable to a high-frequency signal generator. The body of the on-board instrument-simulator emitted an electromagnetic field, which propagated through the aperture of the spacecraft's module model into the outer space, creating interference emissions. The module model of the spacecraft is made in the form of a hollow electrically conductive body with dimensions of 400x300x300x mm. One of the sides of the module model, measuring 300x300 mm, is open. The open side of the module model is the largest aperture. The open side of the module model can be completely covered with conductive plates. For the study, two electrically conductive plates with apertures were used. The first plate has one aperture in the form of round hole with a diameter of 20 mm. The second plate contains nine apertures in the form of nine round holes with a diameter of 5 mm. These plates make it possible to study the interference emissions of the module model with a different number of apertures and with apertures of different sizes. The number and size of apertures were varied in order to compare the levels of electric field strength in the presence or absence of radar absorbing material in the internal space of the module model. The module model of the spacecraft was installed in the transverse electromagnetic wave camera (TEM-camera). The electric field strength of interference emissions was measured using a receiver connected to a TEM-camera. The setup scheme used for the experiment is shown in figure 1.

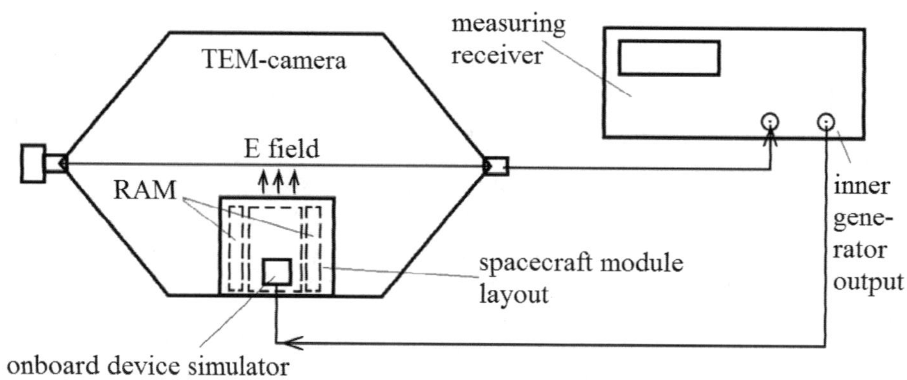

Figure 1. The setup scheme used for the experiment.

Initially, the electric field strength was measured, for the case when the open part of the module model was not covered by conductive plates. The open part of the module model can be considered as a square aperture with a size of 300x300 mm. When the open part of the prototype of the module was covered with conductive plates with different apertures, the electric field strength was measured.

3. Results

Figure 2 shows the frequency diagrams of the electric field strength of interference emissions emitted by the on-board device simulator when one side of the module model is open. Frequency diagram 1 corresponds to the electric field strength of interference emissions without elements of the radio-absorbing material in the internal space of the module model. The frequency diagram 2 of the of the electric field strength of interference emissions is obtained if that the elements of the radio-absorbing material are placed in the internal space of the module model.

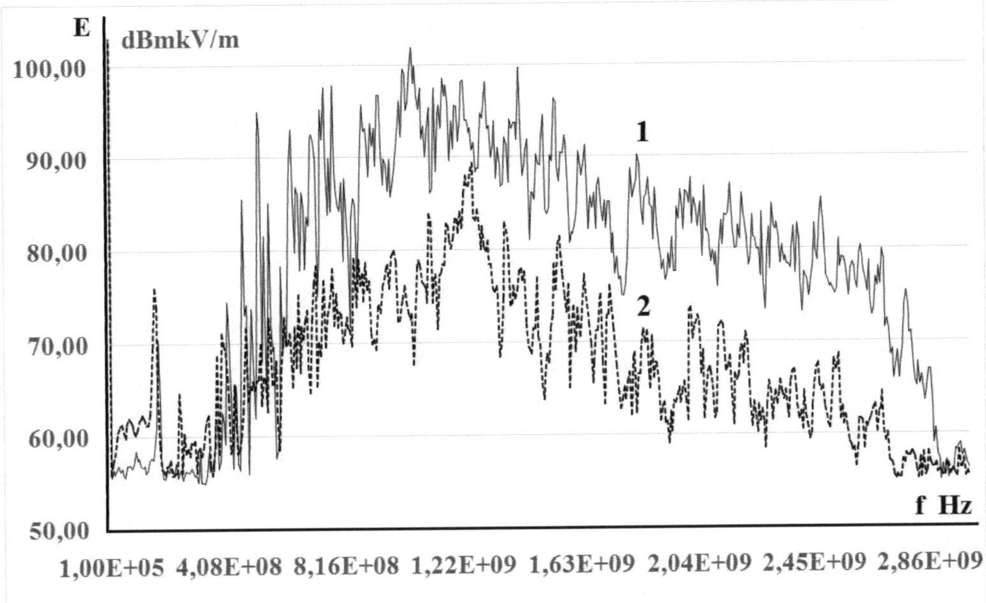

Figure 2. Frequency diagrams of the electric field strength when one side of the module model is open: 1 – without elements of the radio-absorbing material in the internal space of the module model; 2 – the elements of the radio-absorbing material are placed in the internal space of the module model.

The frequency diagram of the difference in the diagrams of the electric field strengths 1 and 2 is shown in figure 3. Figure 4 shows the frequency diagrams of the of the electric field strength of interference emissions of the module model covered by an electrically conductive plate with one aperture in the form of round hole with a diameter of 20 mm. Frequency diagram 1 corresponds to the of the electric field strength of interference emissions without elements of the radio-absorbing material in the internal space of the module model. The frequency diagram 2 of the electric field strength of interference emissions is obtained if the elements of the radio-absorbing material are placed in the internal space of the module model.

Figure 5 shows the frequency diagram of the difference in the diagrams of the electric field strength of interference emissions 1 and 2 of the figure 4.

Figure 6 shows the frequency diagrams of the electric field strength of interference emissions of the module model covered by an electrically conductive plate with nine apertures in the form of round holes with a diameter of 5 mm. Diagram 1 corresponds to the electric field strength of interference emissions without elements of a radio-absorbing material in the internal space of the module model. Diagram 2 of the electric field strength of interference emissions is obtained if the elements of the radio-absorbing material are placed in the internal space of the module model. The frequency diagram of the difference in the diagrams of the electric field strength of interference emission is shown in Figure 7.

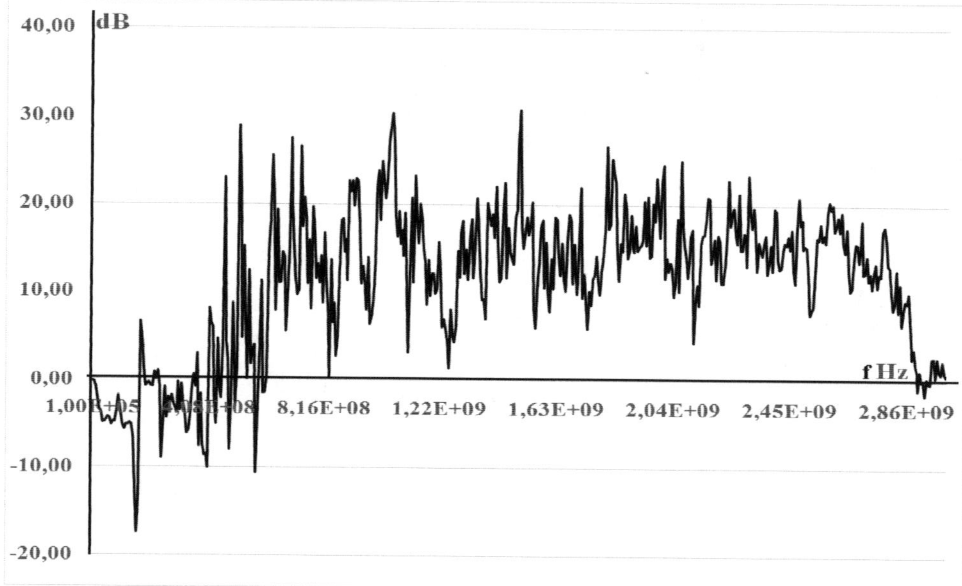

Figure 3. Frequency diagram of the difference in the diagrams of the electric field strength 1 and 2 of figure 2.

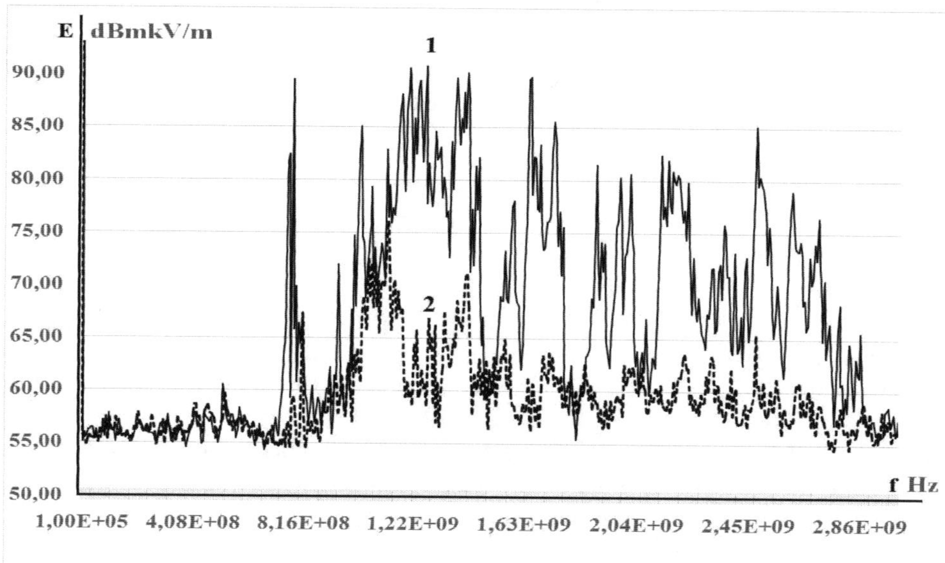

Figure 4. Frequency diagrams of the electric field strength of the module model covered by an electrically conductive plate with one aperture in the form of the round hole with a diameter of 20 mm: 1 – without elements of the radio-absorbing material in the internal space of the module model; 2 – elements of the radio-absorbing material are placed in the internal space of the module model.

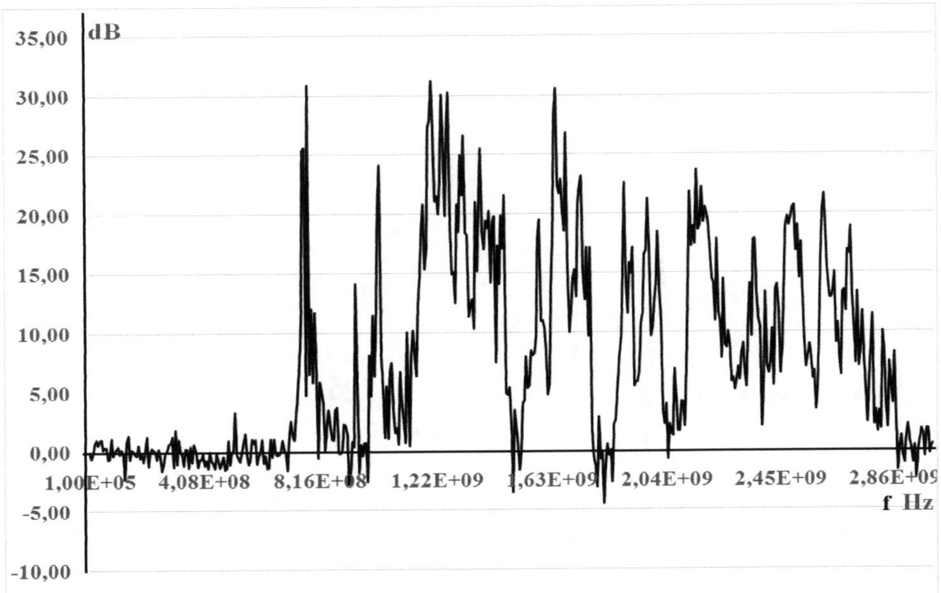

Figure 5. Frequency diagram of the difference in the diagrams of the electric field strength 1 and 2 of the figure 4.

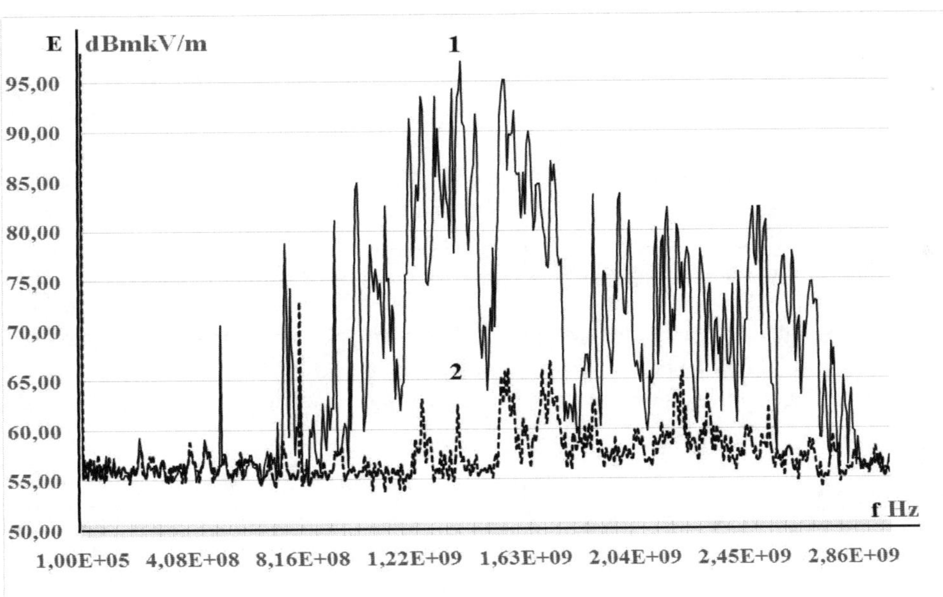

Figure 6. Frequency diagrams of the electric field strength of the module model covered by an electrically conductive plate with nine apertures in the form of round holes with a diameter of 5 mm: 1 – without elements of a radio-absorbing material in the internal space of the module model; 2 – elements of the radio-absorbing material are placed in the internal space of the module model.

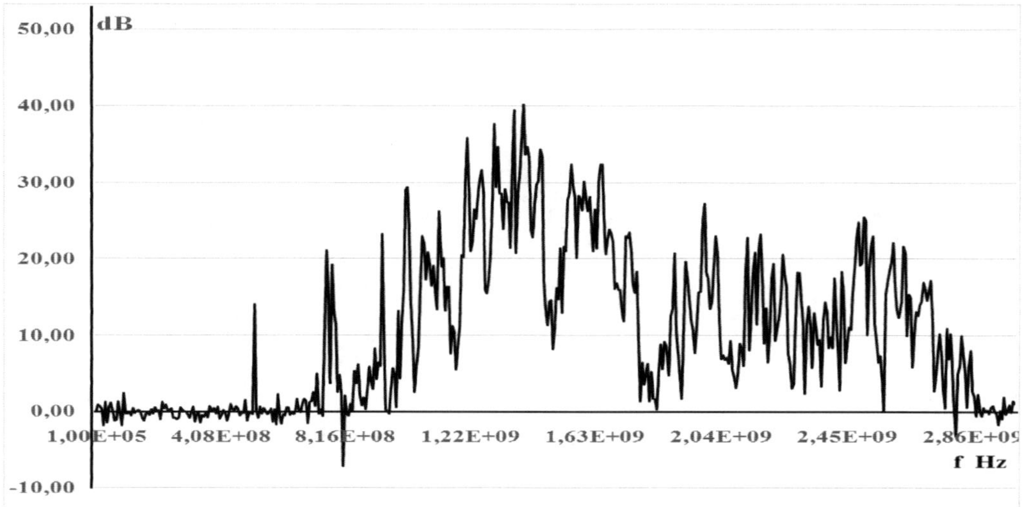

Figure 7. Frequency diagram of the difference in the diagrams of the electric field strength 1 and 2 of the figure 6.

As follows from a comparison of frequency diagrams 1 and 2 in figures 2, 4 and 6 a decrease in the electric field strength of interference emissions occurs in a significant part of the frequency range 100 kHz-3000 MHz. This is confirmed by the frequency diagrams in figures 3, 5, 7 which are the difference of the frequency diagrams 1 and 2 of figures 2, 4, 6 respectively. The positive effect of the use of radar absorbing material is obvious for apertures of various kinds.

4. Conclusion

A comparison of the frequency diagrams obtained because of an experimental study allows us to conclude that a decrease in interference emissions over a significant part of the frequency range occurs when elements of radio-absorbing material are placed in the internal space of the module model. The placement of the elements of the TPMV-1C radio-absorbing material in the internal space of the module mode allows reducing the electric field strength of interference emissions of the spacecraft module model, at a number of frequencies, to 30-40 dBμV/m. The effect of using radar-absorbing materials does not depend on the number and size of apertures, but for different apertures take place different values of the electric field strength of interference emissions. The results of an experimental study allow us to conclude that the use of TPMV-1C radar absorbing material helps to reduce interference emissions and can serve as an effective and useful addition to the shielding of on-board instruments and devices of electrical systems of spacecraft.

Acknowledgments

This work was funded by Russian Foundation for Basic Research, Project number 19-38-90244.

References

[1] Junge A, Wolf J, Mora N, Rachidi F and Pelissou P 2014 Electromagnetic interference control techniques for spacecraft harness Proc. *Int. Sym. Electromagnetic Compatibility* (Tokyo) pp 840-843

[2] Ott H W 2009 *Electromagnetic Compatibility Engineering* (New Jersey: John Wiley & Sons) p 872

[3] Williams T and Armstrong K 1999 *EMC for Systems and Installations* (Elsievier Ltd.) p 312

[4] Liu D-C, Li Z, Li C, Guanghui C, Qing W and Jiang-Feng Z 2003 Study on the failure mechanism of the electronic device in the transient electromagnetic field *IEEE International Symposium on Electromagnetic Compatibility,* (Istanbul) vol 2, pp 1036-1039

[5] Kechiev L N 2019 *Shielding of Electronic Equipment* (Moscow: Griffin) p 720 [in Russian]
[6] Zhang X, Zhang W and Cui X 2010 Research on shielding effectiveness of enclosure with apertures *Proc. Asia-Pacific Int. Symp. Electromagnetic Compatibility* (Beijing) pp 928-931
[7] Yu D, Song H and Liu Y 2008 Estimate of shielding efficiency for double layer enclosure with apertures array *Proc. 8th Int. Symp. Antennas, Propagation and EM Theory* (Kunming) pp 446-449
[8] Zhukov P A 2017 Theinfluence of cable shield terminations and transition resistance on shielding efficiency *Proc.16th Int. Conf. Aviation and Cosmonautics* (Moscow) p 157 [in Russian]
[9] Zhuravlev S Yu, Kirillov V Yu and Zhukov P A 2018 Studies of radar absorbing materials for spacecraft *Electromagnetic Compatibility Technologies* **4 (67)** 32 [in Russian]
[10] Tamminen A, Lonnqvist A, Mallat J and Raisanen A V 2007 Transmittance and monostatic reflectivity of radar absorbing materials for CATR *Prov. 2nd Eur. Conf. Antennas and Propagation, EuCAP 2007* (Edinburgh) pp 1-6
[11] Gevorkyan A V and Yu P T 2018 The Radiation characteristics of 3.43:1 bandwidth dipole antenna with radar absorbing material. *IEEE Radio and Antenna Days of the Indian Ocean*, (Grand Port) pp 1-2
[12] Anyutin N V, Titarenko A V and Elizarov S V 2017 Justification of the conditions of reproducible measured characteristics of radio absorbing materials in free space *Radiation and Scattering of Electromagnetic Waves RSEMW* (Divnomorskoe) pp 165-167
[13] Zhang C-F, Tang W, Mi X-L and Chen L-R 2011 Application of radar absorbing material in design of metal space frame radomes. *Proc. Cross Strait Quad-Regional Radio Science and Wireless Technology Conf.* (Harbin) pp 222-225
[14] Hofmann W, Bornkessel C, Schwind A and Hein M A 2019 Challenges of RF absorber characterization: comparison between RCS- and NRL-arch-methods *Electromagnetic Compatibility - EMC EUROPE* (Barcelona) pp 370-375

IOP Conf. Series: Materials Science and Engineering **868** (2020) 012010 doi:10.1088/1757-899X/868/1/012010

Mathematical model of radio-frequency ion thruster with an additional magnetostatic field

S Kanev, A Melnikov*, **I Nazarenko and S Khartov**

Department of Electric Propulsion and space power plants, *Moscow Aviation Institute* (National Research University), 4 Volokolamskoe shosse, Moscow 125993, Russia

*E-mail: k208@mai.ru

Abstract. The paper presents an engineering simplified mathematical model for estimating integral characteristics of radio-frequency ion thrusters with an additional magnetostatic field. In addition to the integral characteristics calculation, the proposed model allows to obtain the distributions of some plasma local parameters in a gas-discharge chamber of thruster – atoms and charged particles concentrations, and temperature of electrons. In a calculation, axisymmetric geometry of main elements of a thruster structure is considered. Two approaches to taking into account the influence of an additional magnetic field on the distribution of charged particles concentration are implementing. Each of them allows to estimate integral characteristics of radio-frequency ion thrusters with acceptable accuracy. The comparison of the experimental data with simulation results showed that error of thruster's parameters estimation does not exceed 20-25%, which is sufficient for preliminary designing of new radio-frequency ion thrusters models with an additional source of magnetostatic field.

1. Introduction

The increasing of active lifetime of small low-orbiting spacecrafts is one of the relevant trends in the space technology development. For increasing of active lifetime of satellites of this type, it is necessary to take into account the influence of residual atmosphere of the Earth, which leads to aerodynamic drag. For compensation of this effect, the using of ion thrusters is most promising, since they provide highest values of specific impulse and, as a result, more economical consumption of the working medium [1].

Currently, of special interested are development and using radio-frequency ion thrusters (RIT) on board of the small spacecraft. As opposed classic Kaufman type ion thrusters, RIT has not problems associated with a decreasing of its life due sputtering of the cathode surface inside the volume of gas-discharge chamber (GDC). In addition, this type of ion thruster allows to regulate the thrust value by changing both the input RF power and the working medium flow rate. However, RIT has a significant deficiency - high energy consumption for ionization of the working medium. One of the reasons of it is the large loss of charged particles due to their recombination on the walls of the GDC. These losses can be reduced by using additional axial external magnetostatic field in the area of RF discharge.

The first experimental studies in this direction were carried out at radio-frequency technological ion sources at the Department of Physical Electronics of Moscow State University [2-4]. However, the design of technological units and their operating modes differed from the solutions which used in RIT. Therefore, independent experimental studies were carried out at the Moscow Aviation Institute [5]. It also showed a positive result - thruster performance improvement. The studies of radial distribution of plasma local parameters in one of cross-sections of the GDC of the thruster with using electrostatic probe showed that a registered characteristics improvement in the presence of the additional magnetostatic field was related with an increasing of the electron's concentration and redistribution of their temperature along the radius [6]. As a result of complex studies for the laboratory model of RIT with a beam diameter of 8 cm, optimal parameters of the additional magnetic field at which the maximum improvement of characteristics were observed.

For designing of other RIT models with an additional magnetic field source, it is expedient to have possibility of quick estimation of their integral characteristics. In doing so, it is necessary to take into account the geometry of main elements of the thruster structure and its operation parameters (frequency and amperage on the inductor, flow rate of working medium, amperage at an additional winding or properties of permanent magnet).

Currently, a number of mathematical models have been developed, which differ in the detailing and depth of the physical processes' description in the plasma of radio-frequency plants [7-14]. Some of these models doesn't take into account the geometry of plants structure elements and, moreover, no one of them allows to estimate the influence of an additional magnetostatic field on the RIT integral characteristics or on the distribution of local plasma parameters in its GDC. The more appropriate mathematical models were developed at the Moscow State University [15, 16]. They were used to physically explanation the observed regularities during experimental studies of radio-frequency ion sources and plasma generators, as well as in the development of own RIT prototype with an additional magnetic field source [17, 18]. However, these models require a considerable time for calculation, what is not always appropriate it for preliminary design of thrusters, since most often it is necessary to consider and to compare a large number of different design solutions and operating modes.

Thus, the purpose of the presented work is creating the simplified mathematical model which will allow with low time and computational resources costs, but with acceptable accuracy, to estimate the RIT integral characteristics and the distribution of local plasma parameters in its GDC with taking into account the geometry of the main construction elements, the operating mode, and the presence of an additional magnetostatic field.

2. The calculation method of the engineering mathematical model

Calculations were implemented in the COMSOL Multiphysics software package [19]. The axisymmetric geometry of main elements of RIT structure is considering: the gas-discharge chamber, the inductor, a gas inlet and an ion-optical system (IOS). With the purpose reducing calculation time, a description of atoms and charged particles behavior in the model is carried out using a number of analytical expressions.

The influence of an additional magnetostatic field on the distribution of charged particles and integral characteristics of RIT was estimated in two methods. In the first case, a coefficient was used, which had been obtained by considering the motion of a monoenergetic electron flux across uniform magnetostatic field. The second approach assumes the using of a coefficient which approximately describes the Bohm diffusion of charged particles in a magnetic field.

2.1. Assumptions of the mathematical model

In assumptions of the model, characteristic features of an inductive RF discharge, which is realized in the RIT, were taken into account.

The particles mean free path in a RF plasma during electron-atom and ion-atom interactions often exceeds the size of the gas-discharge chamber. Therefore, processes of volumetric charge-exchange weren't taken into account in the model. In addition, volume recombination of charged particles also wasn't considered, since its intensity is much lower than on walls of the thruster gas-discharge chamber.

Based on a fact that a rate of doubly charged ions in the RIT plasma doesn't exceed 1% [20], we can ignore processes of working medium atoms double ionization. A viscous interaction of neutral particles with each other in the model is also neglected.

Since the Debye radius in the RF plasma is of an order of 10^{-5}–10^{-6} m, which is much smaller than the characteristic size of the gas-discharge chamber, in its volume with high precision a quasineutrality condition is satisfied — electrons and ions concentrations are equal ($n_e = n_i = n_{pl}$).

A possibility of transferring momentum and kinetic energy, when elastic collisions of electrons with ions and working medium atoms is taken into account, is neglected, since this effect will be minor due to a large difference in masses.

A close to magnetohydrodynamic approximation is used to describe a behavior of charged particles in plasma [21]. A possibility of applying this approach is due to the fact that a collisional energy transfer mechanism between charged particles prevails in the RIT plasma [22]. Based on this, it is also assumed in the model that the electron energy distribution function corresponds to the Maxwell function.

It is assumed in the model that all charged particles falling onto walls of the gas-discharge chamber recombine and returned into the plasma volume in form of neutral atoms. In doing so, the temperature of neutral particles is assumed constant throughout at the chamber volume and equal to the wall temperature.

The influence of material properties of main RIT structure elements on local plasma parameters and on its integral characteristics isn't considered in the model. Also, the presence of a capacitive component of the discharge, existing due to inter-turn capacitances formed by inductor, isn't taken into account.

To simplify the mathematical model a propagation of waves in the plasma isn't considered.

2.2. Expressions used in the calculation

Initial data required for the calculation are following parameters: frequency f_{RF} and amperage I_{ind} at the inductor, surface temperature of the inner wall of the gas-discharge chamber T_w, mass flow rate of working medium $\dot{m}$, amperage in the additional winding I_{aw} (or magnetic field induction B_{cm} when RIT with a permanent magnet is calculated) and a transparency of emission σ_{ee} and accelerating σ_{ae} electrodes of the ion-optical system. Also, before the calculation, initial values of the concentration n_e and the temperature of electrons T_e, and the concentration of working medium atoms n_a are additionally set.

The distribution of an electromagnetic field of the inductor is determined using the standard module "Magnetic fields" of COMSOL Multiphysics [19]. In doing so, the calculation was made on amplitude value of amperage on the inductor, which is expressed in terms of the current density j_{ind}. It allows with using the Ampere-Maxwell equation to obtain the distribution of amplitude value induction of magnetic field B_{ind} in the volume of gas-discharge chamber:

$$\nabla \times B_{ind} = \mu_0 j_{ind}, \tag{1}$$

where μ_0 – the vacuum permeability.

The distribution B_{ind} is used to find a vector magnetic potential A_{ind}, through which the amplitude strength of vortex electric field E_0 is determined in the model:

$$B_{\text{ind}} = \nabla \times A_{\text{ind}}; \tag{2}$$

$$E_0 = -\omega A_{\text{ind}}, \tag{3}$$

where $\omega = 2\pi f_{\text{RF}}$ – circular frequency of RF field.

The amplitude velocity of electrons u_{e0}, which they acquire in the vortex electric field in the model is determined through the value E_0 [21]:

$$u_{\text{e0}} = \frac{e\omega A_{\text{ind}}}{m_{\text{e}}(\omega^2+{\nu_{\text{ea}}}^2)^{1/2}}, \tag{4}$$

where e – elementary charge; m_{e} – electron mass; ν_{ea} – the frequency of elastic collisions of electrons with working medium atoms, which is defined as:

$$\nu_{\text{ea}} = n_{\text{a}}\sigma_{\text{ea}}\left(\frac{3kT_{\text{e}}}{m_{\text{e}}} + \frac{u_{\text{e0}}^2}{2}\right)^{1/2}, \tag{5}$$

IOP Conf. Series: Materials Science and Engineering **868** (2020) 012010 doi:10.1088/1757-899X/868/1/012010

where σ_{ea} – elastic scattering cross section of electrons by working medium atoms; k – the Boltzmann constant.

In [21], it was shown that time-averaged energy of oscillating motion, taking into account expressions (4) and (5), would be determined as:

$$\frac{m_e u_{e0}^2}{4} = \frac{e^2 \omega^2 A_{ind}^2}{4m_e(\omega^2 + \nu_{ea}{}^2)}. \quad (6)$$

Then the effective temperature T_{eff}, which determines the electrons total energy and includes a thermal component and accumulated in the vortex electric field kinetic energy, can be estimated by the following expression [21]:

$$\frac{3}{2}kT_{eff} = \frac{3}{2}kT_e + \frac{e^2 \omega^2 A_{ind}^2}{4m_e(\omega^2 + \nu_{ea}{}^2)}. \quad (7)$$

The values of the ionization rate coefficient $\langle\sigma_i\vartheta\rangle$ and the scattering cross section σ_{ea} are calculated through the T_{eff}.

Also, the density of induced in the plasma electron current j_e, which reverse to the flowing through the inductor current is determined through the value u_{e0}:

$$j_e = -en_e u_{e0}. \quad (8)$$

Since this current influenced on the distribution of inductor`s electromagnetic field, in the model, when B_{ind} is calculated, it is set and calculated together with j_{ind}. It makes it possible to relatively simple simulate the presence of skin-effect in the plasma.

At first, the distribution of neutral atoms concentration in the volume of the gas-discharge chamber is assumed to be uniform and determined with using the equation of ion balance:

$$n_{a0} = 0.7S_{GDC}\left(\frac{kT_e}{M_i}\right)^{1/2}(\langle\sigma_i\vartheta\rangle V_{GDC})^{-1}, \quad (9)$$

where S_{GDC} – the inner surface area of the gas-discharge chamber; M_i – mass of working medium ion; V_{GDC} – the inner volume of the gas-discharge chamber. The ionization coefficient for used in the RIT working medium — xenon, $\langle\sigma_i\vartheta\rangle$ is determined by the following expression:

$$\langle\sigma_i\vartheta\rangle = 5 \cdot 10^{-14}\exp\left(-\frac{\varphi_{Xe}}{T_e}\right)\sqrt{T_e}, \quad (10)$$

where φ_{Xe} – xenon atom single ionization energy, eV; electrons temperature T_e is set in eV.

An approximate analytical expression, which was obtained by considering the one-dimensional task of gas flow in the tube, is used to describe the distribution of working medium atoms concentration:

$$\mathrm{div}\left(\frac{u_{aT}^2 l}{4(u_{aT} + n_e\langle\sigma_i\vartheta\rangle l)}\,\mathrm{grad}\, n_a\right) = n_a n_e\langle\sigma_i\vartheta\rangle, \quad (11)$$

where $u_{aT} = [8kT_w(\pi M_{Xe})^{-1}]^{1/2}$ – xenon atoms thermal velocity (M_{Xe} – xenon atom mass); l – the characteristic size of the plasma formation (in the model it is equal to the gas-discharge chamber radius).

The distribution of the charged particles concentration n_{pl} over the gas-discharge chamber volume is determined through the ionic component n_i with using the following expression [21]:

$$\nabla^2 n_i = -2.5\frac{M_i}{kT_e}(n_a\langle\sigma_i\vartheta\rangle)^2 n_i. \quad (12)$$

As noted above, the influence of an additional magnetostatic field on the distribution of charged particles concentration was realized in two ways. In the first case, for the plasma concentration the following correction coefficient was introduced:

$$n_{mf} = n_{pl}\exp\left(-\frac{e^2 A_C^2}{18m_e kT_e}\right), \quad (13)$$

where A_C – the vector magnetic potential of an additional magnetostatic field. It is based on the assumption that when electrons move across magnetic field, their momentum changes in proportion to the difference of vector magnetic potential that they pass, which is shown in [23].

In the second approach the coefficient that approximately describes the Bohm's diffusion of charged particles as they move in magnetic field is used:

$$n_{\mathrm{mf}} = n_{\mathrm{pl}} \exp\left(-\frac{16A_{\mathrm{C}}0{,}1u_{\mathrm{i}}}{T_{\mathrm{e}}}\right), \tag{14}$$

where $u_{\mathrm{i}} = \left(kT_{\mathrm{e}}M_{\mathrm{i}}^{-1}\right)^{1/2}$ – Bohm's ion velocity; the coefficient 0,1 before u_{i} characterizes the ion velocity in the plasma volume, which is assumed to be 10% of the Bohm's velocity.

The influence of an additional magnetic field on the effective temperature distribution is estimated using the empirical coefficient K_{MF}, which was obtained by analyzing the results of probe measurements of plasma local parameters in the gas-discharge chamber of RIT laboratory model [6]:

$$K_{\mathrm{MF}} = \left(a + B_{\mathrm{aw}} - \frac{K_{\mathrm{AF}}}{2} B_{\mathrm{ind}}\right)\left(a + B_{\mathrm{aw}} + K_{\mathrm{AF}} B_{\mathrm{ind}}\right)^{-1}, \tag{15}$$

where $a = 10^{-8}$ – the dimensionless value which is necessary to prevent the coefficient K_{MF} from zeroing in the absence of an additional magnetostatic field; K_{AF} – the correction dimensionless coefficient, which characterizes the degree of influence of an additional magnetic field on the effective electron temperature distribution and which is determined as:

$$K_{\mathrm{AF}} = 0.06\left(\frac{B_{\mathrm{av}}}{B_{\mathrm{indo}}}\right)^2 + 0.11\frac{B_{\mathrm{av}}}{B_{\mathrm{indo}}}, \tag{16}$$

where B_{av} – the average induction of an additional magnetostatic field; B_{indo} – the average amplitude value of the magnetic field induction which is generated by the inductor.

Taking into account the coefficient K_{MF} the expression for the effective temperature will take the following form:

$$T_{\mathrm{eff}} = T_{\mathrm{e}} + \frac{e^2\omega^2 A_{\mathrm{ind}}^2}{6km_{\mathrm{e}}(\omega^2 + \nu_{\mathrm{ea}}{}^2)} K_{\mathrm{MF}}. \tag{17}$$

To the electron temperature T_{e} calculation in the model the energy balance equation is used, in which the right part determines the power supplied into the plasma, and the left part – the power consumptions for ionization of working medium atoms and the losses of charged particles on the walls of the gas-discharge chamber:

$$\int_V (\nu_{\mathrm{ea}} + \nu_{\mathrm{ei}}) n_{\mathrm{e}} \frac{\mathrm{m_e} u_{\mathrm{eo}}^2}{4} \mathrm{d}V = \int_S n_{\mathrm{i}} u_{\mathrm{i}} \left(e\varepsilon_{\mathrm{eff}} + \frac{3}{2}kT_{\mathrm{eff}} + kT_{\mathrm{e}} + e\Delta\varphi + \frac{5}{2}kT_{\mathrm{w}}\right) \mathrm{d}S, \tag{18}$$

where ν_{ei} – the frequency of electron-ion interactions, which is determined through the Coulomb's logarithm ($\ln\Lambda$); $\varepsilon_{\mathrm{eff}} = 12.1\{1 + [0.09(T_{\mathrm{eff}} / 12.1)^{-1}]^2\}$ – the effective ionization energy for xenon [21]; $\Delta\varphi$ – the pre-sheath plasma potential drop.

The value of the extractable ion current is determined by the following expression:

$$I_{\mathrm{i}} = 2\pi e \sigma_{\mathrm{ee}} \int_0^R n_{\mathrm{mf}} u_{\mathrm{i}} r dr. \tag{19}$$

The following boundary conditions are used in the model:

- at the boundary of calculation area, the condition for the absence of a magnetic field is set:

$$x_{\mathrm{n}} \times A = 0, \tag{20}$$

where x_{n} – normal to the calculation area boundary; A – the vector magnetic potential;

- on the gas-discharge chamber walls, the condition for the absence pass through walls flow of atoms and ions of working medium is set:

$$n_a u_{a\perp} + n_{mf} u_{i\perp} = 0, \tag{21}$$

where $u_{a\perp}$ and $u_{i\perp}$ – perpendicular components of working medium atoms and ions velocity to the gas-discharge chamber wall; and the reverse flow of neutral particles, which is determined by falling ions onto the walls with a Bohm's velocity, is set:

$$\Gamma_{w\,a} = n_{mf}\left(\frac{kT_e}{M_i}\right)^{1/2}; \tag{22}$$

on the IOS, the condition which determines the number of atoms passing through this thruster assembly, is set:

$$n_x \frac{u_{aT}}{2}\sigma_{ae} = \left[n_a \frac{u_{aT}}{2} - n_i u_i (1 - \sigma_{ee})\right]\left(\frac{\sigma_{ae}}{2-\sigma_{ae}}\right), \tag{23}$$

where n_x – the concentration of atoms which move in the IOS direction,
and the common condition for gas-discharge chamber walls and IOS, which takes into account the charged particles flow:

$$\frac{\partial n_i}{\partial x_{ni}} = -2.5\left(\frac{M_i}{kT_e}\right)^{1/2} n_{mf} n_a \langle\sigma_i \vartheta\rangle, \tag{24}$$

where x_{ni} – normal to the wall surface and the IOS electrode.

The calculation according to the engineering model was carried out using the Nelder-Mead's optimization method. With varying the values of T_e, n_i and n_a with certain accuracy the energy balance condition (18) was verified. With the absence of convergence, a jump to a new iteration is carried out.

3. Calculation results

Some operating modes of RIT laboratory model with a beam diameter of 8 cm (figure 1), were used to verify the mathematical model.

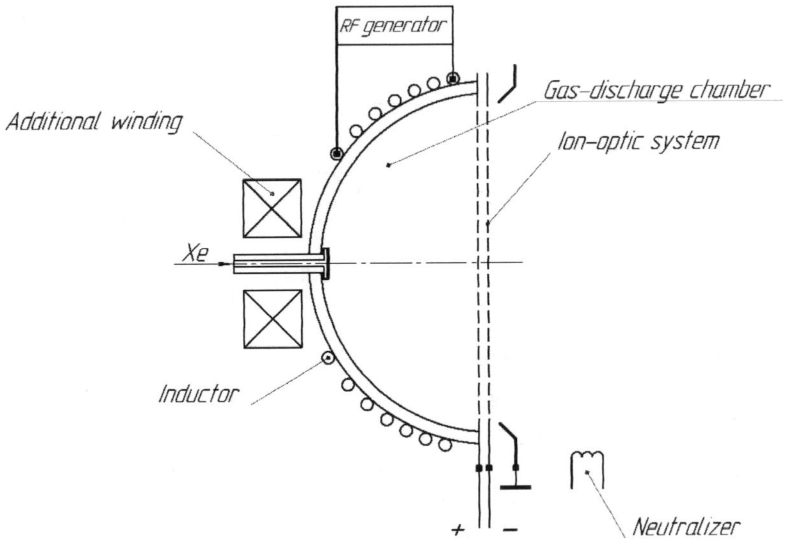

Figure 1. The schematic diagram of the RIT laboratory model with a beam diameter of 8 cm.

In the figure 2 the simplified geometry of the main elements of thruster was given, and in the figure 3 – the created for calculation finite element mesh.

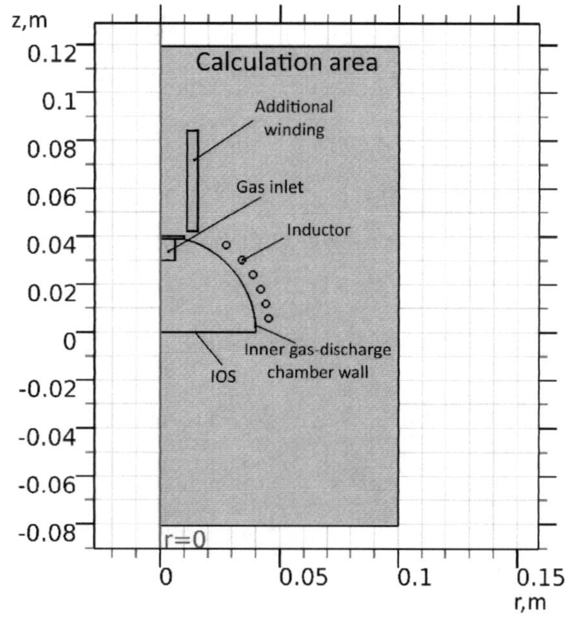

Figure 2. The calculation area and geometry of the main elements of thruster.

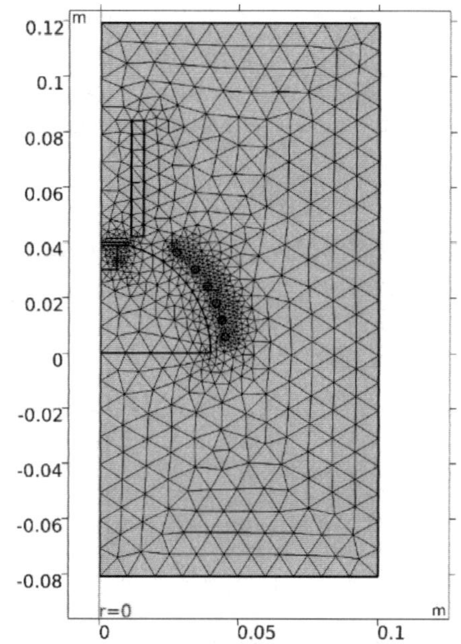

Figure 3. The created in the calculation area mesh of finite elements.

Figure 4 shows the comparison of the calculated dependence of the extractable ion current on the RF power with the experimental data for xenon flow rate 3 and 6 sccm without additional magnetic field. Since the model allows to calculate only the RF power absorbed by the plasma, the efficiency coefficient of energy transfer from the inductor was introduced for comparison with the experiment.

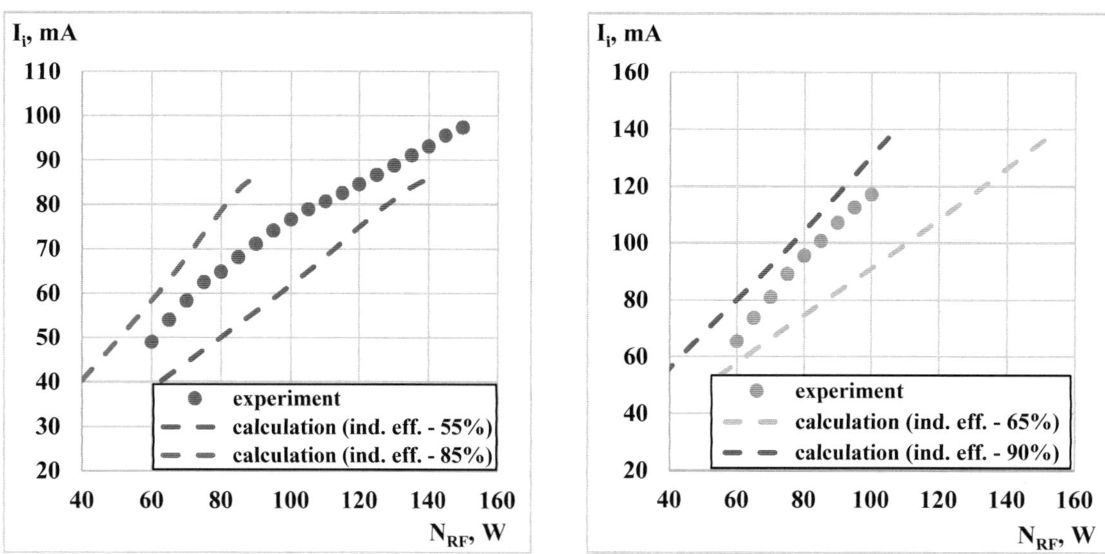

(a) (b)

Figure 4. The comparison of the calculated dependence of the extractable ion beam current on the RF power with experiment: xenon flow rate 3 sccm (a), xenon flow rate 6 sccm (b).

For comparing of two approaches to taking into account the presence of an additional magnetostatic field in the ionization area, similar dependencies were constructed, which are shown in figure 5. Calculation No. 1 was carried out with using coefficient (13), No. 2 – with using coefficient (14).

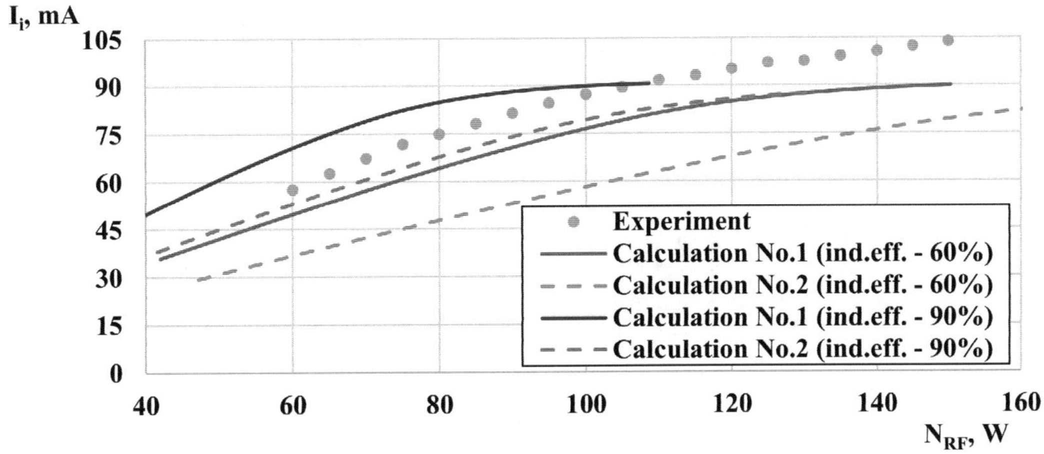

Figure 5. The comparison of calculated dependences of the extractable ion beam current on the RF power with the experiment in the presence of an additional magnetic field. Xenon flow rate is 3 sccm. The amperage in the additional winding is 9 A.

The presented dependences show that coefficient (14) in considered operating modes underestimates the value of the extractable ion beam current.

The comparison of calculated dependences of the extractable ion beam on the amperage in the additional winding with experimental data were carried out to estimate the correspondence of the change integral characteristics character while using an additional magnetostatic field. The comparison result is given in figure 6.

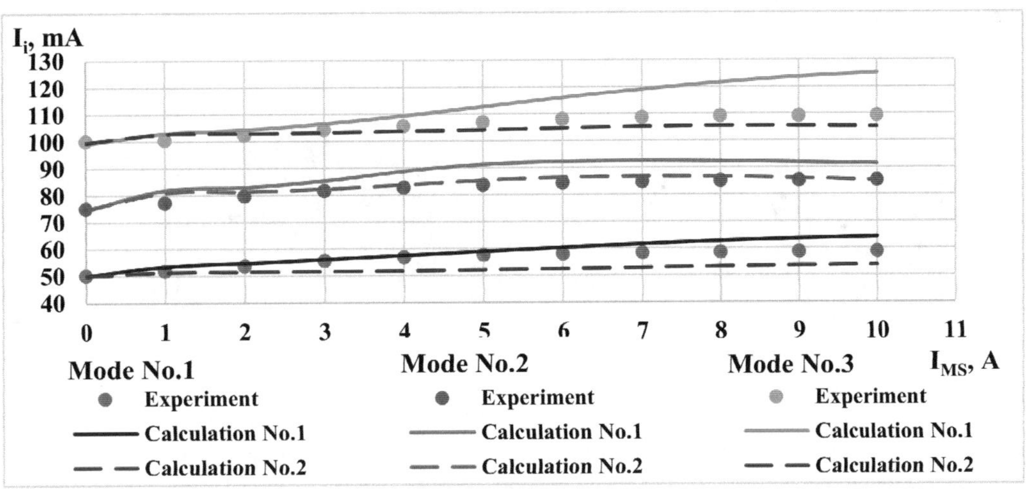

Figure 6. The comparison of calculated dependences of the extractable ion beam current on the amperage in the additional winding with experiment. Mode: No.1 – N_{RF} = 60 W, the xenon flow rate 3.07 sccm; No.2 – N_{RF} = 90 W, the xenon flow rate 3.18 sccm; No.3 – N_{RF} = 90 W, the xenon flow rate 5.14 sccm.

Since the model allows to estimate the plasma local parameters, calculated two-dimensional electron concentration distributions over the gas-discharge chamber volume in the absence and the presence of an additional magnetostatic field were also considered. The calculation result of charged particles concentration distribution for the operating mode of the RIT laboratory model with RF power of 60 W and xenon flow rate 2.24 sccm is given in the figure 7.

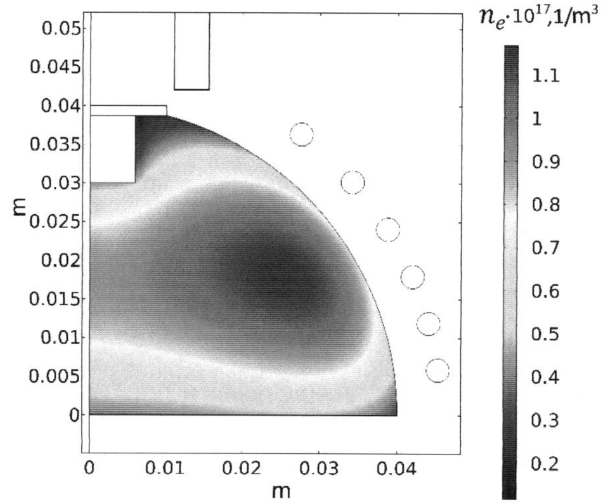

Figure 7. The calculated two-dimensional distribution of electron concentration over the RIT gas-discharge chamber. The RF power 60 W, the xenon flow rate 2.24 sccm.

According to calculation results, without additional magnetic field, the concentration maximum is localized near the gas-discharge chamber wall. However, as a magnetic field is applied, the maximum shifts closer to the axis of a thruster. The comparison of two approaches of the calculation of the electron concentration distribution with the same mode of the laboratory model operation, but in the presence of an additional magnetic field is given in the figure 8 and the figure 9.

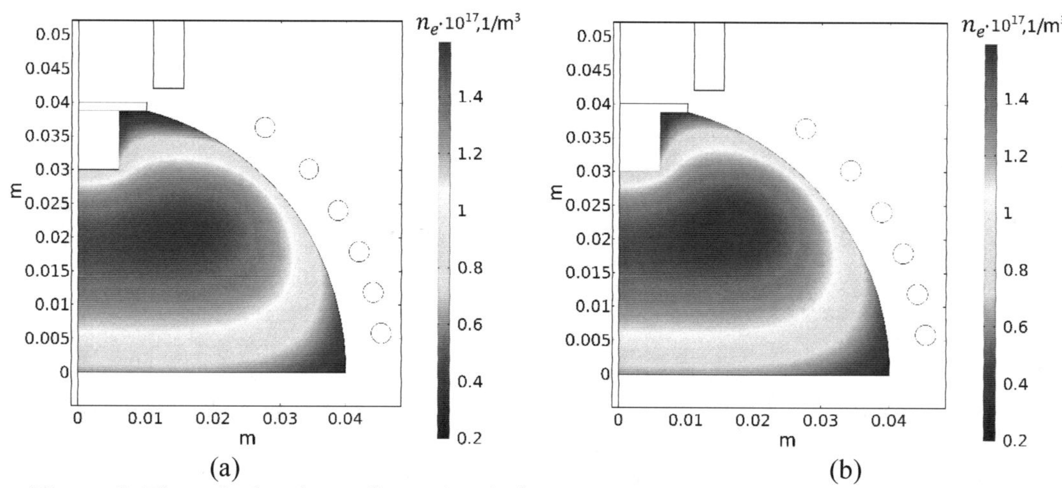

Figure 8. The calculated two-dimensional electron concentration distribution over the RIT gas-discharge chamber volume: the approach No.1 (a), the approach No.2 (b). The RF power 60 W, the xenon flow rate 2.24 sccm, the amperage in the additional winding is 3 A.

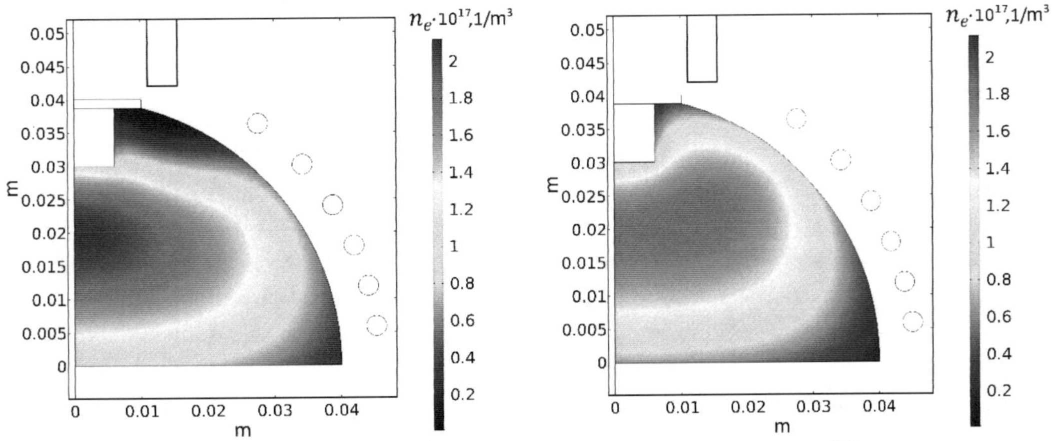

(a) (b)

Figure 9. The calculated two-dimensional electron concentration distribution over the RIT gas-discharge chamber volume: the approach No.1 (a), the approach No.2 (b). The RF power 60 W, the xenon flow rate 2.24 sccm, the amperage in the additional winding is 9 A.

At amperage of 3 A in the additional winding, the difference in the charged particles concentration distribution is minimal. At amperage of 9 A in the distribution, which was calculated using coefficient (13), the stronger influence of an additional magnetic field is observed. Using in calculations the coefficient based on the Bohm's diffusion gives more uniform distribution. At the same time, it somewhat reduces the increasing of charged particles concentration under the influence of an additional magnetic field. Radial dependences of the electron concentration and the effective temperature in the cross-section of RIT laboratory model gas-discharge chamber in which probe measurements were made. were plotted for the comparing of two approaches with the experiment. The comparison of resulting distributions is presented in the figure 10 and the figure 11.

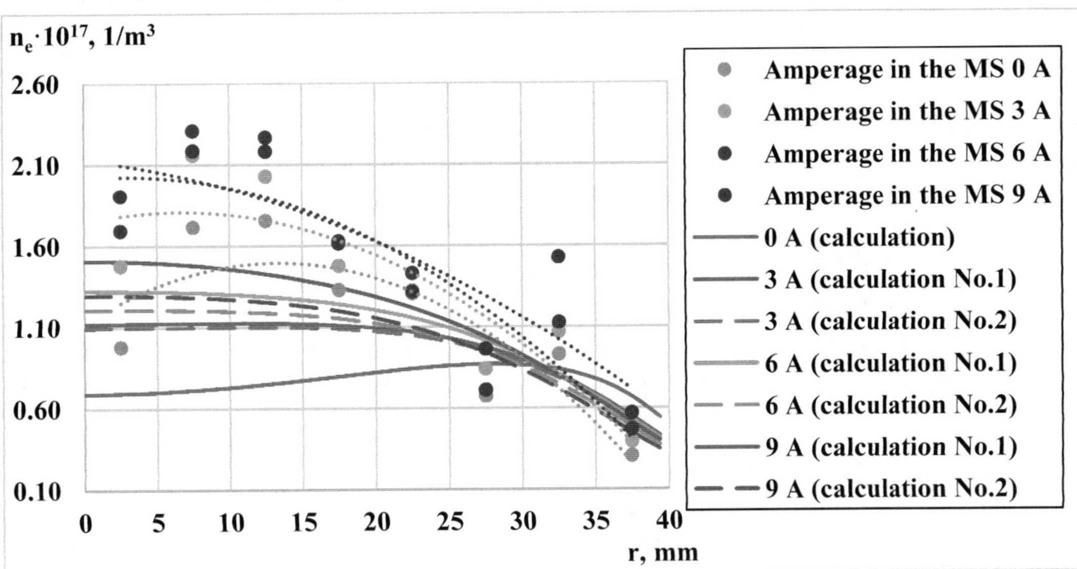

Figure 10. The comparison of calculated radial distributions of electron concentration with the result of probe studies. The RF power 60 W, the xenon flow rate 2.24 sccm.

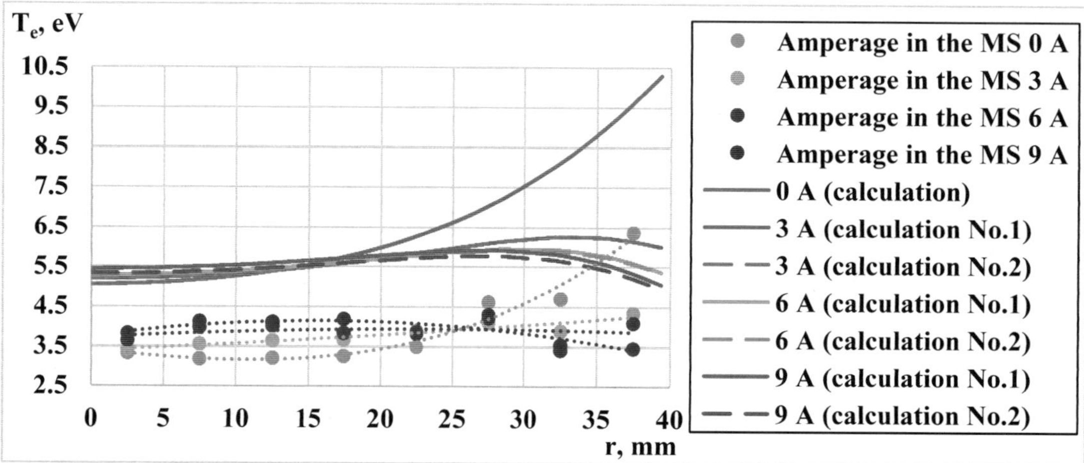

Figure 11. The comparison of calculated radial distributions of electron effective temperature with the result of probe studies. The RF power 60 W, the xenon flow rate 2.24 sccm.

As one can se from the figures, both approaches allow an acceptable estimation for distribution of plasma local parameters in the presence of an additional magnetic field. However, it is worth noting that using the coefficient (14) understates the charged particles concentration.

4. Conclusions

The using of simplified relations in the engineering model allows significantly reduce the required time for estimation of the RIT integral characteristics and the distributions of local plasma parameters in its gas-discharge chamber. For calculation of one operating mode from 200 to 600 iterations are required and on average takes from 1 to 3 minutes. Both considered approaches, which take into account the influence of an additional magnetostatic field on the thruster parameters, with sufficient accuracy allows to simulate of working process. The verification of numerical results with the experimental data has shown that the calculation error doesn't exceed 20 – 25 %. Such accuracy is sufficient for preliminary design of RIT thrusters with the source of an additional magnetic field. However, further improvement of the proposed model is also possible. In particular, additional probe studies of local plasma parameters may allow the more detailed understanding process of changing the electron temperature distribution in the presence of an additional magnetostatic field in the area of the RF discharge.

References

[1] Goebel D M, Katz I 2008 *Fundamentals of Electric Propulsion: Ion and Hall Thrusters* (California Institute of Technology) 486 p

[2] Alexandrov A F, Bugrov G E, Kerimova I, Kondranin S G, Kralkina E A, Pavlov V B, Plaksin V Yu, Rukhadze A A, Vavilin K V 2003 Development of Low Power 13.56 MHz Plasma Source Family. *Proc. of 30th International Conference on Plasma Science* (Jeju, South Korea) DOI: 10.1109/PLASMA.2003.1228688

[3] Vavilin K V, Rukhadze A A, Ri M K, Plaksin V Yu 2004 Low-power inductive RF plasma sources for technological applications. *Plasma Phys. Rep.* **30** 687 https://doi.org/10.1134/1.1788762

[4] Petrov A K, Vavilin K V, Kozlov G P, Kralkina E A, Nekliudova P A, Nikonov A M, Pavlov V B 2015 Plasma parameters in a dual-camera low-power inductive RF discharge with an external magnetic field. *Moscow University Physics Bulletin* **70** 527 https://doi.org/10.3103/S0027134915060144

[5] Melnikov A V, Khartov S A 2018 Radio-Frequency Ion Thruster with Additional Magnetic Field: Experimental Investigation. *Thermal Engineering* **65** 980 DOI: S0040601518130086

[6] Khartov S A, Melnikov A V, Kozhevnikov V V 2019 Characteristics of Radio-Frequency Ion Thruster with an Additional Magnetic Field in the Ionization Area. *Proc. of 36th International Electric Propulsion Conference* (Vienna, Austria) IEPC-2019-240

[7] Celik M, Turkoz E 2013 Optimization of radio-frequency ion thruster discharge chamber using an analytical model. *Proc. of 6th International Conference on Recent Advances in Space Technologies* (Istanbul, Turkey) DOI: 10.1109/RAST.2013.6581262

[8] Goebel D M 2008 Analytical discharge model for RF ion thrusters. *IEEE transactions on plasma science* **36** 2111 DOI: 10.1109/TPS.2008.2004232

[9] Chabert P, Monreal J A, Bredin J, Popelier L, Aanesland A 2012 Global model of a gridded-ion thruster powered by a radiofrequency inductive coil. *Physics of Plasmas* **19** 073512 DOI: 10.1063/1.4737114

[10] Tsay M M T 2010 *Two-dimensional numerical modeling of radio-frequency ion engine discharge*, PhD thesis, Massachusetts Institute of Technology

[11] Bilén S, Mistoco V 2008 Numerical Modeling of a Miniature Radio-Frequency Ion Thruster. *Proc. of 44th Joint Propulsion Conference* (Hartford, CT, USA) DOI: 10.2514/6.2008-5194

[12] Celik M, Turkoz E 2014 2-D electromagnetic and fluid models for inductively coupled plasma for RF ion thruster performance evaluation. *IEEE Transactions on Plasma Science* **42** 235 DOI: 10.1109/TPS.2013.2287903

[13] Heiliger C, Henrich R 2013 Three Dimensional Simulation of Micro Newton RITs. *Proc. of 33rd International electric propulsion conference* (Washington, DC, USA) IEPC-2013-301

[14] Dobkevicius M, Feili D, Müller J 2015 Comprehensive Radio – Frequency Ion Thruster Electromagnetic and Thermal Modelling. *Proc. of Joint Conference of 30th International Symposium on Space Technology and Science 34th International Electric Propulsion Conference and 6th Nano-satellite Symposium* (Hyogo-Kobe, Japan) IEPC-2015-410/ISTS-2015-b-410

[15] Aleksandrov A F, Bugrov G E, Kerimova I F, Kondranin S G, Kralkina E A, Pavlov V B, Plaksin V Yu, Rukhadze A A, Vavilin K V 2003 Self-Consistent Model of an RF Inductive Plasma Source Located in an External Magnetic Field. *Journal of Russian Laser Research* **24** 301 https://doi.org/10.1023/A:1024891824073

[16] Alexandrov A F, Vavilin K V, Kralkina E A, Pavlov V B, Petrov A K, Tarakanov V P 2015 KARAT modeling of the inductive RF discharge placed into an external magnetic field. *Applied Physics* **5** 44

[17] Kralkina E, Zadiriev I, Kharlan A 2017 Exploratory Testing of a Radio Frequency Thruster for Small Satellites. *Proc. of 35th International Electric Propulsion Conference* (Atlanta, Georgia, USA) IEPC-2017-425

[18] Kralkina E A, Vavilin K V, Zadiriev I I, Nekliudova P A, Shvydkiy G V 2019 Optimization of discharge parameters in an inductive RF ion thruster prototype. *Vacuum* **167** 136 https://doi.org/10.1016/j.vacuum.2019.05.041

[19] Official site of COMSOL Multiphysics developers, avaliable at: https://www.comsol.com/

[20] Loeb H W 2010 *Principle of Radio-Frequency Ion Thrusters RIT. RIT-22 Demonstrator Test of Astrium ST at University of Giessen* (University of Giessen)

[21] Kanev S V, Kozhevnikov V V, Khartov S A 2017 Physical and Mathematical Model of Processes in Ionization Chamber of Electric Propulsion Thruster with Atmospheric Gases as a Propellant Proceedings of the Russian Academy of Sciences. *Power Engineering* **3** 21

[22] Mikellides I, Katz I, Goebel D, Polk J 2005 Hollow cathode theory and experiment. II. A two-dimensional theoretical model of the emitter region. *Journal of Applied Physics* **98** 113303 https://doi.org/10.1063/1.2135409

[23] Landau L D, Lifshitz E M 1987 *The Classical Theory of Fields. 4th Edition* (Butterworth-Heinemann) 402 p

IOP Conf. Series: Materials Science and Engineering **868** (2020) 012011 doi:10.1088/1757-899X/868/1/012011

Trends in the development of a digital quality management system in the aerospace industry

E Kovrigin* and V Vasiliev

Institute of Materials Science and Materials Technology, Department of Quality Management and Certification, *Moscow Aviation Institute* (National Research University), 4 Volokolamsk highway, Moscow 125993, Russian Federation

*E-mail: j7eka@yandex.ru

Abstract. Taking into account the analysis of the literature on the development trends of the quality management system of aerospace industry organizations, it was concluded that today there is an acute question of creating effective information-analytical systems for making managerial and other decisions in the development and production of aerospace products, in particular aircraft engines. The purpose of this article is to develop a model of information-analytical system that allows to predict the level of quality in the development and production of aircraft engines. To achieve this goal, modelling methods were used, as well as information analysis and generalization. As a result of the study, a forecasting model and a quality control scheme for the development and production of aircraft engines at different stages of the life cycle were developed and tested, allowing getting a complete picture of it.

1. Introduction

Today's technologies in the Russian aerospace industry do not allow the creation and production of competitive aircraft that meet the high requirements of the use of aircraft. Russian-made aviation equipment has insufficiently high characteristics in terms of compliance with environmental requirements, reliability and cost-effectiveness requirements. Aircraft engines have low performance and reliability. Avionics and general-purpose equipment are inferior in functional completeness, overall weight and weight excellence, reliability and maintainability to the best examples of the world aircraft industry.

However, the whole world is experiencing a digital boom (or as it is commonly called digitalization), which is fundamentally changing the way of life of all mankind. This process is global in nature and in the next 5-10 years will affect almost all spheres of human life and society. Of course, it will not ignore the aerospace industry and such an important component of any organization as a quality management system (QMS).

Digitalization is the process of transformation of the company, aimed at forming a unified information environment throughout the life cycle of the product, combining different methods and tools of data management in the organization. Today, the formation of such an environment in the aerospace industry is ensured by the integration of modern software products: PLM (Product Life-cycle Management), MES (Manufacturing Execution System), ERP (Enterprise Recurse Planning), MDM (Master Data Management), CRM (Customer Relationship Management) [1-3]. The article [4] proposes a PLM structure based on blockchain technology, which is able to facilitate access to certain types of data on the product lifecycle. The concept of integration of industrial blockchain with Internet of things and M2M (Machine-to-machine) technologies is proposed. The results of the study showed that the proposed structure is effective, and therefore, it can be used in industry. In the article [5], the authors first investigated various industries, such as aerospace, semiconductor, and automotive, to adapt the MES system. The authors investigated the modern, so-called Smart MES (in comparison with the usual MES), which can comprehensively and effectively cope with various situations in the workshop in real time. According to the authors of [6], enterprise resource planning (ERP) systems are considered the basis for Industry 4.0. In their article, they try to answer the following question: "Are modern ERP systems ready to integrate into Factories of the Future (FoF)?" The results of the study show that there is such willingness. However, when it comes to the interaction of the ERP system with the "machines" (M2M), certain problems arise due to the lack of a single standard and protocol. The authors of the study [7] are trying to understand (determine) which factors influence the success of CRM. As a result of the study, they came to the conclusion that the success of CRM is greatly influenced by "using information technology", as well as "customer focus", "organizational capabilities" and "customer knowledge management".

The study [8] provides an overview of the benefits of introducing digital technology in the aerospace industry, which raises awareness of all recommended practices and problems that may arise, helping these organizations better implement them. Authors recommend analysing the underlying process before digitizing it so that the organization can avoid introducing bad practices. At the same time, they identified a number of problems that must be taken into account. Firstly, the aerospace industry has characteristics that create some special problems in implementing the digitalization of the organization's management system. Secondly, investing in workflow management tools will not solve the problems associated with the underlying business process. The "digitization" of the process automates existing practices, which is the main reason why the process must be displayed and optimized before it is automated. Third, the challenge for the aerospace industry is how to integrate all of the tools, databases, and systems used. A workflow system must be able to integrate all of these capabilities. Without proper integration, it is very difficult to realize most of the benefits. Implementing a workflow management system requires active management, end-user participation, tools and system integration, and a sound implementation plan. The article [9] explores the possibilities of integrating data between suppliers, manufacturers and developers in the aerospace industry. Methodological issues of creating such knowledge bases are considered taking into account the specifics of the aerospace industry.

Thus, taking into account the development trends of digital technologies, as well as the problems that exist in the aerospace industry, there is a high need for an effective information-analytical system that allows forecasting the level of quality of developed and manufactured aircraft engines, which is the purpose of this article.

2. Methodology

To achieve the goal set in this article, logical, system, functional, and comparative methods of analysis were used; methods of analogies and generalizations. In particular, an analysis was made of the stages of the life cycle of systems given in ISO / IEC 15288: 2008 "Systems and software engineering - System life cycle processes", as well as a number of scientific articles on the product life cycle [10-12]. The following stages of the product life cycle were identified, on the basis of which a model of a quality forecasting system was built: development; mass production; exploitation. In accordance with certain stages of the life cycle, modeling was carried out (a model and scheme (algorithm) were developed), at what stage, where and how specifically, it is possible to obtain information about the quality of developed and manufactured aircraft engines.

3. Model of quality control and forecasting system

One of the important tasks in the quality management of any product, including products of the aerospace industry, is the collection and processing of information about its quality.

Under the quality of products is understood as a set of properties that determine its suitability to meet certain needs in accordance with its purpose.

Of great importance for product quality management are quality indicators laid at the stage of development and coordination of the tactical and technical specifications. The group of quality indicators includes purpose indicators, which determine the main functions for which it is intended, reliability indicators, ergonomic indicators, etc.

At the same time, since it is often not about the finished product, but about assessing the quality of laboratory and developmental prototype at the development stage, it is important to develop quality indicators – the presence of various design errors. Improving product quality largely depends on reducing the number of design errors in the development of schemes and design, manufacturing defects and product failures during acceptance and operation.

Namely, based on this, the sphere of interests of the quality service of the aerospace industry organization should include failures, defects, design errors, i.e. everything that interferes with the product is fully consistent with the purpose. Therefore, the task of forecasting quality becomes the most acute and significant for it. Because, when setting the task of forecasting, it is necessary to answer the question - what could be bad in the future, is everything done in this sense in the present?

The control and forecasting system is a system for assessing the current state of affairs on product quality and an attempt to predict the problems of tomorrow based on the elimination of today's problems.

The general scheme of this system is shown in figure 1. It is important to note that it covers the stages of development (design), mass production, and also partially the period of operation (in the case of complaints and repairs). In other words, throughout the life cycle of the product, data about it should be accumulated, which allows not only to obtain the most reliable and complete information about its quality, but also to evaluate the effectiveness of the measures taken. In essence, the above diagram shows that the development (design), manufacture and operation is a continuous and constant process of finalizing products in order to improve its quality.

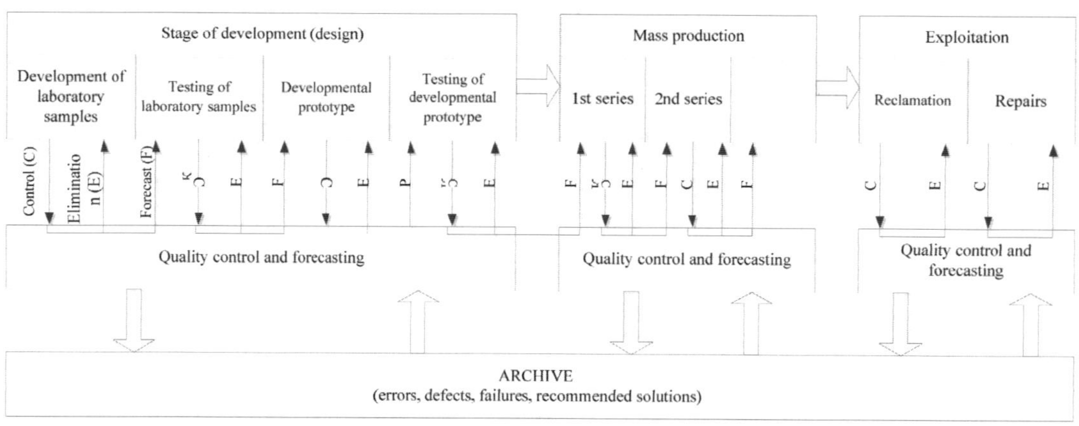

Figure 1. Quality control and forecasting system.

As can be seen from the figure, quality control is carried out at each stage, in the event that developer errors, failures or product defects are detected, they are eliminated and a similar forecast is made in the next stage. Control, elimination and forecasting procedures should be included in any subsystem of the automated quality data management system.

Another important feature of any subsystem is the presence of an archive in which information on the current state of products is accumulated and stored - recorded errors, defects, failures that occurred in the past, as well as some generalized data for long-term storage, information on the causes of failures, defects and measures to eliminate them until they are implemented, generalizing quality indicators, etc.

The archive makes it possible to summarize information on various products and various units, which is important when there is a disconnect and lack of information among developers and employees of supervisory services.

The quality control and forecasting system is a common system and a common name. It represents a set of independent subsystems, determined by the isolation of the individual stages of the product life cycle:

- development of documentation (circuit, design, technological, textual);
- control of purchased materials, semi-finished products, components and elements;
- manufacturing and testing of serial products;
- adjustment of documentation of serial products;
- repair of products.

As control methods can be:

- examination (metrological and reliability) of design, technological or textual documentation;
- quality control of design, technological or textual documentation;
- various types of checks and tests;
- in some cases, at the design stage, calculation methods.

The general quality control scheme is shown in figure 2. As a result of the control, registration is carried out:

- product failures and manufacturing defects during testing and examination;
- design errors detected during the examination or approval of documentation.

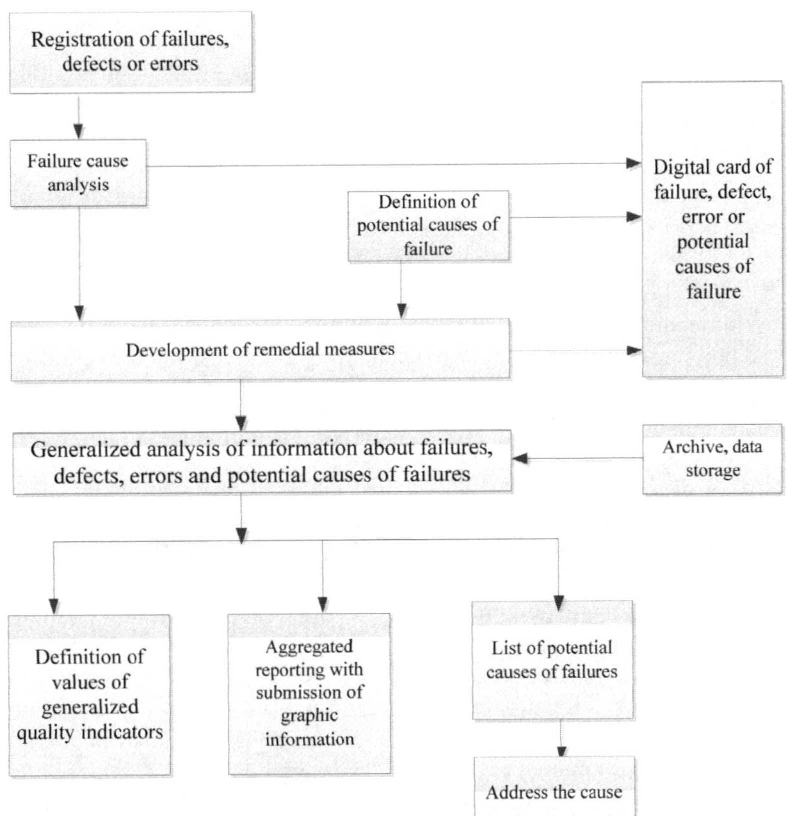

Figure 2. Quality control scheme.

Next, an analysis of the causes of failures, defects and errors, as well as the development of recommendations for their elimination. Obviously, the more thoroughly the analysis is conducted, the

IOP Conf. Series: Materials Science and Engineering **868** (2020) 012011 doi:10.1088/1757-899X/868/1/012011

more reliable the information. In this sense, this stage is important for subsequent decisions and conclusions.

Thus, not only product failures, but also manufacturing defects, design errors, and also the most likely potential causes of failures should get into the field of view of the quality service. It is this information in its totality that can give the most complete picture of the quality of products. And only the inclusion of information on the potential causes of failures with the subsequent development of recommendations and the implementation of measures to eliminate them can allow us to talk about a proactive effect on product quality. Despite all the differences between failures, defects, errors, potential causes of failures (both semantic and origin), there is no fundamental difference in the sense that eliminating the causes that caused them is the most real way to improve quality.

4. Results and discussions

As a result of the study, a quality forecasting model was developed and tested in real practice, as well as a quality control scheme, suggesting where, when and at what stage of development and production of aircraft engines it is necessary to collect information about it. A technique is proposed for collecting information on the quality of aircraft engines being developed and produced during inspections, both technical documentation (for example, design and technological), and a specific product.

At the same time, the analysis of the proposed by a number of researchers [5–9] models of information and analytical systems in the aerospace industry has revealed the following five major unsolved problems:

- it has been found that the data, information and knowledge stored in the company's automated systems, although rich in content, do not have mechanisms that could reduce its very large size and vastness to something more focused, measurable and targeted (by basis of product or process improvements);
- there are problems associated with defining data types that can be useful in preventing defects;
- most research and industry solutions still lack the integration of supplier data and production data that contain knowledge, in particular related to manufacturing defects, which can be used to identify new design decisions;
- given the volume and discreteness of the data and the various tools used to manage it, it is not always clear how to structure diversity in such a way as to ensure its integration into the design process and informing design decisions;
- a significant amount of time is required to access, collect, restructure and reuse any kind of archival data or information to improve new designs.

In part, a number of the problems listed above, within the framework of the study of this article (table 1), were considered and their solutions were proposed.

Also, according to the authors of this article, the authors in their research and proposed models of information and analytical systems, do not fully take into account the risks associated with their implementation and use in the aerospace industry. One of them (and the most important) is the possible leakage of confidential information circulating within organizations (because their systems involve the exchange of data with external organizations, as well as providing access to certain information located in their databases). These risks are associated, including with possible cyberattacks, and as a consequence - loss of data integrity [13]. The information-analytical system of forecasting of quality offered in this article does not assume an exchange of data with external organizations (i.e. has closed character), thus considers collection of volume of information necessary for effective management of quality of production. In view of the above, it should be noted an important direction of further research in this area-the consideration of the possibilities of ensuring information security of information and analytical systems (under the conditions of data exchange with external organizations).

Table 1. Comparative analysis of the results of this article with the studies given by other authors.

Unresolved issues	Results of the study of this article
It has been found that the data, information and knowledge stored in the company's automated systems, although rich in content, do not have mechanisms that could reduce its very large size and vastness to something more focused, measurable and targeted (by basis of product or process improvements)	The problem was not considered / the solution of the problem was not proposed
There are problems associated with defining data types that can be useful in preventing defects	The problem is considered / data types (for example, detected design errors during quality control), which can be useful for preventing defects, are identified / the solution of the problem is proposed
Most research and industry solutions still lack the integration of supplier data and production data that contain knowledge, in particular related to manufacturing defects, which can be used to identify new design decisions	The problem is considered / due to various controls, in particular-control of incoming products from suppliers and control at different stages of the life cycle, including at the production stage, the information and analytical system is able to integrate (accumulate) the necessary data for management decisions / the solution of the problem is proposed
Given the volume and discreteness of the data and the various tools used to manage it, it is not always clear how to structure diversity in such a way as to ensure its integration into the design process and informing design decisions	The problem is considered / figure 2 of this article presents the scheme of information structuring / the solution of the problem is proposed
A significant amount of time is required to access, collect, restructure and reuse any kind of archival data or information to improve new designs	The problem is considered partially/ the model presented in figure 1 takes into account the use of archival data to predict quality, but does not address the issue of time spent to access it / the solution is not fully proposed

5. Conclusions

On the basis of the conducted research it is possible to draw the following conclusions:

- digitalization penetrates into many spheres of human life and activity, including the quality management system of organizations in the aerospace industry;
- digitalization requires a transition from descriptive to prescriptive analytics, respectively, organizations for the successful functioning and increase of competitiveness requires the introduction of effective information-analytical systems;
- To date, studies suggest the development and implementation of information-analytical systems in the organization of the aerospace industry, while the researchers themselves have noted a number of very important problems that are either not fully explored (considered) or specific measures have not been proposed to solve them;
- the model of a quality forecasting system developed by the authors of this article and tested in real practice will help to improve quality manageability in the development and production of aircraft engines in digitalization, and will also solve a number of problems that exist in the aerospace industry. However, a number of issues require further investigation and resolution.

References

[1] Büchi G, Cugno M and Castagnoli R 2020 Smart factory performance and Industry 4.0. *Technolog. Forecast. and Social Change* **150** 119790 doi: 10.1016/j.techfore.2019.119790
[2] Bal H Ç and Erkan C 2019 Industry 4.0 and Competitiveness. *Procedia Computer Science* **158** 626 doi: 10.1016/j.procs.2019.09.096
[3] Dalenogare L, Benitez G, Ayala N and Frank A 2018 The expected contribution of Industry 4.0 technologies for industrial performance. *International Journal of Production Economics* **204** 386 doi: 10.1016/j.ijpe.2018.08.019
[4] Liu X, Wang W, Guo H, Barenji A, Li Z and Huang G 2020 Industrial blockchain based framework for product lifecycle management in industry 4.0. *Robotics and Computer-Integrated Manufacturing* **63** doi: 10.1016/j.rcim.2019.101897
[5] Kim T, Jeong J and Kim Y A 2019 Conceptual Model of Smart Manufacturing Execution System for Rolling Stock Manufacturer. *Procedia Computer Science* **151** 604 doi: 10.1016/j.procs.2019.04.081
[6] Haddara M and Elragal A 2015 The Readiness of ERP Systems for the Factory of the Future. *Procedia Computer Science* **64** 723 doi: 10.1016/j.procs.2015.08.598
[7] Soltani Z, Zareie B, Milani F and Navimipour J 2018 The impact of the customer relationship management on the organization performance. *Journal of High Technology Management Research* **29(2)** 237 doi: 10.1016/j.hitech.2018.10.001
[8] Abollado J, Shehab E and Bamforth P 2017 Challenges and Benefits of Digital Workflow Implementation in Aerospace Manufacturing Engineering. *Procedia CIRP* **60** 85 doi: 10.1016/j.procir.2017.02.044
[9] Souri M, Gao J and Simmonds C 2019 Integrating Manufacturing Knowledge with Design Process to Improve Quality in the Aerospace Industry. *Procedia CIRP* **84** 375 doi: 10.1016/j.procir.2019.04.179
[10] Gardner D 1989 *Product life cycle: A critical look at the literature* (Chicago, IL) pp 162-194
[11] Dhalla N and Yuspeh S 1976 *Forget the product life cycle concept* (Harvard Business Review) pp 102-112
[12] Cox W 1967 Product life cycles as marketing models. *The Journal of Business* 375
[13] Tupa J, Simota J and Steiner F 2017 Aspects of Risk Management Implementation for Industry 4.0. *Procedia Manufacturing* **11** 1225 doi: 10.1016/j.promfg.2017.07.248

IOP Conf. Series: Materials Science and Engineering **868** (2020) 012012 doi:10.1088/1757-899X/868/1/012012

A method of joint high-precision estimation of range and velocity in a radar using ultra-wideband frequency coded waveforms

D Sapronov[1], I Kamenskiy and V Kirdyashkin

Institute of Radioelectronics, infocommunications and information security, *Moscow Aviation Institute* (National Research University), 4 Volokolamskoe shosse, Moscow 125993, Russian Federation

[1]E-mail: danil.sapronov@yandex.ru

Abstract. The article describes a method of joint high-precision estimation of range and velocity in ultra-wideband short-range radar systems. The peculiarity of the method is the two dimensional range-velocity compression function based on ambiguity function of ultra-wideband discrete frequency coded waveforms. Expressions of ambiguity function and compression function such as a block diagram of the radar transceiver for the proposed method are presented. Experimental data confirm the efficiency of the method.

1. Introduction

At the moment, in radar systems, there are various methods for estimating the velocity of small targets, such as small unmanned aerial vehicles (UAVs). The new methods are based on the use of the millimeter wave range and narrow-band step-frequency continuous-wave (SFCW) signals [1-3]. This article proposes a method for joint estimation of range and velocity in radars using ultra-wideband (UWB) signals with SFCW modulation and other waveforms of discrete frequency coded waveforms (DFCW) [4] based on the UWB ambiguity function (AF). A simplified expression for AF of UWB DFCW in range-velocity coordinates with corresponding simplification conditions is given. A distinctive feature of this expression from previously presented [5] is the ability to evaluate the compression characteristics for UWB DFCW with an unequal frequency step and unequal duration of the frequency components. A block diagram of the implementation of the radar for this method is presented, which allows one to work with signals in the band from 50 MHz to 6 GHz, as well as an expression for compressing a digital signal at the ADC output. The results of the velocity estimation for a target with an EPR of the order of 0.125 m^2 are presented by the proposed method compared to the convenience [6] method of estimating the velocity by two successively obtained High Resolution Range Profiles (HRRPs).

2. Materials and methods

2.1. DFCW ambiguity function

In the “classical” form proposed by Woodward in his monograph [7], the ambiguity function of a narrow-band signal is a convolution of the signal with its complex conjugate copy shifted in time and

IOP Conf. Series: Materials Science and Engineering **868** (2020) 012012 doi:10.1088/1757-899X/868/1/012012

frequency, where the time shift is caused by the delay of the signal to propagate and the frequency shift displays the Doppler effect due to the relative radial movement of the target and the radar:

$$|\chi(\tau,\nu)| = \frac{1}{E}\left|\int_{-\infty}^{+\infty} u(t)\overline{u(t-\tau)}\exp\{j2\pi\nu t\}\,dt\right|, \tag{1}$$

where

$u(t)$ – is the complex envelope of the received radio signal;

$\overline{u(t-\tau)}$ – a reference signal, which is a complex conjugate copy of the emitted signal, shifted in time by the value of τ;

$\tau = \frac{2R}{c}$ – signal delay due to the propagation of the signal to the target at a distance R from the radar;

c – the speed of light;

$\nu = -\frac{2\upsilon}{c} f_c$ – the Doppler shift of the carrier frequency of the signal f_c when reflected from a target moving with velocity υ;

E – the total energy of the signal necessary for normalizing the AF.

However, in the case of wideband and ultra-wideband signals (when the frequency bandwidth is higher than 25% of the center frequency [8]), it is incorrect to describe the Doppler effect only by a carrier frequency shift (1). In the general case, the Doppler effect leads to scaling of the signal in time. [9]. In the case of radio signals, when the inequality $\upsilon \ll c$ is true, we can write an expression for the received signal reflected from a moving target, taking into account the temporal scaling of the signal:

$$s_R(t) = s[\alpha t - \tau], \tag{2}$$

where

$s(t)$ – the transmitted signal;

$s_R(t)$ – the expected received signal used as a reference signal;

$\alpha = \frac{c+\upsilon}{c-\upsilon} \cong 1 - \frac{2\upsilon}{c}$ – the time scaling factor due to the Doppler effect;

$\tau = \frac{2R}{c-\upsilon} \cong \frac{2R}{c}$ – the signal delay due to the propagation of the signal to the target at a distance R from the radar;

Taking into account (2) the ambiguity function for broadband and ultra-wideband signals can be written as follows:

$$|\chi(\tau,\alpha)| = \frac{1}{E}\left|\int_{-\infty}^{+\infty} s(t)\cdot\overline{s[\alpha t-\tau]}\,dt\right|. \tag{3}$$

DFCW equation:

$$s(t) = \sum_{n=0}^{N-1} s_n(t) = \sum_{n=0}^{N-1} rect\left(\frac{t - \sum_{m=0}^{n-1} T_m - \frac{T_n}{2}}{T_n}\right)\cdot a_n \exp\{j2\pi f_n t + j\varphi_n\}, \tag{4}$$

N – the number of signal pulses; $s_n(t)$ – the n-th signal pulse; T_n – the duration of the n-th pulse; $\sum_{m=0}^{n-1} T_m$ – the time elapsed before the emission of the n-th pulse;

$$rect\left(\frac{t-\sum_{m=0}^{n-1}T_m-\frac{T_n}{2}}{T_n}\right)=\begin{cases}1, & \sum_{m=0}^{n-1}T_m<t\le\sum_{m=0}^{n}T_m\\ 0, & t\le\sum_{m=0}^{n-1}T_m\cup t>\sum_{m=0}^{n}T_m\end{cases};$$

a_n – the amplitude of the n-th pulse; φ_n – the initial phase of the n-th pulse; f_n – the frequency of the n-th pulse.

The complex conjugate reference signal in (3) corresponding to (4) is written as follows:

$$\overline{s_R(t)}=\sum_{n=0}^{N-1}rect\left(\frac{\alpha\left(t-\sum_{m=0}^{n-1}T_m-\frac{T_n}{2}\right)-\tau}{\alpha T_n}\right)\cdot a_n\exp\{-j2\pi f_n[\alpha t-\tau]-j\varphi_n\}. \quad (5)$$

To simplify further conclusions, let us set the following conditions for the emission and reception of DFCW:

- the radar is monostatic;
- the maximum delay due to the initial range of the target, illustrated in figure 1, does not exceed the duration of the shortest discrete:

$$\tau_{\max}<\min_n(T_n),$$
$$R_{\max}<\frac{\min_n(T_n)\cdot c}{2}; \quad (6)$$

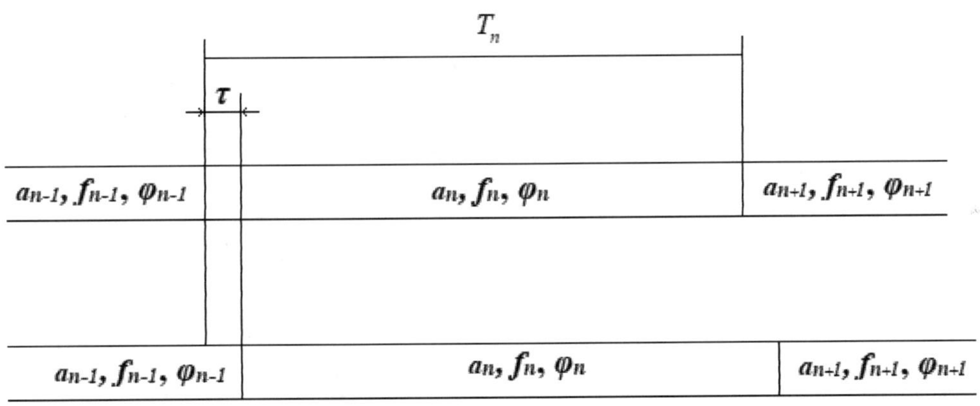

Figure 1. Image of the reference (top) and received delayed (bottom) signals. Low range values lead to the fact that in time only adjacent pulses overlap.

- during the signal emission time, the target moves linearly at a constant velocity, while the time scaling of the reflected signal due to the maximum velocity of the target, shown in figure 2, is much shorter than the average pulse duration:

$$\frac{2\upsilon_{\max}}{c}\sum_{n=0}^{N-1}T_n<0{,}1\cdot\sum_{n=0}^{N-1}T_n\Big/N,\quad \upsilon=const,\quad t\in\left[0;\sum_{n=0}^{N-1}T_n\right]$$
$$\upsilon_{\max}<\frac{c}{20N},\quad \upsilon=const,\quad t\in\left[0;\sum_{n=0}^{N-1}T_n\right]. \quad (7)$$

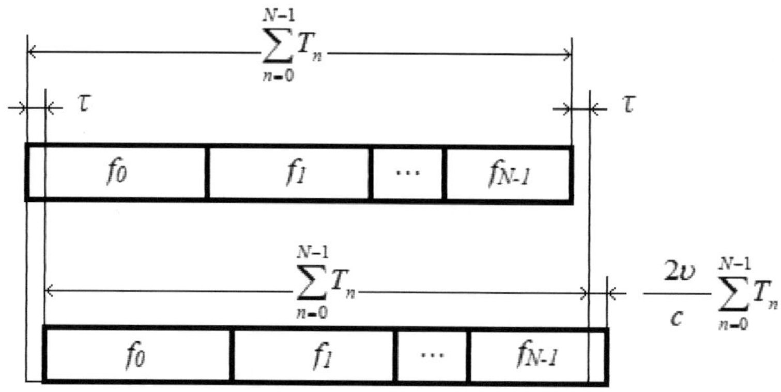

Figure 2. Image of the reference (top) and received delayed (bottom) signals in case of target movement with radial velocity.

These conditions make it possible to simplify expression (5) in such a way that for the reference signal the scaling effect is taken into account only in the phase of the signal and is not taken into account in its amplitude and duration.

Based on the assumption made (7), the (5) can be simplified to the following form:

$$\overline{s[\alpha t-\tau]}=\sum_{n=0}^{N-1}\overline{s_n[\alpha t-\tau]}\cong\sum_{n=0}^{N-1} rect\left(\frac{t-\sum_{m=0}^{n-1}T_m-\frac{T_n}{2}-\tau}{T_n}\right)\cdot a_n\exp\left\{-j2\pi f_n[\alpha t-\tau]-j\varphi_n\right\}. \quad (8)$$

Thus, the convolved emitted (4) and reference (8) signals have the same pulses durations. Under condition (6) overlaps exist only between pulses with the current and previous indices, the remaining products are equal to zero. Then the convolution operation for the sum of the pulses can be replaced by the sum of the convolutions of each pulses with its shifted received copy and with a part of the previous pulse. The integration limits for such a convolution are determined by the duration of the convoluted pulses and the propagation delay from the target:

$$\left|\chi(\tau,\alpha)\right|=\frac{1}{E}\left|\int_{-\infty}^{+\infty}s(t)\cdot\overline{s[\alpha t-\tau]}dt\right|=\frac{1}{E}\left|\int_{-\infty}^{+\infty}\sum_{n=0}^{N-1}s_n(t)\sum_{n=0}^{N-1}\overline{s_n[\alpha t-\tau]}dt\right|,$$

$$\left|\chi(\tau,\alpha)\right|=\frac{1}{E}\left|\sum_{n=0}^{N-1}\int_{\sum_{m=0}^{n-1}T_m}^{\sum_{m=0}^{n}T_m-\tau}s_n(t)\overline{s_n[\alpha t-\tau]}dt+\sum_{n=1}^{N-1}\int_{\sum_{m=0}^{n-1}T_m}^{\sum_{m=0}^{n}T_m+\tau}s_n(t)\overline{s_{n-1}[\alpha t-\tau]}dt\right|,$$

$$\left|\chi(\tau,\alpha)\right|=\frac{1}{E}\left|\sum_{n=0}^{N-1}\chi_{n,n}(\tau,\alpha)+\sum_{n=1}^{N-1}\chi_{n,n-1}(\tau,\alpha)\right|. \quad (9)$$

We substitute (4) and (8) into (9), calculate the integrals and obtain an expression for the ambiguity function of the UWB DFCW recorded for the variables of range and velocity, taking into account the assumptions made (6) and (7):

$$\left|\chi(R,\upsilon)\right|=\frac{1}{E}\left|\sum_{n=0}^{N-1}\chi_{n,n}(R,\upsilon)+\sum_{n=1}^{N-1}\chi_{n,n-1}(R,\upsilon)\right|, \quad (10)$$

where

$$E=\sum_{n=0}^{N-1} a_n^2 T_n;$$

$$\operatorname{sinc}(x)=\begin{cases}\sin(x)/x, & x\neq 0\\ 1, & x=0\end{cases};$$

$$\chi_{n,n}(R,\upsilon)=a_n^2\left(T_n-\frac{2R}{c}\right)\operatorname{sinc}\left(2\pi f_n\frac{\upsilon}{c}\left(T_n-\frac{2R}{c}\right)\right)\exp\left\{j\frac{4\pi}{c}f_n\left[R+\upsilon\left(\sum_{m=0}^{n-1}T_m+\frac{T_n}{2}+\frac{R}{c}\right)\right]\right\};$$

$$\chi_{n,n-1}(R,\upsilon)=a_n a_{n-1}\frac{2R}{c}\cdot\operatorname{sinc}\left(\pi\left(f_n-f_{n-1}+\frac{2\upsilon}{c}f_{n-1}\right)\frac{2R}{c}\right)\cdot$$

$$\cdot\exp\left\{j2\pi\left[\left(f_n-f_{n-1}+\frac{2\upsilon}{c}f_{n-1}\right)\left(\sum_{m=0}^{n-1}T_m+\frac{R}{c}\right)+f_{n-1}\frac{2R}{c}\right]+j(\varphi_n-\varphi_{n-1})\right\}.$$

This equation (10) allows us to analyze the compression characteristics of UWB DFCW with different sequences of frequencies f_n and durations T_n of each pulse. A distinctive feature of this expression from the existing ones [5] is the ability to analyze signals with a non-equidistant step in frequency and non-constant durations of pulses.

Some characteristics of the compressed signal depend of parameters of the DFCW:

- the range resolution: $\Delta R=\dfrac{c}{2F}$, $F=f_{\max}-f_{\min}$;
- the velocity resolution: $\Delta\upsilon=\dfrac{c}{2f_c\sum_{n=0}^{N-1}T_n}$, $f_c=\dfrac{f_{\max}+f_{\min}}{2}$;
- the unambiguous range interval: $R_{\max}=\dfrac{c}{2\Delta f_{\min}}$, $\Delta f_{\min}=\min(f_n-f_{n-1})$, $n=0,1,\ldots,N-1$.

In the most cases, when $\dfrac{\min(T_n)}{\Delta f_{\min}}>1$ the second term $\chi_{n,n-1}(R,\upsilon)$ of (10) becomes negligible.

The implementation of the compression process of the received signal in the radar is proposed in two stages: analog and digital. The analog stage of compression is performed in the receiver circuit of the radar; the digital stage is performed using the expression of the two-dimensional compression function (11) for digital data at the output of the ADC receiver. The block diagram of the radar transceiver circuit and the compression function are proposed below.

2.2. *Radar scheme for UWB discrete coded waveforms*

Figure 3 presents the proposed analog circuit of the radar, which allows one to work with UWB signals in a wide frequency band from 50 MHz to 6 GHz.

2.2.1. *Transmitter*

Combining DDS and PLL in a transmitter synthesizer allows us to take advantage of both circuits. Two DDSs generate N_{DDS} stable signals in bandwidth from 0 Hz to 300 MHz shifted 90° relative to each other with fast T_{DDS} tuning times and a small frequency step Δf_{DDS}. PLL performs tuning of the signal frequency in the band from 50 MHz to 6 GHz in a slow time $T_{PLL}>T_{DDS}$ with a much higher frequency step $\Delta f_{PLL}\gg\Delta f_{DDS}$ ($\Delta f_{PLL}=N_{DDS}\times\Delta f_{DDS}$). The signals from the DDS outputs are mixed with the PLL microwave signals of frequency f_{PLL} in a quadrature modulator (IQM), at the output of which a frequency signal $f_n=f_{PLL}\pm f_{DDS}$ is generated. The signal is then amplified in a wideband amplifier

(WBA). Low pass filters (LPF) in the transmitter are necessary to get rid of harmonics of the DDS clock frequency.

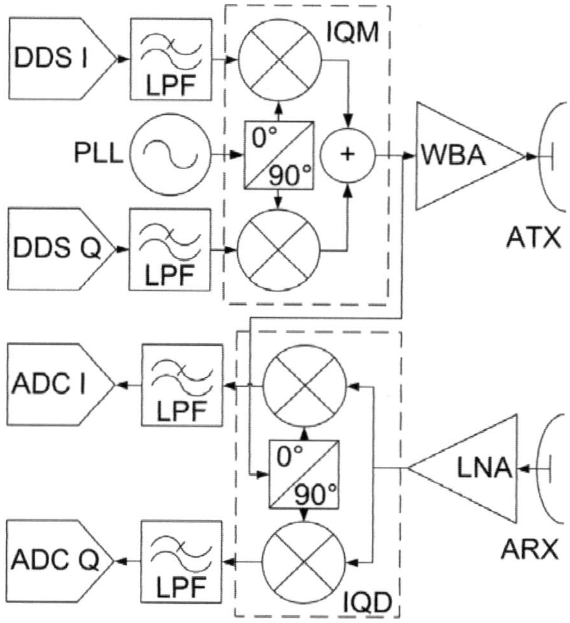

Figure 3. Block diagram of ultra-wideband radar.

2.2.2. Receiver

From the antenna output, the received signals are amplified in a low noise wideband amplifier (LNA). A quadrature detector (IQD) in the receiver channel performs the function of multiplying the received and reference signals. The receiver LPF time constant is consistent with the minimum sampling duration $T_{\min} < T_{\text{DDS}}$. The LPF data has the function of integrating the signal at the output of the IQD. Thus, the joint use of a IQD and LPF acts as a convolution for one received discrete with its original copy $\chi_{n,n}(\tau,\alpha)$ in (9) and $\chi_{n,n}(R,\upsilon)$ in (10).

The narrowband quadrature signals at the output of the receiver are digitized in the ADC. The signals at the ADC output are quadrature digital samples for each frequency f_n in the DFCW:

$$\chi_{n,n}\left(R_{\text{target}},\upsilon_{\text{target}}\right) = CS(n) = I(n) + i\cdot Q(n),\quad n = 0,1,\dots,N-1\,.$$

The next stage of compression is the convolution of the output quadrature digital samples in the coordinates of the range-velocity using the expression of the compression function.

2.3. Range-velocity compression function

The two-dimensional compression function of digital samples of the signal has the form:

$$\left|X\left(R,\upsilon\right)\right| = \left|\frac{1}{N}\sum_{n=0}^{N-1}\left[CS(n) - CS_0(n)\right]\cdot\exp\left\{j\frac{4\pi}{c}f_n\left[R+\upsilon\left(\sum_{m=0}^{n-1}T_m + \frac{T_n}{2} + \frac{R}{c}\right)\right]\right\}\right|. \tag{11}$$

The result of expression (11) is a two-dimensional function with a maximum corresponding to the instantaneous range and velocity of the target. $CS_0(n)$ is the first signal from the accumulated data that is used to background compensation. The form of expression (11) depends on the time-frequency parameters of the signal and can be pre-modeled using the proposed expression of the ambiguity function (10). For example in figure 4 there is the 3D plot of compression function of ideal signal received from target at 3 meters with approach speed 3.8 m/s, signal parameters are the same as in experimental layout from below.

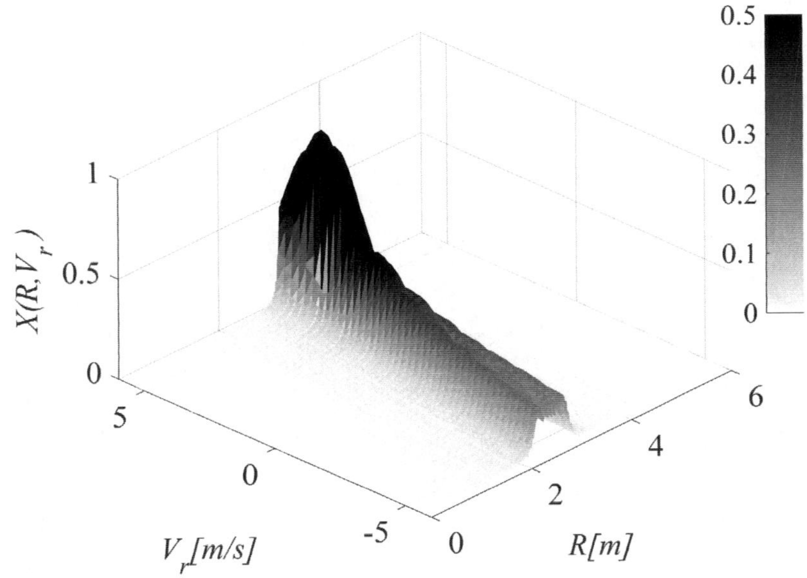

Figure 4. 3D plot of compression function of ideal signal received from target at 3 meters with approach speed 3.8 m/s.

3. Experimental part and results

3.1. Experimental layout

For practical testing of the radar constructed according to the scheme in figure 3 an experimental prototype with an aluminum pendulum of radius 20 cm (RCS is about 0.125 m^2) was built. Figure 5 shows the experimental layout diagram for studies of the proposed method for the joint estimation of range and velocity.

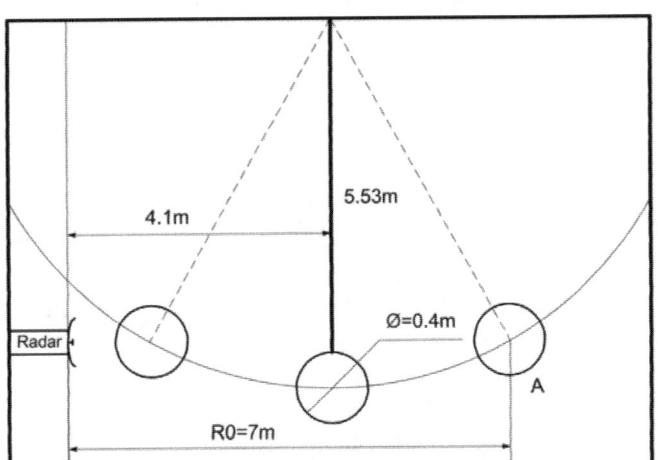

Figure 5. Experimental layout diagram with pendulum: *A* — is the point where the pendulum begins to move.

Data were accumulated for 332 signals with a repetition period of $T_{\text{frame}} = 100$ ms. Parameters of the used DFCW:

- Generally linear frequency code with equidistant frequency step;
- $T_{DDS} = 40$ us, $T_{PLL} = 300$ us, $N_{DDS} = 60$, $N_{PLL} = 18$, $N = N_{DDS} \times N_{PLL} = 1080$;

- $T_{FULL} = N_{DDS}T_{DDS} + N_{PLL}T_{PLL} = 47.88$ ms;
- $\Delta f_{DDS} = 5$ MHz, $f_{DDS} = -153$ MHz, -148 MHz, … ,+147 MHz;
- $\Delta f_{PLL} = 300$ MHz, $f_{PLL} = 900$ MHz, 1200 MHz, … ,6000 MHz;
- $f_n = 752$ MHz, 757 MHz, … 6147 MHz;
- $F = 5395$ MHz.

3.2. Experimental results

The result of compression function (11) normalized to its maximum for 26-th of 332 accumulated signals received from pendulum (figure 5) is shown in figure 6. As can be seen from comprising figure 4 and figure 6 the practical implementation of the compression function from the real target has distortions. This distortion is due to imperfections in the frequency response of the transceiver path. In addition, in the vicinity of 0 meters, there is a certain peak associated with a change in the signal of the proliferating directly from the transmitting to the receiving antenna. However, as in the case of the theoretical function, the position of the maximum of the practical compression function shows the instantaneous values of the range and velocity of the pendulum.

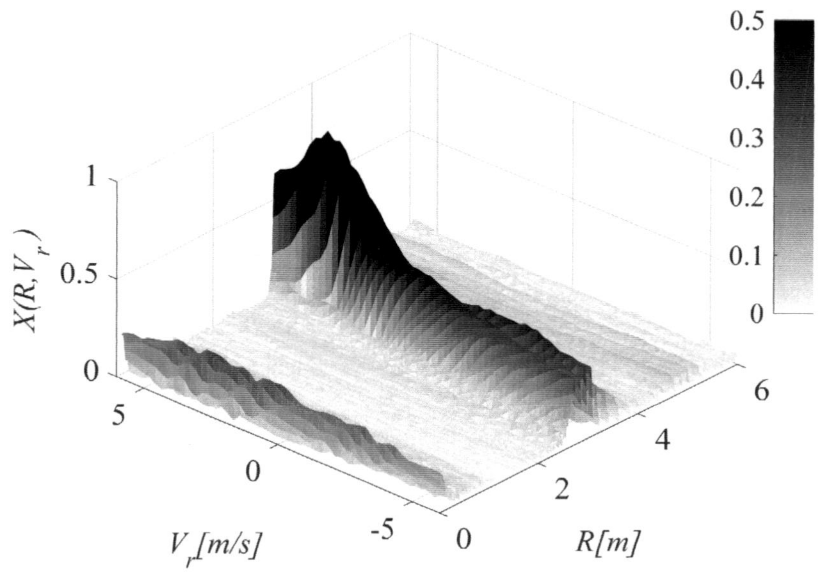

Figure 6. 3D plot of compression function of real signal received from pendulum at 3 meters with approach speed 3.8 m/s.

According to the experimental data, the instantaneous values of the pendulum velocity were calculated in two ways. One way is the proposed method of fixing the maximum of the two-dimensional compression function. The second method is the convenience method for calculating the velocity using Fast Fourier Transform (FFT) from two sequentially received signals and calculating the shift of the maximum of RRRP for signal repetition interval:

$$V_k = \frac{R_k\left(\max\left(\left|FFT\left\{CS_k\right\}\right|\right)\right) - R_{k-1}\left(\max\left(\left|FFT\left\{CS_{k-1}\right\}\right|\right)\right)}{T_{SR}}, \quad k = 0,1,\ldots,K-1, \tag{12}$$

where K – number of accumulated signals; k – index of current signal; $\left|FFT\left\{CS_k\right\}\right|$ – RRRP of k-th signal; T_{SR} – signal repetition interval.

The results of the experiment are shown in figure 7. The upper part of the graph shows the estimates of the pendulum velocity, calculated in two ways.

V_{shift} – is the velocity estimated by the convenience method from two adjacent signals.

V_{AF} – is the velocity estimate obtained by calculating the coordinate of the maximum of the compression function (11).

V_{teor} – theoretical dependence of velocity on time, taking into account the attenuation of oscillations:

$$V_{teor} = A \cdot \omega \cdot \cos(\omega \cdot t + \varphi_0) \cdot \exp(-t/\alpha), \quad \omega = 2\pi\left(g/L\right)^{1/2},$$

where L – the length of pendulum [5.73m]; g – the gravitational acceleration of 9.8 [m / s]; A – swing amplitude 3 [m]; ω – the cyclic oscillation frequency [Hz × rad], depending on the length of the pendulum L = 5.73 [m]; α – some empirical attenuation coefficient of the pendulum 60 [s] to comply with physical quantities.

The lower part of the graph shows the absolute estimation error for the two methods with respect to theoretical values.

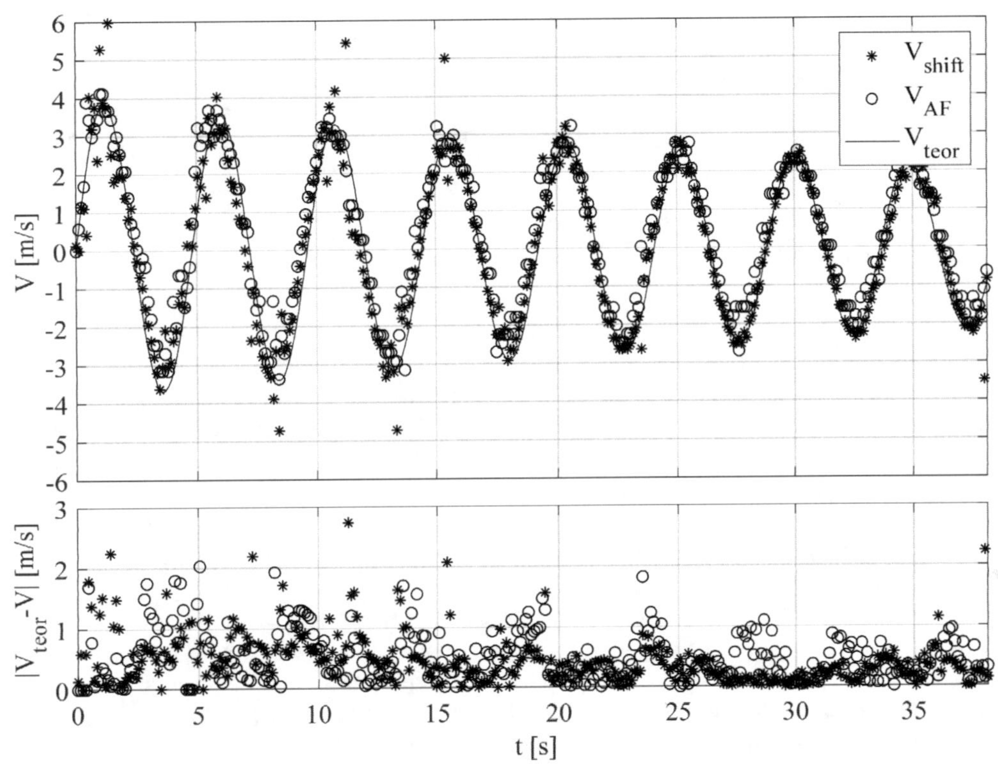

Figure 7. Experimental results. The upper part of the graph shows the estimates of the pendulum velocity. The lower part of the graph shows the absolute estimation error between practical and theoretical values.

The table 1 shows the statistical characteristics of the absolute estimation errors.

Table 1. Statistical characteristics of the absolute estimation errors for 332 accumulated signals.

Estimate value	**Convenience method**	**Proposed method**
Mean [m/s]	1.0698	0.6790
Variance [m^2/s^2]	32.487	1.8037
Median [m/s]	0.3805	0.3855

The experiment repeated three more times with different numbers of accumulated signals.

IOP Conf. Series: Materials Science and Engineering **868** (2020) 012012 doi:10.1088/1757-899X/868/1/012012

Table 2. Statistical characteristics of the absolute estimation errors for 203 accumulated signals.

Estimate value	**Convenience method**	**Proposed method**
Mean [m/s]	0.9188	1.1981
Variance [m^2/s^2]	17.9505	2.8118
Median [m/s]	0.3844	0.4784

Table 3. Statistical characteristics of the absolute estimation errors for 198 accumulated signals.

Estimate value	**Convenience method**	**Proposed method**
Mean [m/s]	0.7305	0.8673
Variance [m^2/s^2]	8.8321	2.0214
Median [m/s]	0.2357	0.2646

Table 4. Statistical characteristics of the absolute estimation errors for 206 accumulated signals.

Estimate value	**Convenience method**	**Proposed method**
Mean [m/s]	0.8864	1.0636
Variance [m^2/s^2]	15.3623	2.5811
Median [m/s]	0.4147	0.3445

It can be seen from the tables 1-4 that the estimation carried out by the convenience method leads to a wider spread of values, despite the fact that the median error value of the convenience method is lower. This is explained by the duration of the necessary process of data accumulation for the convenience method. It is two or more times higher which ensures an improvement in velocity resolution. The proposed method allows one to obtain a similar estimate for the velocity using only one ultra-wideband signal. In addition, local emissions are observed in the graph of the convenience method, which led to a high dispersion of values. These outliers are explained by the fact that as the object's velocity increases, the usual FFT operation will lead to the signal spreading along its range and the jitter of the maximum position (described in [5] and [10]), as shown in figure 8. Since the FFT operation essentially displays the zero velocity cross section of the two-dimensional compression function, the quality of the speed estimation by the classical method using the FFT directly depends on the shape of the compression function. The shape of the compression function depends on both the signal parameters (bandwidth, duration, frequency step, frequency code) and the time-frequency parameters of the radar transmit-receive path. Thus, an accurate assessment of the applicability limits of the convenience and proposed methods requires complex parametric analysis and is not considered as part of this article.

With an increase in the velocity of the object the use of the usual FFT to estimate the velocity will lead to an increase in error.

4. Conclusions

The obtained expression (11) and the proposed radar block diagram allow realizing the ultra-wideband ambiguity function, which makes it possible to carry out a joint estimation of the range and velocity of the target for the time of the one probing. It allows one to reduce the time required to review a given sector of space and improve the quality of the radar. The experimental results obtained from the real radar prototype confirmed the correctness of the proposed method. According to expression (10), the result of this method depends on the time-frequency parameters of the signal. In the following works, it is supposed to study in more detail the influence of more complex forms of DFCW instead of the simplest SFCW signal.

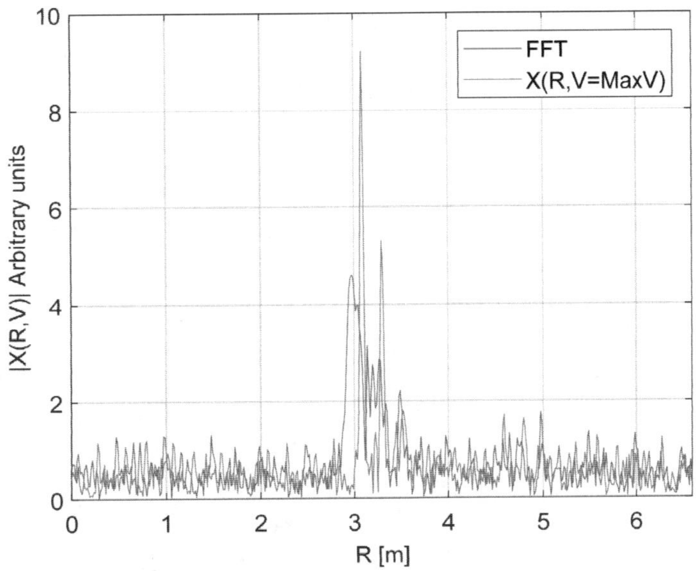

Figure 8. The HRRP of 26-th signal in experimental data compressed by convenience FFT and by proposed range-velocity compression function. The velocity of target is about 3.8 m/s.

For example, signals with pseudo-random frequency coding, signals with variation of a frequency step or time step. As a comparison of the theoretical and experimental two-dimensional compression functions has shown, the frequency response of the transceiver path of the radar affects the result of compression. To improve the accuracy of estimating range and speed, it is required to use some path calibration algorithm, for example, signal predistortion in a DDS and PLL synthesizers, or post-distortion in a signal processor after ADC. A separate area of research is the optimization of finding the maximum in (11) to accelerate the decision on the instantaneous range and speed of the target, for example using the compressive sensing technique.

References

[1] Pieraccini M, Miccinesi L and Rojhani N 2017 RCS measurements and ISAR images of small UAVs. *IEEE Aerospace & Electronic Systems Magazine* 28 doi: 10.1109/MAES.2017.160167

[2] Pieraccini M and Miccinesi L 2017 CWSF Radar for Detecting Small UAVs. *IEEE Int. Conf. on Microwaves, Antennas, Communications and Electronic Systems (COMCAS)* doi: 10.1109/COMCAS.2017.8244710

[3] Pieraccini M, Miccinesi L and Rojhani N 2019 A Doppler Range Compensation for Step-Frequency Continuous-Wave Radar for Detecting Small UAV. *Sensors* **19** 1331 doi:10.3390/s19061331

[4] Cook C and Bernfeld M 1967 *Radar Signals an Introduction to Theory and Application* (New York/London: Academic Press Inc.) p 269

[5] Yimin L, Huadong M, Hao Z and Xiqin W 2008 Motion Compensation of Moving Targets for High Range Resolution Stepped-Frequency Radar. *Sensors* **8** 3429 doi: 10.3390/s8053429

[6] Weiss J M 2009 Continuous-Wave Stepped-Frequency Radar for Target Ranging and Motion Detection. *Proc. of the 42nd Annual Midwest Instruction and Computing Symposium*, available at: http://micsymposium.org/mics_2009_proceedings/mics2009_submission_64.pdf

[7] Woodward P M 1953 *Probability and Information Theory, with Application to Radar* (London: Pergamon Press LTD) p 121

[8] Taylor J D 2012 *Ultrawideband Radar Applications and Design* (New York: CRC Press Taylor

& Francis Group) pp 8–10

[9] Levanon N and Mozeson L 2004 *Radar Signals* (Hoboken, New Jersey: John Wiley & Sons Inc.) pp 2–5

[10] Liu B and Chang W 2013 Range Alignment and Motion Compensation for Missile-borne Frequency Stepped Chirp Radar. *Progress in Electromagnetics Research* **136** 523 doi: 10.2528/PIER12110809

IOP Conf. Series: Materials Science and Engineering **868** (2020) 012013 doi:10.1088/1757-899X/868/1/012013

Study of the specific energy of universal electrode materials for hybrid ultra-high-volume capacitor systems

V V Sleptsov, M N Ushkar, Yu V Zinin, P A Shchur*, A O Diteleva and K Z Lwin

Department of Radioelectronics, Telecommunications and Nanotechnology, *Moscow Aviation Institute* (National Research University), 4 Volokolamskoye shosse, Moscow 125993, Russia

*E-mail: shur-pavel@mail.ru

Abstract. The article discusses specific energy of universal electrode materials based on Busofit for ultra-high-volume capacitor systems and hybrid ultra-high-volume capacitor systems based on $LiCoO_2$ and MnO_2, which can be used in as sources of pulsed power in satellite solar panel deployment systems and emergency door opening systems and emergency platforms in airliners, including the Airbus 380 jumbo jet. The electroimpulse metallization of electrode materials and dependence of specific energy on weight of $LiCoO_2$ and MnO_2 was described. It was shown that hybrid ultra-high-volume capacitor systems increase its energy intensity up to 5,5 times compared with UCS.

1. Introduction

In modern aviation technology, much attention is paid to capacitor structures and electrochemical energy accumulators both in emergency and standby, and in some cases, as the main sources of energy. This is due to the tendency to electrify modern aircraft until the creation of a fully electric aircraft.

Such energy storage are used both separately and jointly in the hybrid power supplies, which are widely used in emergency door and evacuation platform systems in airliners, including the Airbus 380 jumbo jet, and in military aviation as sources of pulsed power, emergency energy supplies, in the stabilization systems of network parameters, in the deployment of solar satellite panels, as well as on-board electronics.

According to the “Bloomberg new energy finance”, in 2040 the electricity generated by burning coal will be reduced from 30% to 13 %, gas – from 24 % to 14 %, fuel oil – from 6% to practically 0%. And also the electricity generated by the hydroelectric and nuclear power station will be reduced slightly. At the same time, the demand for renewable energy sources (RES) will be increased in the market [1].

However, while developing and operating of RES some problems remain to resolve. One of these is the storage of electricity during the period from maximum production to maximum return to consumers.

Today, the traditional solution to solve the problem of accumulation and storage of electricity is the use of lithium-ion accumulators (LIA), providing a maximum energy capacity of 250 Wh/kg, which have some disadvantages [2].

Synthesis of electrode materials of ultra-high-volume capacitor systems (UCS) by using the technology of thin-film metallization on carbon matrix like Busofit or based on ferroelectrics with high dielectric permittivity, allowed to obtain supercapacitor with specific energy of 40-50 Wh/kg, which makes it possible to use them effectively as part of hybrid energy storage, including LIA and UCS. While LIA provides power supply in stationary, UCS will be in transient modes of operation of high-current

IOP Conf. Series: Materials Science and Engineering **868** (2020) 012013 doi:10.1088/1757-899X/868/1/012013

equipment. Nowadays UCS reached only 10-15 Wh/kg with the help of standard tkick-film technology of forming electrode materials [3-5].

Electrode materials are a key element of ultra-high-capacity capacitor structures, determining their energy and performance characteristics. The main requirements to the electrode material: developed specific surface area, high electrical conductivity, availability of porous structure wetting, chemical inertness to the electrolyte, mechanical strength, the possibility of integration with the current collector determines the knowledge-intensive technological process of its creation.

One of the directions of creating supercapacitors is the construction of an electrolytic capacitor structure using carbon materials as electrodes with a developed surface that provides an area of electrodes hundreds or thousands of times larger than that of a conventional capacitor of the same size. Such material is carbon fiber Busofit.

The main drawback of many carbon electrode capacitors is that their performance is often limited due to the high resistance of most carbon electrodes. This high resistance causes large ohmic losses in the capacitor at the discharge stage. The reduction of the mentioned resistance in double-layer capacitors is achieved mainly by reducing the internal resistance in the electrode.

Recently, various manufacturing techniques have been proposed to reduce the internal resistance of carbon composite electrodes. In [6], a method for manufacturing a carbon-aluminum composite electrode by depositing coal powder and other conductivity-enhancing additives on an aluminum substrate is disclosed.

Another similar solution is disclosed in the patent [7], in which it is proposed to introduce metals, in particular aluminum, into carbon fiber electrodes by weaving metal fibers into carbon fiber billets.

Another method is described in [8-10]. In these solutions, it is demonstrated that metal fibers can be intertwined with a coal billet and sintered to produce an electrically conductive matrix with a stable design that can be used as a composite electrode.

In work [11] describes another solution in which the carbon fiber is immersed in an aqueous solution, so that a layer of electrically conductive metal oxide, preferably transition metal oxide, is formed in the pores of the carbon fibers. Nishino et al. the formation of metallic oxides, such as tin oxide or indium oxide, by vapor deposition is also described. In this work plasma spraying of molten metals, in particular, aluminum, on one side of the polarized electrode to form the corresponding layer performing the function of a current collector is offered.

Thus, modern studies show that the deposition of different metals on carbon electrode materials can reduce resistance and increase specific energy, as well as combine the properties of the electrode and the current collector in one material. So it helps us to reduce weight of power supply and increase its capacity.

The purpose of this work was to study the specific energy of the electrode material of UCS and hybrid UCS like Busofit, which is a porous carbon material with a highly developed surface (1000 m^2/g), depending on the addition of chemically active materials as lithium cobalt ($LiCoO_2$) and manganese oxide (MnO_2).

2. Materials and methods

Developed UCS with simple design has ~ 4 Wh/kg of specific energy density of the material [12]. Porous carbon fiber Busofit was used as an electrode material in these cells. The authors developed the following method to increase the energy density of UCS samples, without changing the assembly conditions, measurements, materials and their sizes. The modification of the electrode material is using porous carbon fiber with a high specific surface area (1000 m^2/g) with low surface resistance.

The problem of reducing the internal resistance of the electrode material was resolved by obtaining a metal current collector for flexible electrode with minimal transfer resistance, i.e., applying a continuous layer of metal on the surface of Busofit, which will be the current collector and reduce the internal resistance of the cell, as well as by electro-plating a separate carbon fiber material using such metals as silver (figure 1). Metallization allowed not only to reduce the internal resistance of the UCS cell (ESR), but also contributed to an increase in the specific surface area of the carbon material, and consequently an increase in the specific energy of the material by 63%, which amounted to a specific energy intensity of the UCS material of 6.5 Wh/kg.

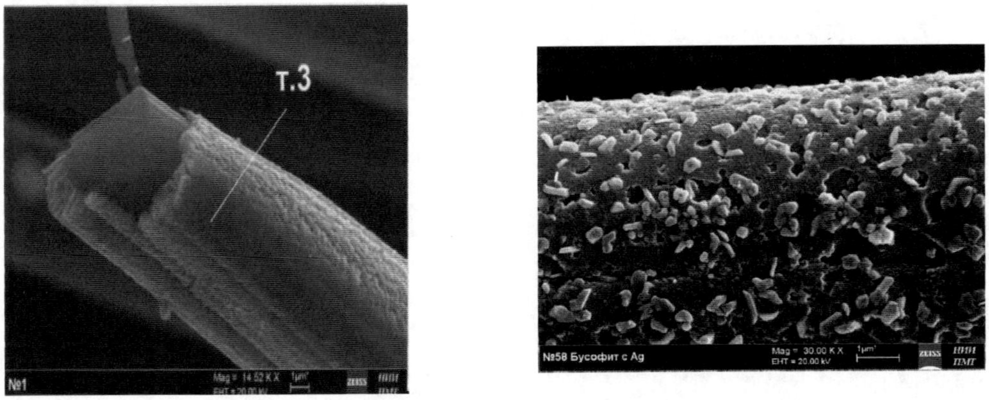

(a) (b)

Figure 1. Metallization of electrode material filament of UCS: (a) a single filament of electrode material with a deposited layer of metal; (b) modified with silver nanoclusters of filament Busofit.

Measurement of electrical characteristics of the samples was carried out under the same conditions on the stand for experimental studies of multicyclic loading of UCS samples. The samples under study are subjected to multicyclic loading in the experimental stand by a set number of charge/discharge cycles and a given current. Charge/discharge cycles are recorded in the stand and automated measurements of parameters of experimental samples of UCS (capacitance, operating voltage, charge and discharge currents, leakage currents) are carried out.

In addition to UCS, hybrid UCS were developed and manufactured based on $LiCoO_2$ and MnO_2. The energy density of chemical current sources made by standard thick-film technology based on $LiCoO_2$ and MnO_2 reaches 160 mAh/g and 150 mAh/g, respectively [13, 14].

A highly porous carbon fabric Busofit with the specific surface of 1000 m^2/g was selected as a matrix for $LiCoO_2$ and MnO_2. The chemically active material filled in the pores of the carbon matrix and covered the filaments of the Busofit with a thin layer [15].

Measurement of electrical characteristics of the samples was carried out under the same conditions on the stand for experimental studies of multicyclic loading of UCS samples. The samples under study are subjected to multicyclic loading in the experimental stand by a set number of charge/discharge cycles and a given current. Charge/discharge cycles are recorded in the stand and automated measurements of parameters of experimental samples of UCS (capacitance, operating voltage, charge and discharge currents, leakage currents) are carried out.

3. Results and discussion

Figure 2 shows that the value of the specific energy density of the cathode is related to the weight percentage of chemically active material (lithium cobalt) of the carbon matrix. The maximum specific energy of the hybrid UCS cathode is achieved when the weight percentage of the chemically active substance of the electrode material is ~100% of Busofit matrix. As a result, the specific energy of the electrode materials is achieved 29 Wh/kg.

And the relation between the specific energy of the cathode material based on $LiCoO_2$ and surface area of the electrode material was studied. It was found that the specific energy density of the electrode material is related to the surface area of the electrode material while hybrid UCS was made based on lithium cobalt ($LiCoO_2$) with the same materials, assembly conditions and measurement. According to the results of measurements, increasing the surface area of the electrode material by 3 times allows to

IOP Conf. Series: Materials Science and Engineering **868** (2020) 012013 doi:10.1088/1757-899X/868/1/012013

increase the energy density of the material by 21%, as well as to reduce the internal resistance (ESR) of hybrid UCS cell by 2.5 times.

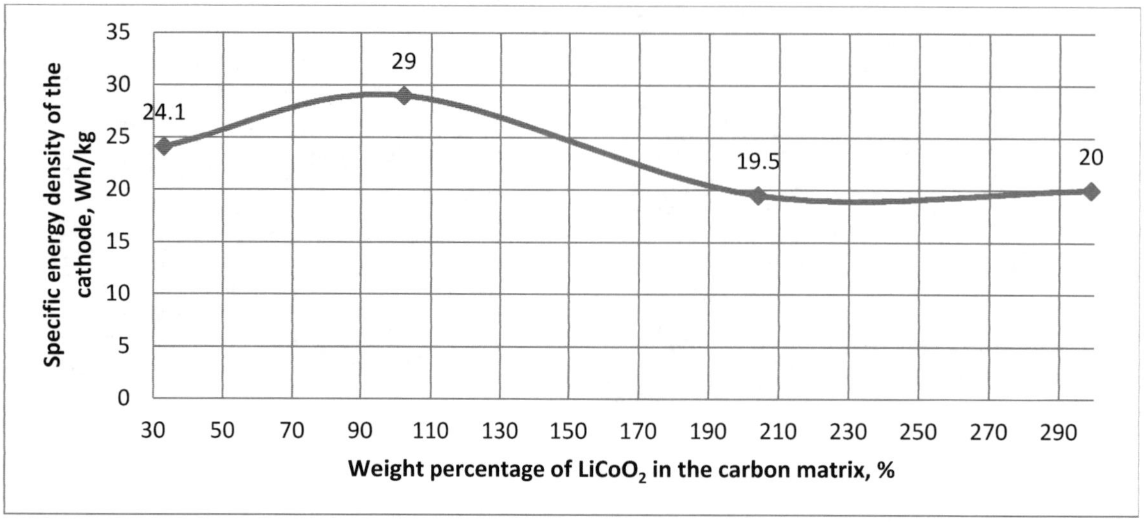

Figure 2. The relation between the specific energy of the cathode and the weight percentage of $LiCoO_2$ in the carbon matrix.

As part of this research, hybrid UCS based on manganese oxide (MnO_2) was constructed and studied. The specific energy of the hybrid UCS primarily depends on the weight percentage of MnO_2 in the cell.

Increasing the specific weight of MnO_2 by 2 times allow to increase the specific energy density of the material by 7 times, as well as to increase the working voltage by 1.8 times. The specific energy density of the material of hybrid UCS reaches 21.2 Wh/kg when MnO_2 is used (figure 3).

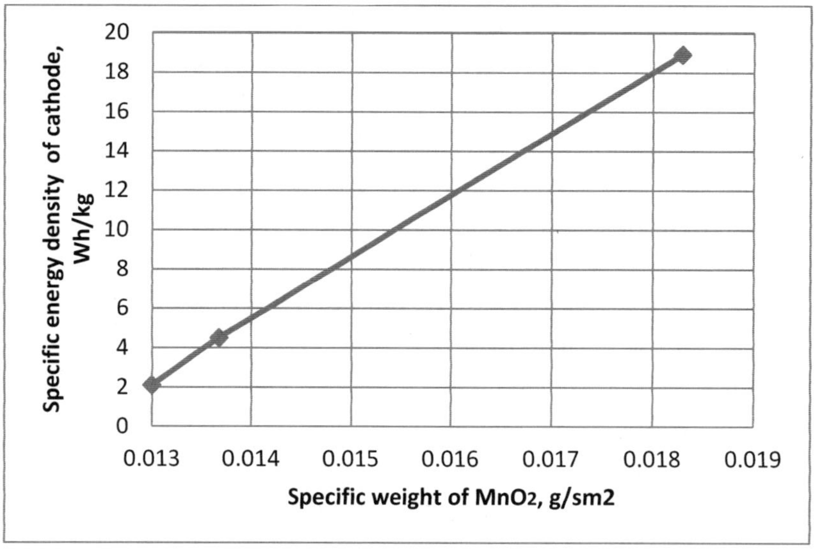

Figure 3. The relation between the specific energy of cathode material and the specific weight of MnO_2.

Figure 4 shows a diagram of specific energy density values of materials of produced samples UCS and hybrid UCS:

- ultra-high-capacity capacitor structures (UCS) - 4.1 Wh/kg;
- ultra-high-capacity capacitor structures with metallization - 6.5 Wh/kg;
- hybrid UCS based on manganese oxide (MnO2) - 21.2 Wh/kg;
- hybrid UCS based on lithium cobalt (LiCoO2) - 22.6 Wh/kg.
-

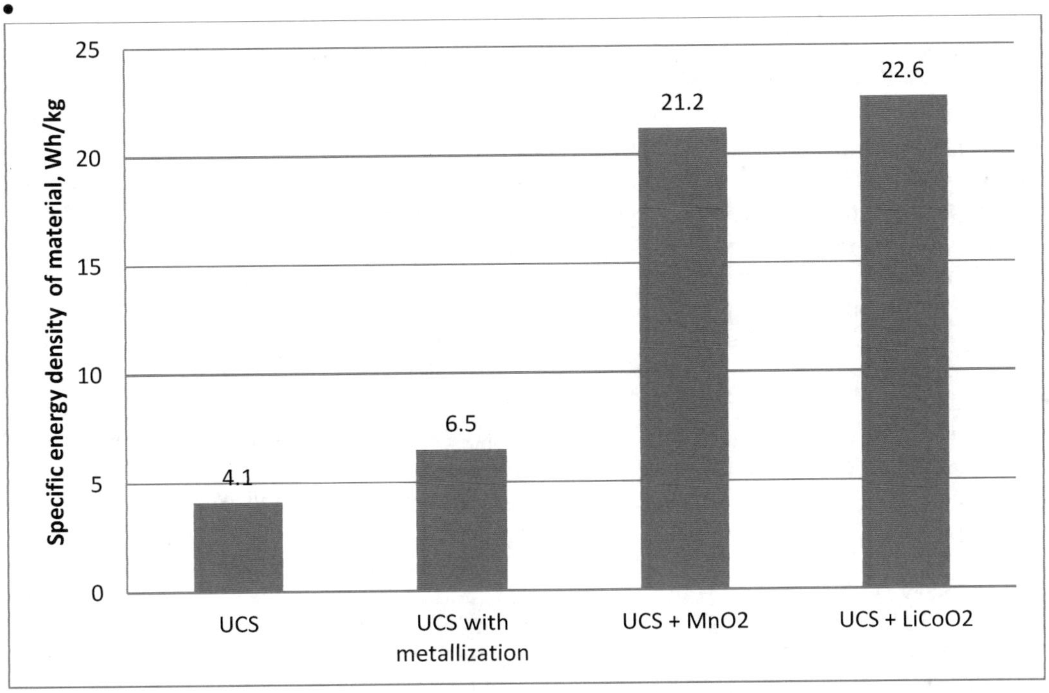

Figure 4. The specific energy of the material of the produced samples UCS and hybrid UCS.

Analysis of the results of the specific energy density of materials of ultra-high-volume capacitor systems, ultra-high-volume capacitor systems (UCS) with metallization, hybrid UCS based on lithium cobalt and hybrid UCS based on manganese dioxide, which are constructed and tested under the same conditions, design and materials, shows that the specific energy density of hybrid UCS based on manganese oxide and hybrid UCS based on lithium cobalt is contributing respectively 5.2 and 5.5 times more than UCS.

4. Conclusions

The obtained results allow drawing the following conclusions:

1. Electro-impulse metallization of electrode materials for ultra-high-capacity capacitor structures allows increasing energy density by 63% compared with the non-metallized electrode material.
2. Increasing the specific weight of manganese dioxide by 2 times is allowed to increase the specific energy density of the material by 7 times, as well as to increase the working voltage by 1.8 times.
3. Increasing the surface area of the electrode material by 3 times allows to increase the energy density of the material by 21%, as well as to reduce the internal resistance (ESR) of the cell by 2.5 times.
4. The developing of hybrid UCS based on $LiCoO_2$ and MnO_2 allows increasing energy density up to 5.5 times compared to UCS and UCS with metallization of Busofit using electro-impulse technology and to reduce the weight of the final power supply, which will be used in aircraft as an emergency power supply, by combining the properties of the electrode and the current collector in one material and increasing the energy intensity due to metallization.

Acknowledgment
The study is carried out with the financial support from the State represented by the Ministry of Education and Science of the Russian Federation. Subsidies Provision Agreement No. RFMEFI57717X0275.

References
[1] Salvatore J 2013 *Bloomberg New Energy Finance; WEC, World energy perspective – cost of energy technologies* (Bloomberg New Energy Finance, New York)
[2] Goodenoug J B Youngsik h K 2010 Challenges for Rechargeable Li Batteries. *Chem. Mater.* **22(3)** 587 https://doi.org/10.1021/cm901452z
[3] Choudhary N, Li C, Moore J, Nagaiah N, Zhai L, Jung Y, Thomas J 2017 Asymmetric Supercapacitor Electrodes and Devices. *Advanced Materials* **29(21)** 1605336 doi: 10.1002/adma.201605336
[4] Lu X, Yu M, Wang G, Tong Y, Li Y 2014 Flexible solid-state supercapacitors: design, fabrication and applications. *Energy Environ. Sci.* **7** 2160 https://doi.org/10.1039/C4EE00960F
[5] Wang F, Xiao S, Hou Y, Hu C, Liu L and Wu Y 2013 Electrode materials for aqueous asymmetric supercapacitors. *RSC Adv.* **3** 13059 https://doi.org/10.1039/C3RA23466E
[6] Yoshida A, Imoto K, U.S. Patent US 5150283A (28 March 2011)
[7] Kucharek A, Marshal J, Lee J, Amdahl C, Yuan L, U.S. Patent US 4597029A (19 March 2004)
[8] Tatarchuk B J, Rose M F, Krishnagopalan G A, Patent US 5102745A (13 November 2009)
[9] Tatarchuk B J, Rose M F, Krishnagopalan G A, Zabasajja J N, David A, Patent US 5304330A (19 April 2011)
[10] Tatarchuk B J, Rose M F, Krishnagopalan G A, Zabasajja J N, David A, Patent US 5080963 A (24 May 2009)
[11] Nishino A, Yoshida A, Tanahashi I, Patent US 4562511 A (30 June 2003)
[12] Sleptsov V V, Kukushkin D Yu and Diteleva A O 2018 Thin-film technologies for producing electrode materials for chemical sources of current (CSC) and ultra-high-volume capacitor structures (UCS). *IOP Conf. Series: Materials Science and Engineering* **387** 012075 DOI: 10.1088/1757-899X/387/1/012075
[13] Xiong F, Yan H J, Chen Y, Xu B, Le J X, Ouyang C Y 2012 The Atomic and Electronic Structure Changes Upon Delithiation of $LiCoO_2$: From First Principles Calculations. *Int. J. Electrochem. Sci.* **7** 9390
[14] Asante J B 2018 Nanostructured $Ni(OH)_2$ and MnO_2 cathode designs for aqueous asymmetric double layer capacitors, Dissertation (Ulm University) DOI: 10.18725/OPARU-13881
[15] Sleptsov V V, Govorov V A, Korobeynikov E V, Kovalev I A and Melnikov A B 2019 Technology of preparation of the $LiCoO_2$ reactive mass for chemical current sources. *Journal of Physics: Conf. Series* **1281** 012076 https://doi.org/10.1088/1742-6596/1281/1/012076

Thin-film coating technologies on porous materials for chemical current sources

V V Sleptsov[1], N K Kulikov[1], Kyaw Zaw Lwin[1], P A Shchur[1*], V A Govorov[2] and D Yu Kukushkin[1]

[1]Department "Radioelectronics, Telecommunications and Nanotechnology", *Moscow Aviation Institute (National Research University)*, 125993 Moscow, Russia
[2]Scientific and educational center "Nanotechnology", *Moscow Power Engineering Institute (National Research University)*, 111250 Moscow, Russia

[*]E-mail: shur-pavel@mail.ru

Abstract. This paper investigates the technologies for the formation of the thin-film electrode material of the chemical current sources. To get the high specific energy density of the electrode we used the grinding of chemically active material ($LiCoO_2$), which the most widely used for cathode materials of chemical current sources to deposit on the materialized carbon-based electrode materials. The tested results show that the specific energy density of the electrode reaches to 210 mAh/g, which is more than thick-film technologies can reach (150 mAh/g).

1. Introduction

Chemical current sources are used in various fields of science and technology: electronics, medicine, aviation, and astronautics. In particular energy storage is used for satellites, spacesuits (extra-ship activities), launch vehicles and portable devices such as computers and video cameras. By comparing with the other type of devices rechargeable lithium-ion battery (LIB) has proven to be a reliable energy storage device since its first market introduction in 1991 because of its high energy density, light weight, low maintenance, long service life, and high efficiency [1].

The characteristics of these energy storage systems largely depend on the electrode materials. Electrode materials must meet the following requirements: high specific energy intensity, good reversibility of electrode processes (cyclability), high conductivity for lithium, the high diffusion rate of lithium.

Among the various cathode materials studied, Layered oxides with the formula $LiMO_2$ (M = Co, Mn, Ni) have been the most widely used/commercialized cathode materials for chemical current sources [1-3]. Lithium cobalt offered many crucial advantages such as its stability in ambient conditions (moisture), a significantly higher Li-ion insertion voltage. The complex oxide $LiCoO_2$ has a layered structure in which the Li and Co ions take place orderly in alternating planes. The presence of planes occupied exclusively by lithium ions provides the possibility of complete deintercalation of the alkali metal and the applicability of this compound as a cathode material in chemical current sources [4, 5]. The material with layered lattice shows better electrochemical performance compared to that with a cubic spinel lattice. The unit cell of the layered form consists of three slabs of edge-sharing CoO_6 octahedra separated by interstitial layers of Li (figure 1). $LiCoO_2$ has a theoretical capacity of 274 mAh/g when one mole of Li was extracted from one mole of $LiCoO_2$, but the practical capacity

IOP Conf. Series: Materials Science and Engineering **868** (2020) 012014 doi:10.1088/1757-899X/868/1/012014

reaches only 150 mAh/g when using standard thick-film technology for applying chemically active material [1, 2].

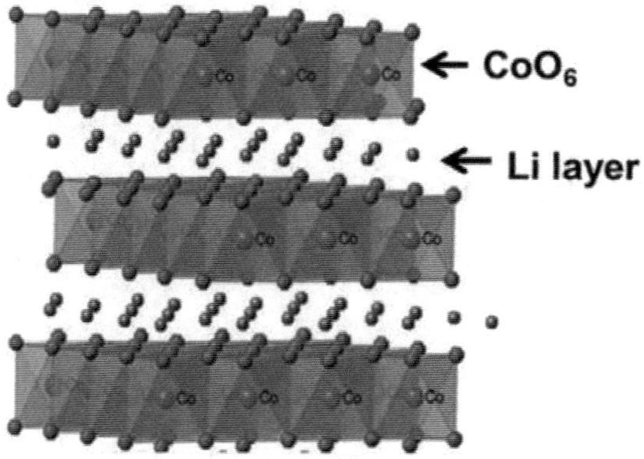

Figure 1. Crystal structure of $LiCoO_2$. In the $LiMO_2$ (M = Co, Mn, Ni) the metal layer is occupied by Mn, Ni, and Co in a random manner [1].

Thin-film electrodes exhibit excellent specific capacitance which could be due to possession of lower contact resistance between film matrix and current collector, and providing a shorter ion diffusion path length [6]. And it points out that the modern thick-film deposition technologies have reached their limit and that need to use the thin-film technology for cathode materials which can use the chemically active material more efficiently [7]. $LiCoO_2$ with a particle size of 8-50 microns is used in the standard thick-film coating technology [8, 9]. There is the technology to possible to disperse the powder to particle sizes of 150-350 nm while maintaining the structure of the material and increasing the effective surface area and energy intensity [10]. In this paper, we will investigate the impregnation of the metallized electrode with 100 to 350 nanometers size of lithium cobalt oxide particles to thin-film electrode material and report the test results of the electrode material.

The purpose of this work is the grinding of chemically active material ($LiCoO_2$) filling the carbon matrix with it in order to create a thin-film cathode material.

2. Materials and research methods

In the research high-porous carbon fiber Busofit is used as an electrode material that meets the requirements for chemical current sources. The use of highly porous carbon as a matrix for $LiCoO_2$ allows switching from a standard thick-film technology using chemically active material to the thin-film using a fine dispersed powder of $LiCoO_2$. And $LiCoO_2$ powder is precipitated in the pore space of carbon fiber type Busofit. The smaller the size of $LiCoO_2$ particles, the greater the effective surface area and the more Li ions can participate in the chemical reaction.

In this paper, the metallized carbon fiber "Busofit-UL-50" is impregnated by using $LiCoO_2$ dispersion in the electrolyte medium based on propylenecarbonate (PC). Metallization of the Busofit before impregnation was carried out to reduce the equivalent resistance between the elements of chemical current sources. Metallization was carried out in a vacuum installation equipped with two magnetrons (figure 2), which can apply titanium coatings at a speed of 0.1 μm per minute. As a result, the 1~3 microlayer of titanium was received and the materialized electrode Busofit with the specific capacity of 200~250 F/g was produced. The advantage of the technology is electro-pulse processing of the Busofit in a liquid medium containing metal nanoclusters. This processing leads to the additional increase of electrical capacitance, the reduction of electrical resistance and stabilization of the parameters of the electrode material.

Figure 2. Vacuum unit UMRM-1.

The deposition of titanium to carbon fabric Busofit increases the specific surface, which allows to increase the energy intensity of supercapacitor systems. Also, titanium can improve lithium intercalation and increase the cyclicity of chemical current sources. Figure 3 shows the comparison between clean single thread Busofit and single thread Busofit coated with a layer of titanium.

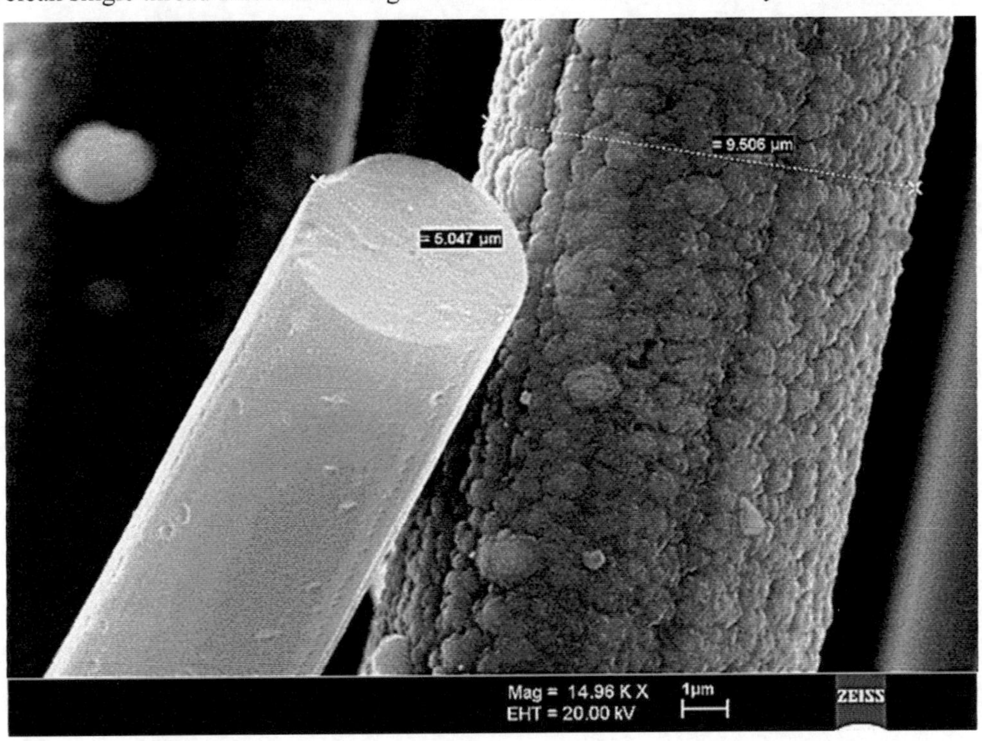

Figure 3. Busofit thread without coating and coated with titanium.

Grinding of lithium cobalt was carried out in a bead mill produced by Netzsch, LabStar type at a speed of rotation of the main shaft 3000 rpm for 8 hours in propylenecarbonate (PC). A grinding chamber with a volume of 600 ml was used, which was 70% filled with yttria-stabilized zirconia beads and with a diameter of (0.4-0.6) mm. 600 ml of PC was poured into the mill, then 500 g of lithium cobaltate was added in portions of 60 g each, after which 200 ml of solvent was added until the mill working volume was completely filled. To measure the size, samples were taken in a volume of 1 ml, and then dried on an electric stove at a temperature of 200 °C until complete removal of the PC. The dried powder was dispersed in ethylene glycol using an ultrasonic bath for 15 minutes. Measurement observed in ethylene glycol. According to the results of the measurement, the 100 to 350 nanometers size of lithium cobalt particles was obtained from grinding (figure 4).

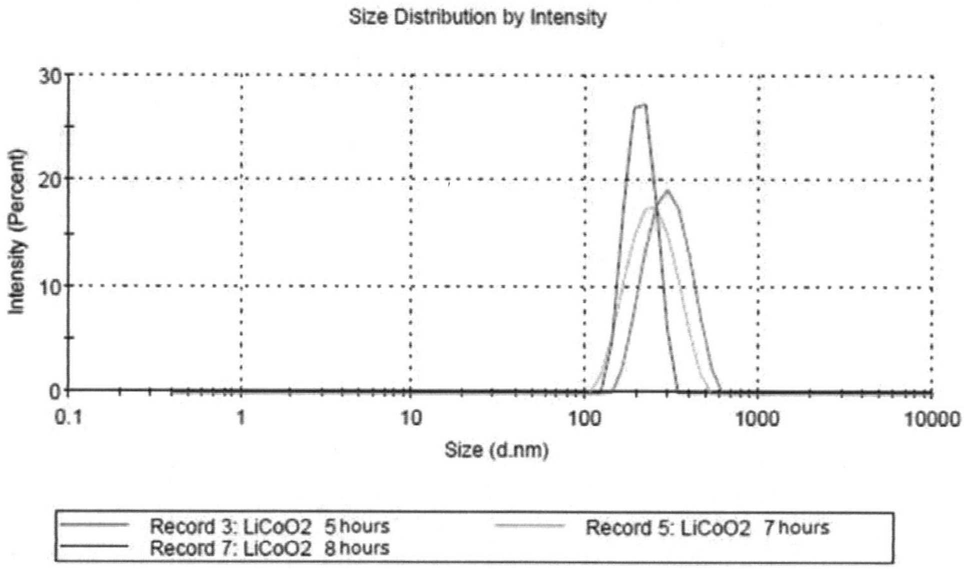

Figure 4. Particle size distribution (three measurements at 30 second intervals).

Impregnation with the prepared dispersion was carried out in a vacuum impregnation unit. Inside the glass cover, the temperature-controlled device is installed to create the required temperature for impregnation.

The carbon fabric should be cleaned of impurities and dried before the vacuum impregnation procedure so that the electrolyte mixed with $LiCoO_2$ can freely penetrate into the empty pores and fill them.

After closing the lid of the tank, a vacuum of at least 10 Pa is created in it by using a vacuum pump, and the air is removed from the open pores. After the "vacuum drying" stage, the electrode material is placed in the electrolyte and impregnated for 30 minutes. The vacuum pump is stopped at the end of the impregnation operation, and atmospheric pressure is restored. The open pores are filled due to the pressure difference between the vacuum and atmosphere.

3. Results and discussion

The relation between surface area of the electrode and the specific energy density of the electrode was studied. According to the measurements, the increase in the area of the electrode material by 3 times allows to increase the energy density of the material by 21% as well as to reduce the internal resistance (ESR) of chemical current source by 2.5 times.

Electrode material with metallization that achieved 670 cm^2 can reach a specific capacity of 17 F/g and ESR 0.1 Om. There is a comparison of results between electrode with metallization and without metallization in the figures 5 and 6. Metallized Busofit has a specific energy similar to other carbon materials of the same area [11], but allows to simultaneously form a chemical current source based on it.

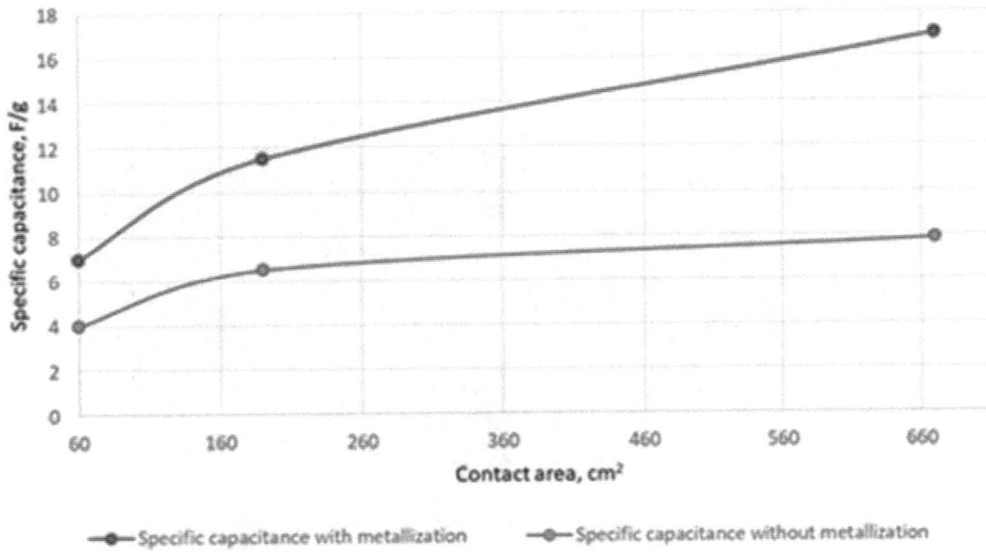

Figure 5. Dependence of a specific electrical capacity on the contact area of electrode materials in the cell: without metallization (blue); with metallization (red).

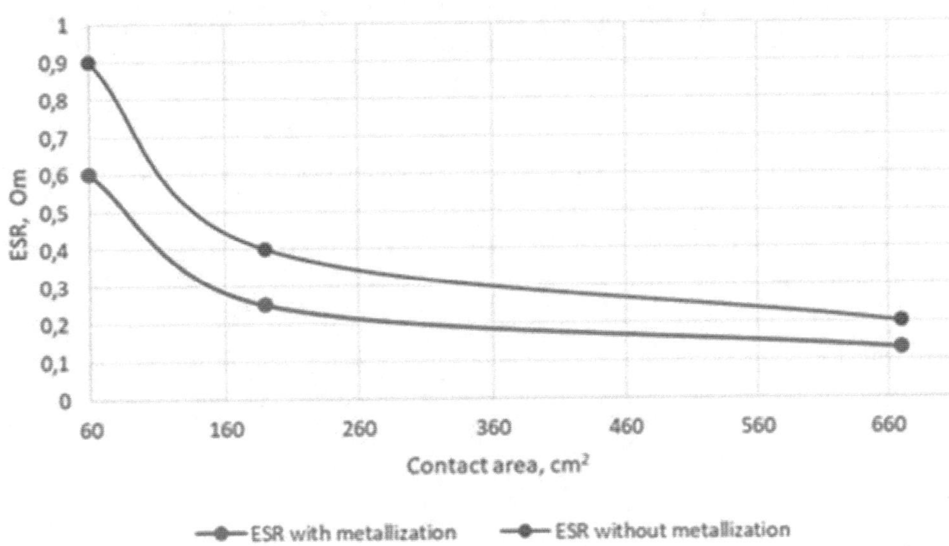

Figure 6. Dependence of a ESR on the contact area of electrode materials in the cell: without metallization (blue); with metallization (red).

The weight of the chemically active cathode material should be 50% to achieve the highest energy intensity in standard cells [11]. By using the technology from this research, the weight of $LiCoO_2$ in the chemical current sources was 50% of the weight of the chemical current cells.

Figure 7 shows a photograph of a filament of carbon material impregnated with a dispersion of $LiCoO_2$ + PC and a graph of the composition of the material deposited on the fiber. The obtained data with SEM show that dispersion made it possible to obtain the required particle size of $LiCoO_2$ less than 300 nm and there are no excess impurities in the formed coating. Figure 7 also shows that the dispersed lithium cobaltate stuck around the Busofit thread 5 microns thick from all sides. Vacuum impregnation made it possible to impregnate Busofit not only in the upper layers, but also in depth, which allowed to increase the entire chemically active material and increase the specific surface area of the electrode material.

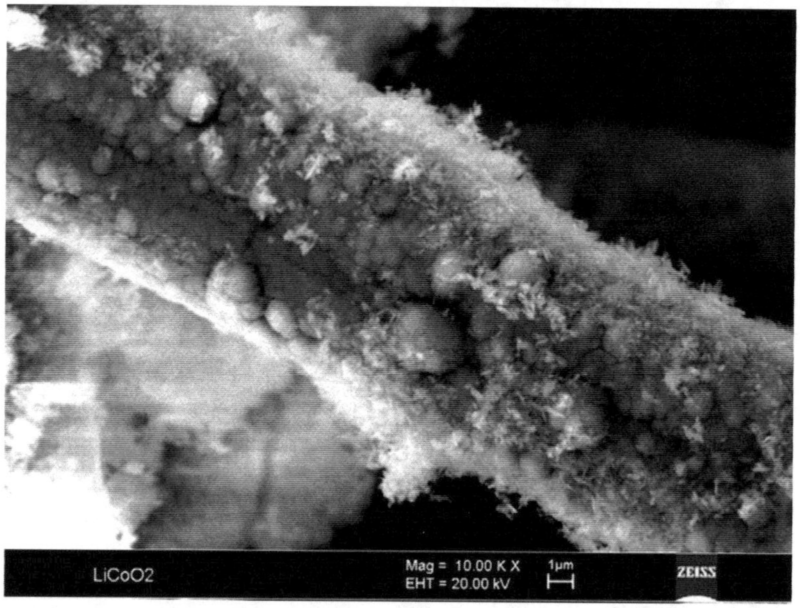

Figure 7. Impregnation of Busofit with dispersion $LiCoO_2$ + PC.

Figure 8 shows that, in addition to titanium and cobalt, there are no other materials when impregnated, which indicates the purity and repeatability of the technology.

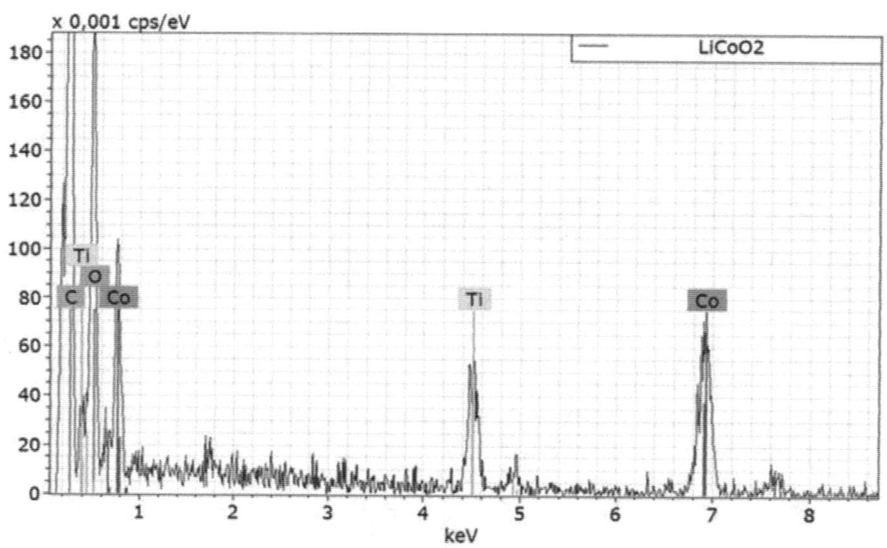

Figure 8. X-ray analysis of the applied $LiCoO_2$ + PC layer on the Busofit filament.

Table 1 shows the specific energy density of the cathode material after impregnation with $LiCoO_2$. Dispersing lithium cobalt and placing it in the pore space allows increasing the specific capacity of the cathode material from working 145-160 mAh/g [1, 2] to 210 mAh/g, which will increase the overall energy density of the chemical current sources. And also we can see from work [12] that the 40 μm thickness thin-film electrode only achieved the charge capacities 141 to 151 mA h/g, and the discharge capacities 140 to 148 mAh/g. Thus, the dispersion of lithium cobaltate and the impregnation of Busofit obtained with the dispersion of carbon fabric made it possible to increase the energy intensity of the electrode material with lithium cobaltate without dispersion up to 30%. However, after 4 charge-discharge cycles, the energy intensity began to decrease, which is apparently due to the movement of dispersed lithium cobaltate inside the cell through a separator and the polarization of the surface layer of the reactive material was occurred.

Table 1. Specific energy density of dispersed $LiCoO_2$ at charge and discharge.

No. of cycle	**1**	**2**	**3**	**4**	**5**	**6**
Discharge (mAh/g)	210	210	146.4	160.7	69.4	87.5
Charge (mAh/g)	239.3	240.3	-	251	149.3	-

In comparison with [1, 2], lithium cobaltate-based electrolytic cells can operate at the same voltage (3V), but the number of charge-discharge cycles is much smaller, which may be due to a lack of information about the chemical processes that occur when using Busofit as a matrix for lithium cobaltate. This phenomenon requires further study.

4. Conclusions

The obtained results allow drawing the following conclusions:

- The dispersion with a particle size of $LiCoO_2$ less than 300 nm was obtained.
- In accordance with the SEM images, it was obtained a carbon fabric material Busofit soaked with dispersed $LiCoO_2$ around each strand into the depth of the material.
- Dispersing lithium cobalt and placing it in the pore space allows increasing the specific energy density of the cathode material $LiCoO_2$ up to 210 mAh/g, which can increase the overall energy density of the chemical current sources.

Acknowledgment

The study is carried out with the financial support from the State represented by the Ministry of Education and Science of the Russian Federation. Subsidies Provision Agreement No. RFMEFI57717X0275.

References

[1] Daniel C, Mohanty D, Li J and Wood D L 2014 Cathode materials review. *AIP Conference Proceedings* **1597** 26 https://doi.org/10.1063/1.4878478

[2] Julien C M, Mauger A and Hussain O M 2019 Sputtered $LiCoO_2$ cathode materials for all-solid-state thin-film lithium microbatteries. *Materials* **12**(17) 2687 doi: 10.3390/ma12172687

[3] Liu C, Neale Z G and Cao G 2016 Understanding electrochemical potentials of cathode materials in rechargeable batteries. *Materials Today* **19**(2) 109 https://doi.org/10.1016/j.mattod.2015.10.009

[4] Ho M Y, Khiew P S, Isa D, Tan T K, Chiu W S and Chia C H 2014 A review of metal oxide composite electrode materials for electrochemical capacitors. *Nano* **09**(06) 1430002 https://doi.org/10.1142/S1793292014300023

[5] Li M, Lu J, Chen Z and Amine K 2018 30 years of lithium-ion batteries. *Adv. Mater* **30**(33) 1800561 https://doi.org/10.1002/adma.201800561

[6] Pang S C, Wee B H and Chin S F 2011 The capacitive behaviors of manganese dioxide thin-film electrochemical capacitor prototypes. *Int. J. Electrochemistry* **2011** 397685 https://doi.org/10.4061/2011/397685

[7] Goodenough J B and Youngsik K 2010 Challenges for rechargeable Li batteries. *Chem. Mater.* **22**(3) 587 https://doi.org/10.1021/cm901452z

[8] Galakhov V R, Karelina V V, Kellerman D G, Gorshkov V S et al. 2002 Elektronnaya struktura, rentgenovskie spektry i magnitnye svojstva nestekhiometricheskih oksidov $LiCoO_2$-delta i $NaxCoO_2$. *FTT* **44** 257 [in Russian] http://journals.ioffe.ru/articles/39378

[9] Xiong F, Yan H J, Chen Y, Xu B, Le J X and Ouyang C Y 2012 The atomic and electronic structure changes upon delithiation of LiCoO2: from first principles calculations. *Int. J. Electrochem. Sci.* **7** 9390 http://electrochemsci.org/papers/vol7/71009390.pdf

[10] Sleptsov V V, Govorov V A, Korobeynikov E V, Kovalev I A and Melnikov A B 2019 Technology of preparation of the $LiCoO_2$ reactive mass for chemical current sources. *Journal of Physics: Conf. Series* **1281** 012076 https://doi.org/10.1088/1742-6596/1281/1/012076

[11] Choudhary N, Li C, Moore J, Nagaiah N, Zhai L, Jung Y and Thomas J 2017 Asymmetric supercapacitor electrodes and devices. *Advanced Materials* **29**(21) 1605336 https://doi.org/10.1002/adma.201605336

[12] Lim C, Yan B, Kang H, Song Z, Lee W C, De Andrade V, De Carlo F, Yin L, Kim Y and Zhu L 2016 Analysis of geometric and electrochemical characteristics of lithium cobalt oxide electrode with different packing densities. *Journal of power source* **328** 46 https://doi.org/10.1016/j.jpowsour.2016.07.119

IOP Conf. Series: Materials Science and Engineering **868** (2020) 012015 doi:10.1088/1757-899X/868/1/012015

Study of high-frequency oscillations and waves in Hall thruster plasma using two-dimensional full kinetic axial-azimuthal model (2D 3V Full PIC)

I Khmelevskoi[1,2,*], D Tomilin[1] and A Lovtsov[1,2]

[1]Department of Electrophysics, *Keldysh Research Centre*, 8 Onezhskaya Street, Moscow 125438, Russia
[2]Department of Aerophysics and Space Research, *Moscow Institute of Physics and Technology* (State University), Dolgoprudny, Moscow region 141700, Russia

[*]E-mail: khmelevskoi@kerc.msk.ru

Abstract. The paper presents the results of Hall thruster plasma simulation using a 2D3V axial-azimuthal Full PIC model. Two types of solution are investigated in detail for two different magnitude of the magnetic field. In the calculations, different types of azimuthal waves were observed. By their characteristics these waves can be corresponded to the electron drift instabilities, gradient-drift waves and ion plasma resonance. The ion plasma resonance was observed in calculations with the magnetic field increase.

1. Introduction

Hall thrusters have long history of various purposes spacecrafts onboard application. Hall thruster (HT) is a coaxial plasma device that uses electrostatic and magnetic forces. Electrostatic forces are used to accelerate an ionized propellant. Magnetic forces are used both to limit the mobility of electrons and to provide an efficient ionization [1-3].

Collisions with neutral particles and interactions with the walls cannot explain the experimental value of electron current. Modelling the Hall thruster plasma allows to investigate the processes responsible for the electron conductivity. Axial-radial geometry is typical for modelling of hall thruster [4-6]. For models in this geometry, it is possible to take into account correctly the interactions with the wall and secondary electron emission. Nevertheless, these models do not give the experimental value of the electron current without artificial additive (Bohm conductivity) [7]. The dependence of Bohm conductivity is inversely proportional to the magnetic field value and the Bohm coefficient. This dependence is a generalization of a large amount of experimental data. Also, azimuthal oscillations and waves in a plasma can significantly affect the electron current in HT. However, this effect can be directly taken into account either in three-dimensional [8] or in 2D axial-azimuthal models [9-13]. Three-dimensional models of real HT devices require high computational resources. Axial-azimuthal models cannot precisely take into account the real configuration of the magnetic field in the HT and the plasma interactions with walls. It is difficult to obtain prediction of HT operating parameters using such axial-azimuthal models. Despite these disadvantages, axial-azimuthal models can be used to research the influence of azimuthal waves on the dynamics of electrons.

In Ref. [9] the authors developed a two-dimensional PIC model for studying the plasma stability in the electron drift direction. The authors suggested that the plasma instabilities could explain the experimentally observed values of the electron current. In Ref. [10–13] the influence of electron drift instability (EDI) on the dynamics of electron component of plasma was studied using the two-dimensional axial-azimuthal PIC models. The influence of the propellant type (Xe, He, Ar, Kr) on the HT discharge in the axial-azimuthal plane was studied using the 2D PIC/MCC model [14]. The authors of 2D3V Full PIC axial-azimuthal model [15] took into account diamagnetic effects. Two types of azimuthal instabilities were observed in the result of modelling: gradient drift waves [16, 17] and electron drift instability. In addition, the influence of the azimuthal component on the dynamics of electrons was studied in the axial-azimuthal plane using a hybrid fluid-particle-in-cell model [18].

This paper presents the results of two-dimensional modelling of a Hall thruster plasma in the axial-azimuthal plane using 2D3V Full PIC model. This model does not require approximation of Bohm diffusion to reproduce correct level of electron transport. The aim of this research is to study the dependence of the solution from the magnitude of the magnetic field. The research is focused on the azimuthal waves and instabilities in the HT plasma.

This work was presented at 18th International Conference "Aviation and Cosmonautics" (Moscow, 18-22 November, 2019).

2. Model description

In Keldysh Research Centre a 2D3V Full PIC model of the Hall thruster plasma has been developed. A rectangle-shaped geometry with periodic boundary conditions in the azimuthal direction and free boundary conditions in the other direction (axial direction) was chosen. The domain scheme is shown in figure 1.

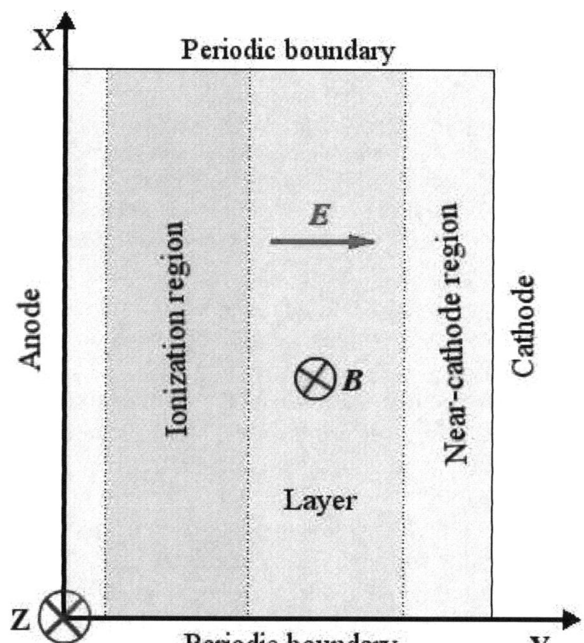

Figure 1. The calculation region scheme.

The region which called "Layer" in the figure 1 is the discharge region, which contains the most of the electric field. This geometry describes a sector of a fixed radius cylinder in the HT discharge channel. The channel height of the HT is considered constant for the entire computational domain. The radial magnetic field is perpendicular to the modelling plane. It corresponds to the B_z component in the model and is considered static. The distribution of the electric field potential is determined at each time step by solving the Poisson equation with periodic boundary conditions along the X axis and the Dirichlet

IOP Conf. Series: Materials Science and Engineering **868** (2020) 012015 doi:10.1088/1757-899X/868/1/012015

boundary conditions along the Y axis using the fast Fourier transform. At the cathode boundary the electron flux is maintained at the level necessary to support plasma quasi-neutrality in the near-cathode region.

Interactions between particles are modelled by the Monte Carlo method. The following types of interactions are taken into account: ionization collisions [19], elastic collisions of electrons with neutral particles [20], excitation collisions [20], Coulomb collisions of electrons with ions [21], electron interactions with walls. Collisions with the walls are modelled with the semi-empirical method. The method implies three steps. First, the probability of electron collision with the walls is calculated. Second, the Debye sheath is calculated with the condition that the ion flux to the wall is equal to the electron flux. Third, electrons with the energy larger than the Debye sheath potential reach the wall.

3. Simulation results

In present research the computational domain is a cylindrical sector with a length of 35 mm and a width of 64 mm. The height of the channel is equal to 14 mm. The calculation grid is 141×256, the grid step is 0.25 mm. The time step is equal to $4 \cdot 10^{-11}$ s. The potential difference between the anode and cathode boundary is equal to 500 V. The xenon consumption is 1 mg/s. To obtain a stable solution, the grid step must be smaller than the Debye radius in the calculation [22]. There is common technique to fulfil this condition [23]. The vacuum permittivity value was artificially increased by a factor $\gamma = 150$. The time step is chosen to resolve plasma oscillations and cyclotron motion of electrons [22]. There are about 10^6 particles of each component in simulation domain. The calculations were carried out on a single CPU.

The computational domain size in the azimuthal direction is approximately equal to $0.5 \cdot \pi$ radian of the circumference of the HT laboratory model with the mean diameter equal to 77 mm. This laboratory model was developed and tested at Keldysh Research Centre [24]. The magnetic field (figure 2a) correspond to the magnetic field values of the HT laboratory model. The magnetic field values were obtained using the Magnet2D software [25]. The line 1 in figure 2a is 30% lower than the line 2. The first calculation was carried out with breathing mode oscillations taken into account. However, the main focus of the paper is the high-frequency oscillations which timescale is much smaller than breathing mode oscillations, consequently, it is possible to use permanent ionization function instead of self-consistent simulation of the ionisation process. Therefore the ionization function is chosen to be equal to the average value during the period of ionization oscillations (breathing mode) for all the further calculations.

The initial distributions of plasma components corresponds to the chosen ionisation function. First, the neutral particles are placed in simulation region in accordance with the permanent ionization function. Figure 2b shows the neutral gas concentration and the normalized ionization function. Second, the plasma components are placed in computational domain. The neutral particles do not move during this phase of the simulation. Particle flows get into the equilibrium when the integral of the ionization function is approximately equal to the difference of the particle fluxes across the *Y* direction boundaries. All the data presented further were obtained after the equilibrium was reached. Hereinafter, the solution with the lower magnetic field (red curve in figure 2a) is called solution 1, and the solution with higher magnetic field (blue line in figure 2a) is called solution 2.

The solutions have significant differences. The graphs in figure 3 and 4 is obtained by averaging the parameters over the time of $8 \cdot 10^{-7}$s. Figures 3 a-d) shows two-dimensional distributions of the electron concentration (figure 3a and figure 3b) and plasma potential (figure 3c and figure 3d). One-dimensional distributions is obtained by averaging the parameters along the azimuthal component (figure 3e and figure 3f). Graphs in the figure 3a, 3c, 3e correspond to solution 1, and graphs in figure 3b, 3d, 3f correspond to the solution 2. For the solution 2, the main potential drop is located near the maximum of the ionization function. A characteristic property of the solution 2 is the local maximum of concentration along the Y axis near the magnetic field maximum. Figure 3a and 3b show the short-wave perturbations (electron concentration). Similar oscillations were obtained in [10–13] and represent as electron drift instability. These waves are characterized by a wavelength in the range from the Debye

IOP Conf. Series: Materials Science and Engineering **868** (2020) 012015 doi:10.1088/1757-899X/868/1/012015

radius to the electron cyclotron radius. The waves propagate in the azimuthal direction at near ion-sound velocity.

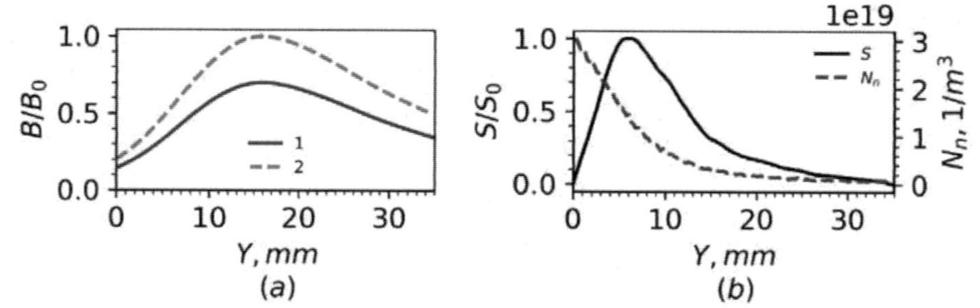

Figure 2. (a) Relative value of the magnetic field (B) for solution 1 and 2 (correspond to the number in legend); (b) Relative value of the ionization function (S) and neutral concentration profile (N_n).

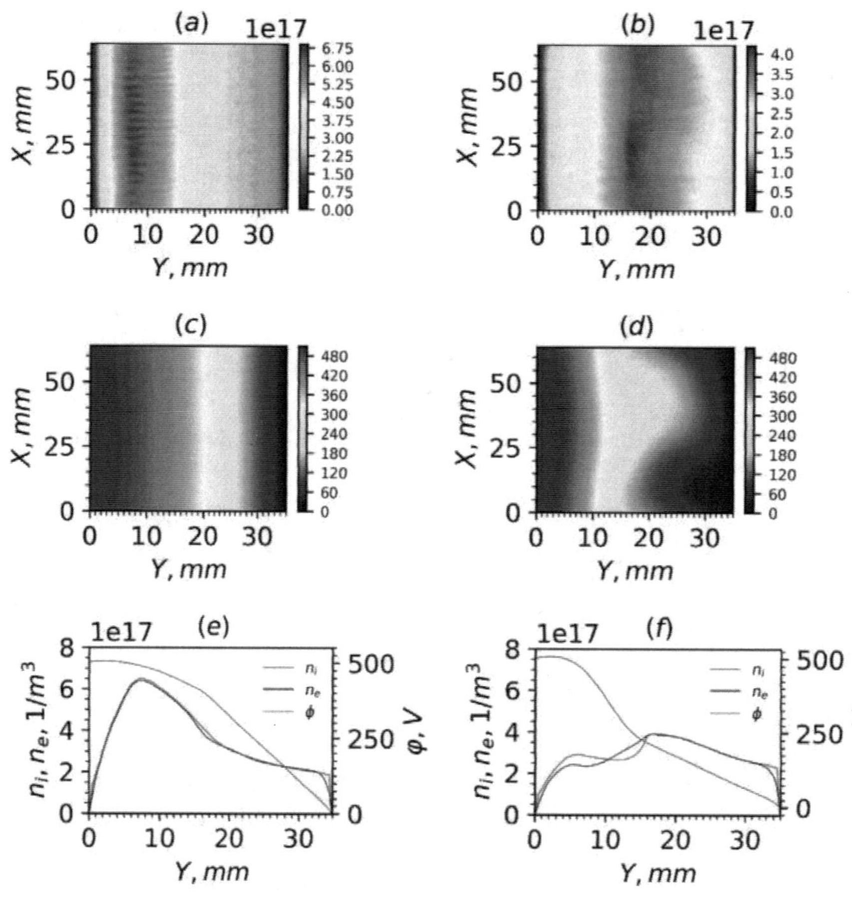

Figure 3. Electron concentration ($n_e, 1/m^3$) for solutions 1 (a) and 2 (b), plasma potential (ϕ, V) for solutions 1 (c) and 2 (d), one-dimensional electron and ion ($n_i, 1/m^3$) concentration and plasma potential for solutions 1 (e) and 2 (f).

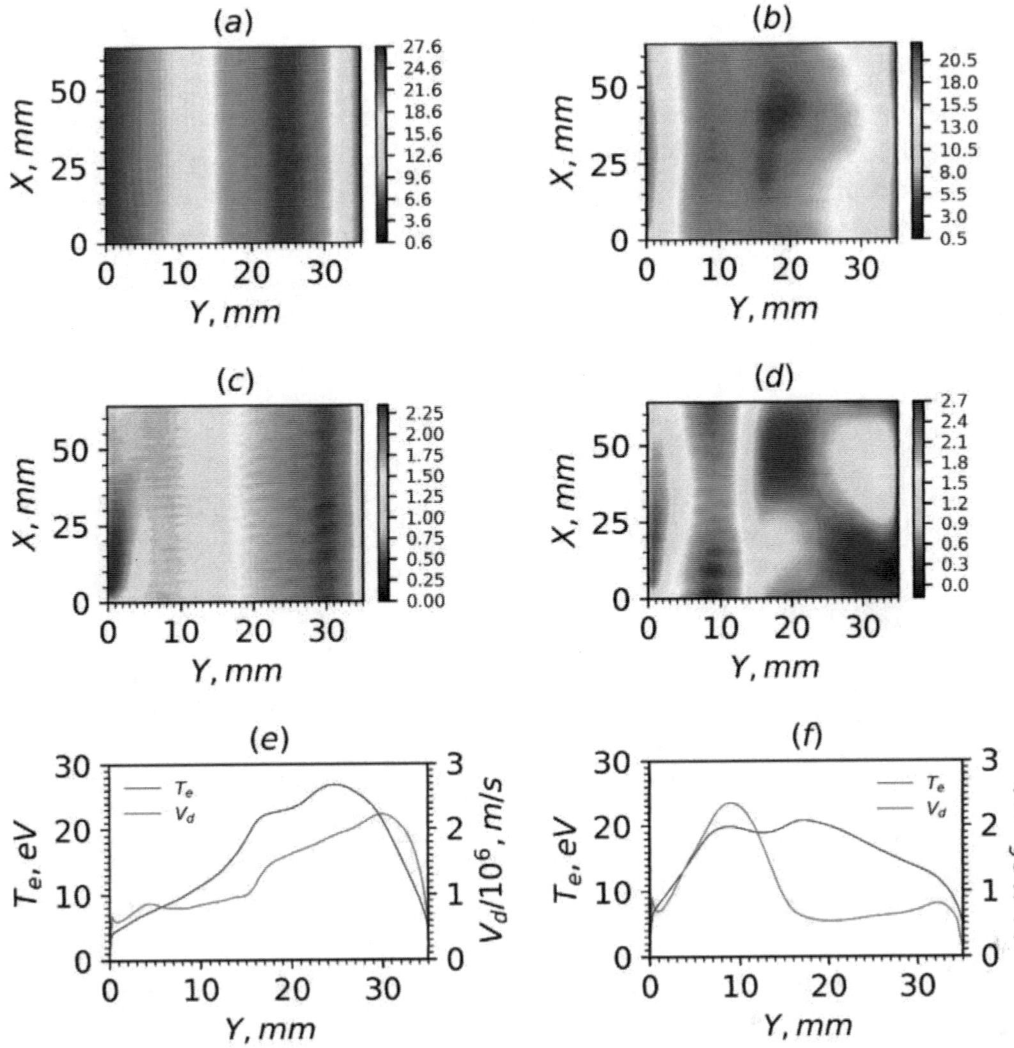

Figure 4. Electron temperature (T_e, eV) for solutions 1 (a) and 2 (b), electron drift velocity ($V_d, m/s$) for solutions 1 (c) and 2 (d), one-dimensional electron temperature and electron drift velocities for solutions 1 (e) and 2 (f).

Figure 4 shows two-dimensional and one-dimensional distributions of the electron temperature and the drift velocity for two solutions. The electron temperature is shown in figure 4a-b and the drift velocity is shown in figure 4c-d. Figure 4e-f contains the one-dimensional distributions of the electron temperature and the drift velocities. The solution 1 (figure 4a, 4c, 4e) is characterized by a maximum of the electron temperature and drift velocities in the region of the magnetic field decrease (right part of computational domain). The maximum values of the parameters in the solution 2 (figure 4b, 4d, 4f) is shifted to the ionization region. The electron temperature and the drift velocity in the ionization region (according to figure 1) are significantly higher for the solution 2 than for the solution 1. Also, the solution 2 is characterized by the presence of a large azimuthal heterogeneity in the region of the magnetic field decrease. This heterogeneity occurs in the distributions of all plasma parameters: concentrations, electron temperature, electron drift velocity and plasma potential.

Next, the dispersion function of azimuthal waves in the region of magnetic field decrease was investigated. Figure 5a-b shows the two-dimensional Fourier transform in time and X coordinate for the fixed $Y = 17.5$ mm. For the solution 1 (figure 5a), the dispersion function is approximately linear. Figure 5c shows the electron azimuthal velocity distribution and the spectral density of waves' phase velocity for the solution 1. The spectral density of waves' phase velocity and the azimuthal velocity distribution are overlapped. It means that the $\frac{\omega}{k} \approx V_d$ condition is satisfied for several frequencies. Therefore, there is an interaction between the azimuthal waves and the motion of the electrons, which leads to an additional transport of electrons across the magnetic field. According to the characteristics of these oscillations, the observed azimuthal waves correspond to gradient drift ones [15-17, 26-28]: the wavelengths exceed the cyclotron radius of the electrons, and the frequencies are in the range from 1 to 200 MHz.

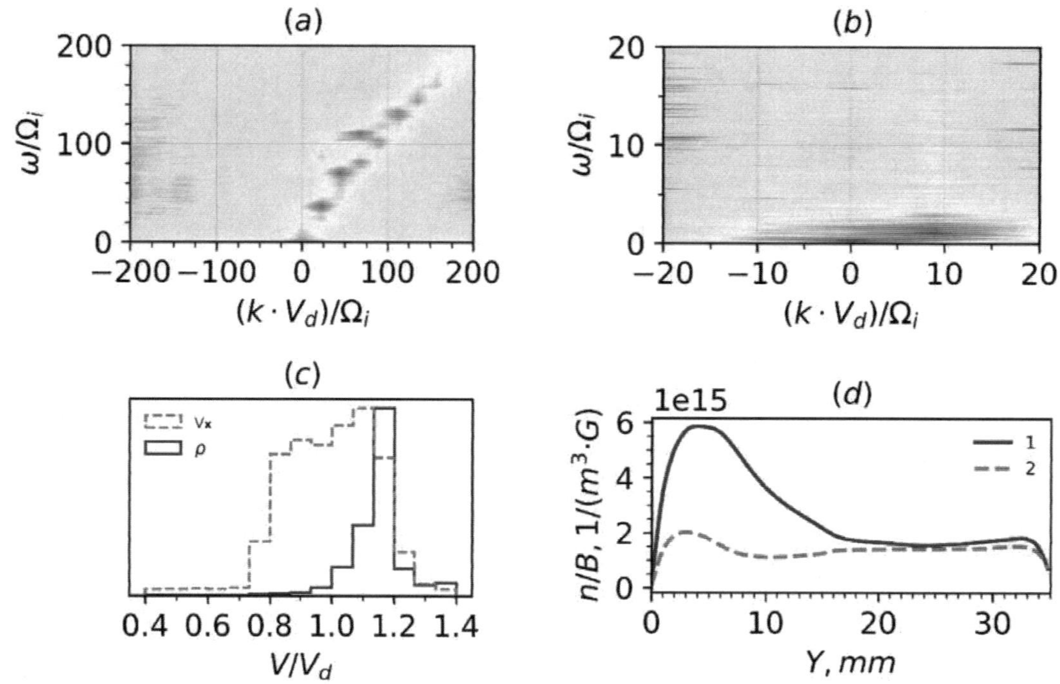

Figure 5. Dispersion functions of the azimuthal waves for solutions 1 (a) and 2 (b); (c) – electron drift velocity distribution and the spectral density of waves' phase velocity (ρ); (d) – ratio of the electron concentration to the magnetic field for solutions 1 and 2 (correspond to the number in legend). Notifications: ω is frequency, k is wave vector in the azimuthal direction, Ω_i is ion plasma frequency, V_X – electron azimuthal velocity distribution function.

For the solution 2 the spectrum (figure 5b) significantly differs from the corresponding spectrum of the solution 1 (figure 5a). The main maximum is found at frequency equal to ion plasma frequency (Ω_i), which matches to ion-plasma resonance. This instability represents a large azimuthal heterogeneity with a wavelength approximately equal to the size of the calculation domain.

In figure 5d the ratio of the electron concentration to the magnetic field is shown. The graph shows that the electron concentration is aligned with the magnetic field in the right part of computational domain. The magnetic field reduces in this part of the domain.

IOP Conf. Series: Materials Science and Engineering **868** (2020) 012015 doi:10.1088/1757-899X/868/1/012015

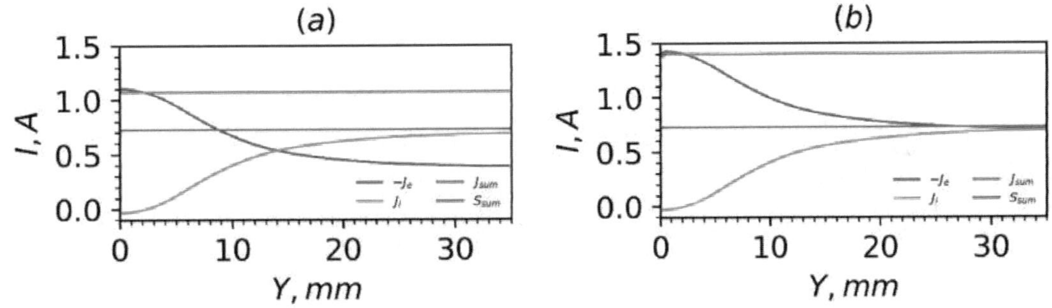

Figure 6. Electron current (J_e), ion current (J_i), total discharge current (J_{summ}) and the integral of the ionization function (S_{summ}) for solutions 1 (a) and 2 (b).

The electron, ion, total discharge currents and the integral of the ionization function along Y axis are shown in figure 6 for the solution 1 (figure 6a) and the solution 2 (figure 6b). The ion current at the exit plane of the thruster almost reaches the value of the integral of the ionization function. The electron current in the solution 2 is about two times higher than in the solution 1. The total discharge current is constant along the Y direction, so the electron current and ion current adjust to each other.

4. Discussion

Table 1 contains the integral parameters of the discharge per length in the azimuthal direction for the laboratory model [29] (experimental value) and for two solutions in the model. The ion current in the experiment and in the calculations is approximately at the same level for the same gas flow rate per length. The discharge current for solution 1 is 20% higher than one observed in the experiment. Such disagreement could be explained by the simplicity of the model of plasma-wall interaction: the channel height is assumed constant, when the discharge chamber has complex shape in the experiment. In addition, the model does not take into account the real 3D configuration of the magnetic field. Thus, only a qualitative agreement between the experimental and model results could be observed.

Table 1. Comparison between model solutions and experimental data

Parameter of operation	**Laboratory model of HT**	**solution 1**	**solution 2**
Gas flow rate per length, mg/s/m	15.625	15.625	15.625
Discharge current per length, A/m	13.7	16.7	21.8
Ion current per length, A/m	11.9	10.7	10.8

Different operational HT modes are observed in the experiments [24, 30-38]. In the experimental works, two of them were called the "jet" and "bell" mode [24, 30]. It was shown that the increase of the magnetic field magnitude could lead to the transition from the "jet" mode to the "bell" mode. The enlarged values of the discharge current are observed in the "bell" mode due to the increase of the electron current component. Also, the waves with azimuthal component in the frequency range 5–150 MHz were found in these two modes [30]. The dispersion function of these waves is close to the linear one. In the "bell" mode, the amplitude of the detected waves significantly exceeds the amplitude in the "jet" mode.

The solutions in the model depends on the magnitude of the magnetic field. In this case, our model could be a promising tool for researching the plasma processes during the transition between different operational modes.

IOP Conf. Series: Materials Science and Engineering **868** (2020) 012015 doi:10.1088/1757-899X/868/1/012015

5. Conclusion

A 2D3V axial-azimuthal Full PIC model of the Hall thruster plasma has been developed. Two different solutions depending on the magnetic field magnitude are investigated using this model. Simulation results provide to research the electron concentration, plasma potential, electron temperature, electron drift velocity and another. The dispersion functions of waves in the azimuthal direction were studied. The solutions show waves, which can be attributed by their characteristic to the electron-drift instability, gradient-drift waves and ion-plasma resonance. The ion current obtained in the model corresponds to the experimental values with good agreement. It is noted that the developed model is a promising tool for researching the processes of the HT transition between different operational modes.

References

[1] Kim V 1998 Main Physical Features and Processes Determining the Performance of Stationary Plasma Thrusters. *Journal of Propul. Power* **14 (5)** 736
[2] Morozov A I, Savelyev V V 2000 Fundamentals of Stationary Plasma Thruster Theory. *Reviews of Plasma Physic* **21** 203 doi: 10.1007/978-1-4615-4309-1_2
[3] Morozov A I, Kislov A Ya, and Zubkov I P 1968 High-current Hall-current plasma accelerator. *JETP Lett.* **7 (172)**
[4] Irishkov S V, Gorshkov O A, and Shagayda A A 2005 Fully kinetic modeling of low power Hall thrusters. *International Electric Propulsion Conf.* (Princeton University) pp 1-10
[5] Parra F I, Ahedo E, Fife J M and Martinez-Sanchez M 2006 A two dimensional hybrid model of the Hall thruster discharge. *J. Appl. Phys.* **100** 023304
[6] Mikellides I G and Katz I 2012 Numerical simulations of Hall-effect plasma accelerators on a magnetic-field-aligned mesh. *Physical Rev.* **E86** 046703 doi: 10.1103/PhysRevE.86.046703
[7] Boeuf J P 2017 Tutorial: Physics and modeling of Hall thrusters. *J. Appl. Phys.* **121** 011101 doi: 10.1063/1.4972269
[8] Taccogna F and Minelli P 2018 Three-dimensional particle-in-cell model of Hall thruster: The discharge channel. *Physics of Plasmas* **25** 061208 doi: 10.1063/1.5023482
[9] Adam J C, Heron A and Laval G 2004 Study of stationary plasma thrusters using two-dimensional fully kinetic simulations. *Phys. Plasmas* **11(295)** doi: 10.1063/1.1632904
[10] Coche P and Garrigues L 2014 A two-dimensional (azimuthal-axial) particle-in-cell model of a Hall thrusters. *Physics of Plasmas* **21** 023503 doi: 10.1063/1.4864625
[11] Boeuf J P and Garrigues L 2018 E × B electron drift instability in Hall thrusters: Particle-in-cell simulations vs. theory. *Physics of Plasmas* **25** 061204 doi: 10.1063/1.5017033
[12] Lafleur T and Chabert P 2018 The role of instability-enhanced friction on 'anomalous' electron and ion transport in Hall-effect thrusters. *Plasma Sources Sci. Technol.* **27** 015003
[13] Lafleur T, Martorelli R, Chabert P, Bourdon A, 2018, Anomalous electron transport in Hall-effect thrusters: Comparison between quasi-linear kinetic theory and particle-in-cell simulations. *Physics of Plasmas* **25** 061202 doi: 10.1063/1.5017626
[14] Croes V, Tavant A, Lucken R, Martorelli R, Lafleur T, Bourdon A and Chabert P 2018 The effect of alternative propellants on the electron drift instability in Hall-effect thrusters: Insight from 2D particle-in-cell simulations. *Physics of Plasmas,* **25** 063522 doi: 10.1063/1.5033492
[15] Chernyshev T, Son E, Gorshkov O 2019 2D3V kinetic simulation of Hall effect thruster, including azimuthal waves and diamagnetic effect. *Journal of Physics D: Applied Physics* **52(44)** doi: 10.1088/1361-6463/ab35cb
[16] Nikitin V, Tomilin D, Lovtsov A and Tarasov A 2017 Gradient-drift and resistive mechanisms of the anomalous electron transport in Hall effect thrusters. *EPL* **117** 45001 doi: 10.1209/0295-5075/117/45001
[17] Tomilin D 2013 Gradient instabilities of electromagnetic waves in Hall thruster plasma. *Phys. Plasmas* **20** 042103
[18] Lam C M, Fernandez E and Cappelli M A 2015 A 2-D Hybrid Hall Thruster Simulation That Resolves the E×B Electron Drift Direction. *IEEE Transactions on Plasma Science* **43(1)** DOI:

10.1109/TPS.2014.2356650
[19] R C Wetzel, F A Baiocchi, T R Hayes and R S Freund 1987 Absolute cross sections for electron-impact ionization of the rare-gas atoms by the fast-neutral-beam method. *Phys. Rev.* **35(2)** 559
[20] Hashimoto T and Nakamura Y 1990 *Electron swarm parameters in xenon and its momentum transfer cross section, Papers of Gas Discharge Technical Committee ED-90-61* (Japan: IEE)
[21] Artsimovich L A and Sagdeev R Z 1979 *Plasma Physics for Physicists* (Atomizdat Publishers, Moscow)
[22] Hockney R W, Eastwood J W 1987 *Computer Simulation Using Particles* (McGraw-Hill Book Company)
[23] Liu H, Wu B, Yu D, Cao Y and Duan P 2010 Particle-in-cell simulation of a Hall thrusters. *Journal of Physics D: Applied Physics* **43(16)** 165202 doi: 10.1063/1.5054009
[24] Khmelevskoi I A, Tomilin D A 2019 Parametric Study of Two Stable Forms of Discharge Burning in a Hall-Effect Thruster. *Techn. Phys.* **64(9)** 1283 doi: 10.1134/S1063784219090068
[25] Shagayda A A 2007 Certificate of Software Application Registration No. 2007614083 p 24 [in Russian]
[26] Lazurenko A, Krasnoselskikh V and Bouchoule A 2008 Experimental Insights Into High-Frequency Instabilities and Related Anomalous Electron Transport in Hall Thrusters. *IEEE Transactions on Plasma Science* **36(5)** 1977 doi: 10.1109/TPS.2008.2000972
[27] Tilinin G N 1977 Experimental study of high-frequency plasma oscillations in the Hall-current plasma accelerator *Sov. Physics-Tech. Journal* **47(8)** 1684
[28] Esipchuk Y B and Tilinin G N 1976 Drift instability in a Hall-current plasma accelerator *Sov. Physics-Tech. Journal* **46(4)** 718
[29] Tomilin D A, Lovtsov A S, Khmelevskoi I A 2019 Parametric studies of 1.5 and 2.5 kW Hall Thrusters with external layer. *The 36th International Electric Propulsion Conf.* (Austria, Vienna) IEPC-2019-342
[30] Khmelevskoi I A, Lovtsov A S, Tomilin D A 2019 Study of two Different Discharge Modes in Hall Thruster. *The 36th International Electric Propulsion Conf.* (Austria, Vienna) IEPC-2019-387
[31] Lovtsov A S, Tomilin D A and Shashkov A S 2014 Distribution of local plasma parameters in a channel of the Hall thruster at two different types of discharge burning. *Tech. Phys. Letters* **40(9)** 754
[32] Azziz Y 2007 Experimental and Theoretical Characterization of a Hall Thruster Plume, Dissertation, Massachusetts Institute of Technology
[33] Conversano R W, Goebel D M, Mikellides I G, Hofer R R, Matlock T S, Wirz R E 2014 Magnetically Shielded Miniature Hall Thruster: Perfomance Assessment and Status Update. *50th AIAA/ASME/SAE/ASEE Joint Propulsion Conf.* (USA, San Diego) DOI: 10.2514/6.2014-3896
[34] Kostin A N, Lovtsov A S, Vasin A I, Vorontsov V V 2013 Development and qualification of Hall thruster KM-60 and the flow control unit. *The 33st International Electric Propulsion Conference* (USA, Washington) IEPC-2013-055
[35] Hara K, Sekerak M J, Boyd I D and Gallimore A D 2014 Mode transition of a Hall thruster discharge plasma. *J. Appl. Phys.* **115** 203304 doi: 10.1063/1.4879896
[36] Ding Y, Boyang J, Sun H, Wei L, Peng W, Li P, Yu D 2017 Effect of matching between the magnetic field and channel length on the performance of low sputtering Hall thrusters. *Adv. Space Res.* **63(3)** 837 doi: 10.1016/j.asr.2017.11.003
[37] Ding Y, Sun H, Li P, Wei L, Xu Y, Peng W, Su H, Yu D 2017 Influence of hollow anode position on the performance of a hall-effect thruster with double-peak magnetic field. *Vacuum* **143** 251 doi: 10.1016/j.vacuum.2017.06.030
[38] Han L, Wei L, Yu D 2016 Mode transition induced by the magnetic field gradient in Hall thrusters. *J. Phys. D: Appl. Phys.* **49** 375203 doi: 10.1088/0022-3727/49/37/375203

IOP Conf. Series: Materials Science and Engineering **868** (2020) 012016 doi:10.1088/1757-899X/868/1/012016

Integration of the inverse dynamics with a reference model technique, and its application for the improvement of the helicopter flying qualities

Z Mbikayi*, A V Efremov and E V Efremov

Flight Dynamics and Control Department, *Moscow Aviation Institute* (National Research University), 4 Volokolamskoe shosse, Moscow125993, Russia

*E-mail: pvl@mai.ru

Abstract. The application of inverse dynamics for the improvement of flying qualities is considered. It is shown that the use of the inverse dynamics in the control loop suppresses the dynamics of the aircraft considerably, leaving only the dynamics of the integral in almost the entire frequency range. The inverse dynamics however could not help in case of atmospheric disturbances or modeling uncertainties. A new scheme using a reference model and its inverse was presented to counter these effects. The results of the mathematical modeling showed that the inverse dynamics allow to improve the tracking performances and all other parameters of the pilot-aircraft system compared to the controller synthesized with feedback gains only. These results were confirmed using ground based simulation. These simulations showed that the use of the inverse dynamics based technique allowed to decrease the variance of error up to 60% during a tracking compensation task. The simulations also showed to that the addition of the reference model and its inverse proved to be able to reduce or suppress atmospheric turbulence, couplings between different control channels and effects of modeling uncertainties, while improving the tracking accuracy and decreasing the pilot workload, thus enhancing the flying qualities of the helicopter.

1. Introduction

One of the major areas in flight control is the provision of necessary flying qualities. Several techniques developed recently allow to change the dynamics of the aircraft considerably. One of them is the dynamic inversion approach introduced in [1].It is a nonlinear control technique based on feedback linearization. This technique has been used successfully over the years for in a variety of application for both airplanes [2] and rotorcrafts [3]. It is used to make an appropriate coordinate transformation of the nonlinear plant, so that any linear control method can be used on the resulting linear plant. In [3] and [4] for example, the non linear dynamic inversion is used in the inner loop, and the outer loop consists of a PID based controller. And in [5], the robustness is provided by an H-infinity based controller. As a linear control technique, it has been used as an addition in the feedforward loop to support the feedback controller. And in this form, its integration with reference models techniques has been studied by [6] and others. In these studies, the desired dynamics from the pilot stick commands are computed via the use of reference models, which are then given to the inverse dynamics in order to compute the actuators positions necessary to achieve those desired dynamics. This method however still requires an exact model of the aircraft to be controlled. Therefore, a PID - type compensator must be added to provide the necessary robustness for modeling uncertainties and disturbances as in [3].

This paper essentially considers a different use of the dynamic inversion technique as a linear control method. A novel scheme using the inverse dynamics, reference models and their inverses is presented in order to enhance the robustness of the controller and provide better tracking performances. The resulting controller that can be used either on a linear plant or on a plant where the feedback linearization based dynamic inversion has been used in the inner loop.

The efficiency of the proposed technique is evaluated by comparing it to the use of feedback gains-based controllers and controllers synthesized with the inverse dynamics and feedback gains that provide robustness.

2. Using the inverse dynamics

The inverse dynamics can be used in the feedforward loop to improve tracking performances. As stated in [7], the inverse dynamics requires a stable system. For the vehicle used, stability can be achieved using a gain of matrix used in full stated feedback, computed using the pole placement method that uses the formula of Ackermann [8] through the controllability matrix [9], in order to compute the stabilization gains.

To achieve zero tracking error for a given reference signal r*(s)* in a controllable and stable system $G(s)$ with a control loop as given in figure 1, the input *u(s)* can be found as:

$$y(s) = r(s) = u(s) \cdot G(s) \tag{1}$$

$$u(s) = \frac{r(s)}{G(s)} = r(s) \cdot G^{-1}(s) \tag{2}$$

Figure 1 shows the use of the inverse dynamics in the control loop, with the stabilizing matrix of gains.

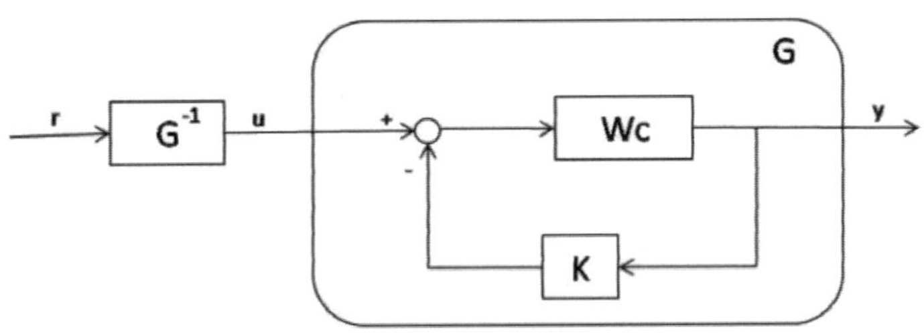

Figure 1. Inverse of closed loop.

Where

$$G = \frac{W_c}{1 + W_c \cdot K}$$

W_c is the aircraft dynamics.

With the pilot in the loop, the goal is to have only one integral between the pilot command and the controlled state. Figure 2 demonstrates the frequency response characteristics of the longitudinal dynamics of the helicopter equipped with the feedback providing the necessary stability, and the same dynamics equipped with the inverse dynamics. It is seen that the inverse dynamics change the controlled element dynamics, approaching it to the integral in practically all the frequency range.

IOP Conf. Series: Materials Science and Engineering **868** (2020) 012016 doi:10.1088/1757-899X/868/1/012016

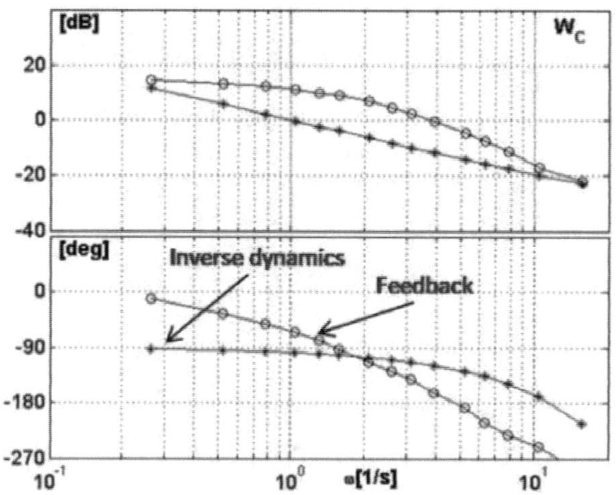

Figure 2. Frequency response inverse dynamics vs. feedback alone.

3. Constructing the Inverse

After making sure that the aircraft is stable, the inverse dynamics can be computed.

For a transfer function $\frac{y(s)}{u(s)}$ the inverse is simply $\frac{u(s)}{y(s)}$.

The stable plant G given in figure 1 is a multiple input multiple output (MIMO) system, and can be representedin state space form as:

$$sX(s) = AX(s) + BU(s) \tag{3}$$
$$Y(s) = CX(s) \tag{4}$$

where *A, B* and *C* are the state matrix, input matrix and output matrix respectively; *U, X* and *Y* are the control vector, state vector and output vector respectively; *s* is the Laplace transform variable.

The inverse can be computed analytically as shown in [7].

From (3) we get

$$X(s) = (sI - A)^{-1}BU(s) \tag{5}$$

Replacing (5) in (4) we get $Y(s) = C(sI - A)^{-1}BU(s)$.

From this, we can get the relationship between the transfer function and the state space representation as:

$$G(s) = \frac{Y(s)}{U(s)} = C(sI - A)^{-1}B \tag{6}$$

Since$A$,B, C and D are matrices, equation (6) is valid only if the number of rows of the matrix C is equal to the number of columns of matrix B. Which means the system needs to be square (Number of inputs = number of outputs), such that each input controls a specific output.

The inputs here are the pilot commands: *Collective, Longitudinal cyclic, Lateral cyclic* and *pedals*. The outputs are: *Climb rate, Pitch rate, Roll rate* and *Yaw rate.*

The inputs and outputs are coupled as follow:

$$\begin{bmatrix} Collective & = & Climb \quad rate \\ Longitudinal & = & Pitch \quad rate \\ Lateral & = & Roll \quad rate \\ Pedals & = & Yaw \quad rate \end{bmatrix} \tag{7}$$

From (6), we get a $(4x4)$ matrix of transfer functions corresponding to couplings of each input and each output as shown in matrix G where G_{ij} represents the transfer function from input i to output$j$.

$$G = \begin{bmatrix} G_{11} & G_{12} & G_{13} & G_{14} \\ G_{21} & G_{22} & G_{23} & G_{24} \\ G_{31} & G_{32} & G_{33} & G_{34} \\ G_{41} & G_{42} & G_{43} & G_{44} \end{bmatrix} \quad (8)$$

Now each transfer function is inversed, and we get a new $(4x4)$ matrix G^* of inversed transfer functions which is put back together in state space form.

However, since all the transfer functions in G are strictly proper, their inverses are improper. Additionally, in the mathematical model used, the actuator dynamics were modeled as a first order system, and there was a time delay in the system that was approximated as a third order system. It was therefore necessary to use a fifth order low pass filter F, to make the inverse dynamics proper. This filter is used not only to proper the system, but also to filter the high frequencies of the inverse dynamics that might lead to non physical actuator rates. And as shown in the next section, the dynamics of this filter is used as the reference model. The frequency of the filter was chosen by experimenting with different frequencies, by trial and error. And the frequency was chosen such that it is high enough not to change the dynamic of the inverse dynamics, but not to high as to keep the actuators rates within acceptable limits.

4. Reference model

As shown by figure 1, the inverse dynamics are calculated for the closed loop system and are used alone in the feedforward. A feedforward loop cannot achieve robustness in case of plant modeling uncertainties, or disturbances. A new scheme using a reference model and its inverse is introduced to provide robustness, as shown in figure 3. The scheme is extended and used for the MIMO system with four inputs and four outputs. This helps suppress the couplings between the aircraft dynamics as well. The reference model is chosen to be the filter F used with the inverse dynamics. This choice was led by the fact that the dynamics of the filter are very close to the dynamics of the entire system. Equation (9) shows that in case of an uncertainty-free system, the dynamics of the entire system will be practically the same as those of the filter used in with the inverse dynamics. Choosing the reference model dynamics to be the same as those of the filer allows it to only act when equation (9) is not true, which means there are modeling uncertainties. Because the inverse dynamics of the filter will also be improper, a second filter $F2$ is introduced in order to make the inverse proper. The frequency of this filter also needs to be chosen carefully such it is not too low as to limit the effect of the reference model, and not to high as to let through high frequency signals that might saturate the actuators.

$$G^{-1}(s) \cdot F(s) \cdot G(s) = F(s) \quad (9)$$

The concept of reference model introduced in this paper is simply a method that the controller uses in order to keep suppressing the aircraft dynamics even in case of uncertainty, only leaving an integral-type response as per flying qualities requirements given in [10] which allows the pilot frequency response to be gain-like.

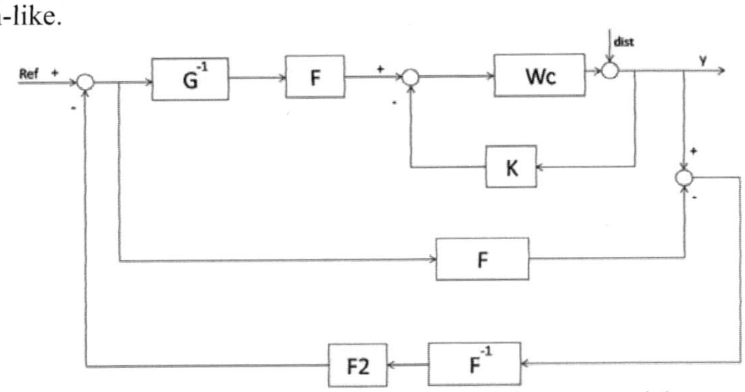

Figure 3. Control structure using a reference model.

IOP Conf. Series: Materials Science and Engineering **868** (2020) 012016 doi:10.1088/1757-899X/868/1/012016

5. The investigation of effectiveness of the proposed principle of inverse dynamics

The effectiveness of the proposed approach was tested by mathematical modeling and in ground based simulation. Modeling uncertainties introducing differences between the inverse dynamics and the plant model are simulated by changing the aerodynamic coefficients, stability derivatives and poles of the linear model of the plant. Additionally, time delays and nonlinear actuator dynamics with rate limits are added to the system in simulations.

5.1. Time domain simulations

Simple time domain verification is done by giving a step input in the longitudinal channel, and observing the behavior of the aircraft.

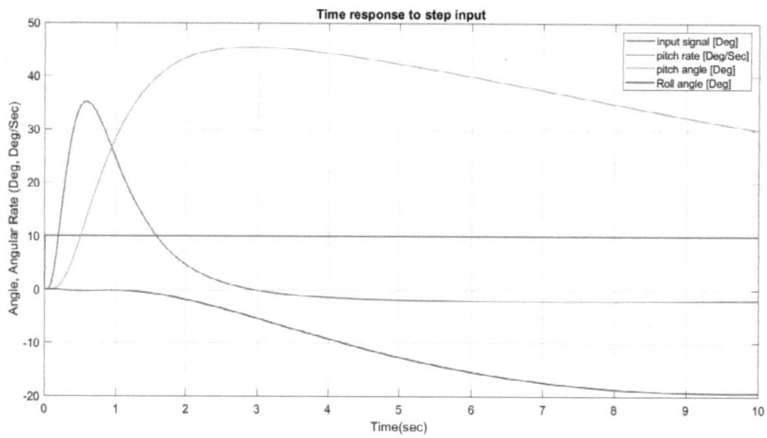

Figure 4. Time response to longitudinal step input (feedback controller).

The time responses to a step input in the longitudinal channel of the helicopter equipped with the feedback controller are show on figure 4. The tracking performances are very bad as neither the pitch rate, nor the pitch angle can track the reference signal. Additionally, the roll angle deviates considerably from the steady state value. This shows strong coupling between the control axes. This will result in a high pilot workload because the pilot needs to be actively controlling the axis in which the task needs to be achieved, and needs to control other axes as well, in order to suppress the coupling effects. The same step input is given to the helicopter equipped with the inverse dynamics and reference model controller, and the time responses are shown on figure 5. The tracking performances are greatly improved and the coupling effects are practically suppressed.

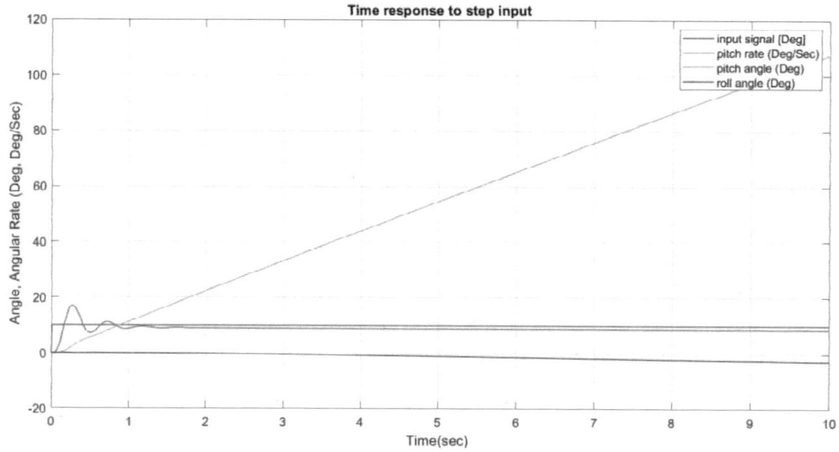

Figure 5. Time response to longitudinal step input (inverse + reference model).

5.2. Mathematical modeling of the pilot-aircraft system with the inverse dynamics

The investigated pilot-aircraft system is shown on figure 6.

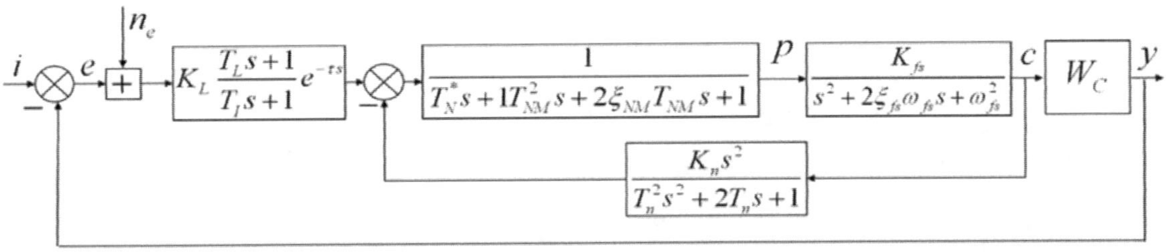

Figure 6. Pilot-aircraft structural model.

For the modeling, the modified pilot-aircraft structural model considered in [11] and [12] was used. This model is based on the well known Hess structural model [13]. The input signal was simulated as a random signal with the spectral density $S_{ii} = \frac{K^2}{(\omega_i{}^2+0.5^2)}$ and variance $\sigma^2 = 4\ cm^2$. As for the model of the vehicle, a middle- sized helicopter was used. The frequency response characteristics of its dynamics are shown on figure 2.

The selection of the parameters K_L, T_L, K_n, T_n, of the pilot-aircraft structural model was done according to the parameter optimization procedure considered briefly in [11] and [12]. As for the parameters $\tau, \xi_{fs}, \omega_{fs}$, they are the same as given in [11] and [12] and are equal to 0.2 sec, 0.25 and 14 rad/sec respectively. The results of the mathematical modeling for the helicopter with feedback and helicopter with feedback + inverse dynamics laws are given on figure 7. The variance of error can be seen with its components: The variances of error correlated with the input signal $\sigma_{e_i}{}^2$, and with the pilot's remnant $\sigma_{e_n}{}^2$. Additionally, the frequency response characteristic of the pilot-vehicle closed loop system is shown.

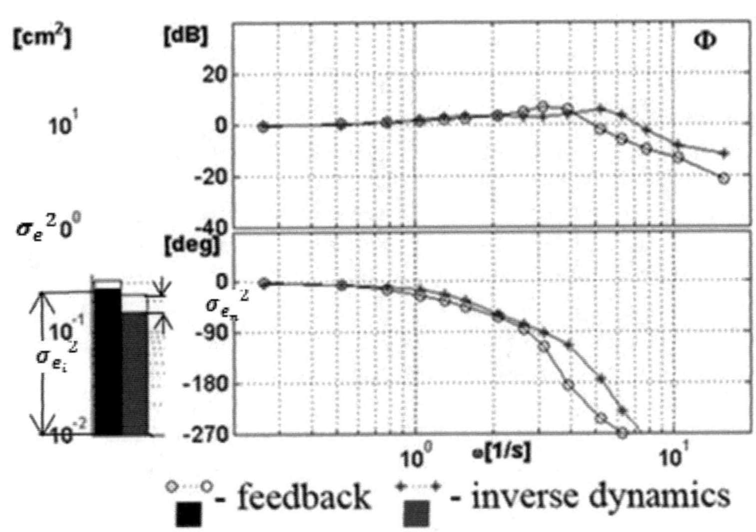

Figure 7. Result of mathematical modeling.

It is seen that the inverse dynamics allow to decrease the variance of error up to 30 %, increase the bandwidth of the closed loop system up to 13 % and decrease the resonant peak up to 12.6 %.

The results of mathematical modeling in case of uncertainties can be seen on figure 8. Uncertainties

IOP Conf. Series: Materials Science and Engineering **868** (2020) 012016 doi:10.1088/1757-899X/868/1/012016

were simulated by changing stability derivatives of the mathematical model of the helicopter up to 40%. It was also simulated by changing the poles of the system, and the results were the same. It can be seen that it was necessary to add the reference model in order to suppress the effects of uncertainties.

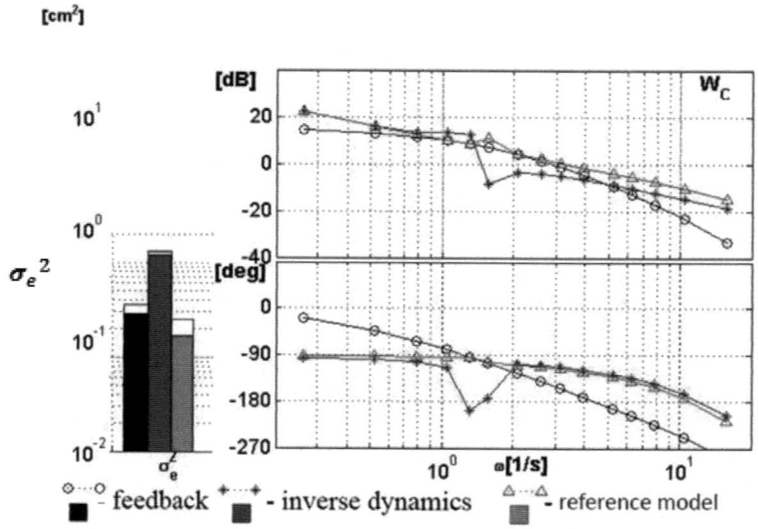

Figure 8. Result of mathematical modeling with modeling uncertainties.

The longitudinal transfer functions of the helicopter are used in the mathematical modeling of the pilot in order the get the results shown above. Additional experiments were carried out with the full linear and coupled model of the helicopter in a ground based simulator, and the results were very close to those obtained above

5.3. Results of ground based simulations

Experiments were carried out on one of the simulators of the pilot vehicle laboratory of MAI. The pilot had to carry out a compensatory tracking task. The experiments were carried out with the same dynamics used in the mathematical modeling. The input signal was the polyharmonic signal $i(t) = \sum_{k=0}^{15} A_k \cos \omega_k t$ where the amplitudes A_k and ω_k were selected according to the technique given in [13].

Figure 9 shows the results of the experiments. In addition to the variance of error which is decreased by up to 80% on figure 9.a, analyses of the pilot performances showed a lower phase lag, which can be translated as less pilot workload during the tracking task.

The variance of error in case of 40 % uncertainties in the knowledge of the aircraft model is shown in figure 9b. It can be seen that when the inverse dynamics are used alone, the error variance is large compared to the use of the feedback alone. It is therefore necessary to use the reference model in order to suppress the uncertainties. It is seen that the use of the reference model allows to decrease the variance of error up to 60 % compared to the feedback alone. The results obtained in ground based simulations showed that the variance of the error decreased even more than it did when running the mathematical modeling simulations.

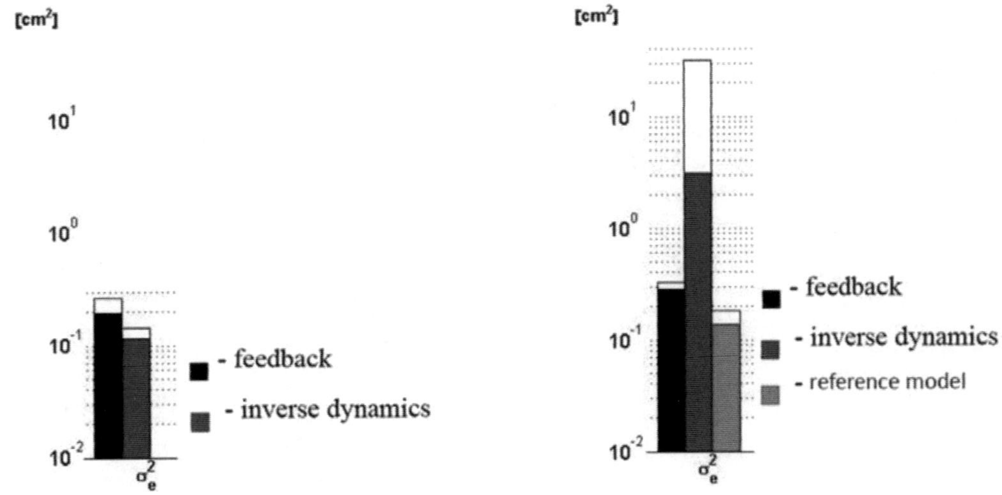

(a) (b)

Figure 9. Results of experiments: (a) experiment without uncertainties; (b) experiment with modeling uncertainties.

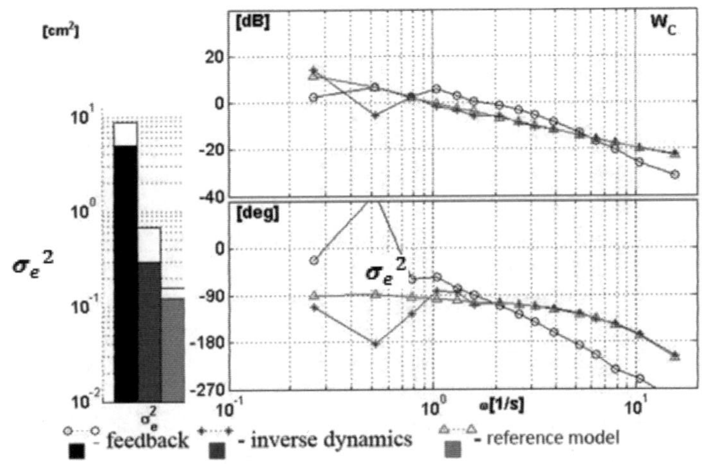

Figure 10. Experiments with atmospheric turbulence.

Figure 10 shows the responses when atmospheric turbulence was added to the simulation (disturbance is added on figure 2 as dist). The turbulence was simulated with a set of harmonics with frequencies non-orthogonal to the input signal. The results showed that the use of the inverse dynamics alone provided good flying qualities when the turbulence had small amplitudes and frequencies.

As the amplitude and the frequency increased, the performances got worse, and eventually, the aircraft was uncontrollable after reaching some threshold, due to periodic instability. From figure 10 it is seen that the turbulence influences considerably on the controlled element dynamics equipped with the feedback only, and the one with the inverse dynamics only. This influence is associated with the couplings between the longitudinal and lateral channels. As for the version equipped with the inverse dynamics and the reference model, it stays practically insensitive to the influence of the turbulence and the variance of error is decreased considerably compared to the feedback alone, or the inverse dynamics alone.

6. Conclusion

A technique was proposed for the integration of the inverse dynamics and reference model. This technique allows to suppress the modeling uncertainties, disturbances, and dynamics couplings. Mathematical modeling demonstrated that the proposed technique allows a considerable improvement of performances and other parameters of the pilot-aircraft system. These results were confirmed with ground based simulations. These simulations demonstrated that the flight control system developed with help of the proposed technique allowed to improve performances and robustness, suppressing the effects of atmospheric turbulence, and decreasing the variance of the error up to 45% when there is no uncertainties, and up to 60% in case of modeling uncertainties. Additionally, it helps decrease the pilot workload considerably, increase the bandwidth of the closed loop system and decrease the resonance peak of the closed loop system, thus, enhancing the flying qualities of the helicopter, placing them in level 1 when measured using the Cooper Harper scale.

Acknowledgments
This work was supported by the Ministry of Education and Science of Russian Federation (project No. 9.7170.2017/8.9).

References

[1] Menon R 1993 Nonlinear Command Augmentation System for a High Performance Aircraft. *Proc. AIAA Conf. on Guidance, Navigation and Control* (Monterey, CA, U.S.A.) doi:10.2514/6.1993-3777
[2] Padhi R, Rao P N, Goyal S and Balakrishnan S N 2007 Command tracking in high performance aircrafts: A new dynamic inversion design. *IFAC Proc.* **4(7)** 79 doi:10.3182/20070625-5-FR-2916.00015
[3] Horn J F 2019 Non-Linear Dynamic Inversion Control Design for Rotorcraft. *Aerospace* **6(3)** 38 doi:10.3390/aerospace6030038
[4] Ansari U and Bajodah A H 2018 Robust generalized dynamic inversion based control of autonomous underwater vehicles. *Proc. of the Institution of Mechanical Engineers, Part M: J. of Eng. for the Maritime Environment* **232(4)** 434 doi: 10.1177/1475090217708640
[5] Peng C, Han C, Zou J and Zhang G 2016 H_∞ Optimal Inversion Feedforward and Robust Feedback Based 2DOF Control Approach for High Speed-Precision Positioning Systems. *Journal of Control Science and Engineering* 7256039 doi:10.1155/2016/7256039
[6] Miller C J 2011 Nonlinear Dynamic Inversion Baseline Control Law: Architecture and Performance Prediction. *Proc. of the 2011 AIAA Conf. Guidance, Navigation and Control* (Portland, OR, USA) doi:10.2514/6.2011-6467
[7] Buchholz J J and Grunhagen W 2008 *Inversion Impossible?*(DLR Braunschweig)
[8] Shinners S M 1998 *Second Edition. Modern Control System Theory and Design* (New York, Wiley-Interscience)
[9] Ogata K 1997 *Prentice-Hall Modern Control Engineering, 3rd Edition* (New Jersey, *Prentice-Hall, Inc.*)
[10] Baskett, Barry 2000 *Aeronautical design standard, performance specification, handling qualities requirement for military rotorcraft* (US Military)
[11] Efremov A V, Aleksandrov V V, Efremov E V and Vukolov M V 2018 The influence of different types of inceptors and their characteristics on pilot-aircraft system. *IFAC Proc.* **51(34)** 372 doi: 10.1016/j.ifacol.2019.01.013
[12] Efremov AV, Efremov E V, Mbikayi Z and Irgaleev IKh 2019 Influence of Inceptors on Pilot-Aircraft System Characteristics and Flying Qualities. *IOP Conference Series: Materials Science and Engineering* **476** 012010 doi: 10.1088/1757-899X/476/1/012010
[13] Hess R 1979 Structural model of the adaptive human pilot. *Journal of Guidance and Control* **3(5)** doi: 10.2514/6.1979-1784

IOP Conf. Series: Materials Science and Engineering **868** (2020) 012017 doi:10.1088/1757-899X/868/1/012017

Analysis of the power part architecture for short-medium-range aircraft control system with local hydraulic systems by reliability criterion

S Postnikov, A Trofimov* and D Smagin

Department of Aircraft Design and Certification, *Moscow Aviation Institute* (National Research University), 4 Volokolamskoye Highway, Moscow 125993, Russian Federation.

*E-mail: trofimovaa@mai.ru

Abstract. This article provides a comparative analysis of the traditional control system power part for a short-medium-range aircraft with a control system power part with local HS. Comparison is based on the criterion of reliability. As initial data, statistics on the probability of failures of the electrical system and hydraulic system and actuators. The considered architecture of the control system power part with local HS fully complies with the required standards for the control system reliability given in regulatory documents, such as AR-25 and CS-25. The considered architecture of the control system power part with local HS is commensurate with the reliability of a typical version of the control system power part architecture. The considered architecture of the control system power part with local HS fully complies with the required standards for the control system reliability given in regulatory documents, such as AR-25 and CS-25. The considered architecture of the control system power part with local HS is commensurate with the reliability of a typical version of the control system power part architecture and in some cases it is an order of magnitude superior.

1. Introduction

When designing new types of civilian aircraft, the developers are faced with the task of increasing fuel efficiency and reducing operational maintenance costs to maintain the competitiveness of the aircraft. The usage of complex digital electronics as part of on-board equipment as well as the development of new design solutions is one of the main principals for increasing fuel efficiency and reducing costs. Also, the main task in the development of aviation systems is to achieve a given level of reliability. Analysis of newly developed systems by the criterion of reliability is an obligatory stage of design. The requirements of ARP 4754 oblige the developer from the very early design stage to prepare materials for the analysis of system fail-safety. Reliability assessment is used not only to confirm that the system meets the requirements of certification bodies in terms of reliability, but also for maintenance planning, calculation of parameters such as mean time between failures (MTBF) or average cycles between failures (MCBF) [1].

At this article a control system (CS) for a short-medium-range aircraft (SMRA) is considered as an object of study. At the moment, most aircraft of this dimension use three centralized hydraulic systems, and electro-hydraulic servo actuators are used as surface actuators. An example of CS typical version is presented in figure 1.

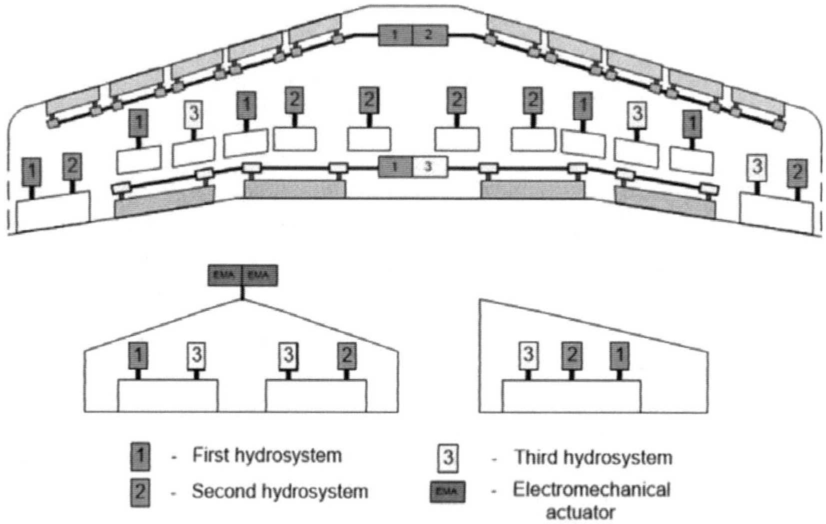

Figure 1. Typical architecture CS.

As an alternative, in this article the control system architecture with three distributed local hydraulic systems (HS) is considered. The use of local hydraulic systems will reduce the length of pressure and drain pipelines and, accordingly, reduce the pipelines weight [2], simplify maintenance and reduce power take-off from aircraft engines [3].

The first local HS is located in the left wing console and is connected to a hydraulic pump mounted on the left engine. The second local HS is located in the right wing console and is connected to a hydraulic pump mounted on the right engine. The third local HS is located in the back of the fuselage and receives hydraulic energy from a pumping station located in the back of the fuselage, as well as in the case of an abnormal situation, from a ram air turbine (RAT).

The block diagram of control system with local HSs is shown in figure 2.

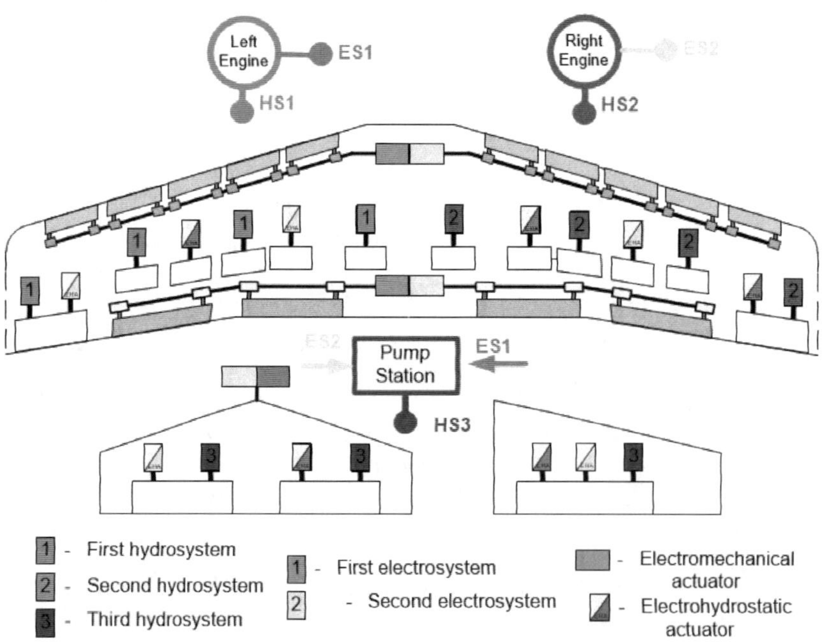

Figure 2. CS with local HSs.

In the proposed architecture version of the control system power part, besides electro-hydraulic servo actuators powered by the corresponding local HSs, EHA powered by two independent power supply systems are used. The use of EHA allows to reduce the control system energy consumption [4], to scale back the number of HSs [5], to simplify HSs installation on aircraft, to simplify HSs testing, as well as to reduce the time of maintenance and operating costs of HSs [2].

Currently, various versions of the control system architecture are offered, for example, TsAGI experts proposed four variants of the control system architecture as part of the "Safety & Energy Efficiency Research on Advanced More Electrical Flight Control Actuation Systems for Short / Middle Range Passenger Aircraft" [6]. These options are based on the use of EMA drives to control ailerons. At the moment, the main drawback of EMA drives is the lack of damping mode. An alternative to EMA is proposed to use electrohydrostatic actuator (EHA). Airbus Corporation engineers also preferred EHA instead EMA for controlling internal ailerons A350 XWB [7]. Also, at the moment, EMA is a more expensive drive compared to hydraulic drives [8], but it is advisable to make a final cost estimate only at the level of the entire aircraft life cycle.

It is worth noting that the assumption is made at this work that roll control is carried out by two surfaces of ailerons and three external surfaces of spoilers on each wing console, all surfaces of spoilers and ground spoilers (GS) participate in braking. The use of interceptors in roll control is a well-known solution and improves not only dynamic characteristics, but also increases the reliability of roll control [1].

2. Calculations

Reliability analysis, as a rule, by the manufacturer using the Fault Tree Analysis (FTA) method built on the basis of the system architecture. In this study, fault trees are replaced by equivalent mathematical expressions.

Based on the data provided in open sources, the average failure rate indicators of HS, ES, and also the main elements of the control system were calculated.

The average probability of one aircraft hydraulic system failure is $4.06 \cdot 10^{-5}$ [9–12].

The average probability of one aircraft power supply system failure $1 \cdot 10^{-4}$ [9–11].

The average failure probability of EHSA is $1.38 \cdot 10^{-5}$ [7, 9, 12].

The average failure probability of EHA is $3.27 \cdot 10^{-5}$ [13, 14].

The average failure probability of EMA is $7.6 \cdot 10^{-5}$ [15, 16].

The average failure probability of high lift power drive unit (PDU) for two electric channels is expertly estimated to be $9 \cdot 10^{-5}$, and for two hydraulic channels $2.7 \cdot 10^{-5}$.

In the framework of this work, failures of the “Jamming” type were not considered as a common cause of control loss of one surface. The SMRA dimension aircraft control system reliability is considered as the sum failure rate for each control channel (roll, pitch, yaw, braking), high lift, THSA and a complete failure (failure of three control channels). Also, as an assumption, the calculation does not take into account the main digital part architecture of the control system. Based on the initial data, the sum failure rate for each channel was obtained, taking into account the considered control system architecture variant, shown in Figure 2. The sum failure rate of the roll, pitch or yaw channel indicates the reliability of the hydraulic and electrical systems in general, the electro-hydraulic servo actuator and electro-hydrostatic actuators involved in roll, pitch or yaw control, respectively.

3. Results and discussion

The main result of this work is to determine the probability of failure of control channels, as well as an analysis of compliance with the power variant of the architecture of the remote control system of aviation rules.

The sum failure rate is:

$$\lambda_{\text{roll channel}}=(\lambda_{\text{HS1}}+(\lambda_{\text{left out.}}^{\text{EHSA ail.}}\cdot\lambda_{\text{left wing}}^{\text{EHSA sp.4}}\cdot\lambda_{\text{left wing}}^{\text{EHSA sp.2}}))\cdot(\lambda_{\text{HS2}}\cdot(\lambda_{\text{right out.}}^{\text{EHSA ail.}}\cdot\lambda_{\text{right wing}}^{\text{EHSA sp.4}}\cdot\lambda_{\text{right wing}}^{\text{EHSA sp.2}}))\cdot \left(\lambda_{\text{ES1}}+\left(\lambda_{\text{right in.}}^{\text{EHA ail.}}\cdot\lambda_{\text{left wing}}^{\text{EHA sp.3}}\right)\right)\cdot\left(\lambda_{\text{ES2}}+\left(\lambda_{\text{left in.}}^{\text{EHA ail.}}\cdot\lambda_{\text{right wing}}^{\text{EHA sp.3}}\right)\right)=9.58\cdot10^{-13} \quad (1)$$

$$\lambda_{\text{pitch channel}}=(\lambda_{\text{HS3}}+(\lambda_{\text{left in.}}^{\text{EHSA elev.}}\cdot\lambda_{\text{right out.}}^{\text{EHSA elev.}}))\cdot(\lambda_{\text{ES2}}+\lambda_{\text{left out.}}^{\text{EHA elev.}})\cdot(\lambda_{\text{ES1}}+\lambda_{\text{right in.}}^{\text{EHA elev.}})=7.15\cdot10^{-13} \quad (2)$$

$$\lambda_{\text{yaw channel}}=(\lambda_{\text{HS3}}+\lambda_{\text{upper}}^{\text{EHSA rud.}})\cdot(\lambda_{\text{ES2}}+\lambda_{\text{middle}}^{\text{EHA rud.}})\cdot(\lambda_{\text{ES1}}+\lambda_{\text{lower}}^{\text{EHA rud.}})=9.58\cdot10^{-13} \quad (3)$$

The proposed architecture with local HSs (figure 2) leads to the organization of control of the first and third section of spoilers from EHA, the control of the second and fourth section of spoilers from EHSA. Ground spoilers in this architecture are controlled by EHA.

$$\lambda_{\text{braking}}=(\lambda_{\text{HS1}}+(\lambda_{\text{left wing}}^{\text{EHSA sp.4}}\cdot\lambda_{\text{left wing}}^{\text{EHSA sp.2}}\cdot\lambda_{\text{left wing}}^{\text{EHSA gsp.}}))\cdot(\lambda_{\text{HS2}}+(\lambda_{\text{right wing}}^{\text{EHSA sp.4}}\cdot\lambda_{\text{right wing}}^{\text{EHSA sp.2}}\cdot\lambda_{\text{right wing}}^{\text{EHSA gsp.}}))\cdot (\lambda_{\text{ES1}}+(\lambda_{\text{left wing}}^{\text{EHA sp.3}}\cdot\lambda_{\text{right wing}}^{\text{EHA sp.1}}))\cdot(\lambda_{\text{ES2}}+(\lambda_{\text{right wing}}^{\text{EHA sp.3}}\cdot\lambda_{\text{left wing}}^{\text{EHA sp.1}}))=1.65\cdot10^{-17} \quad (4)$$

The PDU of the flaps and slats control system is controlled by two electric channels.

$$\lambda_{\text{high lift}}=((\lambda_{\text{slats}}^{\text{PDU}}\cdot\lambda_{\text{flaps}}^{\text{PDU}})+(\lambda_{\text{ES1}}\cdot\lambda_{\text{ES2}}))=1.81\cdot10^{-8} \quad (5)$$

The THSA control is also carried out through two electrical channels.

$$\lambda_{\text{THSA}}=(\lambda_{\text{stab. actuator}}+(\lambda_{\text{ES1}}\cdot\lambda_{\text{ES2}})=1.58\cdot10^{-8} \quad (6)$$

The complete failure of the control system calculation does not take into account the failures of high lift and THSA, since in the conditions of failure of all surface actuators, safe completion of the flight with the working high lift and THSA is impossible.

$$\lambda_{\text{complete failure}}=((\lambda_{\text{HS1}}+(\lambda_{\text{left out.}}^{\text{EHSA ail.}}\cdot\lambda_{\text{left wing}}^{\text{EHSA sp.4}}\cdot\lambda_{\text{left wing}}^{\text{EHSA sp.2}}\cdot\lambda_{\text{left wing}}^{\text{EHSA gsp.}}))\cdot (\lambda_{\text{HS2}}+(\lambda_{\text{right out.}}^{\text{EHSA ail.}}\cdot\lambda_{\text{right wing}}^{\text{EHSA sp.4}}\cdot\lambda_{\text{right wing}}^{\text{EHSA sp.2}}\cdot\lambda_{\text{right wing}}^{\text{EHSA gsp.}}))\cdot(\lambda_{\text{HS3}}+(\lambda_{\text{left in.}}^{\text{EHSA elev.}}\cdot\lambda_{\text{right out.}}^{\text{EHSA elev.}}\cdot\lambda_{\text{upper}}^{\text{EHSA rud.}}))\cdot (\lambda_{\text{ES1}}+(\lambda_{\text{right in.}}^{\text{EHA ail.}}\cdot\lambda_{\text{right wing}}^{\text{EHA sp.1}}\cdot\lambda_{\text{lower}}^{\text{EHA rud.}}\cdot\lambda_{\text{right in.}}^{\text{EHA elev.}}\cdot\lambda_{\text{left wing}}^{\text{EHA sp.3}}))\cdot (\lambda_{\text{ES2}}+(\lambda_{\text{right wing}}^{\text{EHA sp.3.}}\cdot\lambda_{\text{left in.}}^{\text{EHA ail.}}\cdot\lambda_{\text{left out.}}^{\text{EHA elev.}}\cdot\lambda_{\text{left wing}}^{\text{EHA sp.1}}\cdot\lambda_{\text{middle}}^{\text{EHA rud.}})))=6.69\cdot10^{-22} \quad (7)$$

The failure rate calculation of each channel demonstrates that the considered failure conditions leading to the emergence of a catastrophic situation are estimated as extremely improbable events and do not occur due to a single failure of one of the system elements. Thus, the considered architecture of the control system fully meets the reliability requirements presented in Aviation rules number 25 (AR-25 or CS-25) section A-0 [17]. The final conclusion on the power part architecture of the control system with local hydraulic systems can be made by comparing it with a typical version of the control system power part shown in Figure 1. The sum failure rate of the roll, pitch and yaw channel of a typical variant is:

$$\lambda_{\text{roll channel}}^{\text{typical}}=(\lambda_{\text{HS1}}+(\lambda_{\text{left out.}}^{\text{EHSA ail.}}\cdot\lambda_{\text{left wing}}^{\text{EHSA sp.4}}\cdot\lambda_{\text{left wing}}^{\text{EHSA sp.2}}\cdot\lambda_{\text{right wing}}^{\text{EHSA sp.4}}\cdot\lambda_{\text{right wing}}^{\text{EHSA sp.2}})\cdot (\lambda_{\text{HS2}}+(\lambda_{\text{right out.}}^{\text{EHSA ail.}}\cdot\lambda_{\text{left in.}}^{\text{EHSA ail.}}))\cdot(\lambda_{\text{HS3}}+(\lambda_{\text{right in.}}^{\text{EHSA ail.}}\cdot\lambda_{\text{left wing}}^{\text{EHSA sp.3}}\cdot\lambda_{\text{right wing}}^{\text{EHSA sp.3}}))=6.69\cdot10^{-14} \quad (8)$$

$$\lambda_{\text{pitch channel}}^{\text{typical}}=(\lambda_{\text{HS1}}+\lambda_{\text{left out.}}^{\text{EHSA elev.}})\cdot(\lambda_{\text{HS2}}+\lambda_{\text{right out.}}^{\text{EHSA elev.}})\cdot(\lambda_{\text{HS3}}+(\lambda_{\text{left in.}}^{\text{EHSA elev.}}\cdot\lambda_{\text{right in.}}^{\text{EHSA elev.}}))=1.2\cdot10^{-13} \quad (9)$$

$$\lambda_{\text{yaw channel}}^{\text{typical}}=(\lambda_{\text{HS1}}+\lambda_{\text{upper}}^{\text{EHSA rud.}})\cdot(\lambda_{\text{HS2}}+\lambda_{\text{middle}}^{\text{EHSA rud.}})\cdot(\lambda_{\text{HS3}}+\lambda_{\text{lower}}^{\text{EHSA rud.}})=1.61\cdot10^{-13} \quad (10)$$

The calculation of the braking channel is carried out according to the example described for the power part architecture CS with local HS.

$$\lambda_{\text{braking}}^{\text{typical}}=(\lambda_{\text{HS1}}+(\lambda_{\text{left wing}}^{\text{EHSA sp.4}}\cdot\lambda_{\text{left wing}}^{\text{EHSA sp.2}}\cdot\lambda_{\text{right wing}}^{\text{EHSA sp.4}}\cdot\lambda_{\text{right wing}}^{\text{EHSA sp.2}}))\cdot (\lambda_{\text{HS1}}+(\lambda_{\text{left wing}}^{\text{EHSA sp.1}}\cdot\lambda_{\text{left wing}}^{\text{EHSA sp.1}}\cdot\lambda_{\text{right wing}}^{\text{EHSA gsp.}}\cdot\lambda_{\text{right wing}}^{\text{EHSA gsp.}}))\cdot(\lambda_{\text{HS1}}+(\lambda_{\text{left wing}}^{\text{EHSA sp.3}}\cdot\lambda_{\text{right wing}}^{\text{EHSA sp.3}})=6.69\cdot10^{-14} \quad (11)$$

The PDU of the flaps and slats control system is controlled by two hydraulic channels.

$$\lambda_{\text{high lift}}^{\text{typical}}=((\lambda_{\text{slats}}^{\text{PDU}}\cdot\lambda_{\text{flaps}}^{\text{PDU}})+(\lambda_{\text{HS1}}\cdot\lambda_{\text{HS2}}\cdot\lambda_{\text{HS3}}))=7.29\cdot10^{-10} \quad (12)$$

The THSA calculation is carried out according to the example described for the power part architecture CS with local HS.

$$\lambda_{THSA}=(\lambda_{stab.\ actuator}+(\lambda_{ES1}\cdot\lambda_{ES2}))=1{,}58\cdot10^{-8} \quad (13)$$

The control system complete failure of a typical variant is calculated according to the example described for the power part architecture CS with local HS.

$$\lambda^{typical}_{completefailure}=((\lambda_{HS1}+(\lambda^{EHSA\ ail.}_{left\ out.}\cdot\lambda^{EHSA\ sp.4}_{left\ wing}\cdot\lambda^{EHSA\ sp.2}_{left\ wing}\cdot\lambda^{EHSA\ sp.4}_{right\ wing}\cdot\lambda^{EHSA\ sp.2}_{right\ wing}\cdot\lambda^{EHSA\ elev.}_{left\ out.}\cdot\lambda^{EHSA\ rud.}_{upper})\cdot \quad (14)$$

$$(\lambda_{HS2}+(\lambda^{EHSA\ ail.}_{right\ out.}\cdot\lambda^{EHSA\ ail.}_{left\ in.}\cdot\lambda^{EHSA\ elev.}_{right\ out.}\cdot\lambda^{EHSA\ rud.}_{middle}\cdot\lambda^{EHSA\ sp.1}_{left\ wing}\cdot\lambda^{EHSA\ sp.1}_{left\ wing}\cdot\lambda^{EHSA\ gsp.}_{right\ wing}\cdot\lambda^{EHSA\ gsp.}_{right\ wing}))\cdot$$

$$(\lambda_{HS3}+(\lambda^{EHSA\ ail.}_{right\ in.}\cdot\lambda^{EHSA\ sp.3}_{left\ wing}\cdot\lambda^{EHSA\ sp.3}_{right\ wing}\cdot\lambda^{EHSA\ elev.}_{left\ in.}\cdot\lambda^{EHSA\ elev.}_{right\ in.}\cdot\lambda^{EHSA\ rud.}_{lower}))=6.69\cdot10^{-14}$$

More clearly, the calculation results for a typical version CS and CS with local HS are presented in table 1.

Table 1. The calculation results for a typical version CS and CS with local HS.

Control channel	Typical case	Local HS case
Roll channel	$6.69\cdot10^{-14}$	$1.65\cdot10^{-17}$
Pitch channel	$1.2\cdot10^{-13}$	$7.15\cdot10^{-13}$
Yaw channel	$1.61\cdot10^{-13}$	$9.58\cdot10^{-13}$
Braking	$6.69\cdot10^{-14}$	$1.65\cdot10^{-17}$
High lift	$7.29\cdot10^{-10}$	$1.81\cdot10^{-8}$
THSA	$1.58\cdot10^{-8}$	$1.58\cdot10^{-8}$
Complete failure CS	$6.69\cdot10^{-14}$	$6.69\cdot10^{-22}$

As can be seen from table 1, the power part reliability with local HS is comparable in reliability with the typical architecture, and in some cases it is an order of magnitude superior. The increase in reliability during roll channel and in the braking channel is associated with an increase in the number of energy sources for power surface actuators. It is also worth noting a not significant decrease in the reliability of high lift system, this is due to a statistically lower reliability of the high lift electric drives.

The results obtained in this study are comparable in value with other well-known architecture options for the power part of the control system [7, 18], which allows us to conclude that the presented option is competitive

4. Conclusion

The average probability of failure of the control system elements obtained in this study can be used in the development of various versions of the control system architecture, as well as the data obtained can be used in the development of the Preliminary System Safety Assessment (PSSA) and other reporting documents to confirm the reliability of the system. Since the obtained values are based on a large amount of statistical data, we consider them to be more relevant when assessing the architecture of a control system according to the reliability criterion in the early stages of design. The considered architecture of the control system power part with local HS fully complies with the required standards for the control system reliability given in regulatory documents, such as AR-25 and CS-25. The considered architecture of the control system power part with local HS is commensurate with the reliability of a typical version of the control system power part architecture and in some cases it is an order of magnitude superior. The increase in reliability during roll channel and in the braking channel is associated with an increase in the number of energy sources for power surface actuators. The considered architecture requires further comparison with the standard version and optimization by mass and energy efficiency criterion since the implementation of the option with local hydraulic units will reduce the capacity of pumping stations compared to the standard version by reducing the number of consumers and exclude the laying of hydraulic pipes along the fuselage. These measures reduce the weight and increase the reliability of the HS by reducing the number of elements, but lead to an increase in the required power of the power supply system and generators.

IOP Conf. Series: Materials Science and Engineering **868** (2020) 012017 doi:10.1088/1757-899X/868/1/012017

References

[1] Beaverstock C, Maheri A, Richardson T, Lowenberg M and Isikveren A 2009 Methods for Conceptual Flight Control System Design. *47th AIAA Aerospace Sciences Meeting including The New Horizons Forum and Aerospace Exposition* (Orlando, USA) doi: 10.2514/6.2009-1620

[2] Shumilov I S 2014 Hydraulic Actuators with Autonomous Hydraulic Supply for the Mainline Aircraft. *Science and Education Bauman MSTU* **8** 139 doi:10.7463/0814.0724446

[3] Erofeev E V, Kuvshinov V M, Steblinkin A I, Terehov R I and Khaletsky L V 2016 Local hydraulic system with adaptive pressure control for long-haul aircraft. *Conf.: Recent Adv. in Aeropsace Actuation Systems and Compontents (R3ASC)* (France, Toulouse) pp 67-74

[4] Postnikov S E, Trofimov A A and Baikov S V 2018 Architecture options estimate for the near-medium-haul aircraft control system by the reliability, mass and power consumption criteria. *Aerospace Systems* **2(1)** 33 doi:10.18698/2308-6033-2017-12-1711

[5] Ermakov S A, Karev V I and Mitrichenko A.N 2010 Systems of surface actuators of passenger aircraft, their development and ensuring trouble-free operation. *Aerospace MAI Journal*. **17(3)** 18 [in Russian]

[6] Kuvshinov V, Khaletsky L, Steblinkin A, Erofeev E and Skryabin A 2015 Safety & Energy Efficiency Research on Advanced More Electrical Flight Control Actuation Systems for Short/Middle Range Passenger Aircraft. *Conf. More Electric AircraftAt* (Toulouse, France) pp 1-3

[7] Aircraft EHA, MOOG Company, available at: https://www.moog.com/products/actuators-servoactuators/actuation-technologies/electrohydrostatic/aircraft-eha.html

[8] Jianming L, Zhiyuan , Yuping H and Zhiguo L 2016 Review of electromechanical actuation system for more electric aircraft. *IEEE/CSAA International Conf. on Aircraft Utility Systems* (Beijing China) pp 490-497 doi: 10.1109 / AUS.2016.7748100

[9] Baikov S V, Bliznova T B and Obolenskiy Y G 2012 Trends in the development of the architecture of the executive part of the modern aircraft control system. *Scientific Herald State Research Institute of Civil Aviation Collection of Scientific Paper* **2** 15 [in Russian]

[10] Alekseenkov A S 2014 *Improvement of dynamic properties and study of working processes of the aircraft steering hydraulic actuator with combined speed control with increasing external load,* PhD thesis, Moscow [in Russian]

[11] Redko P G 2002 *Increased reliability and improved performance of electro-hydraulic actuators* (Moscow: MSTU Stankin Publishing Center Janus-K Moscow) p 57 [in Russian]

[12] Mare J C 2016 *Aerospace Actuators: Needs, Reliability and Hydraulic Power Solutions* (Wiley-ISTE) p 51 doi: 10.1002/9781119307662

[13] Xue L 2010 *Actuation Technology for Flight Control System on Civil Aircraft.* MSc thesis, Cranfield University

[14] Weiss J 2014 Control Actuation Reliability and Redundancy for Long Duration Underwater Vehicle Missions with High Value Payloads, *Underwater Intervention 2014* pp 3-4

[15] Hyssain Y, Burrow S, Henson L and Keogh P 2016 Benefits Analysis of Prognostics and Health Monitoring to Aircraft Maintenance using System Dynamics. *3rd European Conf. of the PHM Society - PHME16* (Bilbao, Spain) p 11

[16] Actuation-EMA-Series, CIRCOR Aerospace, available at: http://www.circoraerospace.com/downloads/bodet/Bodet-Electromechanical-Actuation-EMA-Series.pdf

[17] Certification Specification. Part 25 (CS-25). *The norms of flight of the aircraft of the transport category. Section A-0 – "General requirements for the airflow of the aircraft when functional system failure"* pp 14-16

[18] Postnikov S E, Trofimov A A and Smagin D I 2017 Options of control system architectures for short-medium haul aircraft. *Eng. J. Sci. Innovat.* **12**·12 doi: 10.18698/2308-6033-2017-12-1711

IOP Conf. Series: Materials Science and Engineering **868** (2020) 012018 doi:10.1088/1757-899X/868/1/012018

Experimental studies of scattering indicatrixes of radar absorbing materials

M Snastin*, E Dobychina and A Solod

Scientific and Production Center of Radio Information Metrology, *Moscow Aviation Institute* (National Research University), 4 Volokolamskoe shosse, Moscow 125993, Russian Federation

*E-mail: mexanizmys@ya.ru

Abstract. There is some kind of materials widely used in microwave engineering that known as radar absorbers. Whether it is the aircraft coating, the anechoic chamber interior or waveguide feed load. Anechoic chamber absorbers are typically characterized by reflectivity at normal incidence of electromagnetic wave. The study of the angular dependence of reflectivity for several absorber samples and the appraising of the impact of their combination on the anechoic chamber performance are the main aims of present work. For this purpose the experimental studies were carried out in an antenna measuring and computing complex based on the anechoic chamber. Obtained results may be useful in the anechoic chamber design and construction stages as well as the modernization of existing ones. It can also be used for modelling of the internal electromagnetic field distribution of indoor field test ranges.

1. Introduction

An anechoic chamber (AEC) allows you to reduce or completely eliminate field tests of antenna systems, which leads to significant costs and time savings for development of complex radio equipment. However, high requirements are imposed on the AEC characteristics to ensure small values of the measurement error [1]. Anechoic chambers are characterized by its dimensions, shield performance and reflectivity level.

AEC is a closed electromagnetic shield, the inner surfaces of which has a regular or special shape and is fully or partially covered with a radar absorber material (RAM). The reflectivity level is primarily determined by the properties of the RAM applied as a part of AEC [2]. Absorbers are typically characterized by reflectivity at normal incidence of electromagnetic wave. But in most applications of absorber in indoor ranges, it is important to know bi-static reflectivity for oblique angles of incidence [3].

Many scientific articles discuss the absorbing and scattering properties of various RAMs, where theoretical relations are derived by various approximate methods for representing complex processes occurring in such materials. These are polynomial equations for predicting reflected energy [4, 5], scaling models [6], developing of electrodynamic models of processes in materials [7, 8], and usage of low-frequency models in case of pyramidal arrays [9]. Also are known the experimental studies of absorbers based on novel materials [10, 11].

This paper outlines comparative tests of several types of RAM. All measurements were carried out in free space with suitable equipment in an antenna measuring and computing complex based on the

IOP Conf. Series: Materials Science and Engineering **868** (2020) 012018 doi:10.1088/1757-899X/868/1/012018

anechoic chamber. The purpose of the research is to measure the transmission and reflection characteristics of various absorbers in a wide frequency range.

2. Radar absorber materials

Regarding functionality, the absorber materials are divided into construction materials and absorber coats. Construction absorbers are used as mounted solution with the aim of backscatter reduction. The absorber coatings in its turn are formed on a reflective surface in the form of paint or foam layers. The main difference between construction RAM and coatings is that the last one are formed directly on the object surface, being held on it by adhesion, whereas constructive materials can be used separately.

According to the method of free space matching absorbers can be divided into three groups. There are resonant, gradient and geometrically matched absorbers. Resonant absorbers attenuate reflected radiation due to destructive interference of reflected wave from the material and reflected wave from the metal backing. These materials are pyramidal, tapered, cellular and others. It should be noted that a generalized complete classification of RAM is impossible due to the wide variety of attributes for their classification [1]. All practically significant absorbers are composite with a micro- or macro-structure.

All absorbers can also be divided into two groups by type of influence: materials that scatter radiation and those that absorb it. Structures and coatings based on radar scattering materials provide attenuation of propagating electromagnetic waves due to the effects of reflection, diffraction and refraction. This type of material absorbs radiation energy not by converting it to other forms, but due to its reallocation in the space. Thus, the reflected signal does not reach the working area and does not distort the measurement results. Among such materials one can distinguish groups of shielding, diffusion and polarizing materials. The main distinctive feature of the absorber materials is that the energy of the electromagnetic radiation entering the material is dissipated with the conversion to other types of energy, heat mainly.

Usually the real materials are multilayer structures that combine various types of functioning. Some materials act as both scattering materials and absorbers at once. For example, the matching surface of uniformly gradient pyramidal materials is capable of effectively conversing specular reflection into diffuse scattering along with absorption of electromagnetic field energy.

3. The study of RAM reflectivity

All measurements were carried out in free space with suitable equipment in an antenna measuring and computing complex based on the anechoic chamber (figure 1). Two measurement setups were established. The first one was designed to measure the reflectivity by the NRL arch method (figure 2) [10]. In front of the metal backed RAM sample (material under test, MUT) placed on base 3 there were two wideband measuring antennas 1 and 2 (figure 3a) mounted on tripods. Antennas were in the far field region relative to MUT at a distance *D*. The measuring antennas moved synchronously relative to the normal of RAM surface at an angle *Az* during the measurements. Using the PNA series vector network analyzer the transmission coefficient was measured over a wide frequency band. A pointwise normalization of measured data to a similar measurement of metal plate only without MUT placing was performed. The result is MUT reflectivity or so-called scattering indicatrix.

The measured samples had an area of about 1 m^2 and were of the following types: pyramidal 'Delta' (figure 3b), carpet 'Ternovnik' (figure 3c) and multilayer flat 'Isoterm' RAM, which is the main absorber in the applied research AEC.

The test results are presented in the form of scattering indicatrixes in the figure 4. Horizontal axis of the graphs shows the deviation angle *Az* of the each measuring antennas position in the range from 10° to 60°. Antennas transmission coefficient (red line – reflectivity from metal plate and blue one – MUT reflectivity) is plotted along the left vertical axis. Attenuation of the signal in the material is shown on the right vertical axis (black line). The graphs are shown for only two frequencies from the measured range: 3.0 GHz and 7.5 GHz.

Figure 1. Inside view of anechoic chamber.

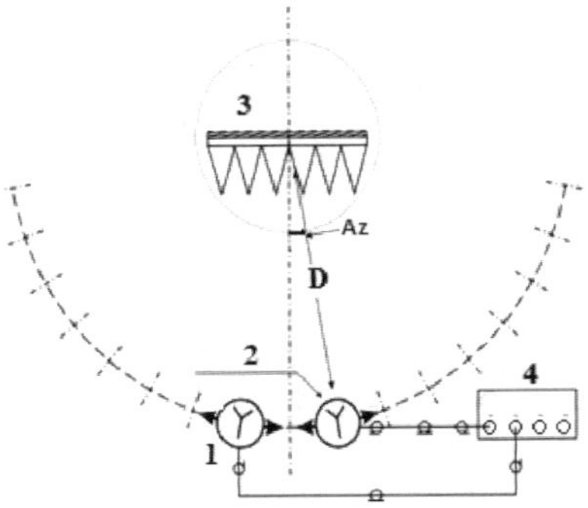

Figure 2. The setup diagram for scattering indicatrixes measurement.

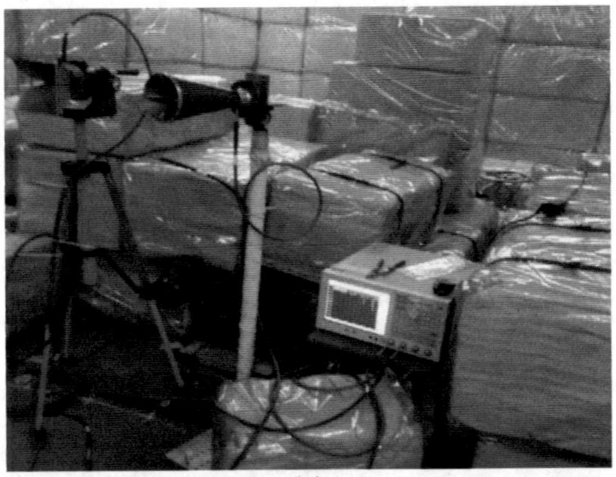

(a)

(b)

(c)

Figure 3. Measurement setup: (a) – Network analyzer and measuring antennas; (b) – 'Delta' absorber with metal backing; (c) – 'Ternovnik' absorber with metal backing.

For example in figure 4a it can be seen that the pyramidal MUT at a given frequency has an absorption of 34 dB at small incident wave angles and it almost uniformly decreases to 6 dB value at higher angles. The other measured materials behave differently but are similar to each other. At small angles the attenuation for both materials is only 20 dB as shown in figure 4c and figure 4e which is worse relative to pyramidal RAM. However, the maximum attenuation occurs at approx. 35° angle and it takes a value of 25 dB for carpet MUT and 37 dB for multilayer flat one respectively.

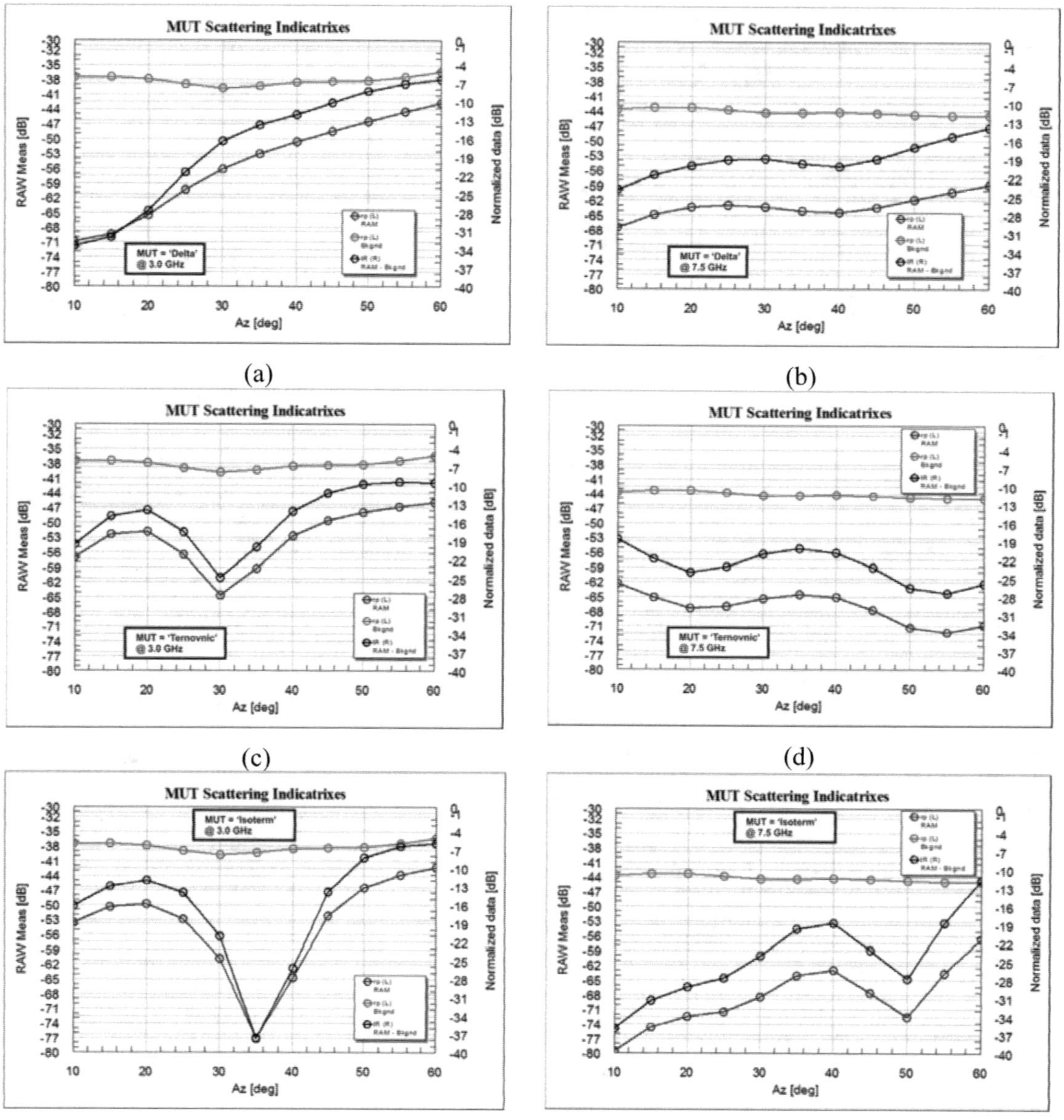

Figure 4. Measured indicatrixes: (a) – 'Delta' at 3.0 GHz; (b) – 'Delta' at 7.5 GHz; (c) – 'Ternovnic' at 3.0 GHz; (d) – 'Ternovnic' at 7.5 GHz; (e) – 'Isoterm' at 3.0 GHz; (f) – 'Isoterm' at 7.5 GHz.

At a frequency of 7.5 GHz the picture is changing. As shown in figure 4f the indicatrix of multilayer flat MUT becomes more uniform, and the maximum attenuation with a value of 35 dB is reached at small incident angles. In turn the indicatrix of carpet RAM becomes less uniform at higher frequency as shown in figure 4d. With a frequency increasing the pyramidal MUT indicatrix does not substantially change its character except its slope as shown in figure 4b.

In this experiment the absorption of the pyramidal MUT does not appear at high frequency so clearly while the reflection from the absorber boundary increases. So it can be assumed that pyramidal MUT is an absorbing material designed for relatively lower frequencies. The other two are materials of scattering type and in a wide frequency band they both demonstrate an improvement in their properties.

The disadvantage of this study is that it does not allow measuring the attenuation of an electromagnetic wave in case of normal wave incidence on material. This is not allowed by the experimental setup and continuous-wave type of measurement.

4. Measurements of RAM attenuation

The second measuring setup (figure 5) was designed to measure the absorber attenuation for normal incidence of wave. The measurement was also carried out in the AEC. Two measuring antennas 1 and 2 mounted on tripods were located opposite each other in the far field region. Then using network analyser the transmission coefficient between antennas was measured. Next, MUT was placed between antennas and transmission coefficient was measured again. Thus, the substitution method was applied to study the MUT properties and free space transmission (without installed MUT) was used as the reference measurement. A comparative analysis of obtained data was carried out.

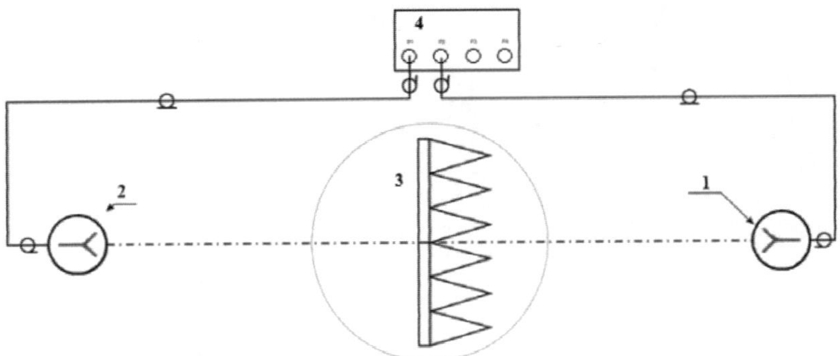

Figure 5. The diagram of MUT attenuation measurement setup.

For a mathematical description of the measurement process, let us write the transmission equation in a logarithmic representation:

$$S_{21} = G_a + G_b + CL_{THRU} + TL_{free} + TL_{MUT} \tag{1}$$

where S_{21} – measured transmission coefficient; G_a, G_b – antennas gain; CL_{THRU} – uncorrected cable loss from antennas to the plane of calibration; TL_{free}, TL_{MUT} – free space transmission loss and MUT attenuation respectively.

In the second experiment the absorbing properties of measured materials were studied depending on the thickness of MUT as well as their combination. Measurements were made with single and double thickness of carpet material and multilayer flat one. Figure 6 show various stages of measurements.

The measurements were carried out in the frequency range 1.0 GHz to 18.0 GHz. The experiment carried out according to the specified scheme allows estimating the MUT absorbing properties in a wide frequency band. In addition to the previous studies of this article, the results allow to reveal the properties of materials more fully.

Figure 7a shows the attenuation of each of three RAMs as a function of the frequency in the case of normal incidence of an electromagnetic wave. The pink line shows the dependence for the 'Ternovnik' (RAM 2), brown for 'Isoterm' (RAM 3) and blue one for 'Delta' (RAM 1). The results of measurements in the frequency sub-bands are combined on one graph with data overlapping; the vertical marker lines show the boundaries of used antennas: 1.0 to 12.0 GHz and 4.0 to 18.0 GHz.

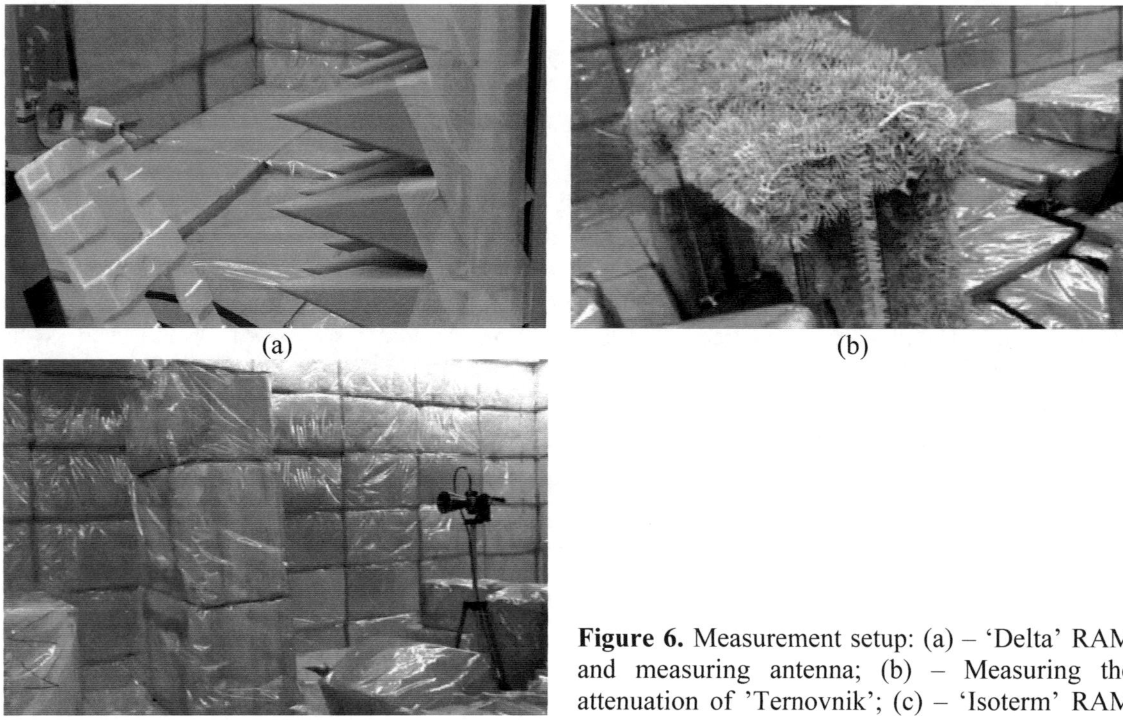

(a) (b) (c)

Figure 6. Measurement setup: (a) – 'Delta' RAM and measuring antenna; (b) – Measuring the attenuation of 'Ternovnik'; (c) – 'Isoterm' RAM and receiving antenna.

The attenuation curves for several MUTs are presented on figure 7b over frequency band: the blue line for one layer of 'Ternovnic' MUT, the brown line for 'Isoterm' and the pink one for double layer of 'Ternovnic'. As can be seen, the doubling of carpet MUT gives a gain in attenuation of up two times at frequencies below 13.0 GHz in comparison with more expensive and broadband multilayer flat material. But for higher frequencies it is inferior in parameters.

Figure 7c shows a comparison of results for compositions of two different material types. Linear interpolation is used to behavior trend tracing of these structures and to evaluate repeatability by differentiate the sub-bands measurements. The pink line shows the frequency response of absorption by a combination of carpet material and multilayer flat MUT. A blue line shows similar characteristic for pyramidal material covered with carpet one. It can be seen that the average attenuation in each of studied MUTs are improved as a result of their combination, reaching values of 50-60 dB at some frequencies. Moreover, the addition of the pyramidal material absorbing properties to the dispersive one carpet seems even more significant than in the case of multilayer flat material. The combination of those materials scattering properties introduces some instability over frequency. However, one can recommend using a RAMs combining to improve their absorbing properties and AEC performance in general.

5. Conclusions

The transmission and reflection characteristics of various RAMs were measured over wide frequency range. The materials under test belonged to the pyramidal, multilayer flat and carpet types. The properties of single materials and their combinations were investigated. If there is necessity to reduce backscatter, for example, in terms of improving the reflectivity level within working area, a technique for combining various types of RAMs should be considered.

The obtained results can be of scientific interest and also used for modelling the internal AEC electromagnetic field distribution.

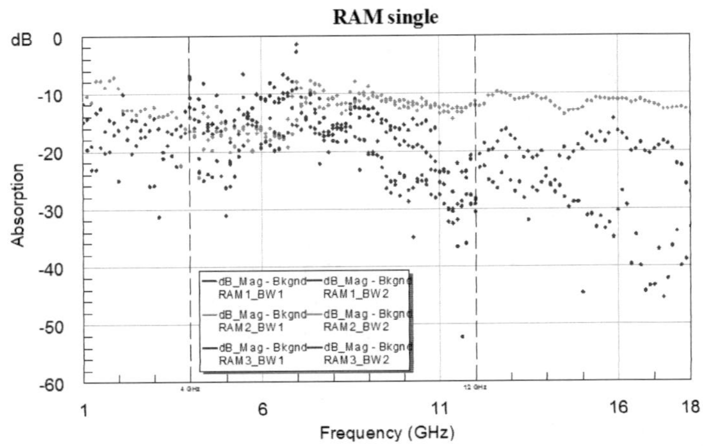

(a)

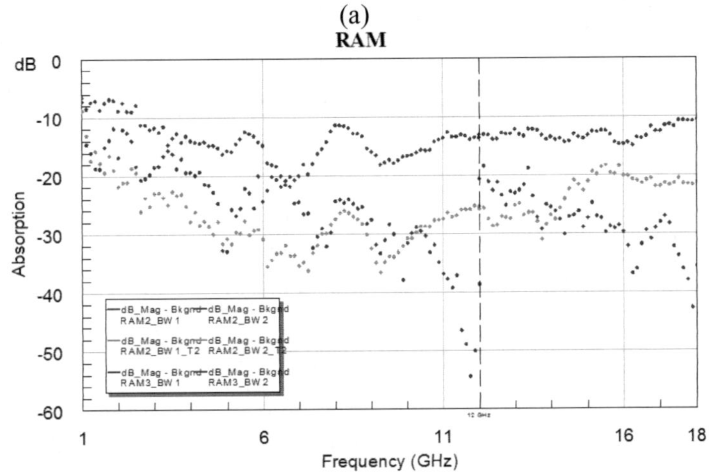

(b)

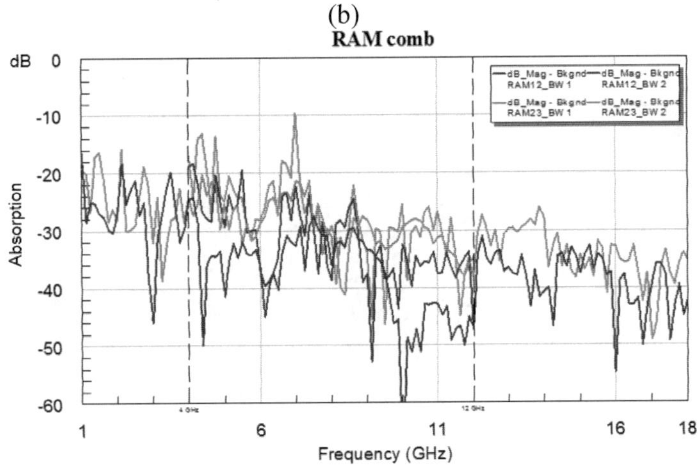

(c)

Figure 7. Measured attenuation: (a) – single MUTs; (b) – carpet RAM thickness doubling; (c) – compositions of different types of materials.

References

[1] Hemming L 2002 *Electromagnetic Anechoic Chambers: A Fundamental Design and Specification Guide* (New York: Wiley-Interscience) pp 57–69

[2] Rodriguez V, d'Abreu G and Liu K 2009 Measurements of Power Handling of RF Absorber Materials: Creation of a Medium Power Absorber by Mechanical Means. *31st Annual Symp. Antenna Measurement Techniques Association* (Salt Lake City, Utah: AMTA) p 218

[3] Rodriguez V, Walkenhorst B and Bruun J 2019 A Method for the Measurement of RF Absorber using Spectral Domain Transformations. *2019 IEEE Int. Symp. on Antennas and Propag. and USNC-URSI Radio Science Meeting* 347 DOI: 10.1109/APUSNCURSINRSM.2019.8889124

[4] Rodriguez V and Barry E 2016 A Polynomial Approximation for the Prediction of Reflected Energy from Pyramidal RF Absorbers. *AMTA 2016 Proc.* (Austin, TX: IEEE) pp 1–6

[5] Rodriguez V 2018 Validation of a Method for Predicting Anechoic Chamber Performance: A Technique That Uses Polynomial Approximations for RF Absorber Reflectivity *IEEE Antennas and Propag. Magazine* **60(4)** 31 DOI: 10.1109/AMTAP.2016.7806302

[6] Hemming L and Sanchez G 2004 Techniques in Scaling the Performance of Dielectric Microwave Absorber for Use in Scale Model Anechoic Chambers. *26th Annual Antenna Measurement Techniques Association Meeting & Symp.* (Stone Mountain, Georgia: AMTA)

[7] Rodriguez V 2016 Basic Rules for Anechoic Chamber Design, Part One: RF Absorber Approximations. *Microwave Journal* **59(1)** 72

[8] Bogaert I, Olyslager F, Arien Y and Pissoort D 2007 Modeling and Optimization of Advanced Multilayered Absorbers. *2007 European Microwave Conference* (Munich, Germany) pp 214–217 DOI: 10.1109/EUMC.2007.4405164

[9] Hofmann W, Bornkessel C, Schwind A and Hein M A 2019 Challenges of RF Absorber Characterization: Comparison Between RCS- and NRL-Arch-Methods. *2019 Int. Symp. on Electromagnetic Compatibility - EMC EUROPE* (Barcelona, Spain) pp 370–375 DOI: 10.1109/EMCEurope.2019.8871983

[10] Rabelsa G, Rianto D, Putra A, Santosa C E, Sri Sumantyo J T and Yohandri 2019 Pyramidal Radar Absorber Based on Coconut Shell Activated Carbon for Anechoic Chamber Application. *2019 PhotonIcs & Electromagnetics Research Symp.* (Rome, Italy) pp 1287–1291 DOI: 10.1109/PIERS-Spring46901.2019.9017897

[11] Mishra N and Chaudhary R K 2019 Design and Development of an Ultrathin Triple Band Microwave Absorber Using Miniaturized Metamaterial Structure for Near-Unity Absorption Characteristics. *Progress In Electromagnetics Res. C* **94** 89 doi: 10.2528/PIERC19043002

IOP Conf. Series: Materials Science and Engineering **868** (2020) 012019 doi:10.1088/1757-899X/868/1/012019

Computational and experimental investigation of a weakly diverging ion beam source

M Cherkasova, A Mogulkin, V Nigmatzyanov*, V Obukhov, O Peysakhovich and G Popov

Research Institute of Applied Mechanics and Electrodynamics, *Moscow Aviation Institute* (National Research University), Moscow 125080, Russian Federation

*E-mail: riame@sokol.ru

Abstract. This paper presents the idea of space debris removal by the ion beam. The main part of this idea is the ion source on the board of the spacecraft. The ion beam should be narrow, with small half-angle of divergence - about 2°. Angle of divergence means that in there are 95% of ions in this angle. Authors show the concept of the slotted ion optic system. Based on calculation, the construction of the ion optic was designed and tested as part of the laboratory model of the ion source. The results obtained during the experimental testing verified the possibility of using an ion source of such type as an ion injector, which can be installed on board a next-generation spacecraft for the removal of space debris objects.

1. Introduction

The idea of active removal of large space debris objects (SDO) from the near-earth space gave rise to the design of weakly diverging ion beam technology. The essence of this idea is to impact of an intense ion beam on SDO (ion beam shepherd concept) [1-3]. The distance between an ion beam shepherd S/C and SDO in the process SDO transportation to the disposal orbit should exceed the sizes of both objects. This distance could be about 20 or 40 meters to exclude accidental collisions between S/C and SDO. However, because of the ion beam divergence [4, 5] only some part of the ion force created by the ion beam transfers to SDO and the transferred force depends on the operating distances. In fact, it is defining the demands to the ion source concerning a divergence angle of the beam on the ion source exit.

The problem of forming a weakly divergent beam of charged particles was considered in detail as applied to electronic tubes [6]. The analytically found geometry of an electrostatic system for obtaining a laminar weakly divergent electron beam for both the ribbon and axisymmetric cases has been called in publications as the "Pierce gun". Using the quasi-Pierce geometry of the ion-extraction system (IES), it was possible to obtain a ribbon beam of ions with energy of 45-60 KeV and a half-angle of divergence of about 1° [7].

The ion beam properties in the near field (at a distance of 5-6 source radii from the source exit) determine the initial conditions in the problem of its subsequent expansion and the beam parameters in the far field in the region of interaction with SDO. When analyzing the divergence of an ion beam upon expansion into outer space, the following factors must be taken into account: 1) the initial formation conditions, 2) the divergence under the influence of electron pressure and an ambipolar electric field, 3) scattering by atoms and ions of the environment. The initial conditions for beam formation in a slotted IOS are the subject of this article and will be discussed below. When a large-sized SDO is removed by an ion beam from the GSO region, where its application has the greatest prospects, the influence of the ionosphere on ion beam propagation is negligible due to the low local density of the ionosphere. An analysis of various models of ion beam propagation taking into account the ambipolar electric field was carried out in [8]. In our estimates of the beam divergence under the action of an ambipolar electric field, we will rely on the data of [9], which gives the final expressions for calculating the ion beam streamtube containing 95% of the ion current using Parks and Katz self-similar model [9].

In the cited paper, a cone beam is considered. In the wedge-shaped beam formed by the slotted IES, the half-divergence angles in the transverse and longitudinal directions with respect to the slits are different. The half-divergence angle along the slits in the near field should be lower than the transverse one, that is determined by the ion's temperature in the discharge plasma which is lower 0.5 eV. According to the results of the experiment with the source of the wedge-shaped ion beam [10], the following half-divergence angles can be accepted with some margin: transverse 4°, longitudinal 2°. The results of the model calculations for the cone beam can serve as an estimate for the wedge-shaped if, as the initial half-angle at the exit from the near field accept in the range of 2°-4°.

Estimates will be carried out for the ion source design parameters, shown in table 1.

Table 1. Source design parameters.

Parameter	Value
Ion beam initial radii, R_0	0.1 m
Propellant	Xenon
Ion energy, W_i	4000 eV
Ion beam current, I_b	0.5 A
Xenon flow rate (mass efficiency η_m 0.8),	0.856 mg/s
Thrust, F	0.52 N
Initial plasma plum number density, n_0	$4\cdot10^{15}$m^{-3}
Initial ion's velocity, v_i	76400 m/s
Specific impulse, I_{sp}	61120 m/s
Ion beam power, N_b	2 kW
Plasma plume Mach number, M_i	51.64
Initial divergence half-angle, α_0:	2°
In transvers	3°
In longitudinal	4°
Distance between IS and SDO, d	Up to 40 m
Ratio N_i/F	3.85 kW/N

The mode of the ion source operation with a high ion energy of 4 keV is accompanied by high specific costs of electric power per unit of force (thrust) created by the beam 3.85 kW/N. However, such a regime, in comparison with regimes with a lower ion energy, can give a lower beam divergence in the far field due to the high Mach number ($M_i \approx 51.64$), higher momentum transfer efficiency and a high specific impulse of 61120 m/s, which gives savings in xenon consumption. Keep in mind that Xenon stock is a non-renewable resource aboard SSC.

The divergences of the ion beam in the far field were calculated using the formula obtained for the cone beam [9].

$$\frac{dh}{d\bar{z}} = \sqrt{\tan^2(\alpha_0) + 12\frac{\ln h}{M_i^2}} \qquad (1)$$

where $\bar{z} = z/R_0$, is the normalized axial coordinate; h is the dimensionless quantity associated with the beam radii $R(z)$ by the ratio:$R(z) = R_0 \cdot h$.

The boundary condition for (1) is: $h(0) = 1$. The calculation results are shown in figure 1 for three values of α_0. On the graph as well as the limiting case, the dependence $h(\bar{z})$ for $M_i \to \infty$ is shown, which follows from (1):

$$h(\bar{z})=1+tan\ \alpha_0.$$

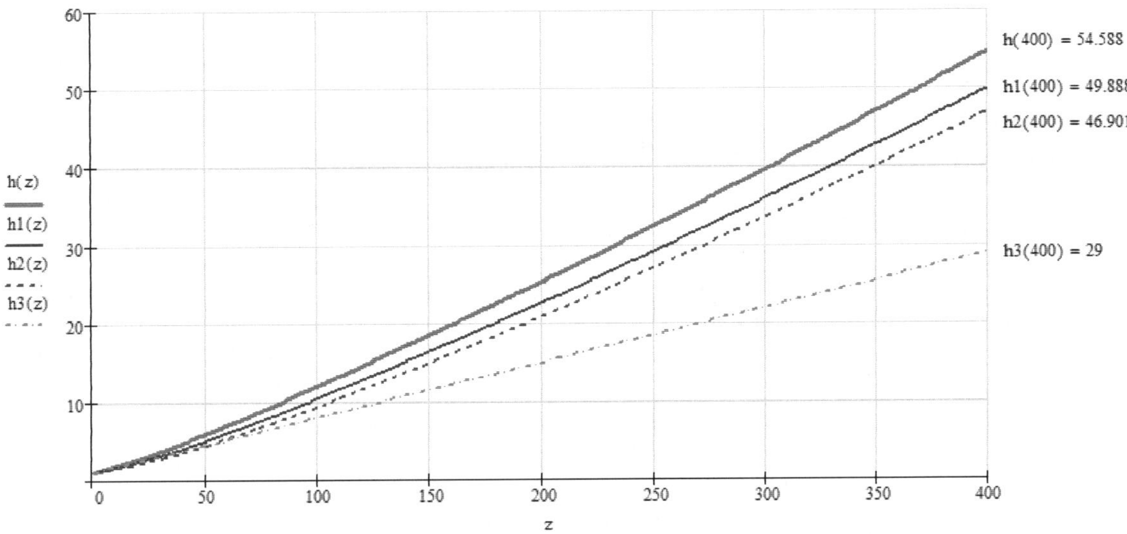

Figure 1. Calculation results.

According to the calculation, at a distance from the ion source $d = R_0\bar{z} = 20\ m$, the beam radii can be between 2 and 2.5 m. In conclusion, it seen that small initial divergency angle is the major factor that increases the momentum transfer efficiency to SDO, and the development an ion source of weekly diverging ion beam is actual.

2. Calculation

2.1. Simulation of the cell of the ion-extraction system

Taking as a basis a three-grid slotted IES, the authors stated the problem of investigating the potential possibilities of such scheme by numerical modeling. The development of the geometry that provides the smallest angle of divergence for the ion beam at a given diameter of the IES (100 mm) is the main task of simulation of the unit cell of the ion-extraction system. The aim was to determine the electrode cell profile, which influences on a plasma meniscus shape. Preferred shape is a convex surface toward plasma boundary for, to focus ion streamlines. Sharp edge of a slit provides the most focused streamlines.

As a result of the simulation by the software IGUN–8.022©, the geometry of IES grids was defined (figure 1) with the following parameters: the width of slit in the screen grid (SG) is 2.5 mm, the SG thickness is 1.0 mm, distance between two grids is 1.7±0.05 mm; the width of slit in the accelerating grid (AG) is 2.0 mm, AG thickness is 1.0 mm. Here is a short description of IGUN–8.022© computations. This program considers a multiple particle tracing method, in terms of solving Laplace's and Poisson's equation. For simplicity lets focus on two particle tracing, at a first step the result of Laplace's equation shows a charged particle parameter, at a second step Poisson's equation result shows a trajectory of a charged particle. These iterations continue until given values of current density. Ion beam divergence is calculated as a geometrical value based on borders of streamlines.

Figure 2 shows the dependence of the half-angle of divergence of the ion beam in the slotted IES of the aforementioned geometry on the normalized perveance. The normalized perveance for the given geometry was determined by the following relationship:

$$\Pi_i = j_i \cdot U_{\Sigma}^{-1.5} = (0.9 \pm 0.05) \times 10^{-8}\ \mathrm{A \cdot cm^{-2} V^{-1.5}} \qquad (2)$$

Here j_i is the ion current density, U_{Σ} is the total accelerating potential.

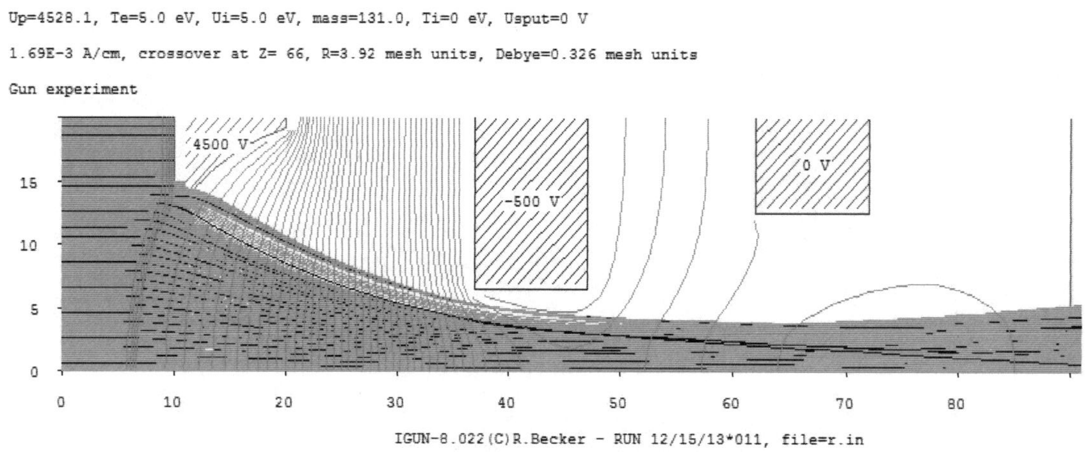

Figure 2. Geometry of the IES elementary slit cell.

Thus, the simulation showed the possibility of creating an IES that provides a minimum beam divergence angle of about 4° (shown in the figure 3).

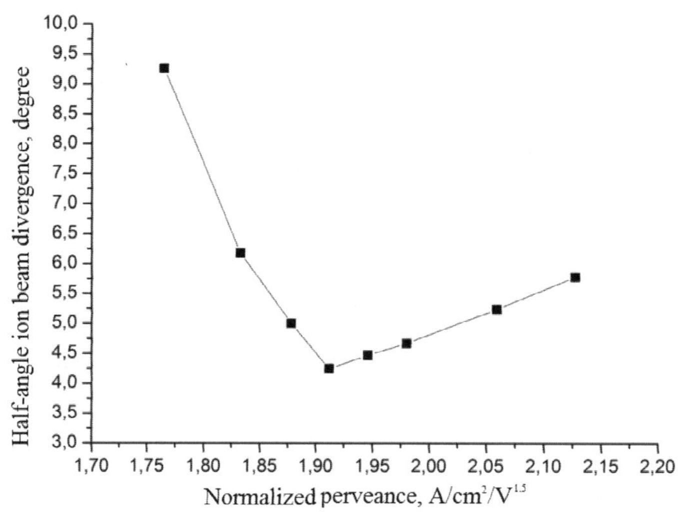

Figure 3. The ion beam divergence half-angle as a function of the normalized perveance.

3. Experimental setup

Based on the results of the simulation, a model of a laboratory prototype of an ion source was developed. The slit ion-extraction system with a diameter of 100 mm (figure 4) was made of Kovar; it comprised two grids – a screen one and an accelerating one. The laboratory prototype, model of the prototype is shown in the figure 5, was developed on the basis of a previously created prototype [10]. As in the prototype, the inductor was located at the top part of the discharge chamber and was surrounded by ferrite to increase the inductive coupling of inductor and plasma. The discharge chamber was made of quartz glass.

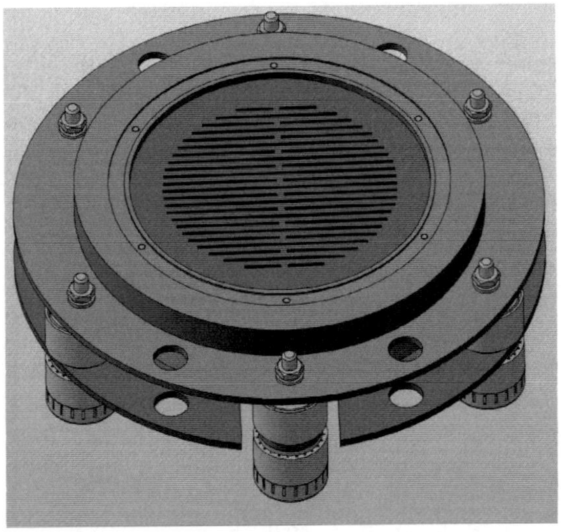

Figure 4. Two-grid ion extraction system.

In order to verify the modeling results, the experiments were conducted to determine:

1. The ion beam current with the ion energy of 1000-4000 eV;
2. The ion current density;
3. The normalized perveance in the indicated ranges of ion energy and ion current density;
4. Impulse transfer by the ion beam to the plate-sensor mounted on the thrust meter.

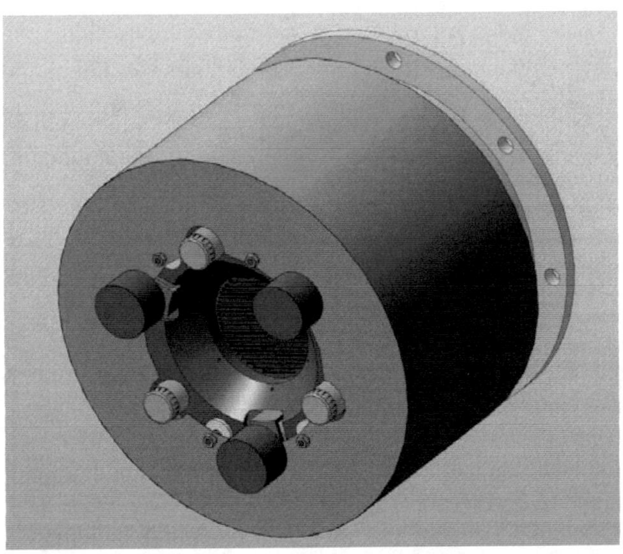

Figure 5. Laboratory prototype.

The experiments were carried out at the test facility comprising vacuum chamber, designed for the investigation of integral characteristics of electric propulsion and of the parameters of plumes of the thrusters with the power of up to 1.5 kW in continuous mode and of up to 4 kW for short-term operation. In addition, the facility comprised the thrust meter for the thrust level determination, the power supply units, and the propellant feed system. The target of thrust meter was mounted normal to a plume at the distance ±600 mm. Target was made from a titanium alloy with a surface of 0.3 m^2. The lowest pressure during experiments was about 10^{-4} Pa.

4. Results and discussion

Experimental work with laboratory prototype included measurements at 24 modes, which differed in the total RF-power, the potentials on the screen grid and accelerating grid. The xenon flow rate remained constant during the experiment. At each mode, the following parameters were measured: the current of the ion beam, the current on the accelerating grid and the thrust F on the plate sensor mounted on the thrust meter. The data obtained during the specified modes of operation of the laboratory prototype is shown in table 2.

Table 2. Experimental results.

Mode	Flow rate, mg/sec	RF-power, W	SG potential, V	AG potential V	Ion beam, mA	Accel. current, mA	Thrust, mN
1		110	1060	100	95	4	4.1
2		120	1000	100	-	-	4.3
3		130	1000	100	-	-	4.5
4		140	1000	100	108.5	17	4.6
5		140	1000	100	119.5	14	4.3
6		**140**	**1200**	**120**	**128.5**	**4**	**5.2**
7		140	1200	160	-	-	5.4
8		140	1200	160	120	4	5.4
9		140	1200	180	123	3.5	-
10		140	1200	150	119.5	4	5.3
11		140	1400	160	131.5	3	6.0
12		140	1600	180	117	2.5	6.5
13		140	1700	180	134	2.5	7.0
14		140	1830	180	132.5	2.5	7.1
15		**140**	**2000**	**200**	**154.5**	**2.5**	**7.6**
16	0.4	140	2500	180	160	4	9.1
17		**140**	**3000**	**180**	**200**	**8**	**10.2**
18		100	3180	180	160	11	7.5
19		100	3400	180	150	-	8.0
20		100	3600	150	150	15	8.0
21		100	4000	160	164.5	18.5	8.2
22		100	2000	200	110	3.5	6.0
23		**100**	**4000**	**180**	**167**	**18**	**8.0**

The bold-type data were chosen for the comparison with the simulation. The thrust F was determined by the following relationship:

$$F=\dot{m}v=\frac{I_{EE}}{e/M_i}\sqrt{2\frac{e}{M_i}U_{EE}}\times K(\alpha) \tag{3}$$

Here $\dot{m}$ is the xenon flow rate, v is the ion velocity, e is the elementary charge, M_i is the xenon ion mass, and $K(\alpha)$ is the coefficient depending on the angle of ion beam divergence in near field. Coefficient $K(\alpha)$ was defined as:

$$K(\alpha)=0.5(1+\cos\alpha) \tag{4}$$

here α is the half-angle of ion beam divergence.

When obtaining the calculated values, it was assumed that the accelerating potential was 10% of the screen grid potential, which differed from the experimental values. In the experiment, the accelerating electrode potential was selected based on the minimum value of the accelerating grid current. Therefore, the calculated values of the normalized perveance differ slightly from the experimental values. The results of comparison of experiments with simulation are shown in table 3.

Table 3. Experimental data analysis.

	$U_{SG}{}^{a}$=1200 V		U_{SG} =2000 V		U_{SG} =3000 V		U_{SG} =4000 V		U_{SG}=4000V
Parameter	Test	Simulation	Test	Simulation	Test	Simulation	Test	Simulation	Simulation
U_Σ, V	1320	1320	2200	2200	3180	3300	4180	4500	4500
Ion current density, A/m^2	34.7	34.7	41.8	41.8	54	54	45	45	60
Ion beam, mA	128.5	128.5	154.5	154.5	200	200	167	167	220
Thrust, mN	5.2	7.18	7.6	11.4	10.5	18	8	16.9	23
Perveance,10^{8} A/cm^{-2}V$^{-1.5}$	7.2	7.2	4.05	4.05	3.0	2.9	1.6	1.5	1.9

[a]Screen grid potential

The value U_Σ, indicated in the table, is equal to the sum of screen grid potential and accelerating grid potential. The total potential defines the value of the normalized perveance in the gap between two grids.

The last column of table 2 contains calculated parameters of the laboratory prototype corresponding to the minimum beam divergence at the perveance of 1.9×10^{-8} A·cm^{-2}V$^{-1.5}$. As can be seen from table 2, the perveance experimental values differ from the optimal value. At that, the lower is the U_Σ value, the stronger is this difference. The experiment, in which U_Σ was 4180 V, corresponded to the perveance of 1.6×10^{-8} A·cm^{-2}V$^{-1.5}$, i.e. less than the optimal.

The conducted experimental and computational study allowed us to outline the main design characteristics of the ion source and the ion-extraction system with an increased diameter of 160 mm, which had improved characteristics compared to the previously developed models. Figure 6 shows the calculation result for the new geometry of the ion-extraction system. The screen grid potential is 4500 V, the accelerating grid potential is minus 500 V, the decelerating grid (the case of the model) has a potential equal to zero. This calculation, provided by the software IGUN–8.022©, showed that the best half-angle of divergence equal to 1.95° is achieved at the current density of 48.6 mA/cm^2, the normalized perveance is equal to $1.324\cdot10^{-8}$ A/cm$^2\cdot$V$^{1.5}$. This geometry of the IES will provide the thrust of the order of 15-40 mN with ion energy of 2-4 KeV and an ion beam current of 200-300 mA.

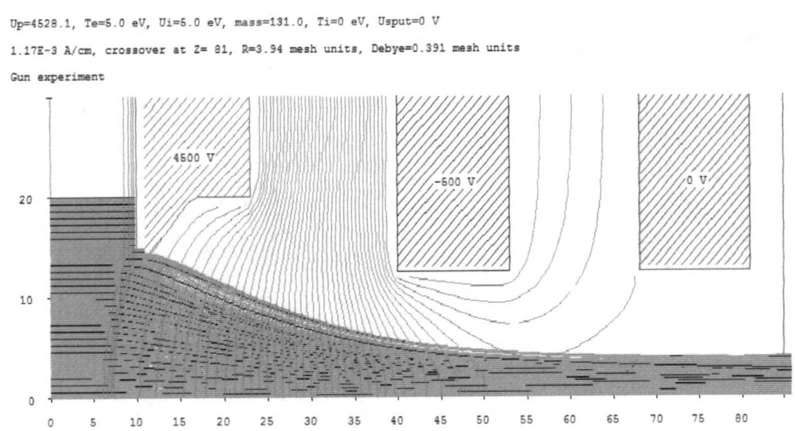

Figure 6. Geometry of the IES elementary slit cell.

5. Conclusion

The paper considers a full cycle of research for the sources of weakly divergent ion beams, including numerical simulation of the ion-extraction system, laboratory testing of models and determination of their optimal parameters.

It is shown that it is possible to develop an IES with the half-angle of beam divergence of 1.95°, ion beam current of 200-300 mA, thrust from 15 mN to 40 mN, and the ion energy of up to 4 KeV. Such parameters of the ion source are within the range of operating values required as applied to the designed spacecraft.

Acknowledgements

This work was supported by the Ministry of Science and Higher Education of the Russian Federation. Agreement Number: 05.604.21.0211. Unique identifier of the project: RFMEFI60419X0211.

References

[1] Bombardelli C and Peláez J 2011 Ion Beam Shepherd for Contactless Space Debris Removal. *Journal of Guidance, Control, and Dynamics* **34(3)** 916 doi: org/10.2514/1.51832

[2] Kitamura S 2014 A reorbiter for large GEO debris objects using ion beam irradiation. *Acta Astronautica* 94(2) p. 725, doi: org/10.1016/j.actaastro.2013.07.037

[3] Bonnal C, Ruault J M, Bultel P and Desjean M C 2010 High Level Requirements for an Operational Space Debris Deorbiter. *61st International Astronautical Congress IAC-10.A6.4.5* (Prague, Czech Republic)

[4] Cichocki F, Merino M and Ahedo E 2014 Modeling and simulation of EP Plasma Plume Expansion into Vacuum. *50th AIAA/ASME/SAE/ASEE Journal Propulsion Conference, AIAA Paper* (Cleveland, OH, USA) 2014-3828

[5] Merino M, Cichocki F and Ahedo E 2015 Collisionless Plasma thruster Plum Expansion Model. *Plasma Sources Science and Technology* **24(3)** 035006 doi: org/10.1088/0963-0252/24/3/035006

[6] Pierce J R 1940 Rectilinear Electron flow in beams. *Journal of Applied Physics* **11** 548 doi: org/10.1063/1.1712815

[7] Forrester A T 1988 *Large ion beams: Fundamentals of generation and propagation* (New York: Wiley-Interscience)

[8] Bombardelli C, Merino-Martínez M, Ahedo Galilea E, Peláez J, Urrutxua H, Herrera-Montojo H, Iturri-Torrea A 2011 *Ariadna Call for Ideas. Active Removal of Space Debris. Ion Beam Shephered for Contactless Removal. Final Report ESA* (Madrid, Spain)

[9] Cichocki F, Merino M, Ahedo E, Smirnova M, Mingo A and Dobkevicius M 2017 Electric Propulsion Subsystem Optimization for "Ion BEAM Shepherd" Missions. *Journal of Propulsion and Power.* Article in Advance DOI: 10.2514/1.B36105

[10] Balashov V, Cherkasova M, Kruglov K, Kudriavtsev A, Masherov P, Mogulkin A, Obukhov V, Riaby V and Svotina V 2017 Radio frequency source of a weakly expanding wedge-shaped xenon ion beam for contactless removal of large-sized space debris objects. *Review of scientific instruments* **88(8)** 083304 doi: org/10.1063/1.4998247

A comparative statistical analysis of global trends in civil helicopter accidents in the U.S., the EU, and the CIS

K Moon* **and A A Yakovlev**

Faculty of Aircraft and Spacecraft Engines and Power Plants, *Moscow Aviation Institute* (National Research University), 4 Volokolamskoe shosse, Moscow 125993, Russian Federation

*E-mail: moonapower@naver.com

Abstract. With helicopter operations increasing worldwide for varying usage and purposes, it is important to identify changes in helicopter accident trends. This paper studies the trends in civilian helicopter accidents in the United States, the European Union, and the Commonwealth of Independent States, and evaluates helicopter flight safety. Comparative analysis was performed, focusing on statistics and accident reports from Federal Aviation Administration, International Helicopter Safety Foundation, United States Helicopter Safety Team, National Transportation Safety Board, European Union Aviation Safety Agency, and Interstate Aviation Committee. The overall helicopter accident trends by flight time, engine type, and phase of flight were considered. The differences among the groups were identified, and the flight safety of helicopter was compared with that of other types of aircraft. In conclusion, gas-turbine engine helicopters showed a significant increase in the number of accidents. Regarding phase of flight, takeoff, landing, maneuvering, and en-route reported the most number of accidents. This study emphasizes the need for more profound research on improving helicopter flight safety by analyzing the global trend of helicopter accidents.

1. Introduction

According to FlightGlobal's October 2018 report, there are a total of 38,536 helicopters in existence worldwide [1]. By December 2018, it was recorded that 32,372 helicopters are currently in operation [2], which is a 14.7% increase from 2012 (28,230). North America leads by accounting for 34.3% of the in-service helicopters, with 81.9% registered in the U.S. Europe accounts for 28% of the total in-service helicopters, with Russia operating the highest number in the region with 26.7% [1]. Airbus's helicopter civil market forecast in 2017 [3] and Geach O's helicopter market update in 2019 [4] both show the global demand for helicopters will continue to grow.

In 2000, Franklin D H et al. analyzed the trends and causes of helicopter accidents in the U.S. from 1963 to 1997 [5]. In conclusion of this work, the research emphasized the importance of the rotary-wing aircraft and called for extended efforts to reduce its accidents [5]. Airworthiness certification system, technical system, pilot education system, and etc. are used in efforts to reduce the helicopter accident rate. In the past, these efforts were carried out domestically, but as the civil helicopter usage is becoming more universal, international efforts with various international organizations, such as International Helicopter Safety Foundation (IHSF) is putting forth more effort.

As mentioned, the global demand for helicopters is growing, and the trend of helicopter accidents are changing alongside. Analyzing the said trend is essential in researching ways to increase the

IOP Conf. Series: Materials Science and Engineering **868** (2020) 012020 doi:10.1088/1757-899X/868/1/012020

helicopter flight safety. Therefore, the paper aims to analyze the trend of accidents of civil helicopters registered and operated by the U.S., the EU, and the CIS.

2. Data

The accident trend analysis was performed based on statistical data and helicopter accident reports from the United States, the EU, and the CIS. In the U.S., the Federal Aviation Administration (FAA)'s data and United States Helicopter Safety Team (USHST)'s data from National Transportation Safety Board (NTSB) were considered; in the EU, the European Union Aviation Safety Agency (EASA)'s data was considered. Lastly, in the CIS, the Interstate Aviation Committee (IAC)'s data and annual aircraft accident reports were considered. Then, using the comparative statistical analysis of accident trends, the total number of helicopter accidents and three different trend groups, including flight time, engine type, and phase of flight, were analyzed.

3. Results and discussion

3.1. Analysis of the total number of civil helicopter accidents

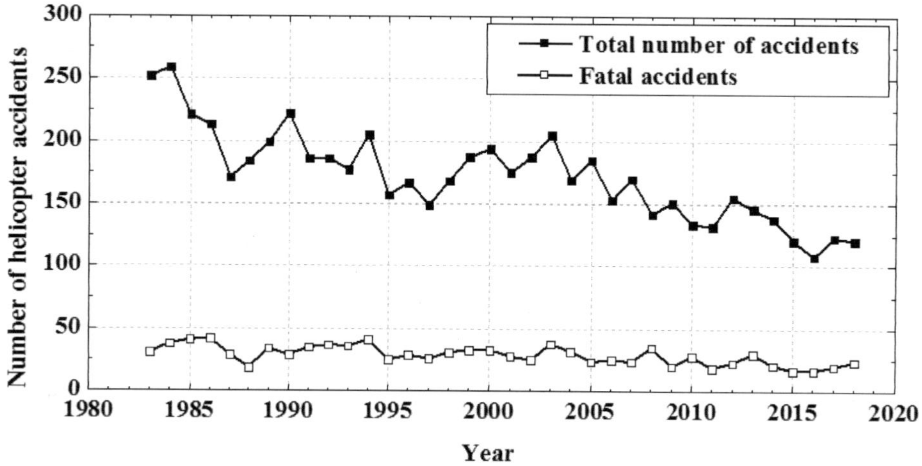

Figure 1. The total number of rotorcraft accidents from 1983 to 2018 in the U.S.

Figure 1 shows the decreasing tendency of helicopter accidents from 1983 to 2018 in the U.S. with data from IHSF and USHST [6, 7]. Comparing the average value for the first and the last three years of this period, it was observed that the total number of helicopter accidents went from 244 to 117.67 — a 51.78% decrease. The number of fatal accidents also decreased from 36.66 to 20.33, which is a 45% decrease. When analyzing the ratio of fatal accidents to the total number of accidents, however, the number had increased from 15% to 17.3%. This shows that present-day helicopter accidents are more likely to be fatal than those in the past. The USHST Safety Report [8] further supports this: from January to October 2019, there were 111 helicopter accidents with 21 of them fatal, marking 18.92% of the total accidents as fatal.

Figure 2a shows the number of helicopter accidents in the EU based on the data collected from EASA [9]. In the EU, the average value of the total number of accidents from the first three years in the figure was compared with that of the last three years, and it was observed that the total number of accidents decreased by 22.5%. The total number of non-fatal accidents was analyzed likewise and showed a 42.33% decrease (71.67 to 41.33). Also, the number of fatal accidents from 2009 to 2015 decreased by 38.07% (14 to 8.67). However, the number began inclining after 2015 all the way to 2018, reaching the same number of fatal accidents as it did in 2008. The average ratios of all accidents to fatal accidents were 15.5% for the first three years and 13.5% for the last three year, which showed an insignificant difference.

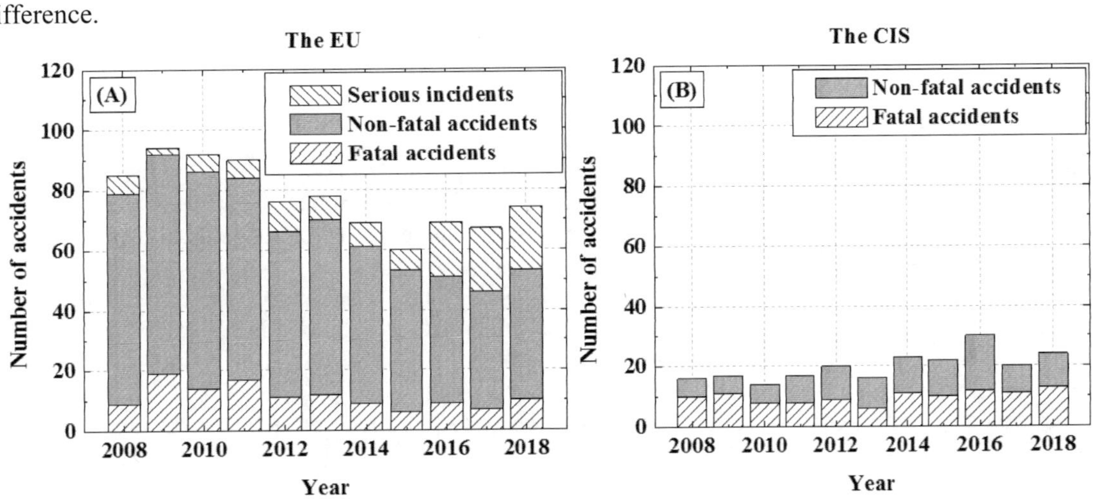

(a) (b)

Figure 2. The number of helicopter accidents in the EU (a) and the CIS (b).

Figure 2b shows the number of helicopter accidents in the CIS based on the data from IAC [10-20]. In the CIS, the average number of accidents from the first three years was 15.67 while the average from the last three years was 24.67 — this is a sharp increase of 57.43%. The number of fatal accidents analyzed likewise also shows an increase of 24.1% (9.67 to 12).

Table 1. Distribution of helicopter accidents in the EU and the CIS W.R.T flight operation.

	EU				CIS	
Year	Section 1[a]	Section 2[b]	Section 3[c]	Section 4[d]	Commercial operation	GA
2008	4(0)[e]	12(2)	16(3)	53(4)	12(8)[f]	4(2)
2009	3(1)	5(2)	32(6)	54(10)	13(7)	4(4)
2010	1(0)	9(2)	25(6)	57(6)	9(5)	5(3)
2011	5(1)	5(2)	27(8)	53(6)	10(4)	7(4)
2012	5(0)	9(1)	23(6)	39(4)	11(5)	9(4)
2013	2(1)	4(0)	19(4)	53(7)	9(2)	7(4)
2014	1(0)	7(1)	14(0)	47(8)	11(5)	12(6)
2015	0(0)	13(2)	9(2)	38(2)	11(5)	11(5)
2016	3(1)	9(3)	17(1)	40(4)	12(3)	18(9)
2017	2(0)	12(1)	17(2)	36(4)	11(4)	9(7)
2018	4(0)	19(2)	18(2)	33(6)	17(10)	7(3)
Total	30(4)	104(18)	217(40)	503(61)	126(58)	92(50)

[a] Offshore Commercial Air Transport Helicopters.
[b] Commercial Operations – Other than Offshore.
[c] Specialized Operations.
[d] Non-Commercial Operations.
[e] Number of total accidents including serious incidents, fatal/non-fatal accidents (Fatal accidents) for EU.
[f] Number of total accidents (Fatal accidents) for CIS.

Table 1 includes the data of helicopter operations in the EU in four main sections: (1) passenger and cargo flights to and from offshore oil and gas installations, conducted by EASA Air Operators Certificate

(EASA AOC) holders with certified helicopters, (2) all passenger and cargo flights not used in offshore oil and gas operations, conducted by EASA AOC holders with certified helicopters, (3) specialized operations involving certified helicopters, such as air ambulance, advertisement, photography, with EASA MS as state of operator or state of registry, and lastly, (4) non-commercial operations involving certified helicopters, with EASA MS as state of operator or state of registry [9]. For the CIS, all commercial flights and general aviation (GA) were included [10-20].

Section 1 showed an irregular trend of decrease and increase. In section 2, the average number of fatal accidents was 1.64 with the number remaining without any significant increase or decrease. The number of serious incidents, however, showed a gradual decrease up until 2013, and then a sharp increase afterwards.

In section 3, the total number of accidents exhibited a sharp decline, going from 32 accidents in 2009 to 9 in 2015. This is a 71.88% decrease. Fatal accidents showed a similar decline from 2011 to 2014. In 2016, the total number of accidents began rising, almost doubling from 2015. Fatal accidents also increased beginning in 2014, and both numbers plateaued since.

The accidents in non-commercial operations demonstrated in Section 4 accounted for 58.9% of all recorded number of accidents from 2008 to 2018. The average value of total accidents in the first three years of the graph was compared with that of the last three years of the graph. For total number of accidents, the analysis showed 33.55% decrease (54.67 to 36.33) comparable to the 30% decrease for the number of fatal accidents (6.67 to 4.67). The ratios of the number of fatal accidents to that of all accidents in each section of the EU are 13.33%, 17.3%, 18.43%, and 12.13%, respectively.

In the CIS countries, 57.53% of accidents in the last 11 years occurred in commercial operations and 42.47% in GA. Fatal accidents accounted for 46% of all recorded commercial operations and 54.84% of all recorded GA. The average number of total accidents in the first three years was compared to that of the last three years: in commercial operations, there was an increase from 11.33 to 13.33; in GA, there was a more significant increase from 4.33 to 11.33. Although fatal accidents in commercial flight showed a slight decrease of 6.67 to 5.67, fatal accidents in GA showed significant increase from 3 to 6.33.

3.2. Analysis of a number of civil helicopter accidents per 100,000 flight hours

Data was collected from multiple sources for figure 3. For all types of helicopter accidents in the U.S. and the CIS, the data for 1999 to 2008 was collected from IAC's 2010 report [19]; Roskop L's report in 2018 [21] was considered for the data from 2009 to 2016 in the U.S., and USHST's 2019 report [7] was considered for the data in 2017 to 2018. For the CIS data, the upper appendage 'b' from 2000 to 2018 considers only the accidents in commercial flights, and not GA [10-12, 17].

For the U.S., the total number of accident average from the first three years compared to that of the last three showed 56.3% decrease (8.2 to 3.58), and the same analysis for fatal accidents showed a 54.5% decrease (1.34 to 0.61). On the other hand, USHST November 2019 report including the data from January to October 2019 revealed the fatal accident rate has increased to 0.69 and the overall accident rate to 3.67 [8]. It is important to note, however, the total active helicopter flight time in the U.S. was approximately 2.1 million hours [5] and the flight time accumulated in 2019 from January to October exceeds 3 million hours [8]. Thus, despite the increase in accident rates this year, it can be assumed that significant progress has been made in improving helicopter flight safety.

The CIS showed a similar trend as the overall accident rate decreased up until 2012, but the rate began to increase again in 2013. Moreover, the rate of fatal accidents in commercial flight showed a sharp increase in 2018. According to IAC's report, the accident rates of all aircraft types other than a helicopter are as follows: the average of non-fatal accidents by flight time from 2000 to 2008 was 1.26 while the average of fatal accidents was 0.52 [19, 22]. Since figure 3 also demonstrates the total number of accidents over each year's entire flight time, the data allows a comparison of accident rates between helicopters and other aircraft types, it is sufficiently demonstrated that helicopter accidents happen more frequently than all the other aircraft accidents. Additionally, the paper considered the accident rate by aircraft weight. From 2009 to 2018, the average accident rate for commercial flight aircraft with a takeoff weight of 10 tons or more was 0.145 and the average rate for fatal accidents was 0.06 [10, 15]. For a takeoff weight of less than 10 t, the average accident rate from 2009 to 2018 was 7.64, while the rate for fatal accident was 3.63 [10, 15]. This shows that although the average accident rate of helicopter is lower than that of heavy-weight aircraft, it is higher than the average for light-weight aircraft.

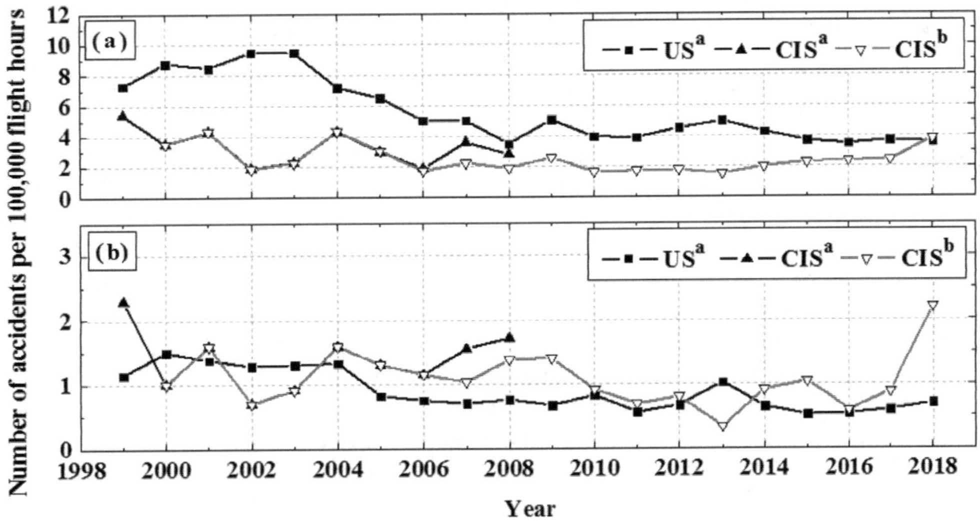

Figure 3. The number of accidents per 100,000 flight hours for total accidents (a) and fatal accidents (b) from 1999 to 2018 ([a] all types of operations, [b] without GA).

3.3. Analysis of helicopter accidents based on engine type

The commercial production of helicopters with gas-turbine engines in the U.S. began in 1958, and increased rapidly [5]. By 1982, it even surpassed the production of reciprocating engines [5]. A similar trend followed suit worldwide, and in 2018, the gas-turbine engine helicopters accounted for 69.1% of all in-operation helicopters, showing a 90% increase from 2004 [2, 23]. Considering this significant increase, an accident analysis by engine type was performed.

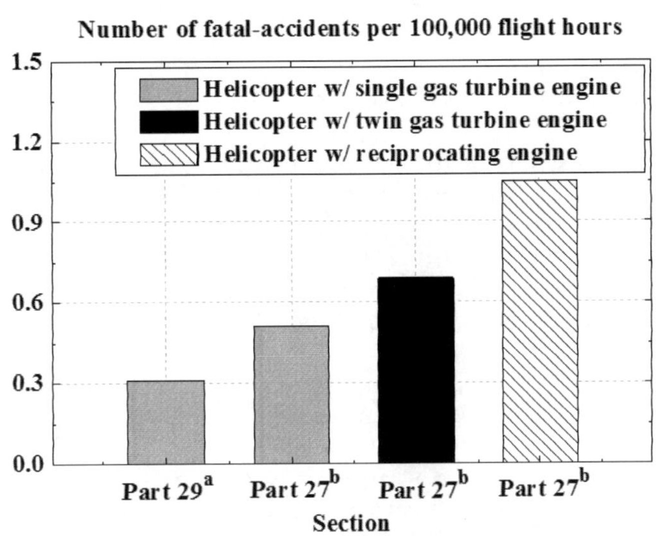

Figure 4. The number of fatal accidents per 100,000 flight hours by the Federal Aviation Regulations and engine installation from Jan 2009 through Aug 2018 [21]. ([a] Airworthiness Standards: Transport Category Rotorcraft; [b] Airworthiness Standards: Normal Category Rotorcraft).

Figure 4 shows the number for reciprocating engines stayed relatively higher than that of gas-turbine engines. This suggests that gas-turbine engines provide higher flight safety. However, after analyzing the US, 2000, 2001, and 2006 accidents of helicopters registered, the percentage of accidents for each engine type was 39~45% for the single reciprocating engine, 45~51% for single gas-turbine engine, 8~10% for the rest [24]. Furthermore, the NASA technical report [5] showed the number of accidents for reciprocating engines has decrease over the years; gas turbine engines also indicated a significant decrease in 1970s, but showed no remarkable decrease since then. Furthermore, figure 4 shows twin gas-turbine engines have lower number of fatal accidents than single turbine engines do, indicating that having an extra engine may increase the chance of survival in the event of an engine failure.

It is mentioned above that gas-turbine engines seem safer in the past, but the report on fatal accidents shows contrasting evidence. When the numbers of accidents per flight time for single and twin gas-turbine engines were combined, it was higher than that of reciprocating engines. The total number of accidents may be lower for the gas-turbine engine, but it is more prone to fatal accidents than the reciprocating engine does. In the past, reciprocating engines were widely used for various flight purposes, but the gas-turbine engines became a more popular choice in the recent years. As the usage for gas-turbine engines further dominate the industry, it is critical to reconsider the flight safety for said engine due to its higher likelihood for fatal accidents.

Table 2 shows the number of accidents according to the installed engine types categorized by operation classifications in the EU. It then compares the data from 2018 to the average data collected from 2008 to 2017 [9]. In 2018, 25 accidents from reciprocating engine helicopters are calculated across all three classifications, while 45 accidents from gas-turbine engines are recorded. For the rate of accidents in different operation classifications in 2018, commercial operations excluding offshore accounted for 27.14% of total helicopter accidents, specialized operations accounted for 25.71%, and non-commercial operations accounted for 47.14%.

Table 2. Distribution of accidents in the EU.

Engine type	Section 1[a]		Section 2[b]		Section 3[c]	
	Avg.[d]	2018	Avg.[d]	2018	Avg.[d]	2018
Reciprocating engine	1.7	0	6	5	30.1	20
Gas turbine engine	6.8	19	14	13	16.9	13

[a] Commercial Operations – Other than Offshore.
[b] Specialized Operations.
[c] Non Commercial Operations.
[d] Average number of helicopter accidents from 2008 to 2017.

Table 2 also shows that section 1 and 2 indicate a much higher accident rate for the gas-turbine engines, while section 3 exhibits contrasting evidence. In comparing the 10-year average for the number of helicopter accidents to the total number of accidents from 2018, it is noted that gas-turbine engines for commercial operations excluding offshore stands alone for exhibiting a significantly higher number of accidents in 2018 than that from a 10-year average. This supports the analysis based on figure 4: the demand for helicopters with gas-turbine engine is increasing, but so is the accident rate for said helicopter type.

Table 3 presents a number of helicopter accidents in CIS from 2008 to 2018, according to engine types, accident types, and then into operation types [10-20, 25]. As of May 2017, a total of 2,560 helicopters were registered in the CIS. Among that number, 93.8% had gas-turbine engines. The ratio is even higher for the active helicopters, with a staggering 98.6% them having gas-turbine engines [2].

When focusing on the total number of accidents, it is observed that those by gas-turbine engine helicopters accounted for 73.52% of all helicopter accidents in the CIS for the last 11 years. Gas-turbine engine helicopters also recorded 84 fatal accidents, which is more than three times that of reciprocal engine helicopters. Comparing the average number of non-fatal accidents in the first and the last three years from the data further confirmed the high risk of gas-turbine engine helicopters: the average value increased from 4.33 to 5.33 for reciprocating engines, but for gas-turbine engines, it increased from 11.33 to 19.33. Similarly, the same comparison for fatal accidents showed a slight increase for reciprocating engines (from 2 to 2.33), while the number for gas-turbine engine showed a greater increased (from 7.67 to 9.67).

Table 3. The number of non-fatal and fatal accidents in the CIS by engine and operation type.

	Reciprocating engine				Gas turbine engine			
	Non-fatal accident		Fatal accident		Non-fatal accident		Fatal accident	
Year	Com. [a]	GA	Com. [a]	GA	Com. [a]	GA	Com. [a]	GA
2008	0	2	0	2	4	0	8	0
2009	1	0	0	2	5	0	7	2
2010	2	2	1	1	2	0	4	2
2011	0	3	0	2	6	0	4	2
2012	0	3	0	2	6	2	5	2
2013	1	1	0	3	6	2	2	1
2014	2	3	0	2	4	3	5	4
2015	2	2	0	3	4	4	5	2
2016	2	3	0	4	7	6	3	5
2017	1	2	2	0	6	0	2	7
2018	0	1	0	1	7	3	10	2
Total	11	22	3	22	57	20	55	29

[a] Commercial flight.

3.4. *Analysis of the number of accidents by phase of flight*

Helicopter operation consists of more flight phases than fixed-wing aircraft operation does. Thus, an analysis of helicopter accidents based on phases of flight was carried out.

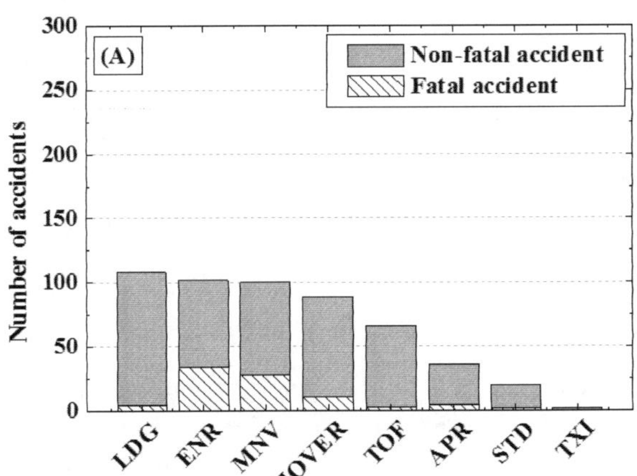

(a)

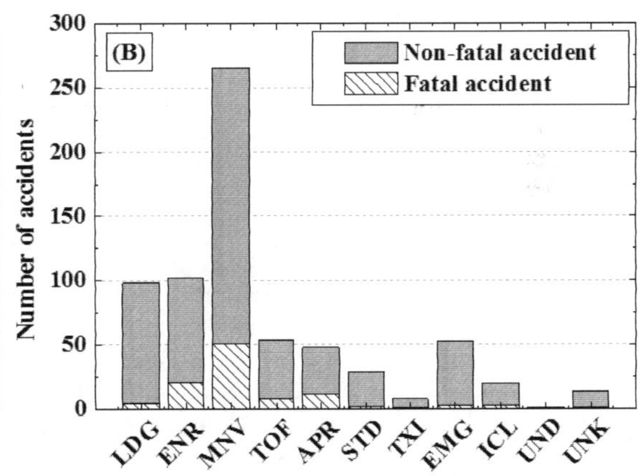

(b)

Figure 5. The number of helicopter accidents registered in the U.S. by phases of flight in 2000, 2001 and 2006 (a) [24], and from 2012 to 2016 (b) [26-30]. (LDG (landing), ENR (en-route), MNV (maneuvering), Hovering, TOF (takeoff), APR (approach), STD (standing), TXI (taxi), EMG (emergency descent), ICL (initial climb), UND (uncontrolled descent), UNK (unknown)).

US-JSHAT's 2011 data reported a total of 523 helicopter accidents from 2000, 2001 and 2006, according to phase of flight the accident occurred [24]. In figure 5a en-route, maneuvering, hovering, and takeoff showed higher number of accidents, with landing showing the highest number. Similarly, maneuvering and hovering showed higher number of fatal accidents, with en-route showing the highest number. For Figure 5b, 2012 to 2016 NTSB database for Part 135 and GA was considered [26-30]. Landing, en-route, and takeoff showed higher number of accidents, with maneuvering dominating as phase of flight with the highest number of accidents. For fatal accidents, maneuvering, en-route, approaching, and takeoff accounted for 86.8%.

Table 4 shows the number of helicopter accidents registered in the EU and Russia by phases of flight [9, 31]. Data are shown separately due to differences in categorization of phase of flight between the EU and Russia. For the EU, section 1 displays a decrease in accidents for all phases, except for approach, which shows a significant increase. Section 2 displays even more significant increases for takeoff and landing. There are no significant differences between the 10-year average and 2018 data for all phases of flight for section 3, but it is noteworthy that 2018 data for standing has increased almost 7 times compared to its 10-year average. In section 4, the number of accidents during landing and en-route has decreased, while during takeoff and maneuvering remained high. Overall, maneuvering, takeoff, landing, and en-route demonstrated to be more dangerous. The data from Russia are comparable to that of the EU: approach, takeoff, landing, and en-route demonstrated to be more dangerous, and together they accounted for 75.9% of all accidents from 2012 to 2016.

Table 4. Number of accidents with helicopter, registered in EU and Russia, by phase of flight.

	EU								Russia	
Phase of flight	Section 1[a]		Section 2[b]		Section 3[c]		Section 4[d]		Phase of flight	2012-2016
	Avg.[e]	2018	Avg.[e]	2018	Avg.[e]	2018	Avg.[e]	2018		
Landing	0.8	0	2	8	2.5	2	12.1	5	En-route	28
En-route	1.2	1	2.7	4	3.1	3	9.8	2	Approach	16
Take-off	0.1	0	1	6	3.1	2	8.4	9	Take-off	11
Manoeuvring	0.1	0	0.9	0	8.8	9	6	8	Landing	8
Approach	0.1	3	0.9	0	1.2	0	3.7	1	Descent	8
Taxi	0.2	0	0	1	0.1	0	2.8	1	Climb	5
Unknown	-	-	0.6	0	0.8	0	2.4	4	Hovering	4
Standing	0.1	0	0.4	0	0.3	2	1.8	3	Low-level flying	3
Total		4		19		18		33		83

[a] Offshore Commercial Air Transport Helicopters.
[b] Commercial Operations – Other than Offshore.
[c] Specialized Operations.
[d] Non Commercial Operations.
[e] Average number of helicopter accidents from 2008 to 2017.

4. Conclusions

The largest number of helicopter accidents occurred in the US, followed by the EU and the CIS. The total number of helicopter accidents has decreased in the U.S. and the EU, but the number of fatal accidents showed no significant decrease. The CIS, on the other hand, showed an increase in the total number of accidents. In recent years, the number of fatal accidents has increased also. The accident analysis according to flight operations based on 2008 to 2018 data indicate that 58.9% of accidents occurred in GA in the EU, while the CIS reported 57.5% of accidents to have occurred in commercial flights. Efforts have been made regarding the increasing trend in helicopter accidents, but further research is necessary to ensure higher flight safety and reduce the number of fatal accidents.

Also, the number of civil helicopter accidents per 100,000 flight hours was analyzed based on the data from the U.S. and the CIS. The analysis confirmed that helicopter flight safety is generally lower than that of fixed-wing aircraft. For commercial flights, helicopter flight safety was in between the safety of heavy and light aircraft.

Helicopter accident analysis by engine type revealed that the accidents with reciprocating engines were more frequent in the early stages of helicopter operation history, but as gas-turbine engines became more popular, the trend changed significantly in the 2000s. Recent data show the number of helicopter accidents with gas-turbine engines are similar and/or more to that of reciprocating engines. This indicates that gas-turbine engine helicopters are increasing in usage and in demand. As a result, further research is required to improve the flight safety of gas-turbine engine helicopters.

Analysis of helicopter accidents by phase of flight confirmed that most of the accidents occurred during takeoff, landing, maneuvering, and en-route. This suggests that although the helicopter's relatively free maneuvering provides many advantages compared to fixed-wing aircraft, but it is also a potential main cause of accidents in flight.

The study shows that the flight safety of civilian helicopters around the world continues to be questionable. The reliability of helicopter would be challenged as well if the number of fatal accidents fails to show a significant reduction. Helicopters with gas-turbine engines are becoming more prevalent in various fields of operation, contributing more complexity and diversity to aviation accident trends. Lastly, flight phases with the highest risk remained almost unchanged. The danger of these high-risk phases leads to either increase the constraints on helicopter activities or accept the risk in flights.

Helicopter operations are increasing worldwide. If more active research and improvements are not conducted to improve flight safety, the reliability of helicopters will inevitably decrease.

References

[1] Flightglobal 2018 *ANALYSIS: Helicopter market report Q3 2018*, available at: https://www.flightglobal.com/news/articles/analysis-helicopter-market-report-q3-2018-452784/

[2] Butov A M 2019 *Civilian Helicopter Market-2019*, Center of Development Institute, available at: https://dcenter.hse.ru/data/2019/06/02/1495288295/Рынок%20вертолетов%20гражданского%20назначения-2019.pdf [in Russian]

[3] Airbus 2017 *20-year helicopter civil market forecast – October,* available at: https://www.airbus.com/content/dam/corporate-topics/publications/backgrounders/Global-Helicopter-Forecast-2017.pdf

[4] Geach O 2019 *IBA's Helicopter Market Update August 2019*, IBA, available at: https://ss-usa.s3.amazonaws.com/c/308462937/media/87315d5bf7d643f0000448375247494/IBA%20 Helicoper%20Webinar%20Slides%20August%202019.pdf

[5] Franklin D H, Eugene F K and Laura E I 2000 *U.S. Civil Rotorcraft Accidents, 1963 Through 1997* Nasa Technical Reports Server, available at: https://ntrs.nasa.gov/archive/nasa/casi.ntrs.nasa.gov/20010038651.pdf

[6] IHSF 2019 *U.S. Registered Rotorcraft Accidents Calendar Years 1983 – 2016*, available at: http://ihsf.aero/Key_Charts/2016%20Accidents83to16.pdf

[7] USHST 2019 *Fatal Accident Rate, Fatality Rate, and Accident Rate Update (CY2019) 1st & 2nd Quarter (Jan-Jun, 2019)*, available at: http://ushst.org.dnn4less.net/Portals/0/2019%20USHST_Metrics%20Update_20190701.pdf

[8] USHST 2019 *Safety Report*, available at: http://www.ushst.org/Portals/0/USHST_SafetyReport_20191108.pdf

[9] EASA 2019 *Annual Safety Review 2019,* available at: https://www.easa.europa.eu/sites/default/files/dfu/Annual%20Safety%20Review%202019.pdf

[10] Interstate Aviation Committee 2019 *Flight safety in civil aviation of contracting states of the Agreement on civil aviation and airspace use in 2018,* available at: https://mak-iac.org/upload/iblock/03e/bp-18-2.pdf [in Russian]

[11] Interstate Aviation Committee 2018 *Flight safety in civil aviation of contracting states of the Agreement on civil aviation and airspace use in 2017*, available at: https://mak-iac.org/upload/iblock/4b6/bp-17-2.pdf [in Russian]

[12] Interstate Aviation Committee 2017 *Flight safety in civil aviation of contracting states of the*

Agreement on civil aviation and airspace use in 2016, available at: https://mak-iac.org/upload/iblock/552/bp-16-2.pdf [in Russian]

[13] Interstate Aviation Committee 2016 *Flight safety in civil aviation of contracting states of the Agreement on civil aviation and airspace use in 2015*, available at: https://mak-iac.org/upload/iblock/ba9/bp15-2.pdf [in Russian]

[14] Interstate Aviation Committee 2015 *Flight safety in civil aviation of contracting states of the Agreement on civil aviation and airspace use in 2014*, available at: https://mak-iac.org/upload/iblock/961/bp14-2.pdf [in Russian]

[15] Interstate Aviation Committee 2014 *Flight safety in civil aviation of contracting states of the Agreement on civil aviation and airspace use in 2013*, available at: https://mak-iac.org/upload/iblock/488/bp13-2.pdf [in Russian]

[16] Interstate Aviation Committee 2013 *Flight safety in civil aviation of contracting states of the Agreement on civil aviation and airspace use in 2012,* available at: https://mak-iac.org/upload/iblock/d89/bp12-2.pdf [in Russian]

[17] Interstate Aviation Committee 2012 *Flight safety in civil aviation of contracting states of the Agreement on civil aviation and airspace use for a 20-year period,* available at: https://mak-iac.org/upload/iblock/b41/bp11-3.pdf [in Russian]

[18] Interstate Aviation Committee 2011 *Flight safety in civil aviation of contracting states of the Agreement on civil aviation and airspace use in 2010,* available at: https://mak-iac.org/upload/iblock/5a4/bp10.pdf [in Russian]

[19] Interstate Aviation Committee 2010 *Flight safety in civil aviation of contracting states of the Agreement on civil aviation and airspace use in 2009*, available at: https://mak-iac.org/upload/iblock/0b1/bp09.pdf [in Russian]

[20] Interstate Aviation Committee 2009 Flight safety in civil aviation of contracting states of the Agreement on civil aviation and airspace use in 2008, available at: https://mak-iac.org/upload/iblock/b34/bp08.pdf [in Russian]

[21] Roskop L 2018 U.S. Rotorcraft Accident Data, available at: http://faahelisafety.org/wp-content/uploads/2018/11/Overview-Accident-Data.pdf

[22] Interstate Aviation Committee 2005 Flight safety in civil aviation of contracting states of the Agreement on civil aviation and airspace use in 2004 and for the period 2000-2004, available at: https://mak-iac.org/upload/iblock/cac/2000-2004.pdf [in Russian]

[23] Fallon A 2019 IBA's Global Helicopter Market Update May 2019, available at: https://ss-usa.s3.amazonaws.com/c/308462937/media/5cdc115f41f82/Global%20Helicopter%20Market%20May%2019.pdf

[24] U.S. Joint Helicopter Safety Analysis Team 2011 The Compendium Report: The U.S. JHSAT Baseline of Helicopter Accident Analysis, vol. I, available at: https://publicsafetyaviation.org/images/Safety_Program_Overview/The_Compenidum_Report_The_US_JHSAT_Baseline_of_Helicopter_Accident_Analysis_Vol_I_09-14-2011.pdf

[25] Interstate Aviation Committee 2012 Flight safety in civil aviation of contracting states of the Agreement on civil aviation and airspace use in 2011, available at: https://mak-iac.org/upload/iblock/461/bp11-2.pdf

[26] National Transportation Safety Board 2018 Aviation: Data & Stats, available at: https://www.ntsb.gov/investigations/data/Pages/AviationDataStats2016.aspx

[27] National Transportation Safety Board 2017 Aviation: Data & Stats, available at: https://www.ntsb.gov/investigations/data/Pages/AviationDataStats2015.aspx

[28] National Transportation Safety Board 2016 Aviation: Data & Stats, available at: https://www.ntsb.gov/investigations/data/Pages/AviationDataStats2014.aspx

[29] National Transportation Safety Board 2015 Aviation: Data & Stats, available at: https://www.ntsb.gov/investigations/data/Pages/AviationDataStats.aspx

[30] National Transportation Safety Board 2014 Aviation: Data & Stats, available at: https://www.ntsb.gov/investigations/data/Pages/2012%20Aviation%20Accidents%20Summa

ry.aspx
[31] Karibov D 2017 Analysis of statistics of accidents with helicopters of the Civil Aviation Administration of the Russian Federation 2012-2016, available at: https://www.aex.ru/docs/4/2017/10/12/2664/ [in Russian]

IOP Conf. Series: Materials Science and Engineering **868** (2020) 012021 doi:10.1088/1757-899X/868/1/012021

Coastal navigation by a solar sail

A V Rodnikov

Moscow Aviation Institute (National Research University), 4 Volokolamskoe shosse, Moscow, 125993, Russian Federation

E-mail: rodnikovav@mai.ru

Abstract. In this paper, we study relative motions of a light spacecraft equipped with a solar sail and tethered somehow to a heavy heliocentric orbital space station. One can say that such spacecraft realizes a kind of a coastal navigation. Formulating a model problem of the spacecraft motion, we state that if the sail can be oriented arbitrary then motions along a rectilinear tether are possible in both directions. Also, we present algorithm of the spacecraft relocation if it is joined with the station by a tether. This algorithm is based on the motion equations solutions that can be interpreted as perturbed motions of a simple pendulum.

1. Introduction

Solar sails are very interesting space propulsions that do not require fuel consumption. Unlike some hypothetical devices, solar sails are elements of spacecrafts in the operating and planned space missions [1,2]. Also, there are hundreds of papers in which the sailing spacecrafts dynamics is studied theoretically. For instance, some proposals for the orbit correction by the solar radiation pressure are offered in [3-12], maneuvers by the solar sail in the Formation Flying is studied in [13], interplanetary spaceflights and flights near asteroids are discussed in [14-17], relative motions of sailing spacecrafts are considered in [18,19], etc. Of course, only a minor part of such papers is cited here.

Note, however, that a solar sail of itself cannot create acceleration directed toward the Sun. Certainly, a solar sail could facilitate to decreasing of the orbit perihelion. But trying 'to move tacks' is pointless for a sailing spaceship. The reason is that the Solar system is not filled with liquid or some other viscous substance. Note, nevertheless, that a sea ship keel factually realizes constraint that restricts the ship motion orthogonally to its course. Tethering the spacecraft to a heavy space station by a cable, one can realize the similar constraint in space.

In this paper we study motions of a light spacecraft equipped with a solar sail with respect to a heavy space station that describes a heliocentric orbit. We assume that the spacecraft motion is restricted by tether(s) which one or both ends fixed on the station. In this case, one might say that the spacecraft makes a kind of a coastal navigation in the orbital frame of reference. Firstly, we study motion of the spacecraft along a rectilinear tether which ends are fixed in two points of the station that is considered sufficiently long. We claim that the spacecraft motion is possible in both directions, excepting the case of the tether that is parallel to the Sun rays, when the motion from the Sun is possible only. We deduce a criterion of the sail and the tether optimal inclinations to the Sun rays to maximize the spacecraft acceleration. Secondly, we study motions of the spacecraft joined with some point of the station by a cable. In this case, if the cable is taut then the spacecraft moves along surface of a sphere with center on the station. We construct an algorithm of the spacecraft displacement between two points of this sphere surface with zero initial and finite velocities. The algorithm is based on the fact that some solutions of the motion equations might be interpreted as perturbed librations of a simple pendulum, where perturbation arises due to the orbital frame of reference is non-inertial.

To achieve these purposes we formulate a model problem of the spacecraft motion. This problem is based on estimations of the forces acting on the spacecraft and on a number of the standard assumptions. We assume that the tether is weightless and inextensible, that the station mass is many times greater than the spacecraft mass, that the spacecraft with the solar sail is a particle equipped with a perfect reflecting flat, that can be placed at any angle to the Sun, and so on. Of course, such assumptions might be revised in future researches as more details might be taken into account.

2. Estimating of acting forces

Using the standard physical units, one seems that the solar radiation pressure is very small. But, there are situations in which the force produced by such kind of pressure is the main force that influence a spacecraft motion. Consider, for instance, a space system consisting of a heavy space station and a light spacecraft *A* of mass m_A near it ($m_A \le 1000$ kg). Let the spacecraft be equipped with a perfect reflecting flat of area *S*, ($S \le 1000\,\mathrm{m}^2$). Assume that the station mass center *B* describes a heliocentric orbit that is close to circular. Let the orbit radius be close to 1AU. Also, let *Bxyz* be right-hand Cartesian coordinate, *Bx* be directed as the sunlight, and *By* be directed as the station velocity with respect to the Sun. Moreover, let the distance between the spacecraft *A* and the station mass center *B* be less than 1 km. Theoretically, it might be elongated up to 20-30 km.

The forces acting on the spacecraft in the orbital frame of reference *Bxyz* are the gravitational force F_G, the force of the Sun radiation pressure F_S, and, also, the force of moving space F^e and Coriolis' force F^c (figure 1). Note that here, at the difference with [20,21], gravitation between the spacecraft and the station is negligible.

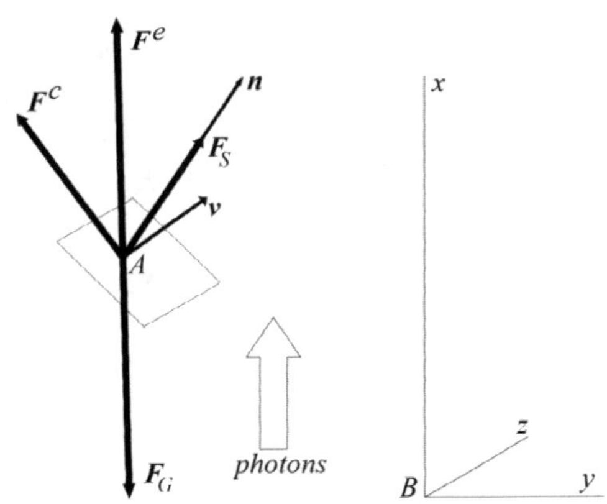

Figure 1. Forces.

One can check easily that for situation under consideration

$$\frac{|F_S|}{|F_G - F^e|} \sim \frac{S}{m_A} \cdot 2.25 \cdot 10^5, \qquad \frac{|F^c|}{|F_S|} \sim \sqrt{\frac{m_A}{S}} \cdot 0.006, \qquad |F_S| = p\frac{S}{m_A}\ \mathrm{N}, \tag{1}$$

where p is the solar radiation pressure. ($p = 9 \cdot 10^{-6}\,\mathrm{N/m^2}$ at 1au from the Sun). In these formulae we assume that $|F_S|$ has maximum value, i.e. the reflecting flat is orthogonal to the solar rays, and that the spacecraft velocity with respect to the station is less than velocity that is reached after moving 1km with acceleration $|F_S|/m_A$ that is less than 0.0009 m/s^2. Let us remark that the spacecraft velocity in *Bxyz* will be less in all situations that will be considered below, and, in any case, less than 1 m/s.

Note that the sail material mass is about 7 g/m^2 presently. So, we can assume $1\mathrm{m^2/kg} < S/m_A < 100\mathrm{m^2/kg}$. For instance, $S/m_A = 6.4$ m^2/kg for LightSail-2 [1]. Thus, in the case under consideration, gravitational force is insignificant, and effect of Coriolis' force is small. Of course, F_S is very small objectively, but note that the force of $9 \cdot 10^{-3}$N can move a mass of 1000 kg by 1km in 4.14 h. (Such situation corresponds to the solar sail of area 1000 m^2). If such displacement is carried out without any costs, this time does not seem too long.

3. The model problem formulation

Let the spacecraft A motion be restricted by one or more than one geometrical constraints, for instance, by tether(s) or by some solid guides. (Of course, the last one seems fantastic these days). These constraints can be written as $f_i(x,y,z) \leq 0$ (x,y,z are coordinates of A with respect to $Bxyz$). Given the above, the spacecraft motion equations could be written as

$$m\ddot{\boldsymbol{r}} = Spn_x^2\boldsymbol{n} - 2m[\boldsymbol{\omega},\boldsymbol{v}] + \Sigma\lambda_i\frac{\partial f_i}{\partial \boldsymbol{r}}, \tag{2}$$

where $\boldsymbol{n} = (n_x, n_y, n_z)$ is a normal to the reflecting flat ($|\boldsymbol{n}| = 1$), n_x, n_y, n_z are projections of $\boldsymbol{n}$ on Bx, By, Bz, respectively, $\boldsymbol{\omega}$ is angular velocity of the orbital frame of reference, $\boldsymbol{r} = \overrightarrow{BA}$, $\boldsymbol{v} = \dot{\boldsymbol{r}}$ is velocity of A in $Bxyz$, λ_i are Lagrange's multipliers. Here we use one of the standard models of a perfect reflecting flat [5,8,15], in which the force of the solar radiation pressure is proportional to n_x^2. Note that if one of the constraints is realized by a tether then this constraint is unilateral and the correspondent multiplier λ_i is sign-definite. In this case, one can assume without loss of generality that $\lambda_i \leq 0$. Note also, that $n_x \geq 0$ by the problem physics.

Let a be some fixed distance, for instance, the tether length. Using transformation $\boldsymbol{r} \to a\boldsymbol{r}$, one might rewrite (2) as

$$\boldsymbol{r}'' = n_x^2\boldsymbol{n} - 2\varepsilon[\boldsymbol{e}_z,\boldsymbol{r}'] + \Sigma\mu_i\frac{\partial f_i}{\partial \boldsymbol{r}}, \tag{3}$$

where $\boldsymbol{e}_z$ is a unit vector of Bz, $\varepsilon = \omega\sqrt{am/(pS)}$ is a dimensionless parameter, μ_i are Lagrange's multipliers of the same signs as λ_i respectively, and the prime $(\,)'$ denotes derivative with respect to dimensionless time $\tau = t\sqrt{pS/(am)}$.

Note that ε is small. Values of ε and units of τ for some values of S/m_A are given in the table 1.

Table 1. Values of ε and units of dimensionless time.

S/m_A (m^2/kg)	ε	$\tau = 1$
1	$4.21 \cdot 10^{-3}$	2.928 h
10	$1.33 \cdot 10^{-3}$	0.926 h
100	$4.21 \cdot 10^{-4}$	0.293 h

So, to simplify the problem under consideration, one can assume $\varepsilon = 0$ and reduce (2) to

$$\boldsymbol{r}'' = n_x^2\boldsymbol{n} + \Sigma\mu_i\frac{\partial f_i}{\partial \boldsymbol{r}}. \tag{4}$$

4. Motion along a rectilinear tether

Let B and C be points of a sufficiently long space station. Moreover, let these points be join by a tether. Suppose that the tether is rectilinear and taut, and that the spacecraft A can move along this tether (figure 2). Assuming $\varepsilon = 0$, one can repute without loss of generality that the tether lays in Bxy. In this case (4) is reduced to

$$s'' = n_x^2(n_x\cos\alpha + n_y\sin\alpha), \tag{5}$$

where $s = BA$ and α is angle between Bx and BC, $0 < \alpha < \pi$. Evidently, right-hand part of (5) can be positive or negative for any α. Hence, the spacecraft can move *in both directions*. Really, projection of $\boldsymbol{F}_S$ on BC can be directed both from B to C and from C to B. As dimensionless mass of the spacecraft A can be assumed 1, this projection coincides with the spacecraft acceleration $\boldsymbol{a}$. Clearly, $|\boldsymbol{a}| = s''$ if $s'' \geq 0$ (this case is depicted in figure 2) and $|\boldsymbol{a}| = -s''$ in the opposite case.

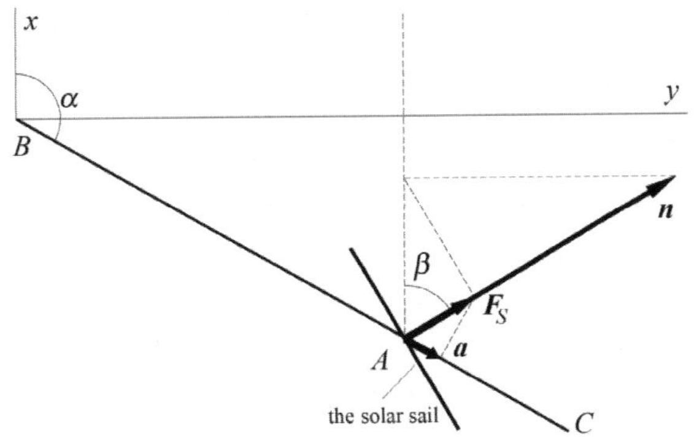

Figure 2. Motions along a rectilinear tether.

Let β be a clockwise angle between Bx and $\boldsymbol{n}$. One can prove that s'' reaches its maximum if $\boldsymbol{n}$ lays in Bxy and β is the maximum point of $\Phi(\beta) = \cos^2\beta\cos(\beta - \alpha)$. This implies that maximum of the spacecraft acceleration directed from B to C is reached if

$$\tan\beta = \frac{\sqrt{9\cot^2\alpha + 8} - 3\cot\alpha}{4}. \tag{6}$$

As it follows from (6), if $\alpha \to \pi$ then $s'' \to 0$, i.e. the 'sailing close to the wind' (α is close to π) is slower than the 'sailing by the wind' ($\alpha > \pi/2$) or the 'sailing wind abeam' ($\alpha = \pi/2$). In the last case we obtain $s''_{max} = 2\sqrt{3}/9$ for $\beta = \arctan\sqrt{2}/2$.

Evidently, if α is acute then the spacecraft can move from B to C even if these points are not tethered, i.e. if the spacecraft moves freely, however, the spacecraft motion along the tether is much faster. The free motion time τ as function of α for $BC = 1$ and zero initial velocity is depicted in figure 3 by the dashed curve. This curve is located much higher than the analogous curve for the time of motion along the tether with maximum acceleration (the solid curve in figure 3). Also, the point curve in this figure is the time of motion along the tether, but if the solar sail is placed in accordance with the usual sailing recommendations. We see that the curves have a common point corresponding to the 'sailing before the wind' ($\alpha = 0$) in which $\tau = \sqrt{2}$ and, also, that $\tau_{min} = 3^{3/4} = 2.280$ for $\alpha = \pi/2$.

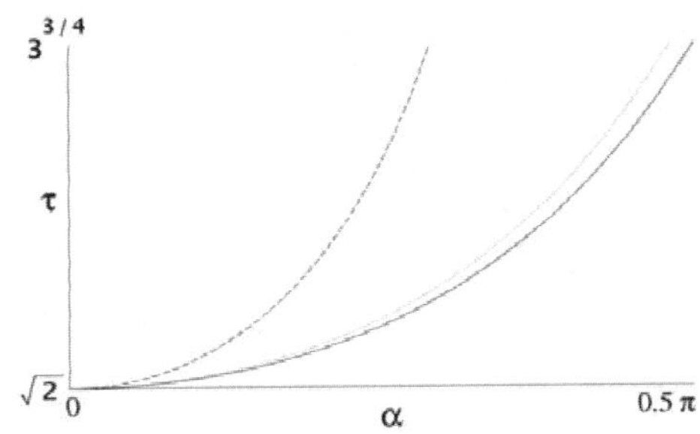

Figure 3. The motion time.

We say that α corresponds to an 'optimal tack' if projection of $\boldsymbol{a}$ on the negative ray of Bx is maximum. To find such value of α we need to maximize $G(\alpha,\beta) = -\cos\alpha\cos^2\beta\cos(\beta-\alpha)$. The maximum point of this function is $\alpha = \pi - \arctan\sqrt{5}$, $\beta = \arctan\sqrt{5}/2$. Substituting this point in G, we get 2/27. This is the maximum acceleration with which the sailing spacecraft can move toward the Sun. Also, let us remark that the tether might be non-rectilinear in reality, as in [22].

5. Navigation on the sphere

Now, let the spacecraft and the station be joined by a tether. In this case the spacecraft motion with respect to the station is restricted by some sphere. Denote, for definiteness, by B the center of this sphere. Using, as above, dimensionless variables, one can rewrite (4) as

$$\boldsymbol{r}'' = n_x^2\boldsymbol{n} + \mu\boldsymbol{r}, \tag{7}$$

where $\mu \le 0$, $(\boldsymbol{n},\boldsymbol{n}) = 1$. Let us restrict to a case of the taut tether, i.e. to motions along the sphere surface. Then, using the tether length as dimensionless unit, one can assume $(\boldsymbol{r},\boldsymbol{r}) = 1$. Combining this equality with (7) we get

$$-\mu = v^2 + n_x^2(\boldsymbol{n},\boldsymbol{r}) \ge 0, \tag{8}$$

where v is the spacecraft velocity with respect to $Bxyz$. This inequality is called a condition of constrained motion. Moreover, eliminating μ from (7), we obtain

$$\frac{d}{d\tau}\left(\frac{v^2}{2}\right) = n_x^2(\boldsymbol{n},\boldsymbol{r}'). \tag{9}$$

This implies that if the form $n_x^2(\boldsymbol{n},\boldsymbol{dr})$ is integrable then the motion equations (7) have the integral of energy.

5.1. A pendulum motion. Algorithms of relocation between two points of the sphere surface

Let now $\boldsymbol{n} = \overrightarrow{const}$. In this case the motion equations (7) have two integrals. Let Bx_1 be directed as $\boldsymbol{n}$. Then from (9) we get $v^2/2 - n_x^2 n_{x1}x_1 = h = const$, where n_{x1} and x_1 are projections of $\boldsymbol{n}$ and $\boldsymbol{r}$ on Bx_1 accordingly. Moreover, from (7) we obtain the integral of angular momentum $(\boldsymbol{r},\boldsymbol{r}',\boldsymbol{n}) = k = const$. Note that if $k = 0$ then equations (7) coincide in fact with motion equations of the simple pendulum with librations with respect to Bx_1.

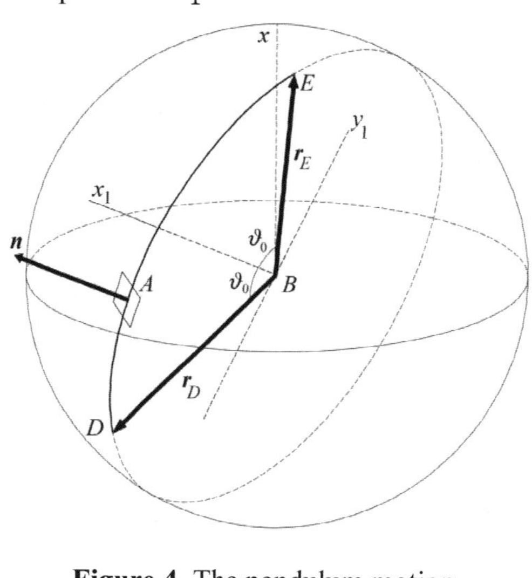

Figure 4. The pendulum motion.

Let $D(x_D, y_D, z_D)$ and $E(x_E, y_E, z_E)$ be a pair of points on the surface $(\boldsymbol{r},\boldsymbol{r}) = 1$ (figure 4). Let the spacecraft A be immovable in the point D initially. Moreover, let Bx_1 be directed as the angle $\angle DBE$

bisector. In this case, if the normal $\boldsymbol{n}$ will be directed as Bx_1 all the time, then the spacecraft A will move from D to E, and will reach the point D with zero velocity. Moreover, if one will continue to hold direction of the normal $\boldsymbol{n}$ then the spacecraft will return into D. Unfortunately, this simple algorithm of relocation is applicable only if two conditions are fulfilled. Firstly, $n_x > 0$ implies $x_D + x_E > 0$. Secondly, (8) implies $\angle DBE \leq \pi$. Particularly, if both points in the pair belong to the semi-sphere $x \leq 0$ then the algorithm is inapplicable.

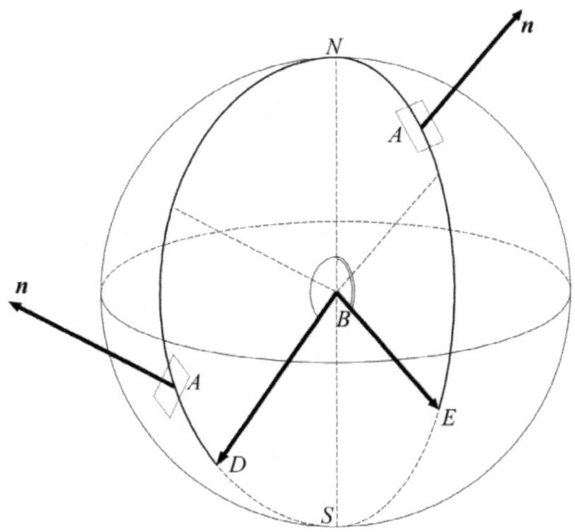

Figure 5. The pendulum motion with an intermediate point.

This problem can be resolved if one use some intermediate point and realize relocation in two steps. Specifically, the pole $x = 1$ (N in figure 5) may be such intermediate point. In other words, if $x_D + x_E \leq 0$ then the spacecraft moves initially from D to the pole N, then from N to the target point D. Evidently, the both conditions are fulfilled for such process.

Theoretically, the modified algorithm is applicable for any pair of points, excepting the pairs including the pole $x = -1$ (S in figure 5). However, the time of the one stage relocation can be computed by the formulae

$$T = \frac{2}{n_x^*} K\left(\sin\frac{\vartheta_0}{2}\right), \qquad \vartheta_0 = \frac{1}{2}\arccos\left(\overrightarrow{BD}, \overrightarrow{BE}\right), \tag{10}$$

where $n_x^* = (x_D + x_E)/(2\cos\vartheta_0)$ and $K(k)$ is the complete elliptic integral of first kind with modulus k. Evidently, if $x_D + x_E \to 0$ then $T \to \infty$. Hence, even one uses the modified algorithm, some vicinity of the pole S is not available in fact.

5.2. Refinement of the algorithm

As ε is of order of magnitude 0.001 actually, the algorithm from the previous chapter has some inaccuracy since it allows the spacecraft deviation of order of magnitude εT from the goal point. Such deviation can reach up to 50 m if the tether length is about 1km. This might not be acceptable.

To correct the algorithm, one might do the following. Let By_1 be directed as $\overrightarrow{DE}$, $Bx_1y_1z_1$ be a right-hand Cartesian coordinate system and x_1, y_1, z_1 are coordinates of the spacecraft A with respect to this system. Also, denote by q_1, q_2, q_3 coordinates of the normal $\boldsymbol{n}$ with respect to $Bx_1y_1z_1$. Let $z_1 \equiv 0$ at all time of motion. Then, from (3) implies that

$$(q_1\gamma_1 + q_2\gamma_2 + q_3\gamma_3)^2 q_3 = 2\varepsilon\left(\omega_x y_1' - \omega_y x_1'\right), \tag{11}$$

where $\gamma_2, \gamma_2, \gamma_2$ are direction cosines of Bx with respect to $Bx_1y_1z_1$; $\omega_x, \omega_y, \omega_z$ are coordinates of $\boldsymbol{e}_z$ with respect to $Bx_1y_1z_1$. Let $x_1 = \cos\vartheta$, $y_1 = \sin\vartheta$. Then one could reduce (3) to

$$\vartheta'' = (q_1\gamma_1 + q_2\gamma_2 + q_3\gamma_3)^2(q_2\cos\vartheta - q_1\sin\vartheta). \tag{12}$$

Assume that q_2, q_3 are of order of magnitude ε. Then, combining (11) and the result of integrating (12), and taking into account the terms of order of magnitude ε only, we get

$$q_3 = \frac{2\varepsilon}{\gamma_1}\sqrt{2(\cos\vartheta - \cos\vartheta_0)}\big(\omega_x\cos\vartheta + \omega_y\sin\vartheta\big). \tag{13}$$

One might rewrite it as

$$q_3 = \frac{2\varepsilon z}{\gamma_1}\sqrt{2(x_1 - x_{D1})}, \tag{14}$$

where x_{D1} is coordinate of D with respect to Bx_1.

Further, one must take into account that the spacecraft velocity in the goal point E must be zero. Basing on this requirement and taking into account only terms of order of magnitude ε, one can express q_2 from the result of integration of (12) as

$$q_2 = \frac{2\varepsilon\gamma_3}{\gamma_1^2\sin\vartheta_0}\int_{-\vartheta_0}^{\vartheta_0}\sin^2\vartheta\sqrt{2(\cos\vartheta - \cos\vartheta_0)}\,d\vartheta, \tag{15}$$

or in explicit form

$$q_2 = \frac{8\varepsilon\gamma_3}{15\gamma_1^2\sin\vartheta_0}\left((2\cos 2\vartheta_0 + 7)E\left(\sin\frac{\vartheta_0}{2}\right) - (2\cos 2\vartheta_0 - 8\cos\vartheta_0 + 7)K\left(\sin\frac{\vartheta_0}{2}\right)\right), \tag{16}$$

where $E(k)$ is the complete elliptic integral of second kind with modulus k. Note that unlike q_3, q_2 is constant. Analyzing (14-16), we see that if the spacecraft moves in the plane of the station orbit Bxy then $q_2 = q_3 = 0$ and correction of the initial algorithm is not required. Also, if $\overrightarrow{BD}$, $\overrightarrow{ED}$ and Bx are coplanar then $q_2 = 0$. Evidently, (14-16) are applicable only if n_x is not small.

One can easily check that the improved algorithm allows the spacecraft deviation of order of magnitude ε^2T^2 from the goal point. For the tether of length 1km this deviation can be estimated as less than 1 meter. It can be considered acceptable for a real situation.

6. Conclusion

In this paper the model problem of constrained motion of a spacecraft with a solar sail near a heavy heliocentric space station is formulated. Factually, a kind of 'the coastal shipping' in space is studied. The motion equations of the spacecraft with respect to station are deduced basing on estimations of the acting forces. Some criterion for the optimal placing of the sail is formulated if the spacecraft moves along a rectilinear tether. Also, motions of the spacecraft tethered to the station are considered. In this case some sphere restricts the spacecraft motion. Algorithm of the spacecraft relocation between two points of such sphere surface is offered. This algorithm provides such relocation as a kind of perturbed motions of a simple pendulum.

Acknowledgments

The author is grateful to Yu F Golubev and A V Karapetyan for useful discussions. This research was carried out at Moscow Aviation Institute (National Research University).

References

[1] *LIGHTSAIL Flight by light for CubeSats*. https://www.planetary.org/explore/projects/lightsail-solar-sailing/#the-lightsail-2-mission
[2] *IKAROS Small Scale Solar Powered Sail Demonstration Satellite* http://www.isas.jaxa.jp /en/missions/spacecraft/current/ikaros.html
[3] Tsander F A 1961 *Problems of Spaceflights with Jet Propulsion Engines: Interplanetary Travels* (Moscow: Nauka) (*russian*)

[4] Beletskii V V 2001 *Essays on the Motion of Celestial Bodies* (Basel Switzerland: Birkhäuser) doi: 10.1007/978-3-0348-8360-3
[5] Polyakhova E N 2011 *Space Flight with a Solar Sail* (Moscow: URSS) (in Russian)
[6] Koblik V, Polyakhova E, Sokolov L and Shmyrov A 2002 Controlled solar sail transfers into the near-sun orbits. *Advances in Space Res.* **29** 2083 doi: org/10.1016/S0273-1177(02)00156-4
[7] Likhachev V N, Sazonov V V and Ul'yashin A I 2004 Evolution of the Orbit of an Earth Satellite with a Solar Sail. *Cosmic Research* **42** 79 doi: org/10.1023/B:COSM.0000017565.85599.0b
[8] Polyakhova E N, Koblik V V and Sokolov L L 2011 solar sail near the sun: point-like and extended models of radiation sources. *Advances in Space Res.* **48** 1717 doi: 10.1016/j.asr. 2011.04.024
[9] Trofimov S P 2015 Dynamically invariant scaling of square solar sail systems. *Keldysh Institute Preprints* **31** [in Russian]
[10] Trofimov S P 2015 Deorbiting of Low-Earth Orbit Spacecraft Using a Sail for Solar Radiation Pressure Force Augmentation. *Keldysh Institute Preprints* **32** [in Russian]
[11] Polyakhova E and Korolev V 2018 The solar sail: Current state of the problem. *AIP Conference Proceedings* **1959**, 040014 doi: org/10.1063/1.5034617
[12] Polyakhova E, Shmyrov A. and Shmyrov V 2018 Solar radiation pressure application for orbital motion stabilization near the Sun-Earth collinear libration point. *AIP Conference Proceedings* **1959**, 040016 doi: org/10.1063/1.5034619
[13] Ovchinnikov M Y and Zaramenskikh I E 2011 Elimination of relative secular drift of orbits of two satellites caused by polar oblatenes of the Earth. *Cosmic Research.* **49** 65 doi: org/10.1134/S0010952511010059
[14] Pereira M C, Melo C F and Meireles L G 2017 Numerical analysis of orbital transfers to Mars using solar sails and attitude control. *J. Phys. Conf. Ser.* **911** 012026 doi: 10.1088/1742-6596/911/1/012026
[15] Petukhov V 2018 Optimal heliocentric trajectories for solar sail with minimum area. *AIP Conference Proceedings.* **1959** 040013 doi: org/10.1063/1.5034616
[16] Y Gao and J Wu 2016 Asteroid rotation control via a tethered solar sail. *Advances in Space Research.* **58** 2304 doi: 10.1016/j.asr.2016.08.016
[17] Heiligers J and Scheeres D 2018 Solar-sail orbital motion about asteroids and binary asteroid systems. *Journal of Guidance, Control, and Dynamics* **41** 1 doi: org/10.2514/1.G003235
[18] Vasily V and Sazonov V V 2011 Calculation of resultant vector and principal moment of light pressure forces acting upon the spacecraft with a solar sail. *Cosmic Research* **49** 56 doi: org/10.1134/S0010952511010072
[19] Chekashov A S, Starinova O L, Alipova B and Gorbunova I V 2018 Modeling of solar sail surface oscillations during interplanetary flight. *J. Phys. Conf. Ser.* **1096** 012064 doi: 10.1088/1742-6596/1096/1/012064
[20] Rodnikov A V 2018 On safe configurations of a natural-artificial space tether system. *AIP Conference Proceedings* **1959** 040018 doi: org/10.1063/1.5034621
[21] Rodnikov A V 2018 On safe orbits of a space station tethered to a precessing asteroid by two cables. *IOP Conf. Ser. Mat. Sci.* **468** 012017 doi: 10.1088/1757-899X/468/1/012017
[22] Rodnikov A V 2018 On librations of a particle on a gantline fixed on a gravitationally stabilized space station *14th International Conference "Stability and Oscillations of Nonlinear Control Systems" (Pyatnitskiy's Conference) (STAB)* doi: 10.1109/STAB.2018.8408393

IOP Conf. Series: Materials Science and Engineering **868** (2020) 012022 doi:10.1088/1757-899X/868/1/012022

Deformed state of viscoelastic bodies in one problem of tidal interaction

A Zlenko

Department of high mathematics, *Moscow Automobile and Road Construction State Technical University,* 64 Leningrad Avenue, Moscow 125319, Russian Federation

E-mail: zalaf121@mail.ru

Abstract. In our previous articles we considered the motion of two viscoelastic bodies in the gravitational field of a massive material point. We studied the evolution of the bodies, stationary solutions and their stability. We solved the equations of translational-rotational motion of the bodies together with the equations of their deformed state, using the method proposed by V G Vil'ke for the systems with infinite numbers of degrees of freedom. In our manuscript we continue our research and our aim is to get the global characteristics of the bodies, deformed by tidal forces: their shape, moments of inertia, the tidal lag angle, the relationship between the viscosity coefficient and the lag angle, tidal torques, internal material properties, depending on elastic constant and viscosity coefficient. It is very important for study of an evolutional motion of the celestial bodies, since tides affect this. In order to achieve this goal we use the tidal potential perturbation and displacements of each point of the bodies. The formulation of the problem, goal set and used methods are the novelty of our research. We applied our theory to the Earth, and it was found that the Earth's substance (in global) behaves like an auxetic.

1. Introduction

This manuscript is continuation of other our papers. In work [1] we formulated problem of motion of two viscoelastic bodies in the gravitation field of massive material point, solved the equations of motion and applied them for investigating the tidal evolution of the Earth – Moon system in the field of the Sun. In work [2] we studied the stationary solutions of these equations of motions and their stability.

As usually in researching of tides are used the classical methods of celestial mechanics or their modifications, for example, generalisation of the Darwin-Kaula theory for tides in a body librating about a spin-orbit resonance [3]. However, we use relatively "young" approach, developed by V G Vil'ke for system with infinite number of degrees of freedom [4], because we solve the equations of translational and rotational motions of the bodies together with equations of their deformation, the equations of the elasticity theory. This is the novelty of our paper for applying this method for such celestial-mechanical problem.

We find the displacements of each point of the bodies. They depend on the separations of the bodies, their orbital and rotational angular velocities, their inner structure, coefficient of viscosity and so on. These displacements change the surface of the bodies and influence in turn on the motion of the bodies. This is the novelty of our paper in comparison with other works on this subject.

We use the Kelvin-Voigt dissipation model. In article [5] it was considered a simple example of a Maxwell material. In work [6] it was used several rheological models, i.e. Andrade, extended Burgers,

Sundberg-Cooper, and power-law to compute the dissipative properties within Mars and calculate the global quality factor. It would be interesting to further apply these rheological models in our approach.

Tidal orbital evolution of the Mars-Phobos system and the Keplerian elements in two-body problem was studied in [6] and [7] respectively without taking into account their rotational motion. But we considered not only the orbital evolution of the Earth-Moon system but also their rotation evolution. We caught the moment when the Moon entered the spin-orbit resonance and continues to remain in it in the future. This is the advantage of the method used.

Papers [1, 2] and our manuscript are a single whole. So, using their results, we are going to get the global characteristics of the bodies, deformed by tidal forces: their shape, moments of inertia, tidal torques, the tidal lag angle, the relationship between the viscosity coefficient and lag angle, *Love number* k_2. The vast majority of works are limited, as a rule, *Love number* k_2, for example, during experiment simulation the Ganimede's tides *Love number* k_2 is determined with an accuracy 10^{-4} [8]. We intend to apply our theory to the Earth and find global internal material properties, depending on elastic constants and coefficient of internal viscous friction. All the goals set are new, because the celestial-mechanical model and methods of its research are new.

2. Main notations and assumptions in statement of the problem

We consider the motion of two homogeneous and isotropic viscoelastic bodies (the Earth and the Moon) with masses m_1 and m_2, respectively, in the central field of massive material point O (the Sun) with mass M and $m_2 \ll m_1 \ll M$ relation. We connect with point O inertial frame $OXYZ$, with points O_i − Koenig's system of coordinates O_iXYZ and $O_iX_iY_iZ_i$ system fixed in the rotating body, where O_i is the centre of mass of the *i*-th body (everywhere in the text $i=1$ is reffered to the first body, $i=2$−the second body). The OX, OY, O_iX_i and O_iY_i axes lie in the orbital plane of motion of the bodies. O_iZ and O_iZ_i axes coincide, and are perpendicular to the orbital plane. *I*-th body rotates about the O_iZ_i axis with ω_{2+i} angular speed. We denote φ_i the angle between O_iX_i and O_iX axes. It is measured from O_iX axis of Koenig's system of coordinates O_iXYZ. Let $\boldsymbol{R}_1$ ($R_1\cos\lambda_1$, $R_1\sin\lambda_1$, 0) denote the position vector of the centre of mass C of the bodies in the inertial frame. Here λ_1 is the angle measure from the OX axis, R_1 is the separation of O and C points. In the O_1XYZ Koenig's system vector $\boldsymbol{R}_2$ $(R_2\cos\lambda_2, R_2\sin\lambda_2, 0)$ denote the position vector of O_2 point, where λ_2 angle is measured from the O_1X axis, R_2 is the separation of O_1 and O_2 points. The centre of mass C moves with respect to origin O of inertial frame with orbital angular velocity ω_1. The second body moves about the first body with ω_2 angular velocity.

In the unperturbed state the bodies are solid, homogeneous and spherical of density ρ_i and r_{i0} radius, and $r_{i0} \ll R_2 \ll R_1$ relation. The centre of mass C moves in circular keplerian orbit with respect to origin O of inertial frame with constant angular velocity ω_1, the second body moves in circular keplerian orbit about the first body with ω_2 constant angular velocity, *i*-th body rotates about the O_iZ_i axis at ω_{2+i} constant rate. All four movements are independent of each other.

If the bodies are viscoelastic then the shape of each body is distorted by gravitational forces of massive point and another body treated as material point. Due to distribution of matter associated with displacements of the points at each body, the tidal potential perturbation (non-central part of the external potential) arises. We call it just perturbing potential. In the perturbed motion the centre of mass C moves in quasicircular orbit with respect to massive point, the second body moves about the first body in quasicircular orbit too. All ω_1, ω_2, ω_3, ω_4 angular velocities are very slow function of time, since perturbations are small.

3. Viscoelastic displacements and perturbing potential

The full statement of the problem and solution of equation of motion is in our work [1]. We got there the displacements $\boldsymbol{u}_i$ of i-th deformed body, potential Π of the system and perturbing potential Π_{p}.

$$\boldsymbol{u}_i(\boldsymbol{r}_i,t)=\boldsymbol{u}_{i11}+\boldsymbol{u}_{i12}+\boldsymbol{u}_{i13} \tag{1}$$

The displacements of points $\boldsymbol{u}_{i11}$ are caused by centrifugal forces, $\boldsymbol{u}_{i12}$ – elastic displacements are caused by massive point and another body in i-th body, $\boldsymbol{u}_{i13}$ – dissipative displacements are caused by massive point and another body in i-th body. We will consider only non-stationary $\boldsymbol{u}_{i12}$ and $\boldsymbol{u}_{i13}$ displacements and $\boldsymbol{u}_{ins}(\boldsymbol{r}_i,t)=\boldsymbol{u}_{i12}+\boldsymbol{u}_{i13}$ their sum.

$$\boldsymbol{u}_{i12}=\sum_{k=1}^{2}\boldsymbol{u}_{ik12}=3\rho_i E_i^{-1}\sum_{k=1}^{2} f_{ki}R_k^{-3}\left[a_{i1}(\boldsymbol{B}_{ki}\boldsymbol{r}_i,\boldsymbol{r}_i)\boldsymbol{r}_i+(a_{i2}\boldsymbol{r}_i^2+a_{i3}r_{i0}^2)\boldsymbol{B}_{ki}\boldsymbol{r}_i\right] \tag{2}$$

$$\boldsymbol{u}_{i13}=\sum_{k=1}^{2}\boldsymbol{u}_{ik13}=-3\rho_i E_i^{-1}\chi_i\sum_{k=1}^{2} f_{ki}R_k^{-3}(\omega_k-\omega_{2+i})\left[a_{i1}\left(\frac{\mathrm{d}\boldsymbol{B}_{ki}}{\mathrm{d}\psi_{\kappa\iota}}\boldsymbol{r}_i,\boldsymbol{r}_i\right)\boldsymbol{r}_i+(a_{i2}\boldsymbol{r}_i^2+a_{i3}r_{i0}^2)\frac{\mathrm{d}\boldsymbol{B}_{ki}}{\mathrm{d}\psi_{ki}}\boldsymbol{r}_i\right], \tag{3}$$

$$a_{i1}=(1+\nu_i)/(5\nu_i+7),\ a_{i2}=-a_{i1}(2+\nu_i),\ a_{i3}=a_{i1}(2\nu_i+3),$$

$$\boldsymbol{B}_{ki}=\frac{1}{6}\begin{pmatrix}3\cos 2\psi_{ki}+1 & 3\sin 2\psi_{ki} & 0\\ 3\sin 2\psi_{ki} & -3\cos 2\psi_{ki}+1 & 0\\ 0 & 0 & -2\end{pmatrix}. \tag{4}$$

Here $f_{1i}=f=\mathrm{G}M$, $f_{2i}=\mathrm{G}m_{3-i}$, G – universal gravitational constant. In the $O_iX_iY_iZ_i$ system vector $\boldsymbol{r}_i(x_i,y_i,z_i)$ denote the position vector of any point of the non-deformed i-th body. E_i – Young's modulus, ν_i – Poisson's ratio, χ_i – the coefficient of internal viscous friction of i-th body, $\psi_{ki}=\lambda_k-\varphi_i$. In formulaes (2), (3) $\boldsymbol{u}_{ik12}$, $\boldsymbol{u}_{ik13}$ – elastic and dissipative displacements are raised in i-th body by k-th material point, index of summation $k = 1$ means – the massive point, $k = 2$ – another body treated as material point different i-th body.

$$\Pi=-fm/R_1-Gm_1m_2/R_2+\Pi_{\mathrm{p}}. \tag{5}$$

We will consider Π_{p} as a function of $\boldsymbol{u}_{ins}$ only.

$$\Pi_{\mathrm{p}}(\boldsymbol{u}_{ins})=\sum_{k=1}^{2}\sum_{i=1}^{2}\Pi_{\mathrm{p}ki}(\boldsymbol{u}_{ins}),\ \Pi_{\mathrm{p}ki}(\boldsymbol{u}_{ins})=f_{ki}R_k^{-3}\int_{V_i}\left[\boldsymbol{r}_i\boldsymbol{u}_{ins}-3(\boldsymbol{\xi}_{ki}\boldsymbol{r}_i)(\boldsymbol{\xi}_{ki}\boldsymbol{u}_{ins})\right]\rho_i\,\mathrm{d}v_i \tag{6}$$

Here $m=m_1+m_2$, $\Pi_{\mathrm{p}ki}$ – is the perturbing potential of k-th material point and i-th body, deformed by massive point and k-th material point; V_i – sphere of radius r_{i0} of the non-deformed i-th body, $\mathrm{d}v_i$ – the element of volume.

$$\boldsymbol{\xi}_{ki}=\tau_{ki}(\cos\psi_{ki},\sin\psi_{ki},0),\ \tau_{21}=-1,\ \tau_{ki}=1\ (k\neq 2,\ i\neq 1). \tag{7}$$

We will transform the sum $\boldsymbol{u}_{ik12}+\boldsymbol{u}_{ik13}=\boldsymbol{u}_{ikns}$ – viscoelastic displacements are raised in i-th body by k-th material point. For this purpose we will introduce a new variable α_{ki} and new matrix $\tilde{\boldsymbol{B}}_{ki}$ as follows

$$\tan 2\alpha_{ki}=2\chi_i(\omega_k-\omega_{2+i}) \tag{8}$$

$$\tilde{\boldsymbol{B}}_{ki} = \boldsymbol{B}_{ki} - \frac{1}{2}\tan 2\alpha_{ki} = \frac{1}{6\cos 2\alpha_{ki}}\begin{pmatrix} 3\cos 2\tilde{\psi}_{ki} + \cos 2\alpha_{ki} & 3\sin 2\tilde{\psi}_{ki} & 0 \\ 3\sin 2\tilde{\psi}_{ki} & -3\cos 2\tilde{\psi}_{ki} + \cos 2\alpha_{ki} & 0 \\ 0 & 0 & -2\cos 2\alpha_{ki} \end{pmatrix},$$

(9)

where $\tilde{\psi}_{ki} = \psi_{ki} - \alpha_{ki}$.

Then $\boldsymbol{u}_{ins}$ and $\boldsymbol{u}_{ikns}$ take the form

$$\boldsymbol{u}_{ins} = \sum_{k=1}^{2} \boldsymbol{u}_{ikns}, \quad \boldsymbol{u}_{ikns} = 3\rho_i E_i^{-1} f_{ki} R_k^{-3}\left[a_{i1}(\tilde{\boldsymbol{B}}_{ki}\boldsymbol{r}_i, \boldsymbol{r}_i)\boldsymbol{r}_i + (a_{i2}\boldsymbol{r}_i^2 + a_{i3}r_{i0}^2)\tilde{\boldsymbol{B}}_{ki}\boldsymbol{r}_i \right]. \tag{10}$$

We substitute expression $\boldsymbol{u}_{ins}$ (10) in $\Pi_{pki}(\boldsymbol{u}_{ins})$ (6) and after cumbersome and tedious calculations obtain tidal potential perturbation

$$\Pi_{pki}(\boldsymbol{u}_{ins}) = -\frac{3}{2}\beta_{ik}\rho_i D_{i2}\sum_{q=1}^{2} f_{qi}R_q^{-3}\left[3\cos(\lambda_q - \alpha_{qi} - \lambda_k) + 1\right], \tag{11}$$

where

$$\beta_{ik} = \rho_i E_i^{-1} f_{ki} R_k^{-3}, \; D_{i2} = 4\pi r_{i0}^7 a_{i1}(9\nu_i + 13)/105. \tag{12}$$

4. The shape of the body, deformed by the viscoelastic displacements

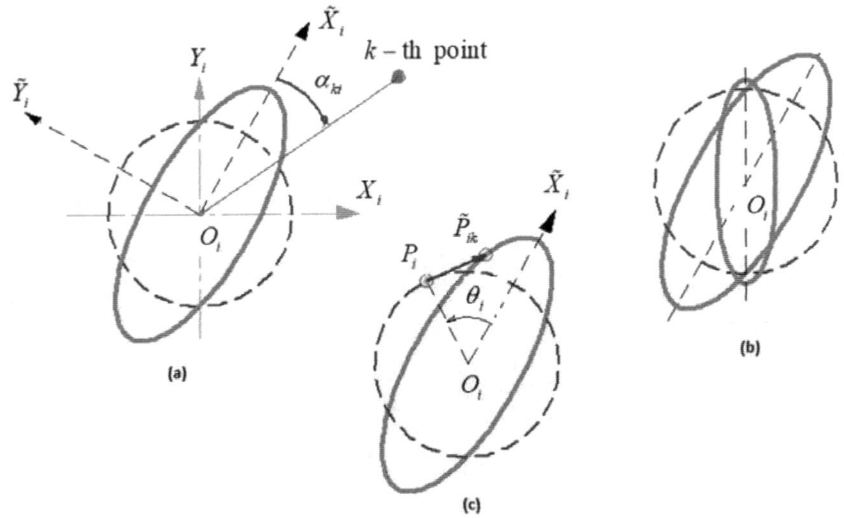

Figure 1. Schematic presentations by dashed circular line the shape of *i*-th non-deformed body with center O_i and by solid line the shape of *i*-th body (oblate spheroid) deformed *k*-th material point **(a)** and **(c)**. On part **(b)** the shape of *i*-th body is deformed by another body as material point and massive point. $O_iX_iY_iZ_i$ – coordinate system, fixed in *i*-th rotating body, $O_i\tilde{X}_i\tilde{Y}_i\tilde{Z}_i$ – system, connected with deformed body **(a)**. The $O_i\tilde{X}_i$ axis is the axis of symmetry of deformed body **(a)** and **(c)**. $|\alpha_{ki}|$ – tidal lag angle **(a)**. The points P_i transit to points $\tilde{P}_{ik}$ in the process of deformation, θ_i – polar angle of points P_i on the surface of *i*-th non-deformed body **(c)**.

We will consider the shape of *i*-th body distorted by the only one *k*-th material point (figure 1a, 1c). The arbitrary point of the non-deformed body with position vector $\boldsymbol{r}_i(x_i, y_i, z_i)$ transits to the point with position vector $\boldsymbol{r}_{ik} = \boldsymbol{r}_i + \boldsymbol{u}_{ikns}$ in $O_iX_iY_iZ_i$ coordinate system. We want to find the shape of the body distorted by $\boldsymbol{u}_{ikns}$ displacements. Let's go to new $O_i\tilde{X}_i\tilde{Y}_i\tilde{Z}_i$ coordinate system. The axes $O_i\tilde{X}_i$ and $O_i\tilde{Y}_i$ lie in the plane $O_iX_iY_i$, the axes $O_i\tilde{Z}_i$ and O_iZ_i coincide. The angle between axes $O_i\tilde{X}_i$ and O_iX_i is equal to $\tilde{\psi}_{ki}$. It is measured from O_iX_i axis. The transformation from $O_i\tilde{X}_i\tilde{Y}_i\tilde{Z}_i$ system to $O_iX_iY_iZ_i$ system is given by the matrix

$$\boldsymbol{\Gamma}_{ik}(\tilde{\psi}_{ik}) = \begin{pmatrix} \cos\tilde{\psi}_{i\kappa} & -\sin\tilde{\psi}_{i\kappa} & 0 \\ \sin\tilde{\psi}_{i\kappa} & \cos\tilde{\psi}_{i\kappa} & 0 \\ 0 & 0 & 1 \end{pmatrix}, \ \boldsymbol{r}_i = (x_i, y_i, z_i)^{\mathrm{T}} = \boldsymbol{\Gamma}_{ik}(\tilde{\psi}_{ik})\tilde{\boldsymbol{r}}_i = \boldsymbol{\Gamma}_{ik}(\tilde{\psi}_{ik})(\tilde{x}_i, \tilde{y}_i, \tilde{z}_i)^{\mathrm{T}}, \quad (13)$$

where $\tilde{\boldsymbol{r}}_i(\tilde{x}_i, \tilde{y}_i, \tilde{z}_i)$ is the vector $\boldsymbol{r}_i$ in $O_i\tilde{X}_i\tilde{Y}_i\tilde{Z}_i$ system.

Let $\boldsymbol{r}_{i0}(\tilde{x}_{i0}, \tilde{y}_{i0}, \tilde{z}_{i0})$ are the coordinates of an arbitrary point P_i on the surface of the non-deformed body (the sphere with radius r_{i0}) in $O_i\tilde{X}_i\tilde{Y}_i\tilde{Z}_i$ system. We introduce the angle θ_i ($0 \le \theta_i \le \pi$) a such that $\tilde{x}_{i0} = r_{i0}\cos\theta_i$, $\sqrt{\tilde{y}_{i0}^2 + \tilde{z}_{i0}^2} = r_{i0}\sin\theta_i$ (figure 1c). In the process of deformation the points $P_i(\tilde{x}_{i0}, \tilde{y}_{i0}, \tilde{z}_{i0})$ transit to $\tilde{P}_{ik}(\tilde{x}_{ik}, \tilde{y}_{ik}, \tilde{z}_{ik})$ points (figure 1c). They form the surface of *i*-th body in $O_i\tilde{X}_i\tilde{Y}_i\tilde{Z}_i$ system distorted by *k*-th material point.

$$\tilde{x}_{ik}(\theta_i) = r_{i0}\cos\theta_i\left\{1 + 4\gamma_{ik}r_{i0}^2\left[a_{i1}P_2(\cos\theta_i) + (a_{i2} + a_{i3})\right]\right\} \quad (14)$$

$$\tilde{R}_{ik}(\theta_i) = \sqrt{\tilde{y}_{ik}^2 + \tilde{z}_{ik}^2} = r_{i0}\sin\theta_i\left\{1 + 2\gamma_{ik}r_{i0}^2\left[2a_{i1}P_2(\cos\theta_i) - (a_{i2} + a_{i3})\right]\right\}, \quad (15)$$

$$O_i\tilde{P}_{ik}(\theta_i) = \sqrt{\tilde{x}_{ik}^2 + \tilde{R}_{ik}^2} = r_{i0}\left[1 + 4\gamma_{ik}r_{i0}^2a_{i1}(2 + \nu_i)P_2(\cos\theta_i)\right], \quad (16)$$

where

$$\gamma_{ik} = \beta_{ik}/(2\cos 2\alpha_{ki}) = \rho_i E_i^{-1} f_{ki} R_k^{-3}/(2\cos 2\alpha_{ki}). \quad (17)$$

So, the deformed body is axisymmetric body of rotation at first approximation, an oblate spheroid, with $O_i\tilde{X}_i$ axis of symmetry. Its long semi-axis $\tilde{x}_{ik}(0)$ is lying along the $O_i\tilde{X}_i$ axis and $\tilde{R}_{ik}(\pi/2)$ short semi-axis gives a circular cross section.

$$\tilde{x}_{ik}(0) = O_i\tilde{P}_{ik}(0) = r_{i0}\left[1 + 4\gamma_{ik}r_{i0}^2a_{i1}(2 + \nu_i)\right], \quad (18)$$

$$\tilde{R}_{ik}(\pi/2) = r_{i0}\left[1 - 2\gamma_{ik}r_{i0}^2a_{i1}(2 + \nu_i)\right]. \quad (19)$$

The largest amplitude $\Delta\tilde{x}_{ik}(0)$ of the tidal bulge at $\theta_i = 0$ or π is equal:

$$\Delta\tilde{x}_{ik}(0) = \tilde{x}_{ik}(0) - r_{i0} = 4\gamma_{ik}r_{i0}^3a_{i1}(2 + \nu_i). \quad (20)$$

The moments of inertia of the deformed body are equal:

$$\begin{aligned} \tilde{A}_{ik} &= \int_{V_i}(\tilde{y}_{ik}^2 + \tilde{z}_{ik}^2)\,\mathrm{d}m_i = 0.4m_i r_{i0}^2 - 8\gamma_{ik}\rho_i D_{i2} \\ \tilde{B}_{ik} &= \int_{V_i}(\tilde{x}_{ik}^2 + \tilde{z}_{ik}^2)\,\mathrm{d}m_i = 0.4m_i r_{i0}^2 + 4\gamma_{ik}\rho_i D_{i2}, \\ \tilde{C}_{ik} &= \int_{V_i}(\tilde{y}_{ik}^2 + \tilde{x}_{ik}^2)\,\mathrm{d}m_i = 0.4m_i r_{i0}^2 + 4\gamma_{ik}\rho_i D_{i2}. \end{aligned} \quad (21)$$

We have calculated the volume V_i of the distorted *i*-th body.

$$V_i = \frac{4}{3}\pi r_{i0}^3\left[1 + 12\gamma_{ik}r_{i0}^2a_{i1}(1 + \nu_i)\sin^2\alpha_{ki}\right] \quad (22)$$

The expressions (14)-(21) are obtained, neglecting the higher order terms γ_{ik}^2 and $\gamma_{ik}\sin^2\alpha_{ki}$. But in the expression (22) we left the term $\gamma_{ik}\sin^2\alpha_{ki}$ to show that the body volume increased. If we take them into account then the body shape will be triaxial. It is caused by viscous forces.

5. Tidal torques

So we have got the perturbing potential $\Pi_{pki}(\boldsymbol{u}_{ins})$ (11) of k-th material point and i-th body, deformed by massive point and k-th material point, in other words, the potential of interaction of k-th material point with the tidal bulge arised by it ($q=k$) and the bulge arised by massive point ($q \neq k$) in i-th body (figure 1b). We can rewrite (11):

$$\Pi_{pki}(\boldsymbol{u}_{ins}) = -6\beta_{i\kappa}\rho_i D_{i2}\sum_{q=1}^{2} f_{qi} R_q^{-3} P_2\left(\cos\left(\lambda_q - \alpha_{qi} - \lambda_k\right)\right). \quad (23)$$

where $P_2(\cos\theta_i) = (3\cos 2\theta_i + 1)/4$ – Legendre polynomial of degree two in $\cos\theta_i$.

In (23) the angle $(\lambda_q - \alpha_{qi})$ is the angle between the axis of symmetry of i-th body deformed by q-th point and the axis O_iX of the Koenig's coordinate system. It is measured from O_iX axis. Here the potential Π_{pki} can give the only one non-zero component $T_{qki\tilde{z}}$ of the $\boldsymbol{T}_{qki}(T_{qki\tilde{x}}, T_{qki\tilde{y}}, T_{qki\tilde{z}})^{\mathrm{T}}$ torque, which is directed along the $O_i\tilde{Z}_i$ (O_iZ_i) axis and is orthogonal to the plane of orbit:

$$T_{qki\tilde{x}} = T_{qki\tilde{y}} = 0,$$

$$T_{qki\tilde{z}} = \partial(-\Pi_{pki})/\partial(\lambda_q - \alpha_{qi}) = -9\beta_{ik}\rho_i D_{i2}\sum_{q=1}^{q=2} f_{qi} R_q^{-3}\sin 2\left(\left(\lambda_q - \alpha_{qi}\right) - \lambda_k\right). \quad (24)$$

We consider two cases for fixed q.

In first case $q \neq k$. Then the torque that acts from the k-th point on i-th body deformed by the q-th point is shortperiodic value and its component is equal

$$T_{qki\tilde{z}} = -9\beta_{ik}\rho_i D_{i2} f_{qi} R_q^{-3}\sin 2\left(\left(\lambda_q - \alpha_{qi}\right) - \lambda_k\right). \quad (25)$$

This torque creates the tidal libration of i-th body and does not affect its evolution.

In second case $q=k$. Then the torque, that acts from the k-th point on i-th body deformed by it, has longperiodic component and is equal

$$T_{kki\tilde{z}} = 9\beta_{ik}\rho_i D_{i2} f_{ki} R_k^{-3}\sin 2\alpha_{ki}. \quad (26)$$

Since $\sin 2\alpha_{ki} \approx 2\chi_i(\omega_k - \omega_{2+i})$, it follows when angular rate ω_{2+i} of i-th body is greater the orbital velocity ω_k of k-th point, we have negative torque acting clockwise and slowing down the rotation of i-th body (a planet). The same, but positive torque (counterclockwise) acts on k-th point (the body) and accelarates its movement in orbit. This leads to the removal of k-th point (satellite) from i-th body (a planet). When angular rate ω_{2+i} of i-th body is lower the orbital velocity ω_k of k-th point, we have positive torque acting counterclockwise and accelarating the rotation of i-th body (a planet). The same, but negaitive torque (clockwise) acts on k-th point (satellite) and slowing down its movements in orbit. This causes k-th point (satellite) to approach to i-th body (a planet).

6. Comparison of results

1. Angle α_{ki} is the angle between the axis of symmetry of i-th body deformed by k-th point (tidal bulge) and the direction to the k-th point (figure 1a). In other words, the module of α_{ki} is the tidal lag angle

$$|\alpha_{ki}| \approx \chi_i|\omega_k - \omega_{2+i}|. \quad (27)$$

In work [9] we can see the tidal lag angle in the form

$$\delta = \Delta t|\omega_p - n|. \quad (28)$$

In this formula Δt is time lag, ω_p is the planetary spin rate, n is the satellite's mean motion. The formulas (27) and (28) are identical. So the coefficient internal viscosity is equal time lag:

$$\chi_i = \Delta\mathrm{t}. \quad (29)$$

This is the mechanical meaning of coefficient internal viscosity χ_i.

2. In work [10] (formulae 4.44) it is supposed the surface of a body deformed by tidal forces is described by equation

$$R(\theta') = C\left[1+\varepsilon_2 P_2(\cos\theta')\right], \tag{30}$$

where $\varepsilon_2 (\ll 1)$ – constant, C – indicates the average radius of the planet. But in our model we not supposed we proved this equality

$$O_i\tilde{P}_{ik}(\theta_i) = r_{i0}\left[1+4\gamma_{ik}r_{i0}^2 a_{i1}(2+\nu_i)P_2(\cos\theta_i)\right]. \tag{31}$$

The expressions (30) and (31) are almost identical. The surface of deformed body is described by Legendre polynomial P_2 in both cases at first approximation.

3. Let us compare the torque $T_{kki\bar{z}}$ (26) and the same torque given by (4.159) in [10].

$$\Gamma = \frac{3}{2}k_2\frac{Gm_s^2}{a^6}C^5\sin 2\varepsilon, \tag{32}$$

where k_2 is *Love number*. In our notations, $m_s = m_k$, $a = R_k$, $C = r_{i0}$, $\alpha_{ki} = \varepsilon$. We rewrite torque (32)

$$\Gamma = \frac{3}{2}k_2\frac{Gm_k^2}{R_k^6}r_{i0}^5\sin 2\alpha_{ki}. \tag{33}$$

We present the torque $T_{kki\bar{z}}$ (26) in the same structure as expression (33)

$$T_{kki\bar{z}} = \frac{3}{2}\left[8\pi G\rho_i^2 E_i^{-1}r_{i0}^2(1+\nu_i)(9\nu_i+13)/\left(35(5\nu_i+7)\right)\right]\frac{Gm_k^2}{R_k^6}r_{i0}^5\sin 2\alpha_{ki}. \tag{34}$$

Equating torques Γ (33) and $T_{kki\bar{z}}$ (34), we can see that in our model k_2 has the following expression

$$k_2 = 8\pi G\rho_i^2 E_i^{-1}r_{i0}^2(1+\nu_i)(9\nu_i+13)/\left(35(5\nu_i+7)\right). \tag{35}$$

7. Global values of Poisson's ratio and viscosity coefficients for the Earth

It is known from observations that the mean value *Love number* $k_2 = 0.299$ for the Earth [10], gravitational constant $G = 6.67\times10^{-11}\,\mathrm{m^3kg^{-1}sek^{-2}}$, density $\rho_1 = 5.52\times10^3\,\mathrm{kg\,m^{-3}}$, mean radius $r_{i0} = 6.37\times10^6\,\mathrm{m}$. But we don't know Young's modulus E_i and ν_i – Poisson's ratio. Let $E_i = 10^{11}\,\mathrm{N\,m^{-2}}$ Young's modulus. This is a reasonable assumption. This modulus is a quarter more than marble and two times less than iron. Then we can find ν_i from equation (35). Its value $\nu_i = -0.73$ is negative. Is the Earth *auxetic*? What it means? How it may be explained? It may be explained from physicochemical point of view. Inside the Earth is a *steam boiler*. When we begin to stretch the globe from the poles, its outer walls will begin to thin out and internal pressure will begin lead to the expansion of the body. Formula (22) confirms this: body volume increased.

We got that coefficient of internal friction χ_i is equal to time lag and we would like to estimate its value for the Earth. It follows from (27)

$$\chi_i \approx |\alpha_{ki}|/|\omega_k - \omega_{2+i}| \tag{36}$$

The tidal lag angle $\alpha_{ki} = 2.52^\circ = 0.044$ rad for Moon's tides on the Earth, orbital angular velocity the Moon $\omega_k = 266\times10^{-8}$ rad/s, angular polar rate of rotation the Earth $\omega_{2+i} = 727\times10^{-7}$ rad/s. Hence we obtain from (36) viscosity coefficient $\chi_i = 629$ s $= 0.007$ d.

8. Conclusion

The very problem of creating an analytical model of the planet (the Earth), deformed by the tide, is determined by the complexity of its internal structure. In our model we find the global characteristics of the bodies. In scientific literature the most popular version of internal structure of the Earth is given by Love [11]. He considered a spherical, homogeneous, incompressible body and showed that its surface must be described the second zonal harmonic $P_2(\cos\theta_i)$. We studied more realistic model of viscoelastic body and found that its surface also is described by the same harmonic at first approximation. But when we take into account viscous forces then its shape is triaxial.

We have discovered the effect of the tidal libration, when on the tidal bulge caused, for example, on the Earth by the Moon is acting the torque from the Sun, formula (25). It means the libration of the tidal lag angle with $T = \pi/(\omega_2 - \omega_1)$ period, where ω_2 is the orbital angular velocity the Moon about the Earth and ω_1 is the orbital angular velocity the Earth about the Sun.

We have obtained the new formula for the tidal lag angle (27), which is equal to the product of the viscosity coefficient of a given body and the absolute value of the difference in the angular velocities of the interacting bodies. The mechanical meaning of viscosity coefficient is that it is equal tidal time lag (29).

We have set very interesting fact that the Earth's substance (in global) behaves like an auxetic. It means when the body is stretched, it becomes thicker perpendicular to the applied force. And from the formula (22) follows that the volume of the deformed body increased due to action of viscous forces.

References

[1] Zlenko A A 2015 A celestial-mechanical model of the tidal evolution of the Earth–Moon system. *Astron Rep* **59(1)** 72 DOI: 10.1134/S1063772915010096
[2] Zlenko A A 2016 Stationary solutions in a model three-body problem. *J Applied Mathematics and Mechanics* **80(4)** 324 https://doi.org/10.1016/j.jappmathmech.2016.09.007
[3] Frouard J, Efroimsky M 2017 Tides in a body librating about a spin-orbit resonance. Generalisation of the Darwin-Kaula theory. *Cel. Mech. Dynam. Astron.* **129** 177 doi: 10.1007/s10569-017-9768-6
[4] Vil'ke V G 1997 *Analytical Mechanics of Systems with an Infinite Number of Degrees of Freedom* Pts I and II (Moscow: Izd Mekh -Mat MGU) p 376
[5] Efroimsky M 2018 Dissipation in a tidally perturbed body librating in longitude. *Icarus* **306** 328 https://doi.org/10.1016/j.icarus.2017.10.020
[6] Bagheri A, Khan A, Efroimsky M, Giardini D 2019 Orbital Evolution of the Mars-Phobos Tidal System. *50th Lunar and Planetary Science Conference 2019* **2132** 3183
[7] Boué G, Efroimsky M 2019 Tidal evolution of the Keplerian elements *Cel. Mech. Dynam. Astron.* **131** 30 https://doi.org/10.1007/s10569-019-9908-2
[8] Cappuccio P, Hickey A, Durante D, Di Benedetto M, Iess L, De Marchi F, Plainaki C, Milillo A, Mura A 2020 Ganymede's gravity, tides and rotational state from JUICE's 3GM experiment Simulation. *Planetary and Space Sci.* **187** 104902 https://doi.org/10.1016/j.pss.2020.104902
[9] Efroimsky M, Lainey V 2007 The Physics of Bodily Tides in Terrestrial Planets, and the Appropriate Scales of Dynamical Evolution. *Astrophysics* 112 DOI: 10.1029/2007JE002908
[10] Murray C D Dermott S F 2010 *Solar System Dynamics* (Cambridge University Press) p 608
[11] Love A E H 1944 *A Treatise on the Mathematical Theory of Elasticity* 4th ed. (Dover, New York) p 670

IOP Conf. Series: Materials Science and Engineering **868** (2020) 012023 doi:10.1088/1757-899X/868/1/012023

Architecture of air conditioning system with electric-motor-driven blowers

K Gubernatorov, M Kiselev, Y Moroshkin and A Chekin[1]

All from Functional Systems Laboratory, *State Research Institute of Aviation Systems,* 7 Viktorenko Street, Moscow 125319, Russian Federation

[1]E-mail: aychekin@2100.gosniias.ru

Abstract. Architectures for air conditioning systems based on electrically driven air pumps for perspective commercial aircraft with a high level of electrification are very diverse. To select the optimal structure of such air conditioning systems for a specific aircraft and to create a scientific and technical groundwork for systems of this type for a wide class of commercial aviation aircraft this article proposes a combinatorial-logic method for design architecture of air conditioning system with electric blowers. The method is based on an oriented graph of combinations of subsystems and units with an assessment the quality of each variant in terms of aircraft fuel efficiency. The method allows to design a concept of on-board equipment in order to create an aircraft with a high level of electrification and to form a knowledge base on air conditioning systems with electric blowers, as well as a hybrid air conditioning systems where electric blowers are combined with other sources of fresh air to supply the aircraft's pressurized cabin.

1. Introduction

The concept of future energy-efficient mainline aircraft address to convert most or all airborne aircraft systems to a single type of energy – electric energy (respectively, the concept of "more electrified aircraft" and the concept of "fully electrified aircraft") [1]. One of the main requirements for air conditioning systems of such airplanes is the rejection of air bleeding for the needs of air conditioning systems from the compressor stage of a gas turbine of jet engine in favor of intake of external air through special air intakes and air compression using electric blowers [2]. The classical principle of air consumption from jet engine is associated with increased fuel consumption and significant loss of jet engine power. The lost power can reach 500-700 kW at a mainline aircraft. The electric air conditioning systems allows reducing the power taken from engines by two or more times [3]. However, even in this case, air conditioning systems remains the most energy-consumption constantly working on-board system, which largely determines the requirements for the power and architecture of the aircraft's power supply system as a whole.

This is the basis for research the most energy-efficient architecture using the electrically driven principle of pressurization of a cabin in the process of forming the concept of the on-board systems of a certain aircraft, as well as for creating a scientific and technical advance of this class of air conditioning systems in order to be prepared for appearing a new more electrified aircraft generation. System designers can not choose the most efficient architecture with new technical solutions because they use mostly the base of past designing skills and they have a lack of information of innovations.

The system integrator of the on-board equipment of a promising aircraft, which does not have sufficient experience in integrating the on-board avionics with systems based on new principles, is in the same situation. In this regard, it becomes urgent to create an approach for the synthesis and analysis of air conditioning systems, as one of the main energy-consuming aircraft systems, which would allow us to consider the widest possible range of system designs with an analysis of their effectiveness. This paper proposes an approach, a combinatorial-logical method for making solutions of multivariate systems based on a directed graph [4] of practicable combinations of the solutions (alternative decision tree, or A-tree [5]) is proposed. Various expressions, as indicators of the effectiveness are used to assess the fuel efficiency of the aircraft [6, 7]. Further, an electric air conditioning system means a system with electric blowers which can be the only source of compressed air, or can work in combination with other sources of compressed external air (hybrid and combined systems).

2. Methods and materials

Air conditioning system with electric blowers for commercial aircraft in its most general form can be represented by a structure consisting of the following components (figure 1): a subsystem supplying the outdoor air with the use of electric blowers, a subsystem of the primary air cooling; a subsystem of temperature and humidity of air control; a subsystem of the target temperature control of heat zones and compartments of the aircraft; an air distribution with the recirculation loop; a cabin pressure control system for automatic regulation of the pressure in the pressurized cabin. In addition, the air conditioning system provides the certain temperature mode of avionics and power electronics. The system also provides an aircraft of neutral gas and pressurization of hydraulic system [8]. There can be added another subsystems apart of the presented subsystems in figure 1 such as liquid coolant and refrigerant coolant systems.

A preliminary review of possible solutions of electrical air conditioning systems or systems with auxiliary using of electric blowers demonstrates a lot of versions of the systems that require scientific and technical and economic analysis. Thus, in the thermodynamic circuits of the air preparation subsystems can present air cycle refrigerating machines with turbochargers of various designs, vapor compression machines of direct cooling/heating or using a secondary coolant, various combinations of air cycle refrigerating and vapor compression machines (combined systems) can be used. In such systems, air cycle refrigerating and vapor compression machines can work in series or in parallel. There are possible systems with air extraction from the fan circuit of the main engine of the aircraft, followed by increased pressure by the auxiliary electric blowers with low power (hybrid systems). Targeted air temperature control can be carried out by the classical method of mixing hot and cold air flows, electric heaters or heat exchangers, as well as combined systems. Hot compressed air is separated to two flows. One flow is a flow of circuit of itself hot air and another is used for preparation of cold air circuit. Figure 1 shows some possible versions of the subsystems that make up the air conditioning system. Various combinations of these subsystems give different circuits and constructive solutions of air conditioning system. These solutions have different weights, volumes, technical complexity, level of energy consumption and the degree of influence on the design of the aircraft and associated on-board systems. These parameters determine the efficiency and technical level of each individual solution. In addition to the shown usual components in circuit solutions there have to be presented innovative developments, which are represented at numerous patents in the field of air conditioning systems.

In the conditions of such a constructive variety, it is very difficult to find the most rational circuit solution for a particular system. It is even more difficult to create a database of rational solutions of such systems for different types of aircraft. The creation of such a database requires the analysis of a very large number of circuit solutions. A designer is not able to analyze it using only own experience. In this case, there is a high probability of loss of non-obvious, but effective technical solutions. An effective tool for structural, parametric synthesis and design of multivariate technical systems, such as, for example, the considered air conditioning systems that is consisted of functionally related objects, is the combinatorial-logical method of directed graphs [4], or “Alternative-trees” [5]. The variety of options for the implementation of the electric air-conditioning systems that is demonstrated above can be visualized in the form of the oriented graph. The graph is shown in figures 2-4.

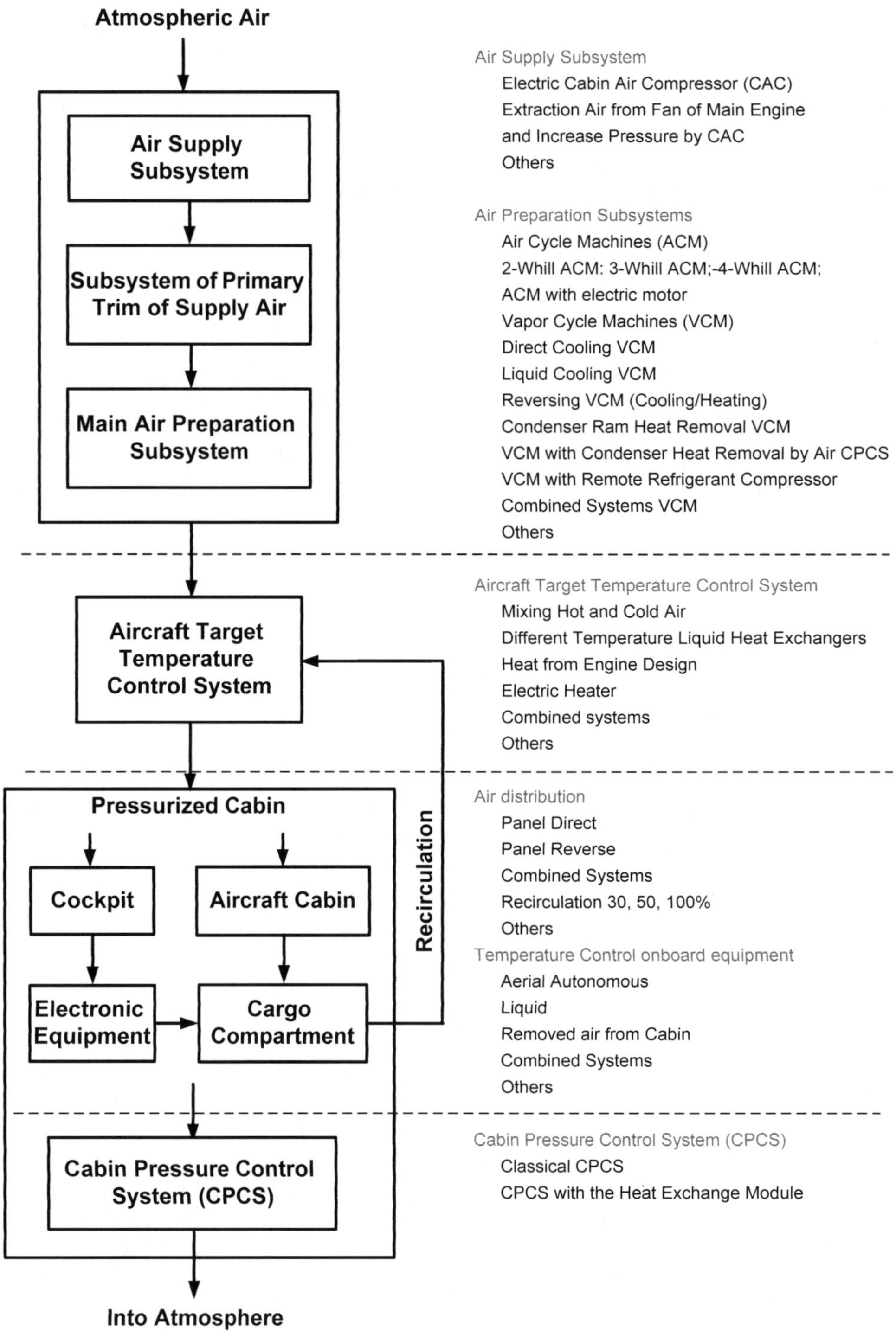

Figure 1. Structure of electric air conditioning system and its subsystems.

The graph is constructed in the form of a "collapsed" alternative-tree, where all decision graphs (air conditioning systems structures) are reduced to one common final vertex - "Performance Indicator". Each vertex has associated vertices of the next functional level (sub-vertices or descendants). The vertices (sub-vertices) of the graph have names of functional modules and subsystems of air conditioning system. The names of the vertices indicate their circuit designation. The composition of the alternative-tree (names of vertices, directions and edges) can be considered as a prototype for the classification of air conditioning systems using an electric compressor in the cabin pressurization system. However, the presented alternative-tree is not limited to this.

In general, the directed graph includes many minimally functionally complete circuitry solutions that are formed from a combination of functionally and technically compatible, technologically feasible components. The alternative-tree of such circuitry solutions is called basic.

The connected components of this set built sequentially into the technological chain implement the process of targeted change of air parameters in the climate control system of the pressurized airplane cabin and the operation of heat-dependent on-board equipment. From a technical point of view, this chain starts from outdoor air intake devices, passes through the functional edges of any one of the possible circuit structures of air conditioning systems, and ends with devices for removing air from the pressurized cabin. This determines a specific technical solution – a decisive graph of a generalized alternative-tree. The final peak – "Performance Indicator" gives a technical and economic assessment of the solution. As such an assessment, the fuel efficiency of the aircraft is used [7], or rather, its change in the implementation of the selected version of air conditioning systems, or other integral indicators of aircraft perfection.

The basic alternative-tree can be considered as a tool for the formation and evaluation of the effectiveness of possible looks of the new generation of air conditioning systems. For this, it must have sufficient "power" so that the search for combinations of functionally connected components of air conditioning system is effective, i.e. the search should give distinct, workable holistic circuitry options that match the specified design parameters in the design operating conditions, i.e. the search should give distinct, workable holistic circuitry options that match the specified design parameters in the design operating conditions. Components included in the base alternative-tree are themselves generalized structures of subsystems with an increased level of detail. Based on such a developed base alternative-tree, a sequential structural, parametric and structural synthesis of the system and analysis of the obtained solution according to the indicated objective indicators can be performed.

3. Results and discussion

In terms of set theory [9], the base alternative-tree is represented by the vector $(\mathbf{G}, \mathbf{S}, \mathbf{y})$, where: $\mathbf{G} = \mathbf{G}(\mathbf{X}, \mathbf{D})$ is the root (in this paper, the base) oriented tree with many vertices $\mathbf{X} = \{\mathbf{x}_i\}$ and sub-vertices (descendants) $\{\mathbf{x}_{ik}\}$ (in Figures 2, 3, 4 the vertices and sub-vertices are indicated by named cells, which mean the corresponding components of air conditioning system, or the logical values "yes", "no" regarding the use of some technical/physical principle); $\mathbf{D} = \{\mathbf{d}_j\}$ is the set of edges (directed numbered or marked arrows); $\mathbf{S} = \{\mathbf{S}_k\}$ is the set of edges emerging from one vertex.

The maps $\mathbf{y(s)}$ establish a logical relationship (logical operations) between the components of air conditioning system connected by the edges and indicated by the signs of the logical relation: $\vee$ – disjunction ("OR"), $\veebar$ – "exclusive OR" ("XOR"), $\wedge$ – conjunction ("AND"), etc. The output parameters of the vertices are the input parameters of the vertices of the lower level. Formally, the parameters are passed through the edges $\mathbf{D} = \{\mathbf{d}_j\}$.

The initial conditions (the parametric part of the design task on the air conditioning system) are laid at the root vertex. The initial conditions include the typical requirements of the technical specifications for the air conditioning systems of the certain type of aircraft. The functions of the vertices $\mathbf{X} = \{\mathbf{x}_i\}$ and the vertices $\{\mathbf{x}_{ik}\}$ are expressed by the operators $\mathbf{A}_i = \{\mathbf{A}_{ij}\}$ and $\mathbf{A}_{ik} = \{\mathbf{A}_{ijk}\}$, where $\mathbf{i}$ – the serial number of the vertex, which in our case is replaced by an identifier – the name of the vertex function; j and k are, respectively, identifiers denoting the $\mathbf{j}$-th function in the sub-vertex $\mathbf{k}$ of vertex $\mathbf{i}$. The operators $\mathbf{A}_i$ and $\mathbf{A}_{ik}$ are, in most cases, algebraic relations that reflect the thermodynamic, mechanical, and other processes in their static model (calculated operating modes) that occur within this vertex/sub-vertex – a component of air conditioning system based on the parameters transmitted from the connected vertex/sub-vertex (flow rate, temperature, pressure, humidity or enthalpy of air, or liquid coolant (LC) parameters, as well as the total mass of the previous components in the circuit).

IOP Conf. Series: Materials Science and Engineering **868** (2020) 012023 doi:10.1088/1757-899X/868/1/012023

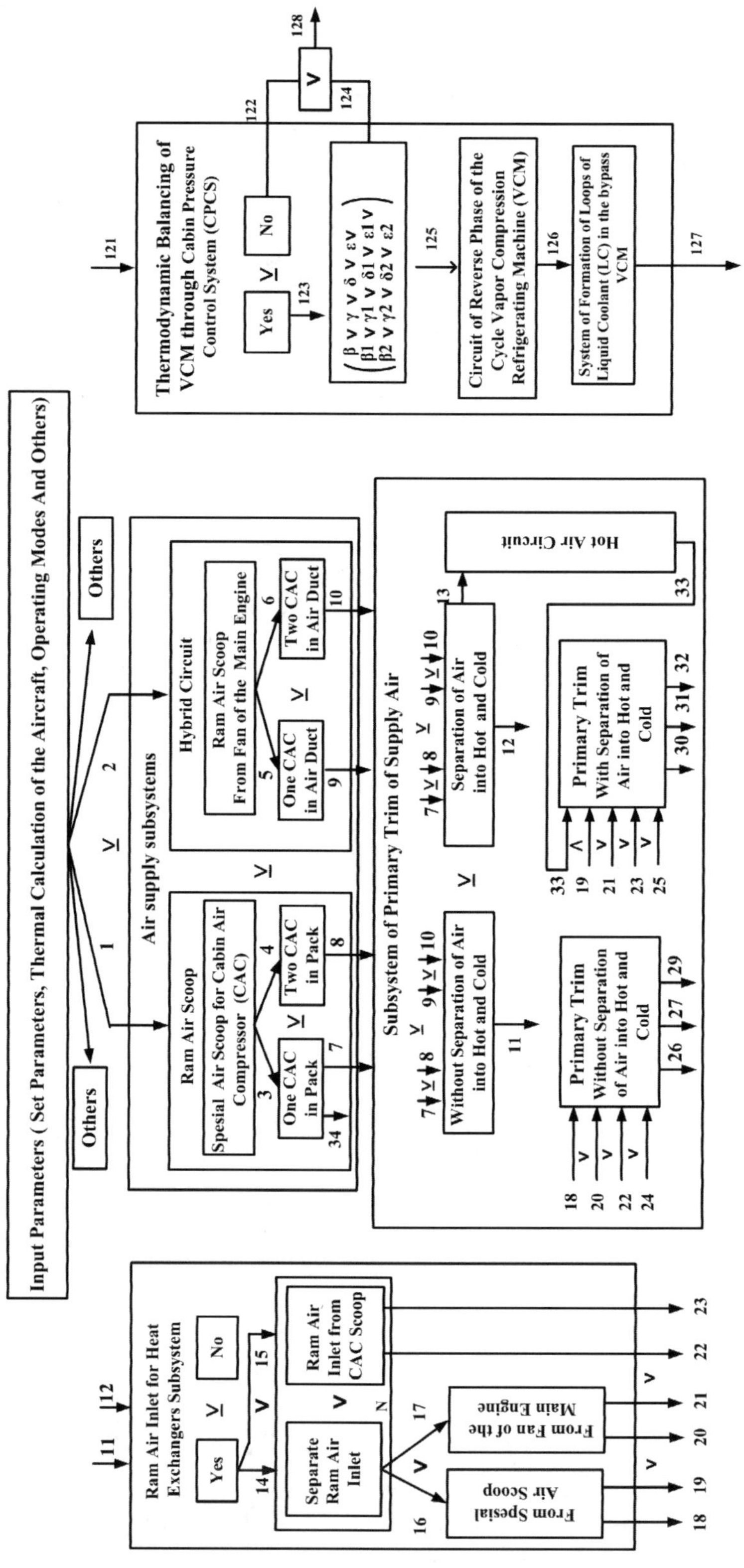

Figure 2. Upper part of alternative-tree.

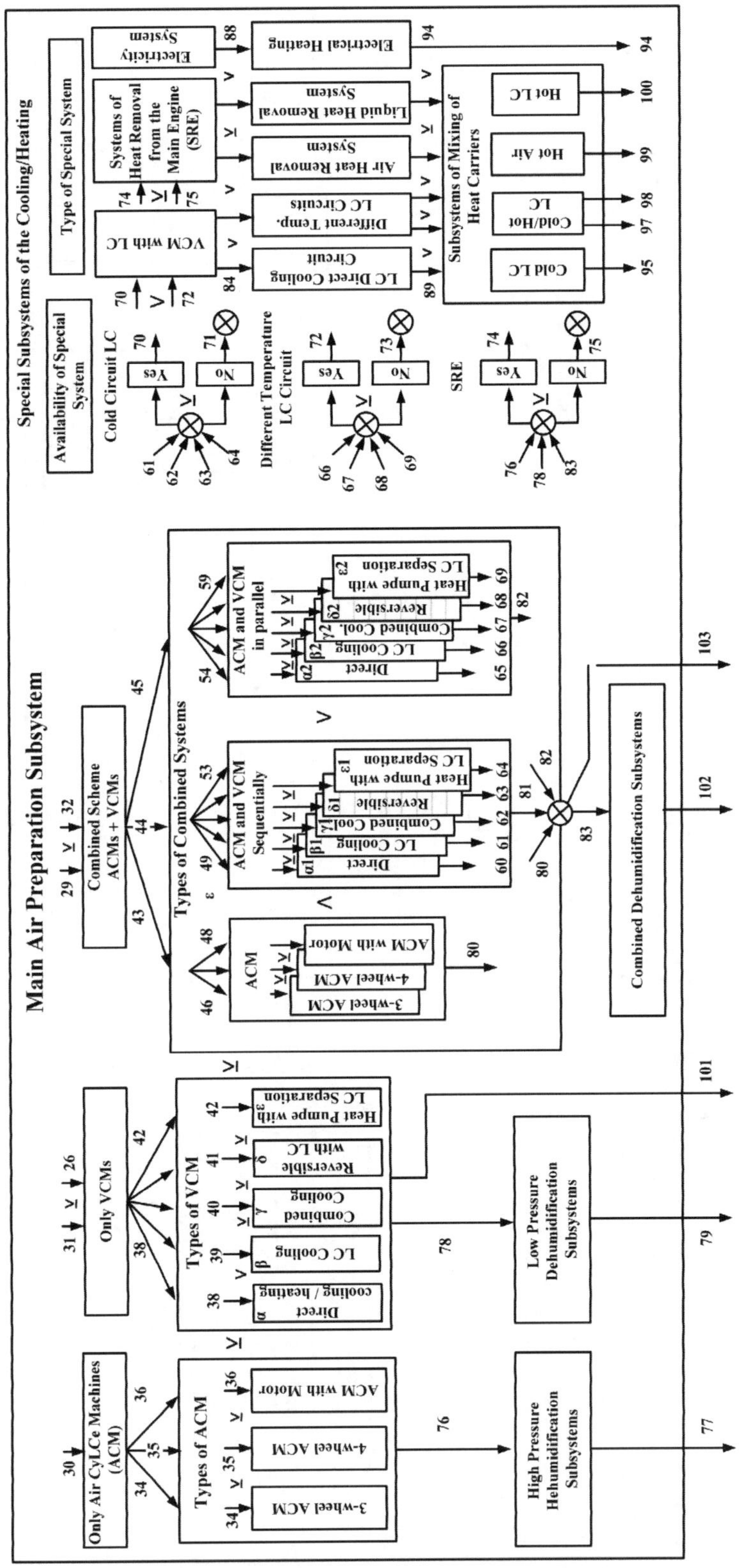

Figure 3. Middle part of alternative-tree.

IOP Conf. Series: Materials Science and Engineering **868** (2020) 012023 doi:10.1088/1757-899X/868/1/012023

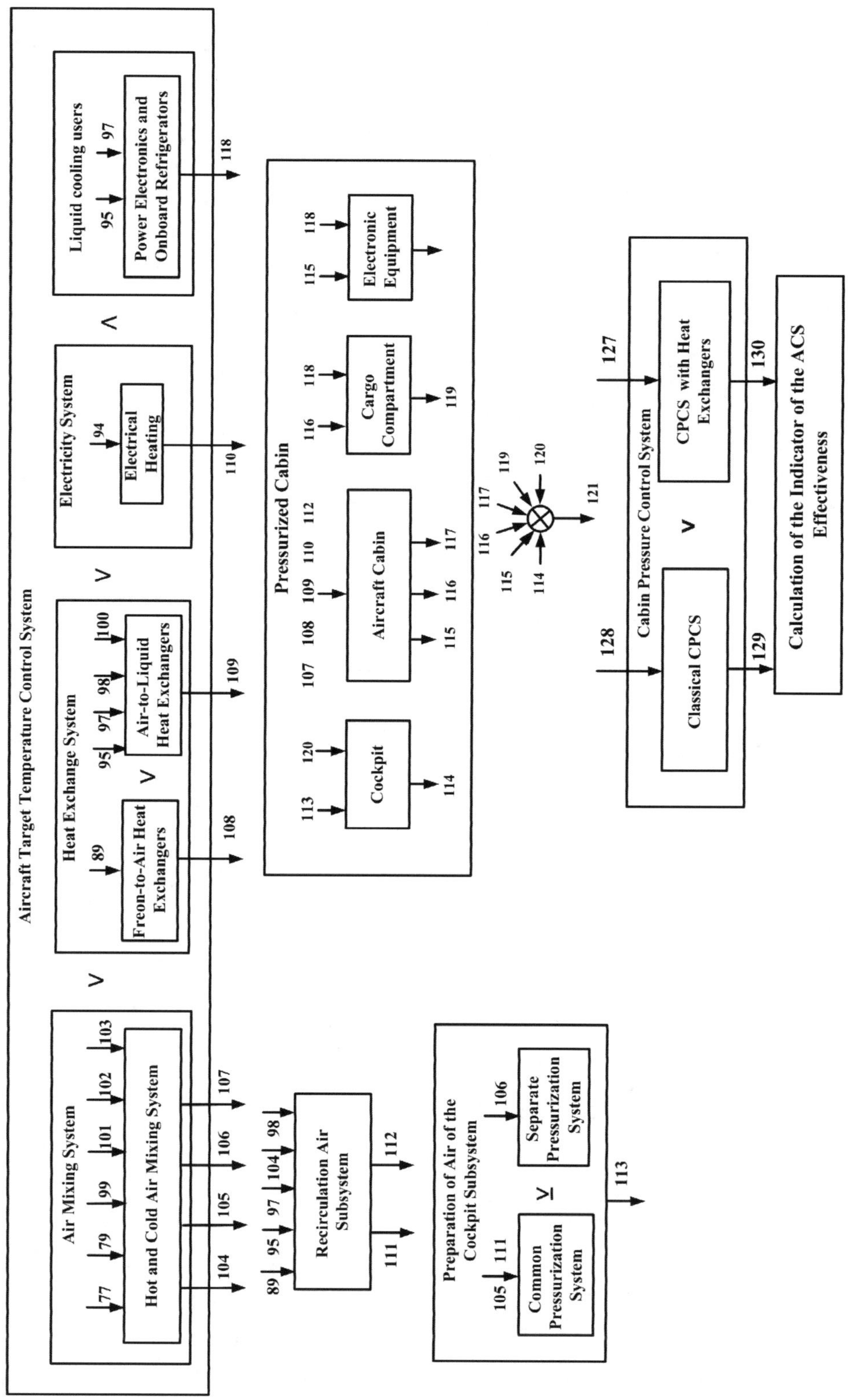

Figure 4. Bottom part of alternative-tree.

IOP Conf. Series: Materials Science and Engineering **868** (2020) 012023 doi:10.1088/1757-899X/868/1/012023

The operators $\mathbf{A_i}$ and $\mathbf{A_{ik}}$ calculate the components of the performance indicator of air conditioning system, which is the mass of the structure that implements the functions of this vertex, or other additive technical and economic indicators. As the decision graph is formed, the components of these indicators are transmitted through the edges $\mathbf{D} = \{\mathbf{d_j}\}$ of the decision graph of the alternative-tree and accumulate during their growth. The end nodes "Performance Index" on the base alternative-tree are combined into one component, where, $\mathbf{g_n}$ value of mass of air conditioning system which calculates the increment of the take-off mass of the aircraft $\mathbf{\Delta g_{ac}}$. The value of mass is based on the accumulated for each **n**-th solution ($\mathbf{n = 1...N}$, where **N** is the total number of productive decisions of air conditioning system under given initial conditions). The mass $\mathbf{\Delta g_{ac}}$ takes into account the effect on changes the weight of associated aircraft systems and the required amount of fuel. The complete enumeration of all **N** solutions gives a pair $\mathbf{R} = \{\mathbf{G_n}, \mathbf{r_n}\}$, where $\mathbf{r_n}$ contains a sequential list of edge numbers $\mathbf{r_n} = \{\mathbf{d_{jn}}\}$, which forms the **n**-th decisive graph. Priority solutions on the value of $\mathbf{\Delta g_{ac}}$ are determined during the process of analyzing the vector **R**. The solutions can be restored, detailed and developed to the required design level using the list of edges $\mathbf{r_n}$. Since the values of $\mathbf{g_n}$ are an additive quantity from the indices $\mathbf{g_{jn}}$, $\mathbf{g_n} = \mathbf{\Sigma g_{jn}}$, which depend on the methods of thermodynamic conversion of air flow, secondary heat carrier, or energy conversion in the components of the air conditioning systems circuit (vertices of decision graphs), we can use as the initial base alternative-tree as a basic graph whose operators $\mathbf{A_i}$ and $\mathbf{A_{ik}}$ be only the weight or mass-dimensional functions of the corresponding technical devices. In this interpretation of the base alternative-tree, it is advisable to operate with the mass-dimensional parameters of the components related to the reference values corresponding to any electrified air conditioning system of an aircraft of a similar type, for example, the air conditioning system of long-range aircraft Boeing-787 [10].

The installation weights of most typical units of air conditioning systems are known from sources as [7] or are established on the basis of modern developments. They are statistical functions of changes in the state of the working fluid or energy conversion in the unit (the ratio of the characteristic parameters of the unit at the input and output). These changes are necessary and, in our case, sufficient characteristics of the unit to determine the mass of unit. The mass formulas of atypical devices included in the base alternative-tree (patent components) characteristic of inventive products are to be determined. The last can be based on the analysis of analogues in the aircraft design or general industrial area. This is one of the most time-consuming and controversial tasks of forming a basic alternative-tree, since patent descriptions in many cases are limited only to explaining the principle of the invention.

In a more developed structure of the alternative-tree, the operators $\mathbf{A_i}$ and $\mathbf{A_{ik}}$ can be finite-state machines, transfer functions, matrix operations, and, ultimately, CAD components (fragments of the thermal engineering scheme) with interfaces to the CAD components of the next vertex.

The solution to the synthesis problem from the basic alternative-tree consists in performing the natural steps of constructing complete chains of related components (decision graphs), starting with the selection of the edge coming from the root vertex and the sequential passage of variants of its descendants to the end vertex of each option with the final calculation of the efficiency indicator obtained decision graph.

Moreover, depending on the functional content of the vertices of the base alternative-tree, structural synthesis can be performed simultaneously, i.e. the formation of specific and isolated circuitry looks, parametric synthesis of the formed structure and, ideally, the structural synthesis of the system.

The structural synthesis applications based on non-planar graphs can help in the formation of the base alternative-tree. However, the specifics of the base Alternative-tree with operator filling of vertices makes it advisable to use special programming of combinatorial-logical enumeration of solutions. After the formation of a graphical representation of the alternative-tree and its operator filling, programming enumeration of solutions is not a significant difficulty. The dimension of the searching problem increases exponentially with increasing number of edges in the alternative tree and can become unreasonably large. This requires a rational organization of the process and filtering of productive solutions. A natural limitation of searching is the output of the calculated parameters at any vertex beyond physically or technically feasible limits, including those associated with the high-speed range of operation of air conditioning system. The calculated parameters need to be introduced at the stage of operator filling of vertices. The second limitation is the reduction of the functional chain of vertices, after which the remainder of the tree is the same for the solutions found earlier. A preliminary assessment shows that productive solutions for a certain alternative tree, passing through similar filters and restrictions, lie within a few hundred searches.

4. Conclusion

An approach is proposed based on the combinatorial-logical apparatus of oriented graphs (alternative-trees) for circuitry, parametric synthesis and synthesis of the designs of the electrical air conditioning systems at the stage of conceptual design of the aircraft and its on-board systems.

The practical goal of creating this method is based on the need to form a scientific and technical reserve in the field of air conditioning systems using electric blowers for promising aircraft with a high level of electrification and to ensure the synthesis of rational architectures of air conditioning systems for specific projects of new aircrafts.

The developed tool will be useful not only to the developers of air conditioning systems, but also to the system integrator of the on-board equipment set.

It is advisable for the system integrator to have a complete set of alternative-trees of all aircraft systems.

Moreover, a number of these tools should have the property of mutual influence, similar to the interconnections of the onboard systems themselves, for example: air conditioning system and power supply systems, air conditioning system and air signal systems, air conditioning system and aircraft digital computer system, etc.

Such mutual influence can be realized through the connection of the root vertices of alternative-trees and "cross-pollination" at the level of the sub-nodes with interdependent parameters.

References

[1] Kaushik Rajashekara 2014 More Electric Aircraft Trends (Technology Leaders). *IEEE Electrification Magazine* **2(4)** 4 doi: 10.1109/MELE.2014.2365621

[2] Kalliopin A K, Savel'ev R S, Smagin D I 2017 The main trends in the development of air conditioning systems promising aircraft. *Engineering Journal: Science and Innovation* **6** doi: 10.18698/2308-6033-2017-6-1627

[3] Bozhko S, Hill C I, Yang T 2018 *More-electric aircraft: systems and modeling* (Wiley Encyclopedia of Electrical and Electronics Engineering) pp 1-31

[4] Tom Leighton, and Marten Dijk 2010 *Mathematics for Computer Science* (Massachusetts Institute of Technology: MIT OpenCourseWare) pp 121-127

[5] Bozhko A N and Tolpachov A C 2019 Structural synthesis on elements with limited compatibility, available at: https://www.metodolog.ru/00562/00562.html

[6] Lim D, Nam T, Kirby M and Burdette G 2014 Measurement of Aircraft Fuel Efficiency. 2014 *McGraw Hill Yearbook of Science and Technology* (Atlanta, Georgia)

[7] Shustov Y M and Bulaevsky 1978 *Aviation air conditioning systems* (Moscow: Mashinostroenie)

[8] Gubernatorov K N, Kiselev M A, Moroshkin Y V, Chekin A Y, RU Patent No. 2,686,609 (29 April 2019)

[9] Beckmann A and Dimitracopoulos C, Lowe B 2008 *Logic and Theory of Algorithms. Fourth Conference on Computability in Europe, Local Proceedings* (University of Athens)

[10] Mike Sinnett 2007 *787 No-Bleed Systems: Saving Fuel and Enhancing Operational Efficiencies* (787 Systems)

IOP Conf. Series: Materials Science and Engineering **868** (2020) 012024 doi:10.1088/1757-899X/868/1/012024

Mechanical characteristics of short fiber composite samples located behind circle, rectangle, triangle obstacles

E I Kurkin*, M O Spirina, V O Chertykovtseva and Ya V Zakhvatkin

Joint Russian-Slovenian Laboratory Composite Materials and Structures, *Samara National Research University,* 34 Moskovskoe shosse, Samara, 443086, Russian Federation

*E-mail: eugene.kurkin@mail.ru

Abstract. The article investigates the influence of the weld line on the mechanical characteristics of flat samples on tensile strength according to ISO-527-2, cut from a plate with an obstacle, molded from polyamide-6, reinforced from 30% short glass fibers. There are considered circular, rectangular, triangular obstacles. Melt front, weld line location and reinforcing fiber orientation are calculated in Moldex3D. The stiffness and strength of the samples were predicted in ANSYS using an anisotropic nonlinear composite material model obtained in the Digimat by reverse engineering method. The transversely isotropic criterion of Tsai Hill in the first pseudo-grain failure statement was used for predict strength. The paper presents the influence of the weld line on the stress, deformation and first pseudo-grain failure strength criterion. The greatest contribution to the samples mechanical characteristics was made by the rectangular obstacle.

1. Introduction

Injection technologies are used to produce large batches of spatially loaded products of complex shape. The addition of short fibers to binders significantly changes the mechanical characteristics of the material, allowing it to increase its stiffness [1, 2]. Short-reinforced composite materials can be used to create spatially-loaded structures of complex shape, for example, aerospace joints, as well for creation tips of lattice structures [3]. The flow of material inevitably leads to a change in the orientation of the reinforcing fibers. The stiffness and strength of composite materials is largely determined by the orientation of the reinforcing fibers, therefore, when designing products from short-reinforced composite materials, it is necessary to simulate all stages of their production [4]. Studies of the mechanical characteristics of composite material reinforced by short fibers are based on the work of Eshelby, who studied the elasticity of a field with an ellipsoidal particle [5-7]. Tandon and Weng based on Eshelby's theory determined the elasticity characteristics of a composite material reinforced by unidirectional short fibers [8]. The calculation of the characteristics of a composite material reinforced by short fibers with an arbitrary orientation tensor is presented in [9, 10]. Finite element modeling of microstructure of a representative volume is used for calculate the mechanical characteristics of the composite material reinforced by short fibers [11]. The study of the stiffness and strength of short-reinforced composites on temperature is presented in [12]. The weld lines are formed at molding products of complex topology in the case of combining flows after flowing around obstacles [13]. Weld lines, that occur when two mix flows are connected in the mold, influence the strength of the product in this area [14]. Takayama [15] investigate weld lines in samples from short fiber reinforced polypropylene and shows that weld strength tended to decrease as the fiber content increased. The study

[14] is devoted to studying the influence of the weld line when flowing around a triangular and rectangular obstacle on the mechanical material characteristics. A feature of [14] is the comparison of the mechanical characteristics of samples cut out behind an obstacle with samples located near the gate. The aim of the work is to compare the stiffness and strength of samples located on the weld line when flowing around circular, rectangular, triangular obstacles, with the stiffness and strength of samples located on the plate in the region of uniform flow far from the weld line along and across the molding direction.

2. Materials and research methods

The calculation of plate molding is carried out in the Moldex3D system. The size of the plate is 200x150x4 mm. The calculation was carried out on a mesh with solid elements obtained in ANSYS Meshing and transferred to Moldex3D by means Rhinoceros. The technique of mesh creation in ANSYS Meshing for Moldex3d presented in [16]. The meshes consists of 29...35 thousand hexagonal elements located in 8 layers (figure 1).

The plate material is polyamide-6 reinforced by 30% glass fibers – PA6 Armamid PA SV 30-1ETM (figure 2a). Melt temperature is 250 °C, mold temperature is 110 °C. Injection-molding machine is the Negri Bossi VE210-1700 with screw diameter 60 mm. Flow rate is 33 $cm^3 \cdot s^{-1}$. Hexagonal elements and sweep meshing technique allow to increase calculation accuracy of fiber orientation along the plate thickness. An example of melt front calculation is shown in figure 2b. The values of the reinforcing fibers orientation tensor are calculated in Moldex3D and exported in the o2d file format, supplemented by mesh file in the cdb file format.

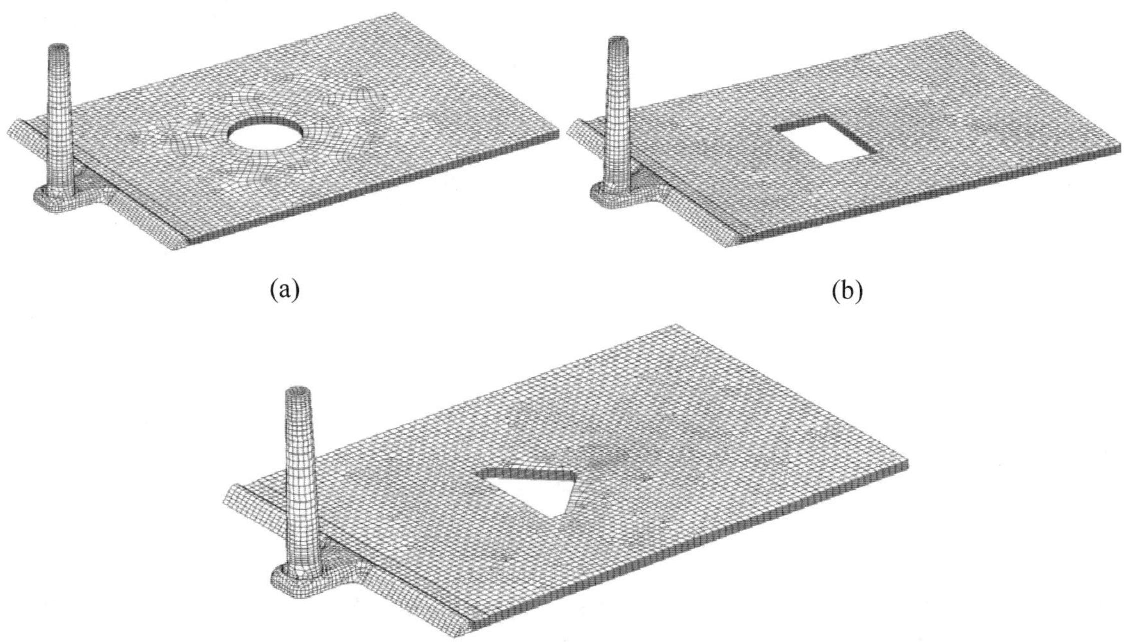

(a) (b) (c)

Figure 1. Meshes on panes with obstacles: (a) circle (29 127 elements), (b) rectangle (29183 elements), (c) triangle (35 759 elements).

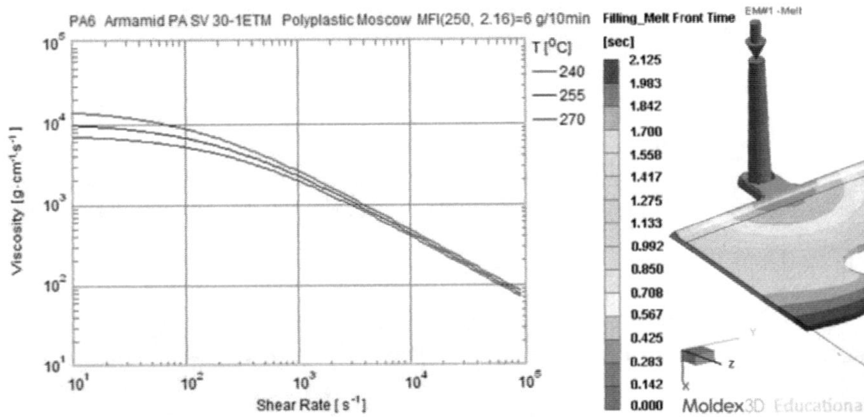

(a) (b)

Figure 2. Molding calculation in Moldex3D: (a) material model, (b) melt front in plane with circular obstacle.

We study the influence of the weld line based on the estimation of the stiffness and strength of material samples cut from a plate according to ISO 527-2 [17] (figure 3a), numbered as shown in figures 3b-d. The samples location on the plates is shown in figure 3.

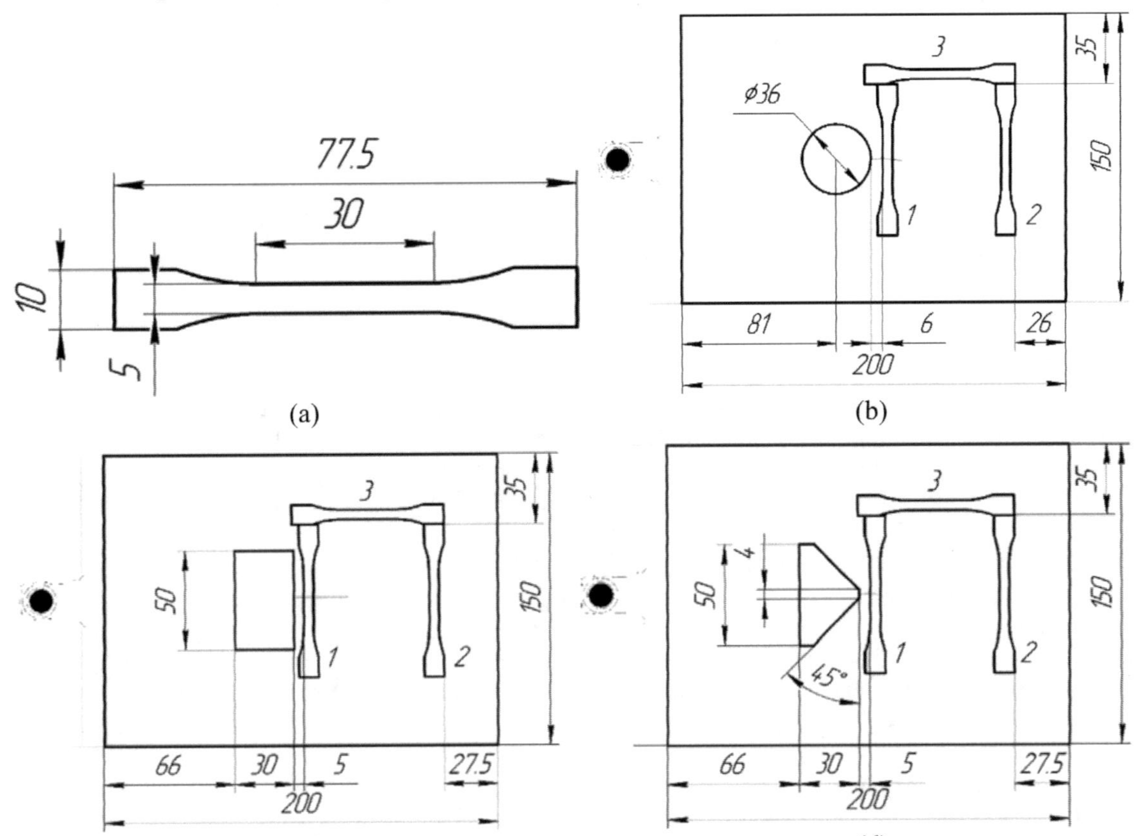

(a) (b) (c) (d)

Figure 3. Samples on the plates: (a) ISO 527-2 type 1 BA sample; plates with obstacle: (b) circle, (c) rectangle, (d) triangle.

Using the Digimat MAP, the components of the fiber orientation tensor are transferred from the hydrodynamic calculation mesh to the stress-strain state calculation mesh (figure 4). The mesh for calculating the stress-strain state can include not all the entire molding area, that way simulating the process of cutting products and removing technological elements of the half-finished product (figure 4a). The mesh for calculating the stress-strain state of ISO 527-2 tensile specimens is structured and consists of 2550 hexagonal elements. The fibers orientation data on the mesh for calculating the stress-strain state are obtained from the fibers orientation data calculated on the Moldex3D mesh by interpolation (figure 4b).

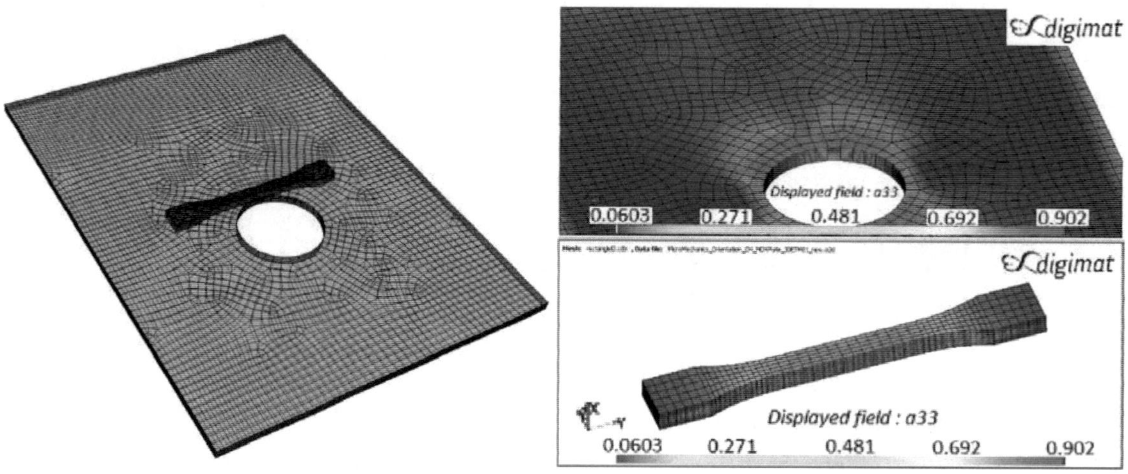

(a) (b)

Figure 4. Transferring the orientation tensor values from the molding calculation mesh to the static structural mesh using Digimat MAP. Example of plate with circle obstacle, sample 1.

For the static structural calculation, an anisotropic material model is used, previously obtained in Digimat MX by reverse engineering, so that the properties of the composite as a whole correspond to the experimentally obtained dependences of stresses on strains for the images cut at an angle of 0°, 45° and 90° to the flow direction in without obstacle. The material consists of two phases – 30% mass fraction of linearly elastic glass fibers with density 2600 kg·m^{-3}, Young modulus 74 GPa, Poisson's ratio 0.22, fiber aspect ratio is 10.888 and 70% mass fraction of J2_plasticity polyamide binder model with density 1400 kg·m^{-3}, Young modulus 2.82 MPa, Poisson's ratio 0.36726, yield stress 25 MPa, power law of hardening model with 122 MPa hardening modulus and 0.396 hardening exponent. For failure prediction Tsai Hill 3D transversely isotropic model with first pseudo-grain failure (FPGF PGA) formulation was used. The representative volume cell was divided into 12 pseudo-grains, and destruction occurred when 75% of all pseudo-grains were destroyed. Axial tensile strength limit is 155.66 MPa, in-plane tensile strength limit is 57.385 MPa, transverse shear strength limit is 44.644 MPa.

3. Results and discussion

Using the meld flow calculation in Moldex3D, we investigated the distribution of the melt front isoline (volume share), the size and location of the weld line, and obtained the orientation tensor of the reinforcing fibers at each point of the plate (figure 5). Weld line at figure 5 is defined as the line where meeting angle between two melt fronts is in the range from 0° to 135° (0° corresponds to head-on converge of two melt fronts).

The fibers behind the obstacle (figure 6) have more random orientation, while the fibers in the external flow are mainly orient in the flow direction, with the exception of fibers in the center of the plate thickness and in the wall region, which is in accordance with the results of work [18].

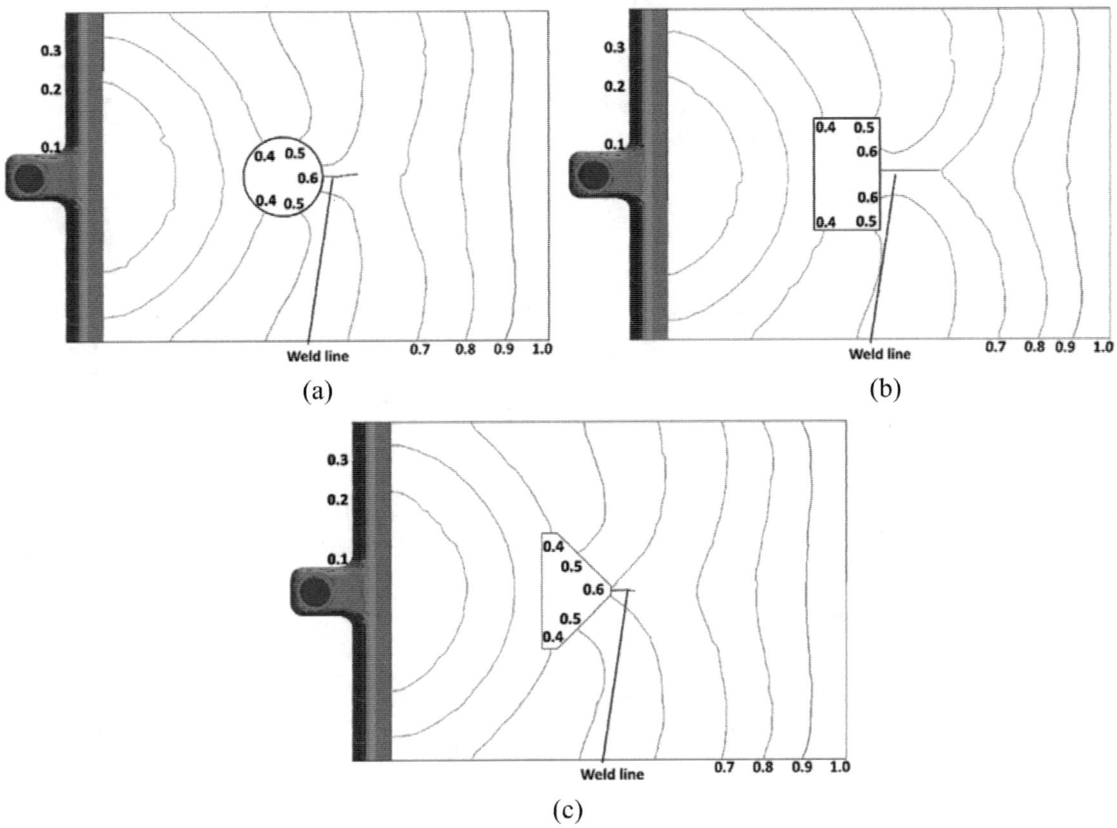

Figure 5. Melt front isoline (volume share) and weld lines on panes with obstacles: (a) circle, (b) rectangle, (c) triangle.

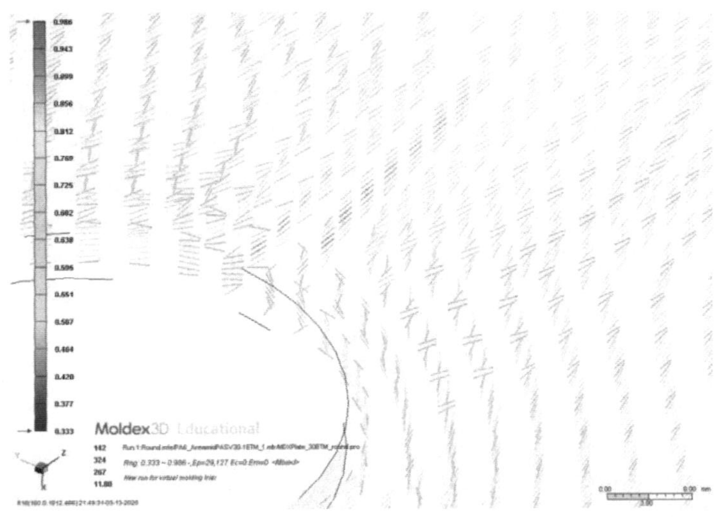

Figure 6. Fiber orientation on panes with circle obstacles – vector is the predominant direction of the fiber, color is the probability of fiber orientation in this direction.

IOP Conf. Series: Materials Science and Engineering **868** (2020) 012024 doi:10.1088/1757-899X/868/1/012024

Samples cut from plates are loaded by tension by fixed support of one of the ends and force condition to the other end (figure 7a). Calculated samples tension, where stress of samples is force divided by area of cross section (20 mm^2), strain is differences between sample working area deformation divided by length of this area (30 mm) (figure 7b) shown in figure 8.

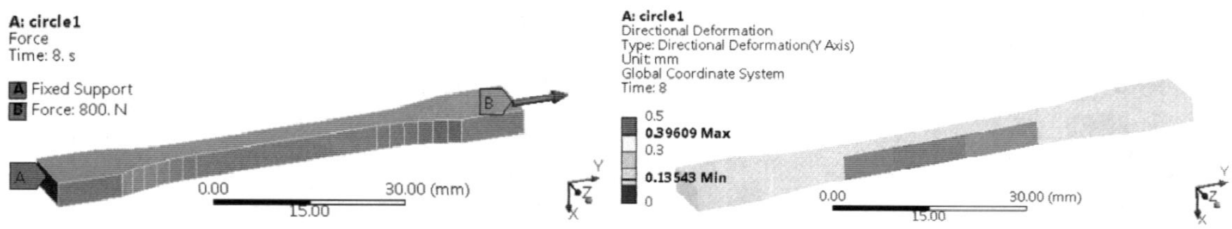

(a) (b)

Figure 7. Stain-stress task: (a) loading scheme, (b) average strain calculation.

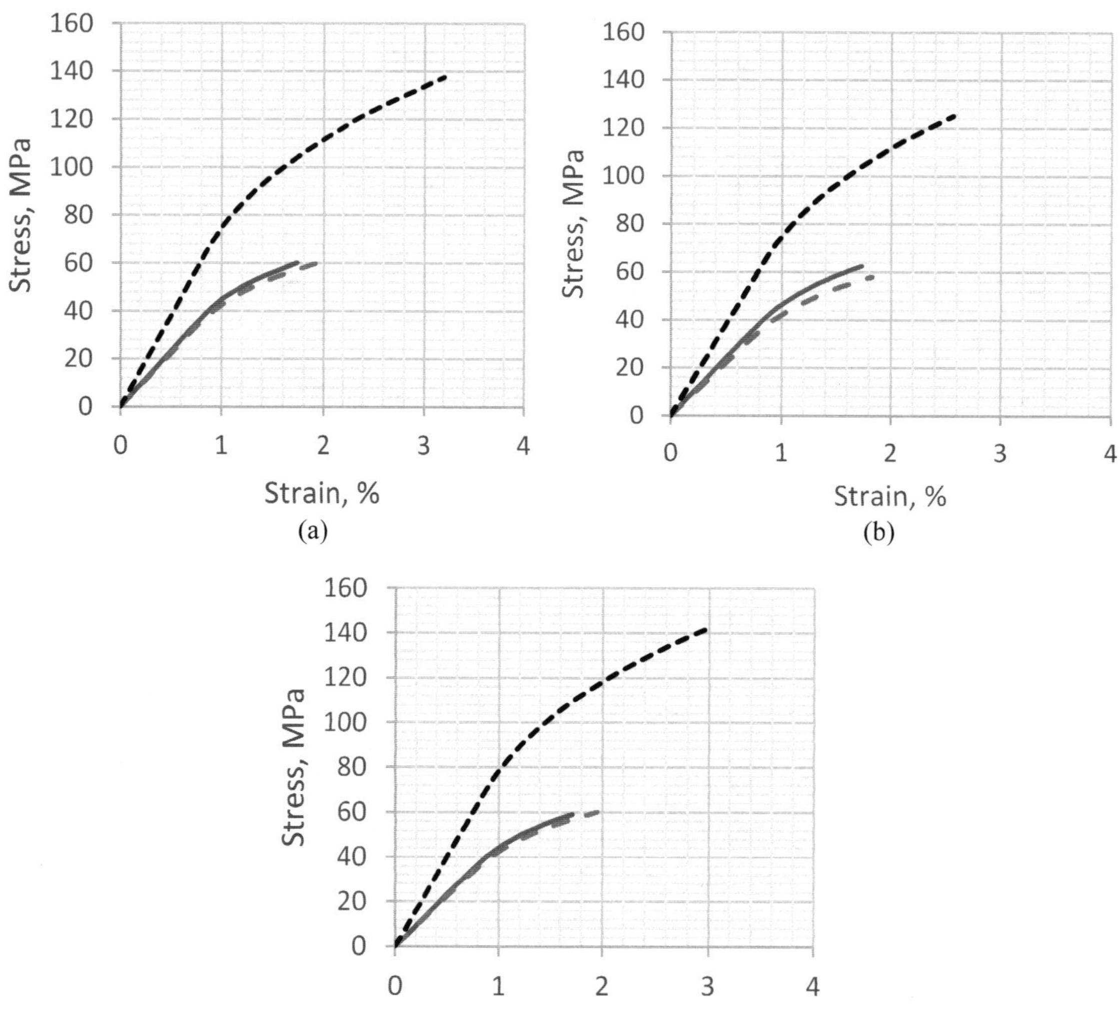

(c)

Figure 8. Samples tension calculation (1 – blue solid, 2 – rad dash, 3 – black dot) on a plate with an obstacle: (a) circle, (b) rectangle, (c) triangle.

IOP Conf. Series: Materials Science and Engineering **868** (2020) 012024 doi:10.1088/1757-899X/868/1/012024

End of lines indicate the fracture samples beginning – first time when maximum value of FPGF criterion become equal to one. In sample 3 (along flow direction), the fibers are oriented mainly along the sample, in sample 2 - mainly perpendicular to the direction of the sample. The fibers in the sample 1, located after the weld line, are oriented more randomly than in sample 2, so its stiffness is slightly higher than the stiffness of sample 2, but significantly lower than the stiffness of sample 3.

Stress and deformation fields at tensile samples with a force of 800 N are shown at figure 9. Stress field of samples 2 and 3 correspond to the fiber orientation layers, shown at figure 6 and correspond to the results of work [18] – the most stressed regions in the perpendicular sample 2 are near the plate surface and in the center of the thickness, while the most stressed regions in sample 3 are located in the middle layers of the plate — the places where the fibers are most likely to be located along the flow. Violation of the fibers orientation field in the weld line after the obstacle leads to a stress field in the sample 1 shown in the figure 9a.

The stiffness of the longitudinal sample 3 is significantly higher than the stiffness of the transverse samples 1 and 2, which is reflected in smaller total deformations of the sample 3 (figure 9f). The unevenness of the stress-strain state of the specimen 1, containing the weld line, leads to its bending under tension (figure 9b).

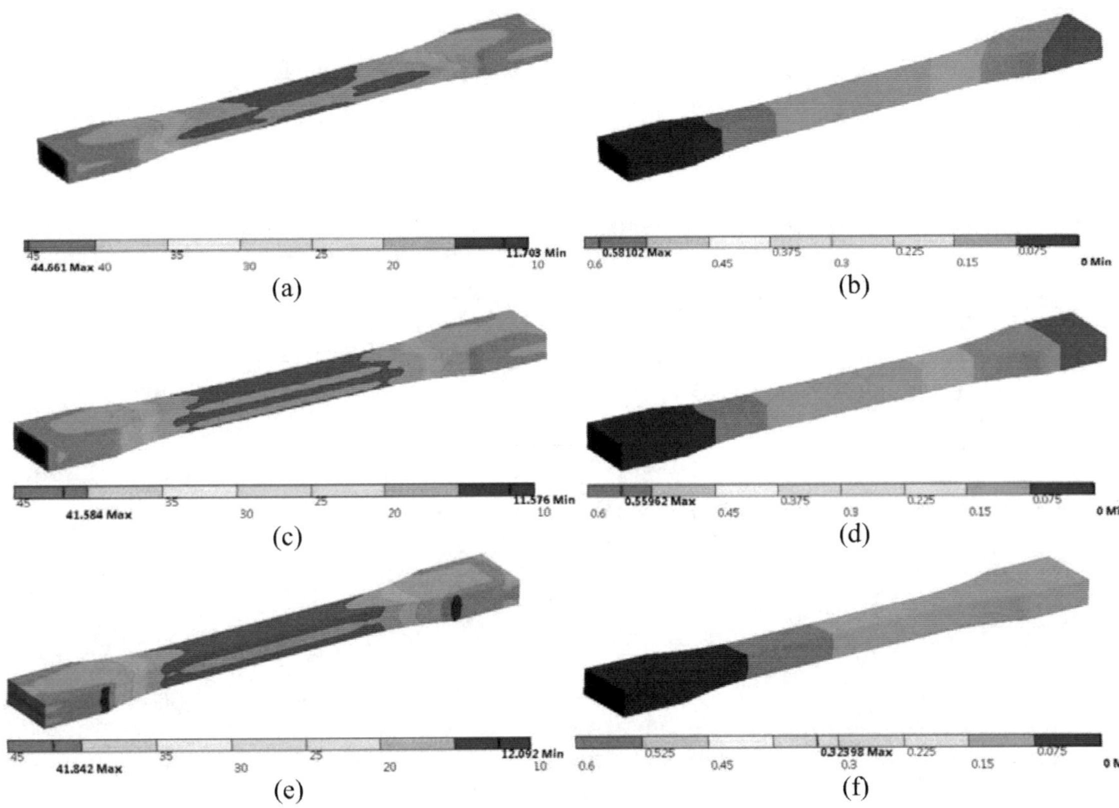

Figure 9. Results of stress-strain analysis (samples on a plate with a circular obstacle): (a), (b) sample 1; (c), (d) sample 2; (e), (f) sample 3; (a), (c), (e) equivalent (von Mises) stress, MPa; (b), (d), (f) total deformation, mm.

The distributions of the Tsai Hill criterion in a transversely isotropic formulation using the model of the first pseudo-grain failure at ultimate load are presented in figure 10. The sample 1 failure begins from weld line location. The sample 2 does not contain a weld line and its failure can begin evenly over its entire working area. The longitudinal sample 3 failure begins near the extension location.

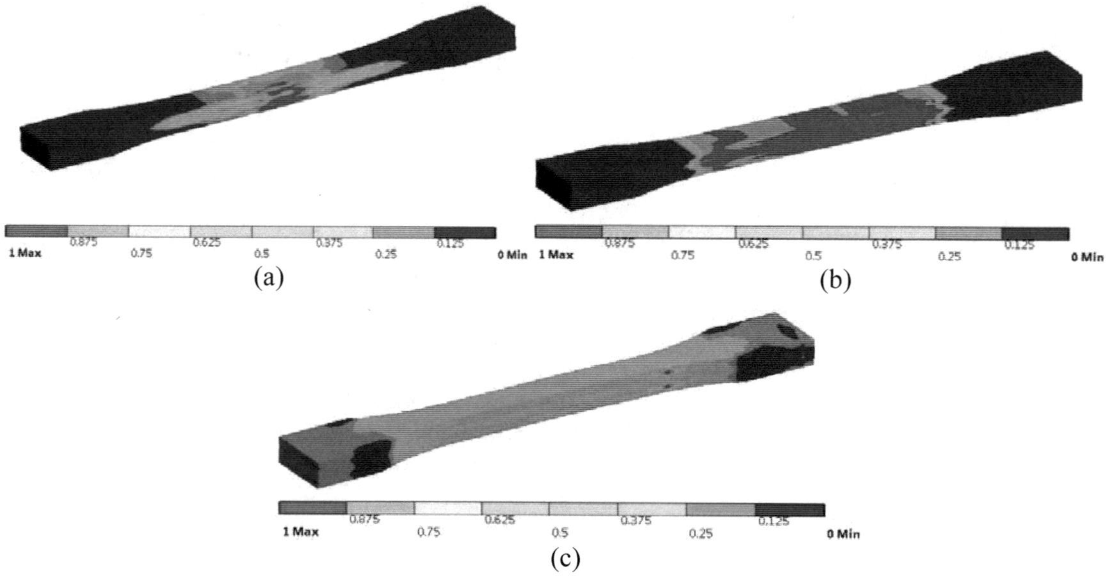

Figure 10. Value of FPGF criterion for samples from plate with a circular obstacle, numbered: (a) 1, (b) 2, (c) 3.

4. Conclusion

The influence of the weld line on the stress-strain state of short fiber reinforced composite material samples was investigated. Samples containing the weld line are slightly stiffer than samples cut across the molding of plate in a place that does not contain the weld line. The contribution of weld line to the samples stiffness is less than the change in the sample orientation with respect to the molding direction of the plate. The greatest contribution to the samples mechanical characteristics was made by the rectangular obstacle. Using an anisotropic model of short fibers reinforced composite material allows to take into account the influence of the weld line on the mechanical characteristics of thermoplastics products.

Acknowledgments

The reported study was funded by Russian Science Foundation according to the research project No. 19-79-10205.

References

[1] Alizadeh V and Amirkhizi A V 2020 Milled Carbon Fiber Polyurea Elastomeric Composites. *Conf. Proc. of the Society for Experimental Mechanics Series. Challenges in Mechanics of Time Dependent Materials, Fracture, Fatigue, Failure and Damage Evolution* (Cham: Springer) vol 2 p 170 doi: 10.1007/978-3-030-29986-6_26

[2] Alizadeh V, Morris J and Amirkhizi A V 2019 Cavitation erosion of polyurea composite coatings. *Proc. of the American Society for Composites–Thirty-fourth Technical Conf.* (Lancaster: DEStech Publications) USB flash drive or CD-ROM doi: 10.12783/asc34/31337

[3] Liu X, Alizadeh V and Hansen C J 2020 The compressive response of octet lattice structures with carbon fiber composite hollow struts. *Compos. Struct.* **239** 111999 doi: 10.1016/j.compstruct.2020.111999

[4] Matthews F L and Rawlings R D 1999 *Composite Materials: Engineering and Science* (Cambridge: Woodhead Publishing) p 480 doi: 10.1016/C2013-0-17714-8

[5] Eshelby J D 1951 The force on an elastic singularity. *Philosophical Transactions of the Royal Society of London. Series A, Mathematical and Physical Sciences* **244** 87 doi: 10.1098/rsta.1951.0016

[6] Eshelby J D 1957 The determination of the elastic field of an ellipsoidal inclusion, and related problems. *Philos. T. R. Soc A.* **241** 376 doi: 10.1098/rspa.1957.0133

[7] Eshelby J D 1959 The elastic field outside an ellipsoidal inclusion. *Proc. R. Soc. Lond. A.* **252** 561 doi: 10.1098/rspa.1959.0173

[8] Tandon G P and Weng G J 1984 The effect of aspect ratio of inclusions on the elastic properties of unidirectionally aligned composites. *Polym. Composite.* **5** 327 doi: 10.1002/pc.750050413

[9] Laspalas M, Crespo C, Jimenez M A, Garcia B and Pelegay J L 2008 Application of micromechanical models for elasticity and failure to short fibre reinforced composites. Numerical implementation and experimental validation. *Computers and Structures* **86** 977 doi: 10.1016/j.compstruc.2007.04.024

[10] Gusev A A 2017 Finite element estimates of viscoelastic stiffness of short glass fiber reinforced composites. *Compos. Struc.t* **171** 53 doi: 10.1016/j.compstruct.2017.03.021

[11] Jain A, Lomov S V, Abdin Y, Verpoest I and Paepegem W V 2013 Pseudo-grain discretization and full Mori Tanaka formulation for random heterogeneous media: Predictive abilities for stresses in individual inclusions and the matrix. *Composites Science and Technology* **87** 86 doi: 10.1016/j.compscitech.2013.08.009

[12] Li Y et al. 2018 Temperature-dependent longitudinal tensile strength model for short-fiber-reinforced polymer composites considering fiber orientation and fiber length distribution. *J Mater Sci* **53** 12190 doi: 10.1007/s10853-018-2517-8

[13] Tadmor Z 1974 Molecular orientation in injection molding. *J. Appl. Polym. Sci.* **18** 1753 doi: 10.1002/app.1974.070180614

[14] Ozcelik B, Kuram E and Topal M M 2012 Investigation the effects of obstacle geometries and injection molding parameters on weld line strength using experimental and finite element methods in plastic injection molding. *Int Commun. Heat Mass.* **39** 275 doi: 10.1016/j.icheatmasstransfer.2011.11.012

[15] Takayama T 2019 Weld strength of injection molded short fiber reinforced polypropylene. *Mechanics of Materials* **136** 103064 doi: 10.1016/j.mechmat.2019.103064

[16] Kurkin E, Chertykovtseva V and Zakhvatkin Y 2019 Construction of Aerospace Bracket Manufactured from Short-Reinforced Composite. *9th Int. Conf. on Recent Advances in Space Technologies* (New York: IEEE) p 606 doi: 10.1109/RAST.2019.8767797

[17] ISO-527-2-2012. *Plastics. Determination of tensile properties. Test conditions for moulding and extrusion plastics* (ISO, Geneva, Switzerland) https://www.iso.org/standard/56046.html

[18] Tseng H C, Chang R Y and Hsu C H 2017 Numerical prediction of fiber orientation and mechanical performance for short/long glass and carbon fiber-reinforced composites. *Composites Science and Technology* **144** 51 doi: 10.1016/j.compscitech.2017.02.020

Influence of the discharge circuit inductance on the ablative pulsed plasma thruster performance

G A Diakonov, N V Lyubinskaya* and S A Semenikhin

Department of Low Power Electric Propulsion Systems, *Research Institute of Applied Mechanics and Electrodynamics of the Moscow Aviation Institute (RIAME MAI)*, 125080, 5 Leninrgadskoye sh., Moscow, Russian Federation

*E-mail: riame3@sokol.ru

Abstract. The most promising option for motion control of small spacecrafts with weight less than 50 kg is a pulsed plasma thruster with an energy of less than 20 J (micro-PPT), which has such advantages as low weight, structure simplicity, precise pulse control, sufficiently long life time, constant operational readiness and low inertia. It was believed that PPT with low discharge energy (less than 20 J) needed reduction of the initial inductance, despite the fact that the low capacity of the capacitor bank would not allow to achieve the optimal C/Lo ratio. However, the results of calculations and experiments show that a slight increase in the initial inductance of the discharge circuit, contrary to the assumptions, has a positive effect on the micro-PPT characteristics, reducing the propellant consumption and mainly increasing the specific impulse without significant reduction of the thrust impulse (unit impulse). This feature of the discharge circuit of the PPT with a low discharge energy allows us to discover new solutions in micro-PPT designing and improve the thruster specific characteristics.

1. Introduction

Currently, there has been a steady increase in the number of small spacecrafts with masses of up to several tens of kilograms. Russia and other countries consider the use of spacecraft in orbital groups consisting of up to several hundred units. To control the spacecraft motion, it is necessary to develop a highly efficient small-sized thruster for correcting and maintaining orbits. In this regard, the ablative pulsed plasma thruster (APPT) is seen as the most promising option among the wide range of spacecraft thrusters. Its advantages are precise pulse control, sufficiently long lifetime, constant operational readiness, low inertia and low cost. The area of use of these thrusters is expanding, so it is necessary to increase their specific characteristics, in particular specific impulse.

The ablative pulsed plasma thrusters (APPTs) continue to be developed in Russia and abroad. These thrusters are intended mainly for small spacecraft with limited power supply system capacity [1-6]. Figure 1 demonstrates the schematic diagram of the APPT discharge circuit with the electromagnetic mechanism of plasma acceleration. The main characteristics of the APPT discharge circuit are the constant capacitance Co, resistance Ro and inductance with the constant Lo and the variable component ΔL [7].

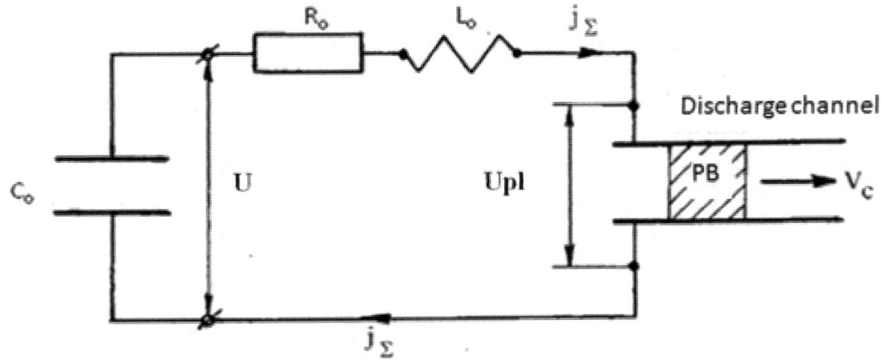

Figure 1. Schematic diagram of the APPT discharge circuit.

The electrodynamic efficiency is largely determined by the ratio of the constant capacitance to the initial inductance of the discharge circuit. Therefore, in order to increase the electrodynamic efficiency for APPT with discharge energy of more than 50 J the initial inductance Lo has to be reduced [8-10]. It was believed that the same approach was necessary for APPT with small discharge energy (less than 20 J), despite the fact that the bank capacity is much lower and thus the optimal Co/Lo ratio for such thrusters could not be achieved.

This being said, the search for the optimal ratio is a necessary, but not sufficient condition for obtaining high specific characteristics. A significant part of the mass of the plasma-forming substance enters the channel after the discharge is almost over. This part of the low ionization plasma flow is accelerated only to thermal speeds, which leads to a decrease in both specific impulse and propulsion efficiency of the thruster. In order to improve the coordination of input of energy and mass into the channel it is necessary to study how inductance affects the propulsion and specific characteristics of the thruster.

To conduct experiments on the influence of the discharge circuit inductance on the characteristics of the APPT, we took the previously studied laboratory model of rail geometry with side feed of the propellant with energy of 6.6 J (figure 2) [11]. During the vacuum bench tests, the initial voltage Un on the capacitor bank and oscillograms of discharge current of the thruster were registered, its average thrust was periodically measured, and the amount of discharge pulses were recorded. The discharge current and the thrust were measured using a dual-traced digital storage oscilloscope, one input of which was reading the signal transmitted via an RC integrator from the Rogowski belt located on the cathode bus of the model, while the other was receiving the signal from the inductive sensor of the thrust-measuring device.

To measure the average thrust of the APPT, we used a direct string pendulum thrust-measuring device fitted with an inductive displacement sensor and a calibration system with low-friction joints. Figure 3 displays the device diagram.

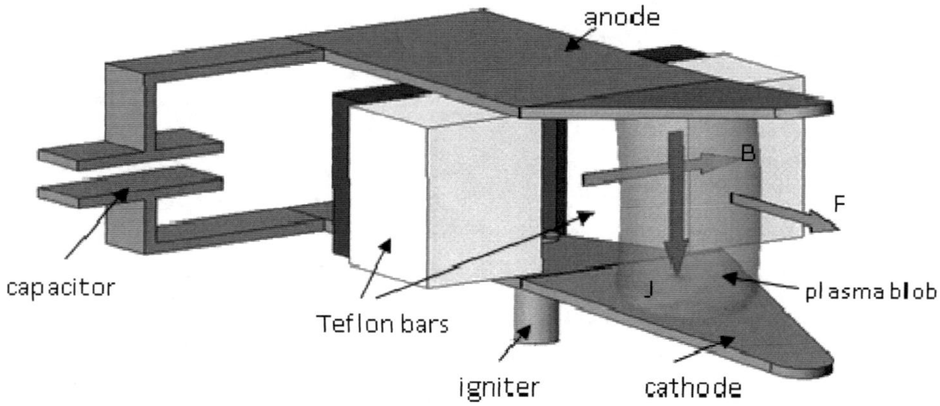

Figure 2. Principle diagram of the pulsed plasma thruster discharge channel [9, 11].

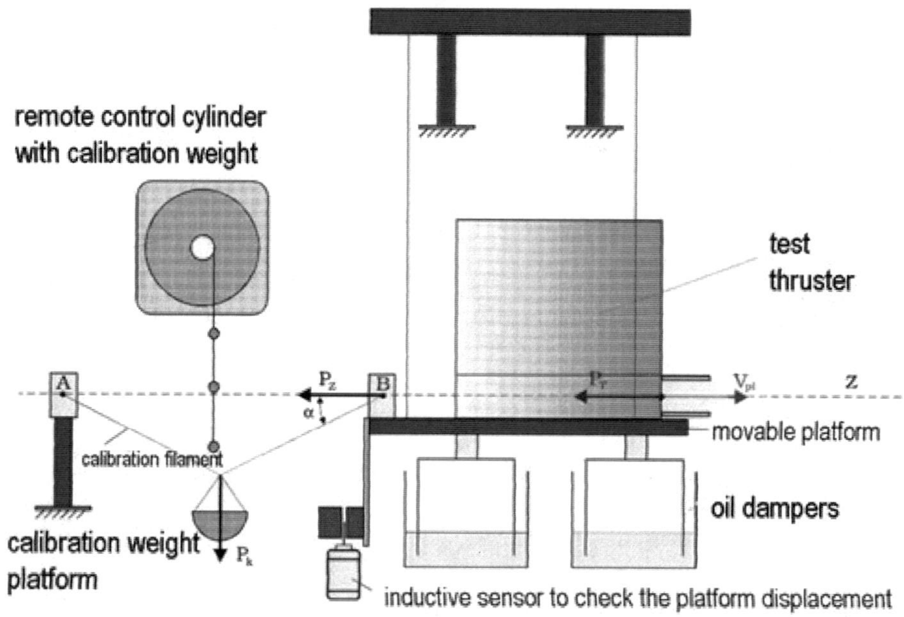

Figure 3. Diagram of the thrust measuring device.

Measurement of the flow rate of the propellant during the operation of the APPT was carried out by weighing the Teflon bars on highly sensitive analytical balances before and after a series of $(3...8)\cdot10^3$ pulses, the results of which also determined its flow rate per pulse. With a total number of pulses of $\sim5\cdot10^3$, the relative error in the mass of the bars did not exceed ±0.5%.

The integral characteristics of the APPT prototype were determined based on the measured APPT thrust and average flow rate of the propellant per pulse. The errors did not exceed ±5% for thrust and ±1% for discharge voltage.

2. Materials and research methods

2.1. Physical-and-mathematical modeling of plasma acceleration processes in APPT with low discharge energy

A physical-and-mathematical model of the plasma acceleration process in the APPT discharge channel was previously developed [11] in order to study the impact of the initial inductance of the discharge circuit on the characteristics. The model is a system of ordinary differential equations and one integral equation. Plasma blob on the one hand is considered as a material point (it has zero dimension) and on the other hand is described as an extended object that changes its shape and characteristics. In this model, we used the ordinary differential equations of motion of the center of mass of the PB and equations describing the evolution of the plasma blob averaged characteristics, such as temperature and pressure. The plasma blobs accelerate in the channel mainly as the result of electromagnetic forces, with their mass replenishing due to the Teflon bars ablation. For simplicity, it is stated that all the radiation energy originates from the blob center and falls on the surface of the Teflon bars that are facing the acceleration channel. The absorption of energy by the bars results in ablation of the propellant entering the channel. A moving PB absorbs the propellant that ablated inside the PB, and “scoops” the ablated propellant in the channel in front of it. The propellant entering the channel behind the PB is lost for this particular PB, but the next one can pick it up. The propellant ablated behind the last blob is completely lost. The proposed scheme is simplified, but at the same time, it describes all the basic processes occurring in the discharge channel. The model contains correction factors that are chosen in accordance with the condition of the best match of the calculated current curve and the physical experiment.

Despite the fact that the model mainly describes the motion of the center of mass of a plasma blob, we used in it the two-dimensional spatial models of plasma radiation, radiation transport and energy absorption by the channel walls. This allows us to describe adequately these processes in the model.

In APPT with energy up to 20 J there is an oscillatory discharge with a significant number (4...6) of half-periods of the discharge current oscillations, leading to wave-like generation and acceleration of the plasma. This quasi-one-dimensional physical-and-mathematical model takes these features into account. The periodicity of the plasma generation process is modeled by the formation of several plasma blobs moving one after the other in the acceleration channel. The dynamics of all plasma blobs is calculated similarly to the first blob, while the density and pressure are determined by the amount of propellant that has ablated by this time.

The initial inductance of the discharge circuit Lo varied during the calculations. The results are summarized in table 1.

Table 1. Calculation results for the thruster characteristics at different initial inductance.

Initial inductance, nH	33	40	46	50	52
Discharge energy, J	6.62	6.62	6.62	6.62	6.62
Mass flow rate, μg per pulse	23.7	20.8	17.5	15.8	14.9
Single thrust impulse, mN·s	0.112	0.118	0.134	0.150	0.154
Specific thrust impulse, m/s	4741	5648	7672	9436	10346

The above calculation results let us conclude that an increase in the initial inductance of discharge circuit for an APPT with the low discharge energy positively affects the characteristics of the APPT, contrary to the generally accepted principles. It reduces the flow of the propellant and increases the thrust specific impulse, due to better coordination of energy and mass input in the discharge channel.

IOP Conf. Series: Materials Science and Engineering **868** (2020) 012025 doi:10.1088/1757-899X/868/1/012025

2.2. Experimental study of the impact of the discharge circuit initial inductance on the characteristics of APPT with low discharge energy

In order to verify the results of the calculation by the physical-and-mathematical model, we conducted experimental studies of the impact of the discharge circuit inductance on the characteristics of APPT with low discharge energy. For this purpose, we made a laboratory model of APPT with energy of 6.6 J, which consisted of the following main elements: a discharge channel, a capacitor bank consisting of two capacitors, and a system for feeding the propellant (figure 4). The discharge channel was formed by two flat copper electrodes (cathode and anode), two Teflon bars installed on the sides between them and an end insulator. High-voltage igniter was inserted into the hole in the cathode electrode.

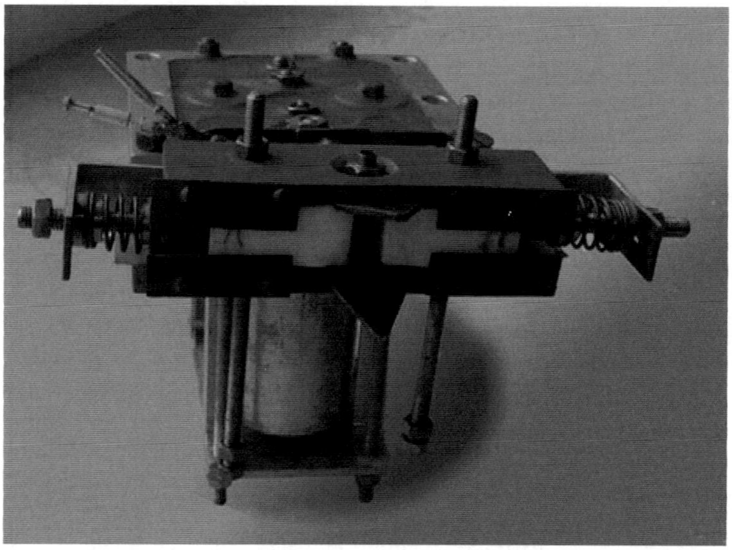

Figure 4. The laboratory model of the APPT.

We installed additional flexible copper conductors between the capacitor block and the discharge channel in order to vary the inductance in the APPT (figure 5). It was possible to achieve different values of the initial inductance of the discharge circuit by changing the distance between these conductors.

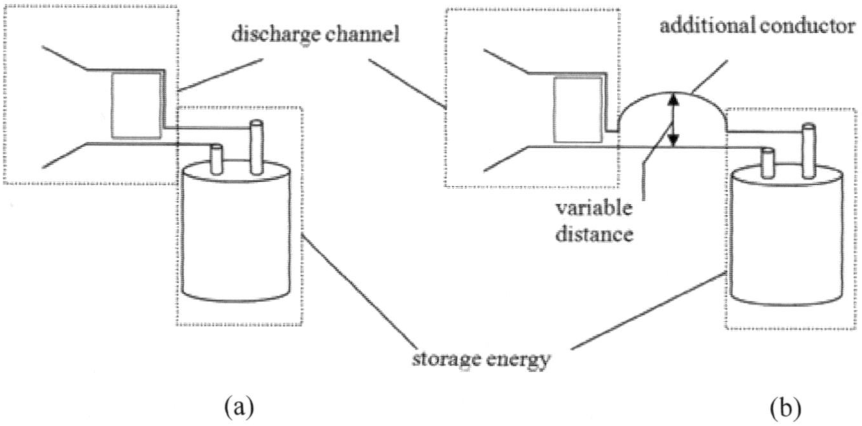

Figure 5. Modification of the laboratory model of APPT in order to increase the initial inductance of the discharge circuit: a) before the change, b) with additional conductors.

IOP Conf. Series: Materials Science and Engineering **868** (2020) 012025 doi:10.1088/1757-899X/868/1/012025

Figure 6 shows the oscillograms of the discharge current at different Lo values.

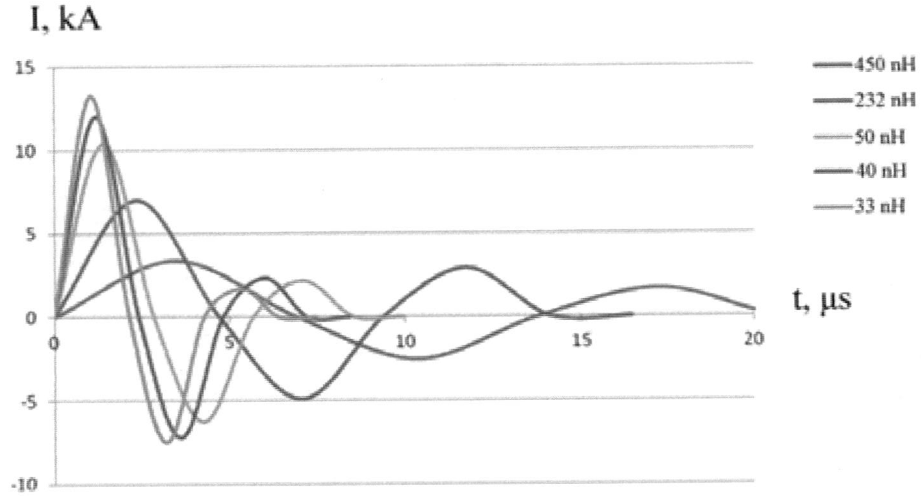

Figure 6. The oscillograms of the current at different values of APPT discharge circuit inductance.

To determine the value of the initial inductance, a current oscillogram was taken at the short circuit of the electrodes at the very beginning of the discharge channel. Then the value of the discharge circuit initial inductance Lo was calculated by the following formula:

$$L_o = \frac{T^2}{\pi^2 \cdot C},$$

where C is the bank capacity and T is the duration of the half period of discharge current.

During the tests, we measured the average thrust and flow rate of the propellant as well as the discharge circuit initial inductance. In addition, we calculated the characteristics of the test sample, such as discharge energy, power consumption, thrust impulse and mean – mass plasma flow rate (specific thrust impulse). The results of tests of APPT with energy of 6.6 J with various inductance values are given in table 2. Comparison of the experimental data with the calculated data is shown in table 3 and figure 7.

Table 2. Test results for a 6.6 J APPT at different initial inductance.

Characteristics	Lo=33	Lo=41	Lo=50
Discharge energy, J	6.6	6.6	6.6
Power consumption, W	13.2	13.2	13.2
Single thrust impulse, mN·s	0.11	0.109	0.095
Mass flow rate, µg per pulse	21.4	18.9	13.5

IOP Conf. Series: Materials Science and Engineering **868** (2020) 012025 doi:10.1088/1757-899X/868/1/012025

Table 3. Comparison of experimental and calculated data.

Experiment		Calculation	
Inductance, nH	Specific thrust impulse, m/s	Inductance, nH	Specific thrust impulse, m/s
33	4966	33	4741
36	5070	36	5142
41	5777	41	6074
46	6850	46	7672
50	7058	50	9436
51	6976	-	-
103	7797	-	-

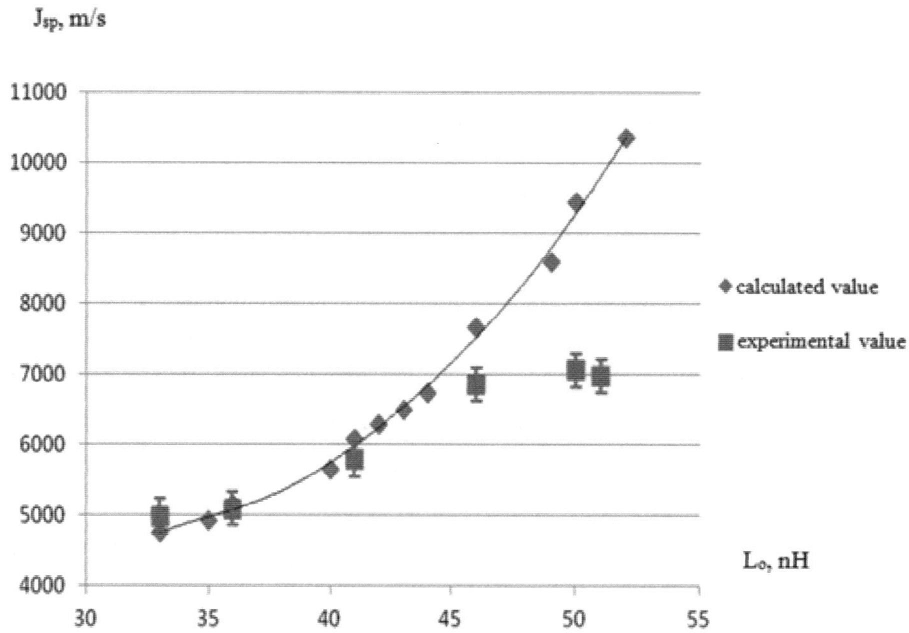

Figure 7. Dependence of the specific impulse on the initial inductance of the discharge circuit.

It is evident that the tendency of specific impulse growth with an increase in the initial inductance of the thruster discharge circuit coincides well with the preliminary calculations. At the same time, the experiments have shown that with a significant increase in inductance the duration of the discharge increases so much that it is carried beyond the discharge channel. This leads to a decrease in the energy inside the APPT channel, which in turn results in such a negative phenomenon as Teflon bar carbonization. It also happens due to the low degree of ionization of the propellant that evaporated at that moment, which settles on the surface of the discharge channel. The calculation model does not take into account the carbonization and reduction of the working surface area of the Teflon bars. This explains the observed discrepancy between theoretical and experimental results when the inductance exceeds 45 nH.

3. Results and discussion

3.1. Modernization of the sample propulsion system PPT-120 experimental model

The results of computational, theoretical and experimental studies have allowed us to finalize the PPT-120 propulsion system (PS PPT-120) in order to improve its specific characteristics. The PS PPT-120 experimental model has been developed on demand of the Space Systems Research and Development Institute (NII KS) and is intended to maintain and correct the orbit of a small low-orbit spacecraft weighing less than 100 kg. The PPT-120 propulsion system is shown in figure 8.

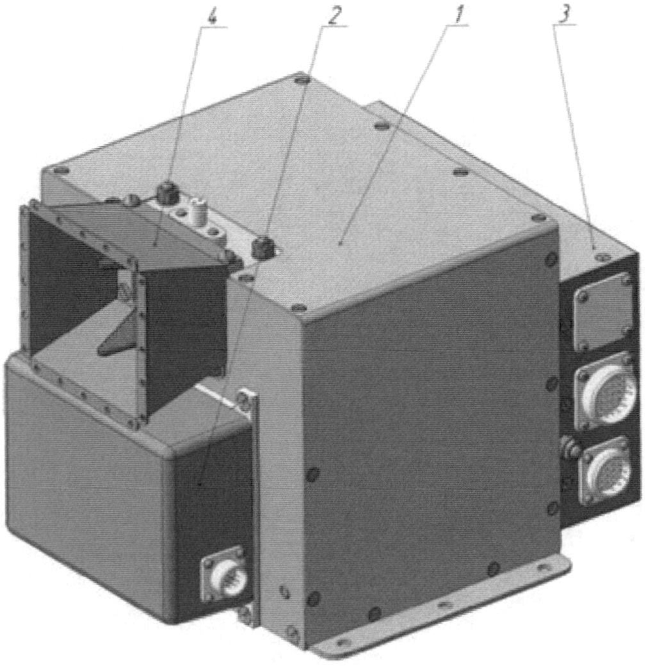

Figure 8. PPT-120 propulsion system.

The PPT-120 propulsion system is made as a monoblock comprising an energy storage unit (ESU) (1) with an electrodes system forming a discharge channel, a discharge initiation unit (DIU) (2), a power propulsion unit (PPU) (3) and a screen shield (4) intended to protect the thruster elements and adjacent spacecraft systems from the effects of peripheral flows of the thruster plasma jet (figure 8).

Capacitors located in the energy storage unit are connected to the electrodes of the discharge channel by means of flexible conductors. A series of conductors of different lengths was designed to change the initial inductance of the discharge circuit. After installing each set of the conductors, we measured the inductance of the model and conducted the experiment. Figures 9-11 show the experimental dependences of the specific thrust impulse, the mass flow rate per pulse and the single thrust impulse on the initial inductance of the discharge circuit.

IOP Conf. Series: Materials Science and Engineering **868** (2020) 012025 doi:10.1088/1757-899X/868/1/012025

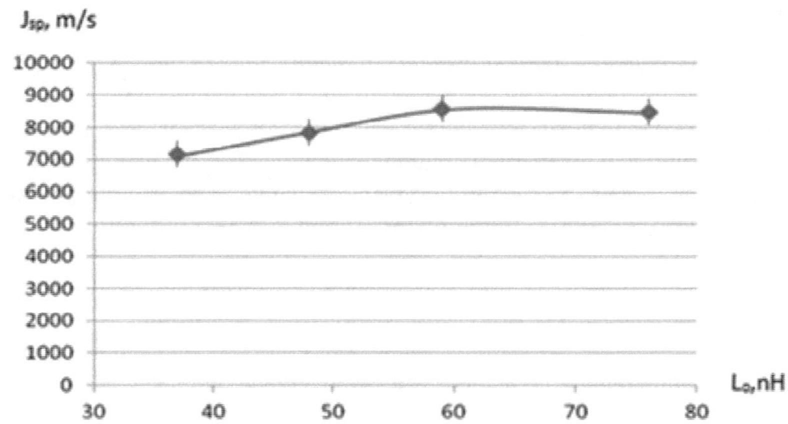

Figure 9. Dependence of the specific thrust pulse of PPT-120 on the initial inductance of the discharge circuit.

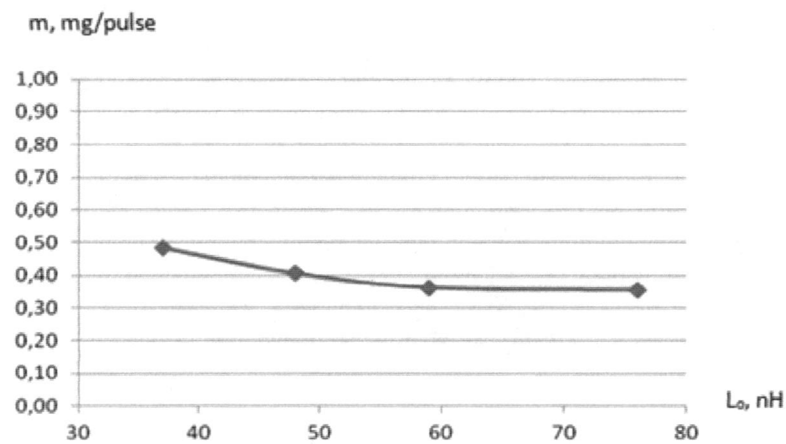

Figure 10. Dependence of the mass flow rate in PPT-120 per pulse on the initial inductance of the discharge circuit.

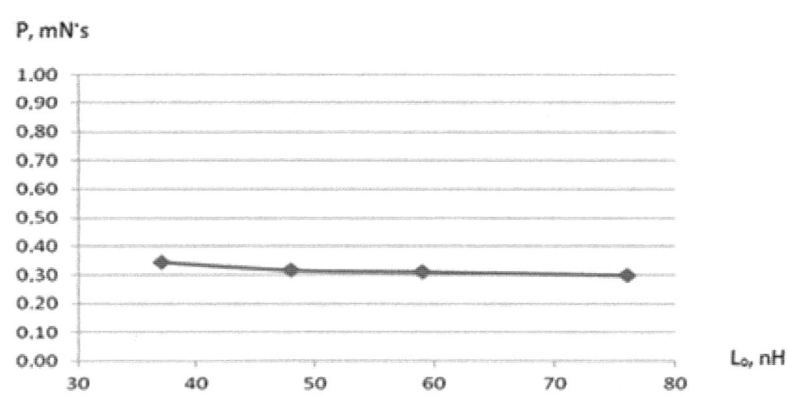

Figure 11. Dependence of a single thrust pulse on the initial inductance of the discharge circuit of PPT-120.

The graphs demonstrate that the increase in the initial inductance of the electric circuit in the range of ~ 35 ... 60 nH leads to a significant increase in the specific impulse (20%), from 7125 m/s to ~8500 m/s. This happens due to a decrease in the mass flow rate per pulse (25%) with a significantly smaller decrease in the thrust single impulse (10%). The characteristics of the original and upgraded propulsion system are given in table 4.

Table 4. Characteristics of the original and upgraded PPT-120 propulsion system.

Characteristics	PS PPT-120 with Lo=37 nH	PS PPT-120 with Lo=59 nH
Discharge energy, J	20	20
Single thrust impulse, mN·s	0.34	0.31
Mass flow rate, μg per pulse	48	36
Specific thrust impulse, km/s	7.1	8.5
The operating frequency of the impulses, Hz	2.65	2.65
Power consumption, W	53	53
Average thrust, mN	1.1	0.82
Propellant	Teflon	Teflon
Propellant store, kg	0.12	0.12
Number of pulses (cycle life)	$2 \cdot 10^6$	$3 \cdot 10^6$
Total thrust impulse, kN·s	0.8	0.95
PPT total mass, kg	3.5	3.5

The table shows that the specific and total thrust impulses of the thruster after its completion significantly increased by 20% and 19%, respectively. At the same time, there is a decrease in the single thrust impulse and, consequently, the average thrust, which can easily be compensated by increasing the frequency of the propulsion system pulses.

A reduction in consumption and, consequently, an increase of the specific thrust impulse are affected by a decrease in the effect of propellant bar post-streaming after completion of the process of main Teflon mass acceleration. With the increase of the inductance, this happens due to an increase in the discharge duration and a decrease in the peak value of current, which in turn leads to a corresponding decrease of the plasma radiation energy and the positioning of the radiating plasma blob at a greater distance from the Teflon bars.

The efficiency of the PPT-120 propulsion system was verified by long-term tests, which resulted in the absence of carbonization traces on the Teflon bar working surfaces. Figure 12 shows the state of the discharge channel after long tests.

Figure 12. The PS PPT-120 discharge channel after longer tests.

IOP Conf. Series: Materials Science and Engineering **868** (2020) 012025 doi:10.1088/1757-899X/868/1/012025

4. Conclusion

The calculations by the above-mentioned physical-and-mathematical model allow us to understand the reasons for the increase in the specific impulse of the thrust with an increase in the initial inductance of the discharge circuit. Figure 13 displays graphs of ablated mass in the discharge channel as a function of time under different initial inductance of discharge circuit, taken from the calculation model. There are also graphs of the square of the discharge current as a function of time, characterizing the power released in the discharge channel in the first two half-cycles.

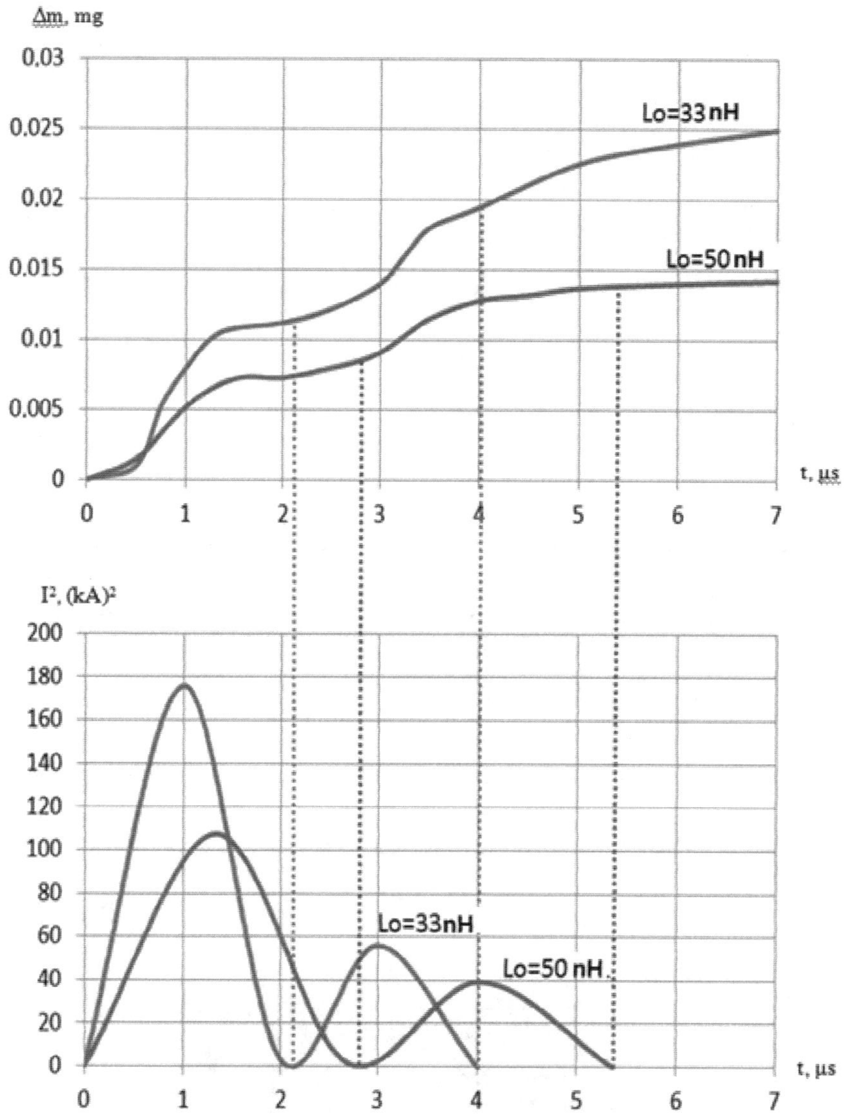

Figure 13. Calculated curves of ablated mass and square of discharge current as a function of time.

The figure shows that the duration of the discharge increases with an increase in the initial inductance of the discharge circuit, while the value of the evaporated mass fed into the channel decreases. At the same time, the mass flow of the plasma-forming substance stops at the end of the second half-period, while at a lower inductance it goes on. If the ablation mass decreases with increasing Lo, this leads to an assumption that the flux of radiant energy hitting the ablating surface decreases, both due to the fact that the discharge burns farther from the surface of the Teflon bars and due to a decrease in the current amplitude inversely proportional to the root of Lo. That is, with Lo increasing, the so-called after-evaporation effect is reduced when the ablated mass is not accelerated efficiently in the discharge channel. However, with a further increase, the positive effect of increasing the inductance Lo is lost. Also, the optimal Lo inductance value can be different for each APPT model.

The results of the experiments allowed discovering the nature of the influence of the discharge circuit Lo inductance on the characteristics of a low-energy APPT.

We can therefore say that with an increase in the initial inductance of the circuit, the APPT with low discharge energy sees improvement in the coordination of the input of energy and mass into the acceleration channel.

Acknowledgment

The work was fulfilled within the framework of implementation of the Federal Purpose-Oriented Program "Researches and Developments in the Priority Directions of Growth of the Scientific and Technologic Complex of Russia for 2014–2020" (unique identifier of works project: RFMEFI60719X0308).

Reference

[1] Schönherr T, Guiza Cepeda J L, Skalden J, Ilic D, Herdrich G and Komurasaki K 2018 Coaxial air-fed pulsed plasma thruster research and development for RAM-EP application. *Conference "Space Propulsion 2018"* (Seville, Spain, 14–18 May, 2018) SP2018-116 p 6

[2] Liang Ling W Y, Zhang Z and Tang H 2017 Progress in fundamental pulsed plasma thruster research. *The 35th International Electric Propulsion Conference* (Georgia Institute of Technology, Atlanta, Georgia, USA, October 8 – 12, 2017) IEPC-2017-140 p 8 http://electricrocket.org/IEPC/IEPC_2017_140.pdf

[3] Fujita R, Kanaoka K, Ono K, Morikawa N, Ryuho K, Enomoto K, Tahara H, Takada K and Wakizono T 2018 Research and development of electrothermal pulsed plasma thruster systems onboard the Osaka Institute of Technology 2nd PROITERES Nano-Satellite. *Journal of the Japan Society for Aeronautical and Space Sciences* **66**(4) 91 doi.org/10.2322/jjsass.66.91

[4] Woodruff C A, King D M, Burton R L and Carroll D L 2019 Fiber-fed Pulsed Plasma Thruster (FPPT) for small satellites. *The 36th International Electric Propulsion Conference* (University of Vienna, Austria, September 15-20, 2019) IEPC 2019-A899 p12 http://electricrocket.org/2019/899.pdf

[5] Vaupel M, Hey F G, Moneva I G, Papendorf D, Braxmaier C, Tajmar M, Sell A, Eckert K, Weise D, Saks N and Johann U 2017 The Next Generation milli-Newton μHEMPT as potential main Thruster for small satellites. *The 35th International Electric Propulsion Conference* (Georgia Institute of Technology, USA, October 8-12, 2017) IEPC-2017-275 p 8 http://electricrocket.org/IEPC/IEPC_2017_275.pdf

[6] Hey F G, Vaupel M, Braxmaier C, Tajmar M, Sell A, Eckert K, Weise D, Saks N and Johann U 2017 HEMPT Downscaling, way forward to the first EM for CubeSat applications. *The 35th International Electric Propulsion Conference* (Georgia Institute of Technology, USA October 8-12, 2017) IEPC-2017-274 p 9 http://electricrocket.org/IEPC/IEPC_2017_274.pdf

[7] Kolesnikov P M 1971 *Elektrodinamicheskoe uskorenie plazmy* (*Electrodynamic Acceleration for Plasma*) [in Russian] (Moscow: Atomizdat) p 392

[8] Grishin S D, Leskov L V and Kozlov N P 1983 *Plazmennye uskoriteli (Plasma Accelerators)* [in Russian] (Moscow: Mashinostroenie) pp 132-133

[9] Diakonov G A, Lyubinskaya N V, Semenikhin S A and Khrustalev M M 2014 Ablyatsionnyi impulsnyi plazmennyi dvigatel dlya malorazmernykh kosmicheskich apparatov (Ablative Pulsed Plasma Thruster for Small Spacecraft) [in Russian] *Trudy MAI* **73** http://trudymai.ru/upload/iblock/b05/b051f8402f5dc4801ca5cadf0d455e35.pdf?lang=ru&iss

ue=73
[10] Nawaz A, Albertoni R and Auweter-Kurtz M 2010 Thrust efficiency optimization of the pulsed plasma thruster SIMP-LEX. *Acta Astronautica* **67** 440 doi: 10.1016/j.actaastro.2010.03.006
[11] Antropov N N, Bogatyy A V, Dyakonov G A, Lyubinskaya N V, Popov G A, Semenikhin S A, Khrustalev M M, Yakovlev V N and Tyutin V K 2012 A new stage in the development of Ablative Pulsed Plasma Thrusters at the RIAME. *Sol Syst Res* **46** 531 doi.org/10.1134/S0038094612070064

IOP Conf. Series: Materials Science and Engineering **868** (2020) 012026 doi:10.1088/1757-899X/868/1/012026

Criterion for evaluation of erosion of aviation engine parts from polymer composite materials

A Y Ravikovich, D P Kholobtsev, A N Arkhipov and A S Shakhov*

Department of Gas Turbine Engine Design, *Moscow Aviation Institute* (National Research University), 4 Volokolamskoe shosse, Moscow, Russian Federation

*E-mail: shakhov_alexander@mail.ru

Abstract. A criterion is proposed for assessing the reliability of parts made of polymer composite materials in the event of sand and dust entering the compressor during take-off and landing. Conducting erosion tests with a concentration of abrasive particles corresponding to a conventional dust stream is practically impossible due to their long duration. When conducting tests with a high concentration that is several orders of magnitude higher than the concentration of a conventional stream, the reliability of the estimate is reduced due to the need to extrapolate the results to a large resource. Therefore, to improve the accuracy of the assessment of the erosion resistance of engine parts from polymer composite materials, it is proposed to conduct periodic measurements of the thickness of the part during operation. By calculating the dynamic eigenfrequencies and oscillation shapes and maximum static stresses, the critical thickness of the part is established, below which its eigenfrequency approaches the dangerous rotor harmonics.

1. Introduction

The implementation of new polymer composite materials (PCM) [1] in the design of an aircraft engine is an important task to improve its efficiency. PCMs are mainly used in the outer air path of aircraft engine fan [2], which are not subjected to high loads, but during operation these structures are exposed to various external factors. Therefore, the problem of assessing the reliability and durability of PCM parts in operation is important. It is necessary to choose the criteria for such. One of the important conditions for the use of PCM in aircraft engines is to maintain operability when sand and dust get into the compressor during take-off and landing.

The issues of assessing the reliability and durability of the PCM internal panel, which is part of the low-pressure compressor of the Sam146 aircraft engine, under the influence of a gas and dust flow, are considered in this paper. Figure 1 shows the location of the panel in the engine.

Dust and gas mixture that is sucked into the engine leads to erosion of parts and deterioration of its characteristics. This effect is inherent in the take-off and landing stages when sand and dust are lifted from the runway by an air stream, which greatly affects the durability of PCM parts.

The erosion resistance of composites depends on many factors. Thus, matrix materials, reinforcement fibers, impact angle and particle velocity have an effect on the solid particle erosion behavior of thermoplastic resins reinforced by short fibers with special attention on an incubation period of erosion [3]. The results of the experiments showed that the mass of an incubation period was strongly dependent on initial surface roughness. The erosion rate at a steady state depended on the matrix material and on the volume content of fibers, regardless of fiber type. Evaluation of composites erosion at different impingement angles and impact velocities of sand particles was done in [4]. Different materials showed peak erosion at different impingement angles as well as significant influence on erosion behavior at different impingement velocity and mass of erodent. Then, alumina filling had an effect on the erosion wear performance of glass fiber-reinforced polyester composites [5]. Pure glass-polyester composite without filler showed greater erosion rate whereas a significant improvement in the erosion resistance was observed with alumina fillers. Test of blades with TBC was performed to determine the erosion rates and particle restitution characteristics under different impact conditions [6]. The experimental results show that the erosion rates increase with increased impingement angle, impact velocity, and temperature. In [7] the authors evaluated solid particle erosion behavior of unfilled and graphite particulate filled carbon fabric reinforced epoxy composites. Maximum wear rate was observed at 45° impingement angle. The erosion rates of plain polymers and these composites was studied at various impact angles (α=15-90°) and showed that the erosion behavior did not depend on fabric orientation [8]. The authors of [9] studied influence of the concentration, angles of attack and speeds of abrasive particles acting on PCM parts at simulation of the sand and dust throw into the intake of engine fan. A model of the movement of abrasive particles in a compressor of a gas turbine engine was used.

Figure 1. Location of the internal panel on the Sam146 aircraft engine.

Thus, the criteria of the erosion resistance for PCM proposed in the most references is experimental test at injection of air with a concentration of abrasive particles and evaluation of erosion rates by measurements of erosion depth or loss of material weight. The main task of experimental research is to reproduce operating conditions during the test as accurately as possible. Due to the relatively low concentration of abrasive particles in the air during operation, the creation of such a concentration during the experiment will lead to a long test duration. When conducting tests with a high concentration that is several orders of magnitude higher than the concentration in a conventional flow, the reliability of the estimate is reduced due to the need to extrapolate the results to a large resource. Besides, engine operation at different environment condition does not allow to estimate erosion damage in dependence on usual parameters of material erosion but could be done indirectly on the base of assessment of the technical condition of the panel during operation, namely residual thickness of panel.

Therefore, the new indirect criterion of the erosion resistance is proposed for PCM internal panel instead of the usual criteria of erosion rates evaluation. This new criterion is based on estimation of critical thickness when statics and dynamics of panel still satisfy design criteria. When the abrasive flow acts on the panel, the material is carried away and panel thickness is reduced, which changes the strength characteristics of the parts during long operation. Thus, periodical measurement of panel thickness will allow to estimate panel statics and dynamics indirectly if their dependence on thickness is well known.

A dynamic and static calculations of a panels with different thicknesses made of a polymer composite material have been carried out and critical panel thickness when dynamic frequencies and static stress do not exceed the permissible level has been determined.

2. Evaluation method

2.1. Criterion for assessing the erosion resistance of engine parts from PCM

In this paper, we study the PCM internal panel, which is part of the low-pressure compressor of an aircraft engine (figure 2). To increase the accuracy of evaluating the erosion resistance of the panel, it is proposed to periodically measure the thickness h of the panel (figure 3) in operation and then compare it with an acceptable value.

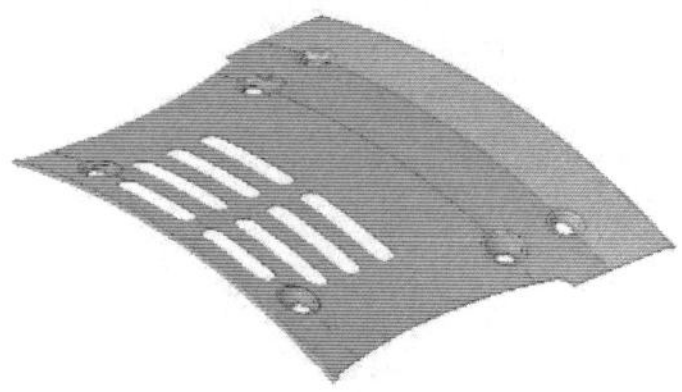

Figure 2. General view of the studied internal panel from PCM.

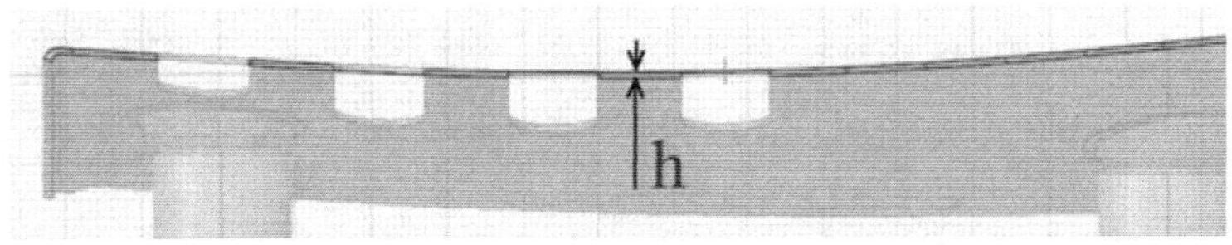

Figure 3. Longitudinal section of panel: h – is panel thickness.

Estimates of the natural frequency of oscillations in the first form and maximum static stresses for panels with different thicknesses after the expected erosion are proposed as criteria for reliable operation.

2.2. Model

To study the effect of erosion on the panel strength, several design models with different thicknesses were created. Static and dynamic calculation of models was carried out in the ANSYS software package. An example of a model is shown in figures 4a and 4b.

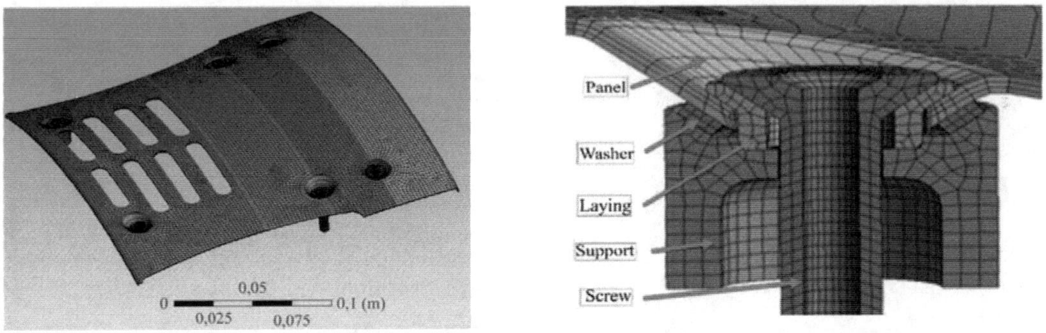

(a) (b)

Figure 4. Finite element model of the interior panel of the PCM: general view (a) and attachment parts (b).

Detailed mathematical modeling of all factors affecting the product during operation, allows to achieve high reliability of the calculation results. An important issue is the modeling of the boundary conditions, on which the accuracy of the obtained results depends significantly [10]. Modeling of contacts and panel assembly on the engine are shown below.

Figure 5 shows all types of contacts simulating real operating conditions. The tightening of the screws was set to 5.25 N·m. A tension of 0.1 mm was set between the bushing and the seal washer.

IOP Conf. Series: Materials Science and Engineering **868** (2020) 012026 doi:10.1088/1757-899X/868/1/012026

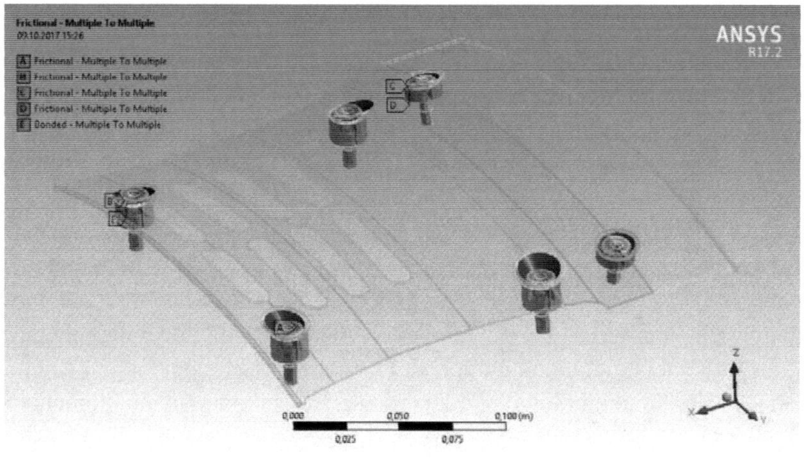

Figure 5. Types of contacts in assembling of panel.

Figure 6 shows the model fixing areas corresponding to the product fixing areas on the engine. The screws and bushings were secured at all degrees of freedom. Friction contacts is simulated with a coefficient of friction of 0.15. The first stage of the calculation is to check the static strength of the installation of the product, for this, the bolt tightening is calculated, shown in the figure 7.

Figure 6. Fixing of panel.

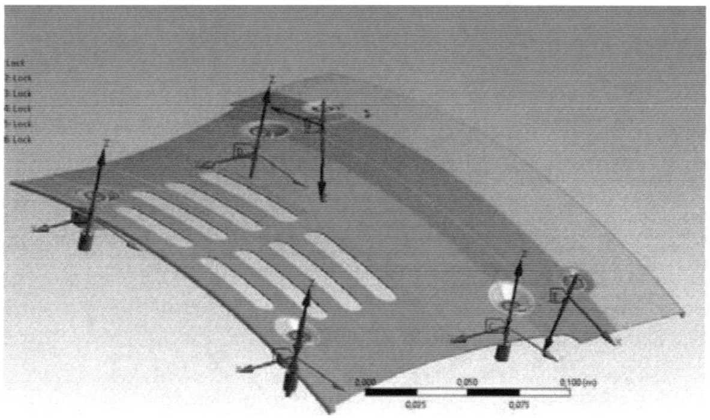

Figure 7. Tightening of screw-bolts.

From the previously performed aerodynamic calculations of the engine under study, pressure distributions (figures 8 and 9) and temperatures (figure 10) over the panel surface were obtained.

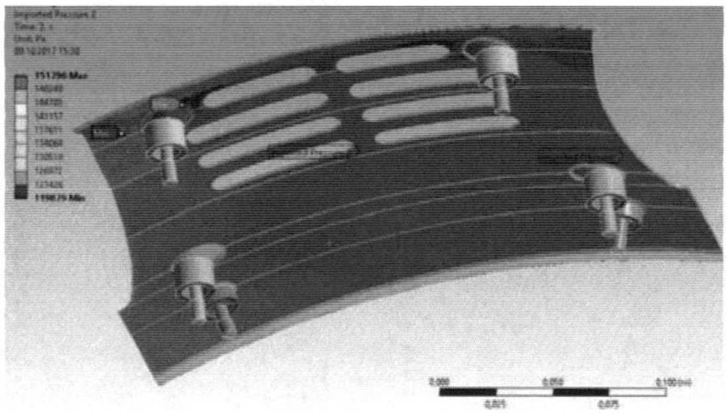

Figure 8. Distribution of pressure on the inner side of panel.

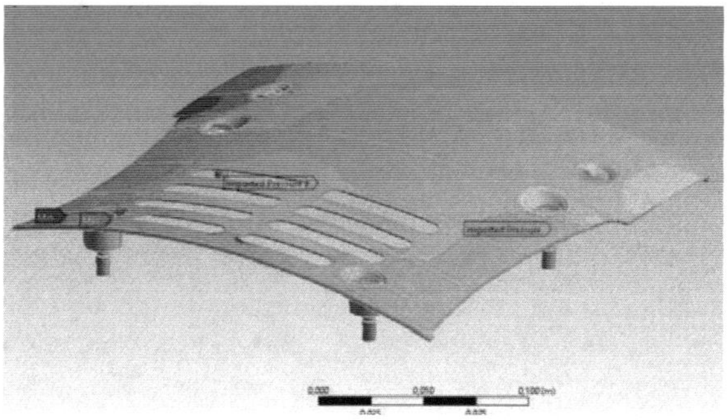

Figure 9. Distribution of pressure on the outer side of panel.

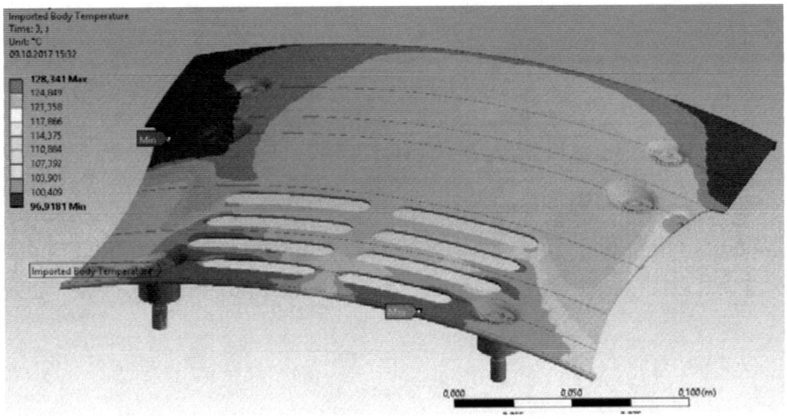

Figure 10. Distribution of temperature on a panel.

2.3. Material properties

Toho TENAX material has been used as panel material with transversely isotropic properties. Titanium alloy VT6 has been used for support and washer (figure 4b). The screw was made of EP99 material. VITONE material was used for laying between the panel and support (table 1).

Table 1. Properties of materials.

Properties	PCM	Titanium	EP99	VITONE
Modulus of elasticity E11, E22, MPa	59,400	114,000	190,300	50
Modulus of elasticity E33, MPa	5714	114000	190300	50
Poisson ratio μ12	0,07	0.3	0.3	0.45
Poisson ratio μ23, μ13	0.15			
Shear modulus G12, MPa	4,500	472	732	
Shear modulus G23, G13, MPa	4,100			
Density, kg/m^3	1,530	4,510	8,440	1,500

2.4. Calculation results and discussion

An example of the calculation results of the first three forms of vibration and static stresses for the ready panel with a thickness of 1.2 mm is shown in figures 11a-c, 12a, and 12b.

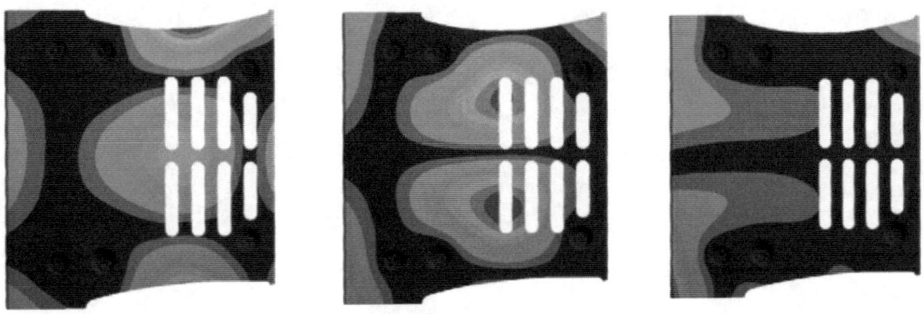

(a) (b) (c)

Figure 11. Example of results of calculation of three first forms of vibrations: first (a). second (b), and third (c) forms.

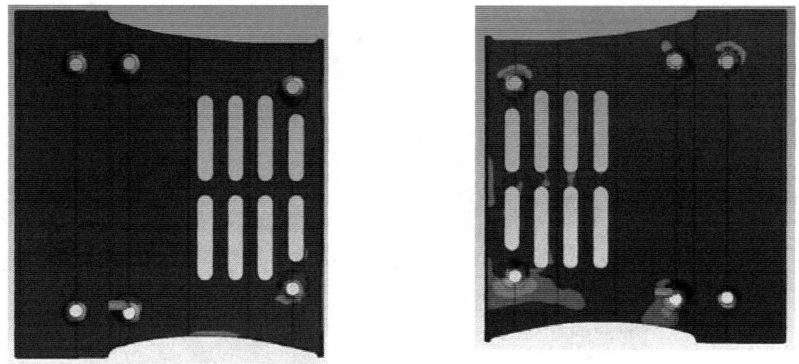

(a) (b)

Figure 12. Example of results of calculation of static stresses at inner (a) and outer (b) sides.

The results of calculating the panel natural frequencies are shown in the table 2. The obtained frequencies are compared with the frequencies of the harmonics of the engine rotor under study. The critical thickness in our calculations is the interval between 0.9-0.8 mm, with this change, the natural frequency of the oscillations of the panel drops from 442 to 371 Hz, which crosses the 4th harmonic value of the engine rotor.

Table 2. Results of calculation of eigenfrequencies and vibration modes of the internal panel.

Forms of vibrations	**Panel thickness, mm**				
	1.2	1.1	1.0	0.9	0.8
	Frequency of vibrations, Hz				
1	517	498	475	442	371
2	575	552	526	496	457
3	602	578	548	514	458

Table 3. Results of calculating the static stresses of the internal panel.

Stress component	**Panel thickness, mm**				
	1.2	1.1	1.0	0.9	0.8
	Maximum stresses, MPa				
σx+	61.4	66.0	64.6	64.3	67.1
σx-	-83.0	-83.4	-83.6	-83.6	-87.4
σy+	78.1	80.9	83.4	85.8	108.1
σy-	-101.7	-103.6	-105.2	-106.7	-106.2
σz+	6.4	6.8	7.2	7.6	8.7
σz-	-57.2	-57.0	-57.0	-56.8	-56.4
τxy	19.8	20.3	20.8	21.2	22.0
τyz	30.3	30.1	30.4	30.6	30.7
τxz	23.0	23.0	22.9	22.8	22.7

The results of the static stress calculation are shown in the table 3. Static stresses at the selected thicknesses are increased up to 40% maximum and do not exceed the permissible level. Maximum radial displacements do not exceed 1 mm that is also acceptable from aerodynamics point of view. Therefore, allowable reduction of panel thickness due to erosion can be down to 1 mm on the base of dynamic and static calculations and taking into account accuracy of measurement during inspection in field.

Taking into account that maximum calculated stresses are a few times lower that minimum yield limit of panel material, an influence of reduction of panel thickness due to erosion is less significant than it influences on dynamics. Therefore, allowable reduction of panel thickness down to 1 mm is also acceptable from static loading point of view.

3. Conclusion

The new indirect criterion of the erosion resistance is proposed for polymer composite materials internal panel instead of the usual criteria of erosion rates evaluation. This new criterion is based on estimation of critical thickness when statics and dynamics of panel still satisfy design criteria.

Series of dynamic and static calculations of a panels with different thicknesses have been carried. Critical panel thickness of 0.9 mm was revealed at which the vibration frequency in the first form decreases to 442 Hz and approaches the dangerous rotor harmonics of the engine, at which the probability of the work of the part in the critical region increases.

Static stresses at the selected thicknesses are increased up to 40% maximum and do not exceed the permissible level. Maximum radial displacements do not exceed 1 mm that is also acceptable from aerodynamics point of view. Allowable reduction of panel thickness can be down to 1 mm on the base of dynamic and static calculations and taking into account accuracy of measurement during inspection of aviation engine in field. Proposed criterion can be used by aviation companies for assessing the reliability of parts made of polymer composite materials in the event of sand and dust entering the compressor during take-off and landing.

Acknowledgement
The work was carried out with the financial support of the Ministry of Education and Science of the Russian Federation as part of the comprehensive project No. 2017-218-09-172. The authors thank the staff of PJSC UEC-Saturn, Rybinsk for their help and support with carrying out this work.

References

[1] Jones R M 1999 *Mechanics of Composite Materials*. 2nd ed. (Boca Raton: Tailor & Francis Group), p 538 https://doi.org/10.1201/9781498711067
[2] Anoshkin A N, Zuiko V Yu, Shipunov G S and Tretyakov A A 2014 Technologies and problems of composite materials mechanics for production of outlet guide vane for aircraft jet engine. *PNRPU Mechanics Bulletin*. **4** 5 doi: 10.15593/perm.mech/2014.4.01
[3] Miyazaki N and Hamao T 1994 Solid particle erosion of thermoplastic resins reinforced by short fibers. *J. Compos. Mater.* **28** 320 https://doi.org/10.1177/002199839402800905
[4] Harsha A P, Tewari U S and Venkataraman B 2003 Solid particle erosion behavior of various polyaryletherketone composites. *Wear* **254** 453 https://doi.org/10.1016/S0043-1648(03)00143-1
[5] Patnaik A, Satapathy A, Mahapatra S S and Dash R R 2008 Parametric optimization of erosion wear of polyester-GF-alumina hybrid composites using Taguchi method. *Journal of Reinforced Plastics and Composites*. **27** 1039 https://doi.org/10.1177/0731684407086867
[6] Swar R, Hamed A, Tabakoff W and Miller R A 2012 Surface deterioration of thermal barrier coated turbine blades by erosion. *International Journal of Rotating Machinery*. **2012** 601837 https://doi.org/10.1155/2012/601837
[7] Sudarshan Rao K, Varadarajan Y S and Rajendra N 2015 Erosive wear behaviour of carbon fiber-reinforced epoxy composite. *Materialtoday Proceedings*. **2** 2975 doi: 10.1016/j.matpr.2015.07.280

[8] Liu B, Bao L and Xu A 2016 Effect of fabric orientation and impact angle on the erosion behavior of high-performance thermoplastic composites reinforced with ductile fabric. *Wear*, **352–353** 24 https://doi.org/10.1016/j.wear.2016.01.016

[9] Zinin A V, Dobrovolsky S V, Lebedev A K, Krupennikov V A and Shevyakov A O 2019 Wear and erosion resistance in a gas-abrasive stream of composite elements of an aircraft engine. *Russian Aeronautics* **62** 696 https://doi.org/10.3103/S1068799819040226

[10] Arkhipov A N, Volgina M V, Matushkin A A, Ravikovich Ya A and Kholobtsev D P 2018 Effect of a model and boundary conditions on the results of analyzing the stress-strain state of a low-pressure compressor rotor. *Russian Aeronautics* **61** 509 https://doi.org/10.3103/S1068799818040025

IOP Conf. Series: Materials Science and Engineering **868** (2020) 012027 doi:10.1088/1757-899X/868/1/012027

Estimation of fatigue margins for composite parts of aircraft engine on the base of vibration tests

Y Ravikovich*, D Kholobtsev, A Arkhipov and G Popov

Research department of the chair 203, *Moscow Aviation Institute* (National Research University), 4 Volokolamskoe shosse, Moscow 125993, Russian Federation

E-mail: *yurav@mail.ru

Abstract. During the modernization of the low-pressure compressor of the regional aircraft engine, a flow separator made of polymer composite material was developed, which has less weight, is easier and cheaper to manufacture. However, the separator must undergo a series of tests before being installed on the engine, in particular the endurance tests, necessary to assess the fatigue strength margin. A method for determining the endurance limit of a large part of an aircraft engine is developed in the presented work and consists in replacing the tests of the entire separator with tests of individual structural elements cut from the part. The shape, dimensions and fixation of the elements have been chosen in such a way that their own test frequency was close to the natural frequency of the flow separator on the engine. The natural oscillation frequencies and places of maximum dynamic stresses have been determined before testing on three-dimensional models of the flow separator and elements cut from it. As a result, the structural elements have been tested until appearance of fatigue cracks which allowed to assess the endurance limit and evaluate fatigue strength margin of the separator.

1. Introduction

Currently, a large amount of work is being carried out to study the possibility of using polymer composite materials (PCM) for the manufacture of aircraft engine parts. The main advantage of PCM is a lower density than titanium, as well as lower cost of manufacture, so the use of PCM in aircraft and other engineering industries increases [1].

Use of PCM for the manufacture of parts, which are exposed by high loads and temperatures [2], requires more expensive preliminary investigations. Therefore, with regard to aircraft engines it is advisable to use PCM for the external fan circuit, where main damage could be accumulated due to static and dynamic loading at relatively low external forces and temperatures.

During the modernization of the low-pressure compressor of the SaM146 engine, a flow separator (FS) made of a polymer composite material was developed for the regional aircraft. The location of the flow separator in the low-pressure compressor is shown as figure 1.

However, the separator must pass a series of tests before installation on the engine, including vibration tests, which should evaluate the margin of fatigue strength K_v. K_v evaluation is carried out by the results of endurance limit evaluation during fatigue tests and FS strain measurement on the engine.

The problem of fatigue strength of PCM is important for practice and there are many publications on this topic. For example, I K A Widi et al [3] showed that the automotive industry is already utilizing many composite materials for its products and investigated the effect of fiber orientation (unidirectional, woven and random) on several resin matrices (epoxy and polypropilene). Fatigue test of the specimens showed the optimum value with the orientation variation of fiber.

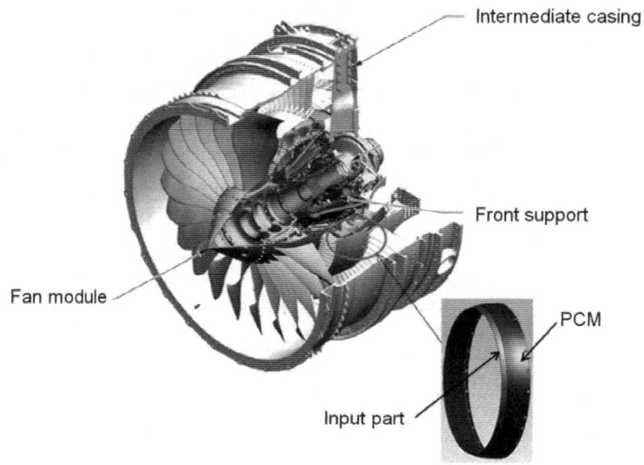

Figure 1. The separator flows into the low-pressure compressor.

A J Brunner [4] considered fatigue mechanism for fiber-reinforced polymer (FRP) composites and concluded that there is still no full understanding of the effects that require consideration when testing FRP composite materials under cyclic fatigue loads with the aim of using the data for design limits.

A J Brunner et et al [5] investigated fracture mechanics tests on carbon fiber composites reinforced with polymer matrix and focused on cyclic fatigue at tensile rupture, plane shift, and mixed loading. The paper analyzes the data using different approaches in an attempt to reduce the spread and determine the parameters for the structural design of carbon fiber. In addition, selected test data comparing load and displacement control in cyclic fatigue tests are discussed.

R C Alderliesten et al [6] discussed various aspects related to the physics of fatigue fracture testing of fiber-reinforced polymers (FRP) composites based on experimental observations. The authors recommended further developing and evaluating approaches that quantify the contribution of fibre bridging unambiguously to use data obtained from unidirectional FRP for structural design with multidirectional lay-ups and interfaces.

L Molent and C Forrester [7] reviewed the cyclic fatigue strength of damaged aircraft composite structures under operating loads and proposed a framework to solve the problem of assessing the damage resistance of these structures based on the slow growth approach.

Yu Kashtalyan and C Soutis [8] reviewed experimental and theoretical studies into the internal-damage mechanisms of cross-polycomposite laminates made within the framework of damage micromechanics and into their effect on the behavior and properties of laminates.

N Lisin et al [9] developed statistical strength theories to describe the results of fatigue tests and to predict the strength of machine components with variable loads. Experiments show that the strength of materials depends significantly on structural defects, and the limiting stress is a statistical quantity.

N S Azikov, A V Zinin [10] performed strength analysis of the structure based on the model of mechanical property degradation for composite material layers taking into account the process of damage accumulation in the course of loading.

T D Karimbaev and D V Matyukhin [11] showed that the variety of forms of fatigue damage in composite materials products and their development are not allowed so far to form a composite material commonly accepted laws such as the law of Paris, the linear summation of Miner and other. They proposed to carry out a fatigue tests of cantilever-fastened specimens at first bending form until vibration frequency decreased to an acceptable value, which is determined from the analysis of the Campbell diagram.

Therefore, the problem of fatigue strength of PCM used for the manufacture of aviation engine parts is also important. However, the most investigations are related to fatigue test of specimens and don't considered the test of real parts used in aviation engines, especially. Besides, these investigations are studied crack propagations mainly. At the same time, it is not advisable to use crack propagation as criteria for fatigue damage of aviation engine parts, such as flow separator, taking into account higher exciting frequencies in comparison with automotive and aircraft parts. It is better to determine a moment of crack growth until vibration frequency decrease down to acceptable value as criteria of fatigue failure.

Vibration tests of the entire separator didn't allow to get fatigue failure that was associated with limit of existing stand power. A method for determining the endurance limit of a large part of an aircraft engine is developed in the presented work and consists in replacing the tests of the entire separator with tests of individual structural elements cut from the part.

This article describes the procedure of the flow separator endurance limit determination based on:

- calculation of natural frequencies and mode shapes for whole separator,
- unsuccessful fatigue test of separator,
- choice of shape, dimensions and fixation of the elements that could be cut from the separator in such a way that their own test frequency was close to the natural frequency of the flow separator on the engine,
- fatigue tests of elements to determine the endurance limit of separator

2. Calculation of flow separator

2.1. Model

The flow separator is a ring made of polymer composite material with an attached input part made of titanium alloy.

Special attention was paid to the simulation of fastening on the engine, when choosing the design model and boundary conditions [12].

The model of the composite ring and titanium part has been made using three-dimensional hexahedral twenty-node quadratic SOLID186 elements. Number of nodes in the model: 373200; total number of elements: 84597; PCM elements: 50399 (HEX20); titanium part elements: 25499 (HEX20); contact elements: 8699 (QUAD8).

The elements of the composite are connected to the titanium part using contact elements CONTA174/TARGE170, the contact type is bonded.

The model was fixed to the right end by fixing the nodes belonging to the face plane in the three directions in the global coordinate system.

Temperature has not been applied because heating of flow separator is insignificant for dynamics.

Pressure has not been applied because its influence on dynamics is insignificant.

2.2. Material properties

Properties of materials are presented in table 1.

Table 1. Properties of materials.

Propeties	**PCM**	**Titanium**
Modulus of elasticity E11, E22	65900 MPa	114000 MPa
Modulus of elasticity E33	65900 MPa	114000 MPa
Poisson ratio μ12, μ23, μ13	0.3	0.3
Shear modulus G12	3500 MPa	
Shear modulus G23, G13	3200 MPa	
Density	1500 kg/m^3	4510 kg/m^3

PCM consists of epoxy polymer matrix and carbon fibers with usual orientation along separator axis (direction 11) and in circumferential direction 22.

2.3. Calculation procedure and results

The calculation has been made in the finite element package ANSYS APDL.

The solver type is SPARSE non-linearity is taken into account large model displacements (nlgeomoption).

Dynamic calculation of the vibration frequencies and mode shapes has been performed when the separator is fixed rigidly around the circle face plane to the right (to the air exit). The titanium part mounted on the PCM is free on the left.

The results are presented as figure 2 and table 2.

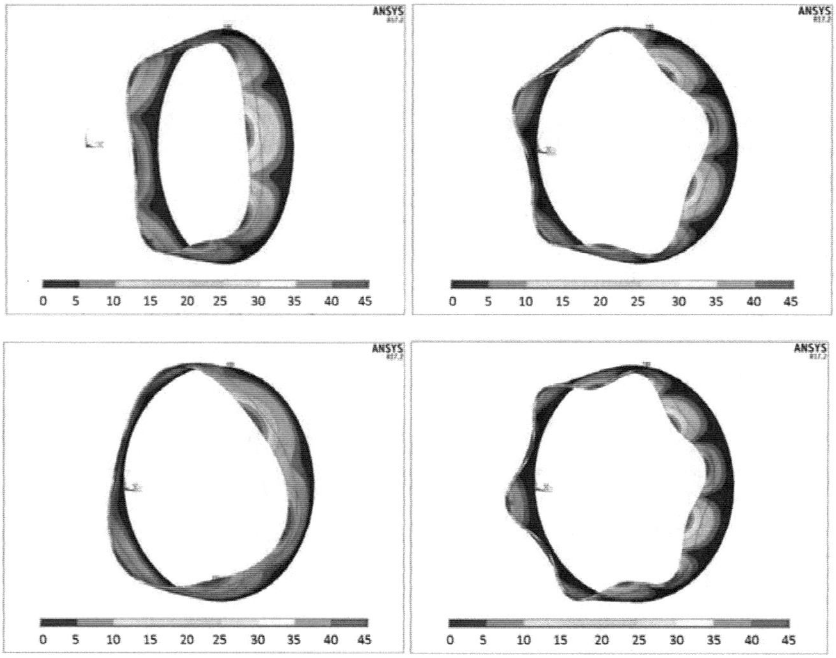

Figure 2. The first 4 vibration mode shapes of the flow separator

Table 2. Natural oscillation frequencies of the flow separator.

Frequency	Value, Hz
1	308.321
2	333.402
3	416.871
4	448.265
5	604.310
6	665.11
7	780.38
8	965.36
9	1120.9
10	1145.1

Calculation showed that one of the modes 1-3 frequencies should be choosen for fatigue test of separator.

3. Fatigue test of the separator

3.1. Test procedure

Fatigue test of separator has been done on the high power electro-dynamic stand Data Physics LE-612/DSA10-40K (USA) in fatigue test laboratory of PJSC UEC-Saturn, Rybinsk.

A special tool has been developed to fix the separator on the electro-dynamic stand, simulating its fastening on the engine (figure 3).

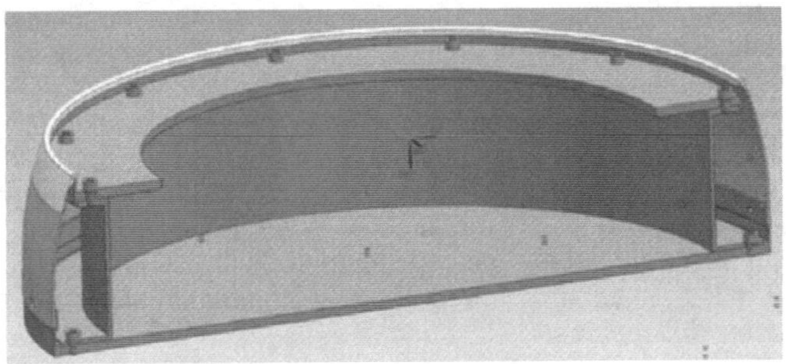

Figure 3. Installation device for FS on the vibration table.

Strain gauges, accelerometers and laser vibrometer have been installed on the separator and table of stand to measure dynamic frequencies, relative strains, vibrospeeds and displacements.

First of all, calibration test has been done to choose vibration mode with the best excitation.

The best excitation has been measured by strain gauges, accelerometers and laser vibrometer at mode 3 with three nodal diameters that are minimum from the first 5 modes.

Excitation at mode 6 with 2 nodal diameters was worse than excitation at 3rd mode due to higher frequency.

3.2. Test results

Tests at 3rd mode frequency were conducted by stepwise increase of the exciting force beginning with maximum stress level of 0,7 of the expected endurance limit σ_{-1}^{*} that is determined as 30-40% of material yield limit. After test at this level during base number of 10^7 cycles the level of maximum stress/displacement measured by strain gauges/laser vibrometer was increased and test during 10^7 cycles were repeated again.

After some tests with increased level of stresses during the resonance trials failed to achieve the required for the appearance of the fatigue damage the stress level in the separator flows, which was associated with limit of existing stand power. Therefore, the separator passed a base number of 10^7 cycles without cracking at a stress level below its real endurance limit.

In order to achieve an endurance limit of separator it was proposed to replace the tests of the entire separator with fatigue tests of individual structural elements cut from the part, which require energy to excite to a destructive level of stress within the capabilities of existing vibration stands in the industry.

4. Fatigue testing of structural elements

4.1. Modesl and calculations

The scheme of the structural elements (SE) cutting from the flow separator is shown as figure 4.

Width and length of SE models were varied in the ranges 20-40 mm and 150-250 mm. Thickness of SE has not been changed.

Fastening the structural element on the table of low power vibration stand ETS Solutions L215M (China) is carried out in Moscow Aviation Institute (National Research University) laboratory using the original fastener of separator on the engine so that the maximum dynamic stress of structural elements coincide with the location of maximum stress in the separator. The use of this approach makes it possible to obtain the values of the endurance limit of the separator taking into account its manufacturing technology and design features.

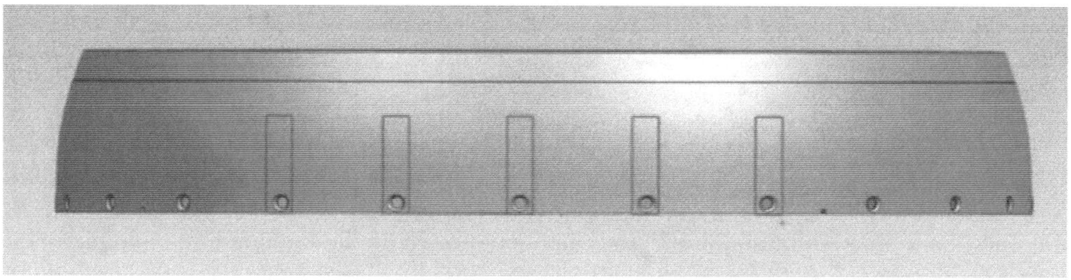

Figure 4. The scheme of the SE cutting from the flow separator.

For this purpose, a number of 3D finite element models were made and a series of their dynamic calculations was carried out. Finite-element meshes, fixation and calculation procedure are similar to ones for separator (subsections 2.1, 2.2).

The final shape and dimensions of the structural elements were chosen 170×22 mm in such a way that their natural frequency of testing 253 Hz is slight lower then 1st mode frequency of the flow separator, taking into account a necessity of element failure. An example of the results of the calculation of dynamic displacements and stresses is shown as figure 5.

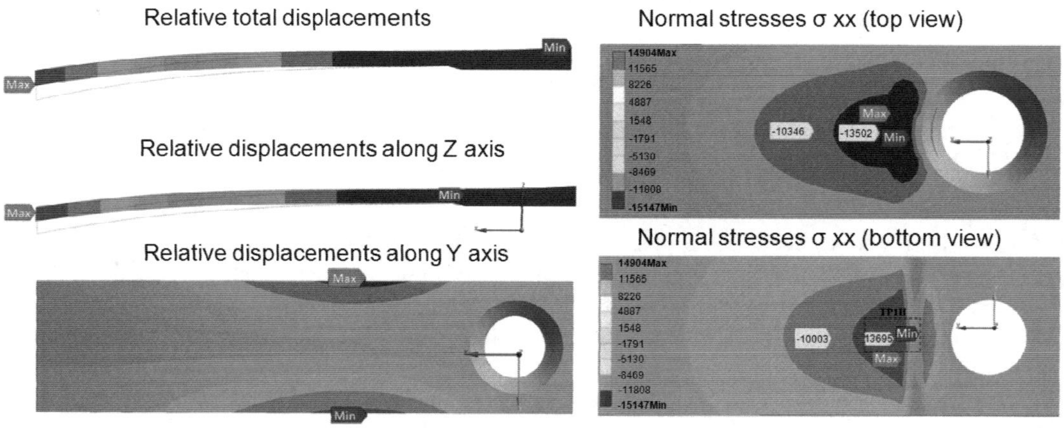

Figure 5. Dynamic displacements and stresses in SE.

4.2. Calibration

The SE of the flow separator is cut according to the size determined as a result of the calculation, and strain gauges are installed in the area of calculated maximum dynamic stresses.

Then the specimen is fixed on the table of the vibration stand with the help of a specially designed device that simulates the fastening of the flow separator on the engine using a standard fastener (figure 6), and its calibration is carried out by strain gauges. The excitation mode at the resonant frequency is set by means of a triangulation laser, which measures the amplitude of the SE tip displacement, and the relative strain corresponding to the obtained value of the amplitude is determined using the values measured by the strain gauge. The excitation is changed stepwise with a relative strain step of $20 \cdot 10^{-5}$ to a maximum relative strain of $70 \cdot 10^{-5}$ in order to avoid fatigue damage of strain gauge during calibration. The data are entered in the table and a calibration graph is built on them (figure 7), which is used in the fatigue test to assess the stress level using amplitude of specimen vibration.

4.3. Test procedure and results

Fatigue tests for the elements of flow separator were carried out using the 'staircase' method, the essence of which is a stepwise change in the load on the next tested element after test of previous one during 10^7 load cycles. SE are tested starting from the stress level exceeding the expected endurance limit σ_{-1}^{*} by at least 20%. At destruction of SE before life 10^7 cycles, the following element is tested at the level of stresses on a step below (-20 MPa). The criterion of failure in determining the limits of endurance is a drop in the resonant frequency of more than 10%. If after operating time 10^7 cycles SE has not reached the state of destruction, the next element is tested at a step higher (+20 MPa). For the endurance limit of the flow separator, a stress of 80% of the minimum alternating stress at which the tested elements have passed the test base 10^7 cycles without failure is accepted. The test results are presented in table 3.

Figure 6. Device for fixing the SE on the vibration stand table.

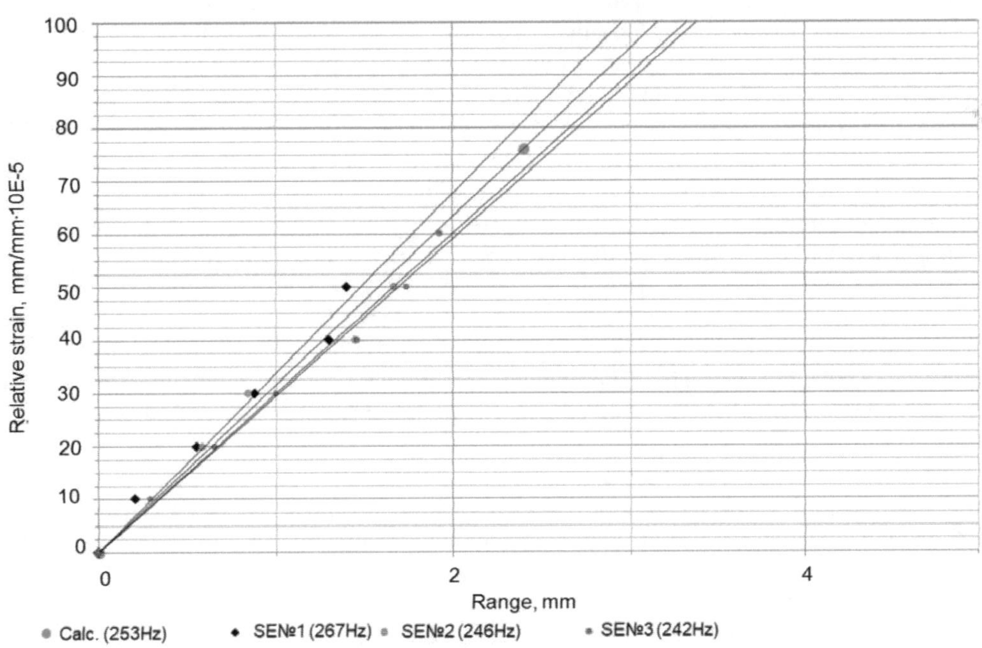

Figure 7. Calibration graph of the dependence of the SE displacement range on the relative strain.

IOP Conf. Series: Materials Science and Engineering **868** (2020) 012027 doi:10.1088/1757-899X/868/1/012027

Table 3. The results of the fatigue tests.

Number SE	stress	Notice
SE No.1	$1.2\sigma_{-1}^{*}$	Not cracked
SE No.2	$1.4\sigma_{-1}^{*}$	Cracked
SE No.3	$1.2\sigma_{-1}^{*}$	Not cracked

The destroyed element No.2 was examined using optical and electron microscopes to determine the location and nature of the damage. The results are shown as figure 8.

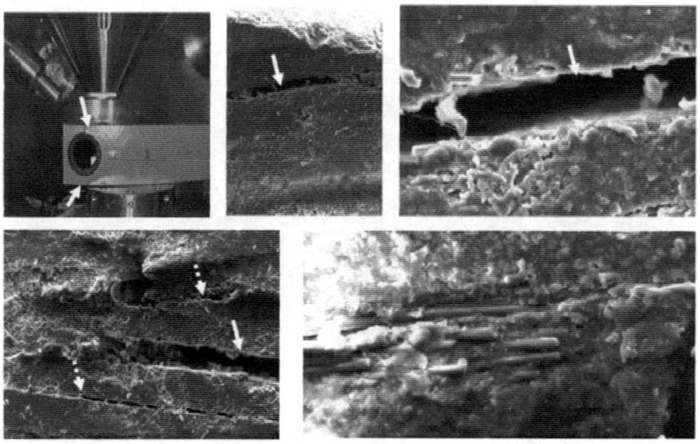

Figure 8. Examination of destroyed samples with a microscope.

The main cracks are located on the lateral sides near hole for element fixation (see white arrows). Cracks are propagated in the epoxy polymer matrix in axial direction. Some additional cracks are visible in parallel with main crack (dash arrows). The broken carbon fibers available on the lateral sides (see right bottom photo) were appeared due to the cutting of element from separator.

Presented photos show that real stresses on the location of crack initiation and propagation don't correspond to location of calculated maximum stress. More complex finite-element model is necessary for calculation of real stress. However, this academic task is not important for applied task of endurance limit and fatigue strength margin estimation. It is possible to use proposed procedure of calculation and test taking into account that both flow separator and structural element are calculated using similar anisotropic finite-element models and stresses are measured using similar strain gauges.

The obtained values of the endurance limits were compared with the dynamic stresses measured during the strain gauging of the flow separator during the preliminary tests of the engine, and the fatigue strength margin was estimated from the obtained results.

Proposed procedure will allow to avoid expensive and unsuccessful test of whole separator.

5. Conclusion

A method for determining the endurance limit of a large part of an aircraft engine such as flow separator made of polymer composite material is developed and consists in replacing the tests of the entire separator with tests of individual structural elements cut from the part.

The procedure of the flow separator endurance limit determination is based on:

- calculation of natural frequencies and mode shapes for whole separator,
- choice of shape, dimensions and fixation of the elements that could be cut from the separator in such a way that their own test frequency was close to the natural frequency of the flow separator on the engine,

- fatigue tests of elements to determine the endurance limit of separator

The developed technique allowing estimation with minimal cost the fatigue strength margin of aircraft engine parts, for which the local fatigue resistance significantly depends on the design features and manufacturing technology.

Acknowledgement
The work was carried out with the financial support of the Ministry of Education and Science of the Russian Federation within the framework of the complex project No. 2017-218-09-172.

The authors thank the staff of PJSC UEC-Saturn, Rybinsk for their help and support with carrying out this work.

References

[1] Jones R M 1999 *Mechanics of Composite Materials*, 2nd edition, CRCPress, p 538 doi 10.1201/9781498711067

[2] Ibeh C C 2011 *Thermoplastic Materials: Properties, Manufacturing Methods, and Applications*, CRCPress, p 640

[3] Widi I, Pohan G, Sujana W and Prihatmi T 2019 Effect of Ramie Fibers on Fatique Strength Automotive Product. *IOP Conf. Ser.: Mater. Sci. Eng.* **694** doi: 10.1088/1757-899x/694/1/012038

[4] Brunner A J 2018 Scatter, Scope and Structures: What fatigue fracture testing of fiber polymer composites is all about. *IOP Conf. Ser.: Mater. Sci. Eng.* **388** 012003 doi: 10.1088/1757-899X/388/1/012003

[5] Brunner A, Stelzer S, Pinter G, Terrasi G P 2015 Cyclic fatigue delamination of carbon fiber-reinforced polymer-matrix composites: Data analysis and design considerations. *Int. J. Fatigue* **83** 293 doi: 10.1016/j.ijfatigue.2015.10.025.

[6] Alderliesten R, Brunner A, Pascoe J 2018 Cyclic fatigue fracture of composites: What has testing revealed about the physics of the processes so far? *Eng. Fract. Mech.* **203** 186 doi: 10.1016/j.engfracmech.2018.06.023

[7] Molent L, Forrester C 2017 The lead crack concept applied to defect growth in aircraft composite structures. *Compos. Struct.* **166** 22 doi: 10.1016/j.compstruct.2016.12.076

[8] Kashtalyan Y and Soutis C 2002 Mechanism of internal damage and their effect on the bechaviour and properties of crossplay composite laminates. *Int. Appl. Mech.* **38(6)** 3 doi: 10.1023/A:1020456726805

[9] Lisin A, Nabokov I, Mozalev V 2017 Predicting the fatigue strength of structural elements. *Rus. Eng. Res.*, **37** 388 https://link.springer.com/article/10.3103/S1068798X17050161

[10] Azikov N, Zinin A 2018 A Destruction Model for an Anisogrid Composite Structure. *J Machinery Manufacture and Reliability* **47(5)** 427 doi: 10.3103/S1052618818050035.

[11] Karimbaev T, Matyukhin D 2016 To the estimation of sample fatigue from polymer composite materials. *J. Mech. Compos. Mater. Str.* **22(3)** 329 https://www.elibrary.ru/item.asp?id=27216367

[12] Arkhipov A, Volgina M, Matushkin A, Ravikovich Y, Kholobtsev D 2018 Effect of a Model and Boundary Conditions on the Results of Analyzing the Stress-Strain State of a Low-Pressure Compressor Rotor. *Russian Aeronautic* **61** 509 doi: 10.3103/S1068799818040025

IOP Conf. Series: Materials Science and Engineering **868** (2020) 012028 doi:10.1088/1757-899X/868/1/012028

The influence of the weld line location on the mechanical characteristics of lugs from short fibers reinforced composite material

E I Kurkin*, M O Spirina, Ya V Zakhvatkin and V O Chertykovtseva

Joint Russian-Slovenian Laboratory Composite Materials and Structures, *Samara National Research University,* 34 Moskovskoe shosse, Samara, 443086, Russian Federation

*E-mail: eugene.kurkin@mail.ru

Abstract. The work is devoted to assessing the influence of the weld line on the mechanical characteristics of lug from short-reinforced composite material PA6 Aramid SV 30-1ETM. Lug molding process was calculated in Moldex3D, including prediction of weld line position and fiber orientation. Static structural analysis was done in ANSYS Workbench with nonlinear anisotropic material model, constructed in Digimat MX by reverse engineering. For failure prediction Tsai Hill 3D transversely isotropic model with FPGF (PGA) formulation was used. The ultimate load is taken as the minimum load at which the maximum value of the FPGF criterion reaches unity. Three cases of melt entrance point position was compared. The smallest stiffness and strength were shown by lug in which the weld line located closest to the maximum values of the strength criterion.

1. Introduction

Lugs are a typical structural elements and they contain stress concentrators. The calculation of stresses in the lug presented in [1]. Lug is an important part of a pin-joints. Stress and strength analysis of pin-joints from laminated composite submitted in [2]. The study of the features of contact stresses in the laminated composite pin-joints is presented in [3, 4]. Review of stress analysis and strength prediction of fastened joints in composite laminates is made by Camanho and Mathews [5]. Short-reinforced composite materials find their application in cases where the shape of the product is difficult to manufacture from a laminate composites or in cases when loads are applied in such a way that the laminates are faced with the problem of delamination. It's showed by Airbus-designed A350XWB's door bracket from short-reinforced thermoplastic material Victrex PEEK 90HMF40 that injection-moulded polymer solution can reduce to 40% in weight and cost in comparison from bracket machined from an aluminum block [6]. Failure models for unfilled PMMA thermoplast lug tension presented in [7]. The research of lugs allows to evaluate the behavior of the anisotropic material and give recommendations on the correct simulation of products with a complex stress state [8]. The stiffness and strength of lugs from short-reinforced composite materials depends on the fibers orientation in the product [9]. The strength in weld lines decreases with increasing filler content [10]. Molding of lugs is associated with the flow around the axis of lug, which leads to the formation of a weld lines [11]. Weld lines affect the orientation of the reinforcing fibers and can lead to a change in stiffness and strength of the product [12]. The aim of the work is assessing the influence of the weld line on the mechanical characteristics of lug from short-reinforced composite material.

2. Molding with weld line model

Moldex3D was used for calculate the molding of lugs (figure 1). Lugs size is 65 x 24 x 4 mm. The diameter of the hole is 12 mm. The plate material is polyamide-6 reinforced by 30% glass fibers – PA6 Aramid SV 30-1ETM (figure 1b). Melt temperature is 250°C, mold temperature is 110°C. Injection-molding machine is the Negri Bossi VE210-1700 with screw diameter 60 mm. Flow rate is 24 cm^3/s. The calculation was carried out on a mesh with solid elements obtained in ANSYS Meshing and transferred to Moldex3D by means Rhinoceros.

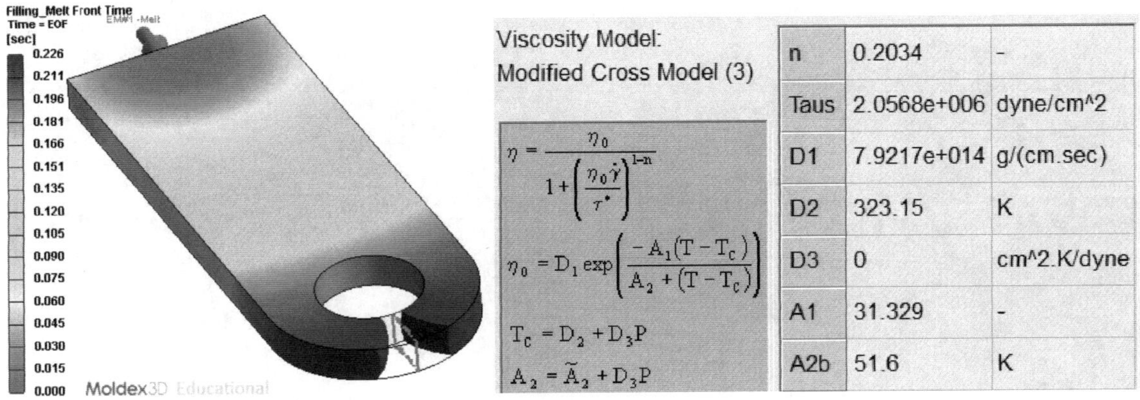

(a) (b)

Figure 1. Molding calculation in Moldex3D: (a) melt front and weld line, (b) material model.

The work investigates the dependence of the melt front and the location of the weld line on the melt entry point (figure 2). The values of the orientation tensor of the reinforcing fibers at each point of each lug molding case were obtained. The values of the orientation tensor of the reinforcing fibers are exported in the o2d format, supplemented by mesh file in the cdb format. The paper is considered three cases for molding lugs, ensuring the location of the weld line along the axis of lug (1st case), diagonally to the axis of lug (2nd case) and across the axis of lug (3rd case).

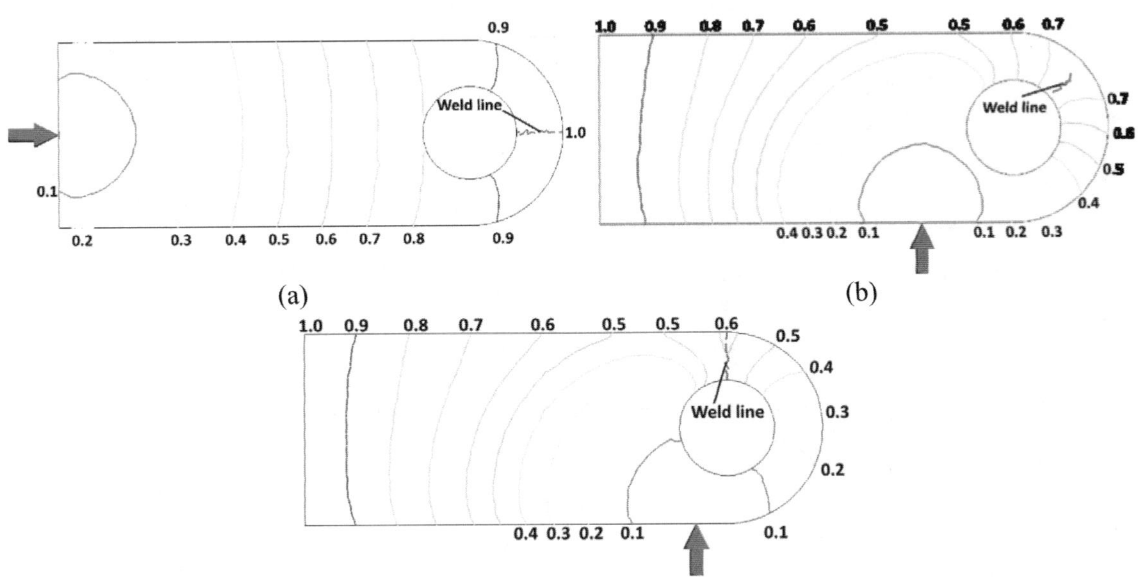

(a) (b)

(c)

Figure 2. Melt front isoline (volume share) and weld lines on lugs at three cases of melt entry point position: (a) 1st case, (b) 2nd case, (c) 3rd case.

IOP Conf. Series: Materials Science and Engineering **868** (2020) 012028 doi:10.1088/1757-899X/868/1/012028

3. Stress-strain state investigation

The stress-strain state is calculated taking into account the anisotropy of the composite material and the nonlinear nature of its deformation. Values of orientation tensor components were transferred to static structural mesh using Digimat MAP (figure 3a). Material properties obtained in Digimat MX by reverse engineering, so that the properties of the composite as a whole correspond to the experimentally obtained dependences of stresses on strains for the samples, cut at an angle of 0°, 45° and 90° (figure 3b).

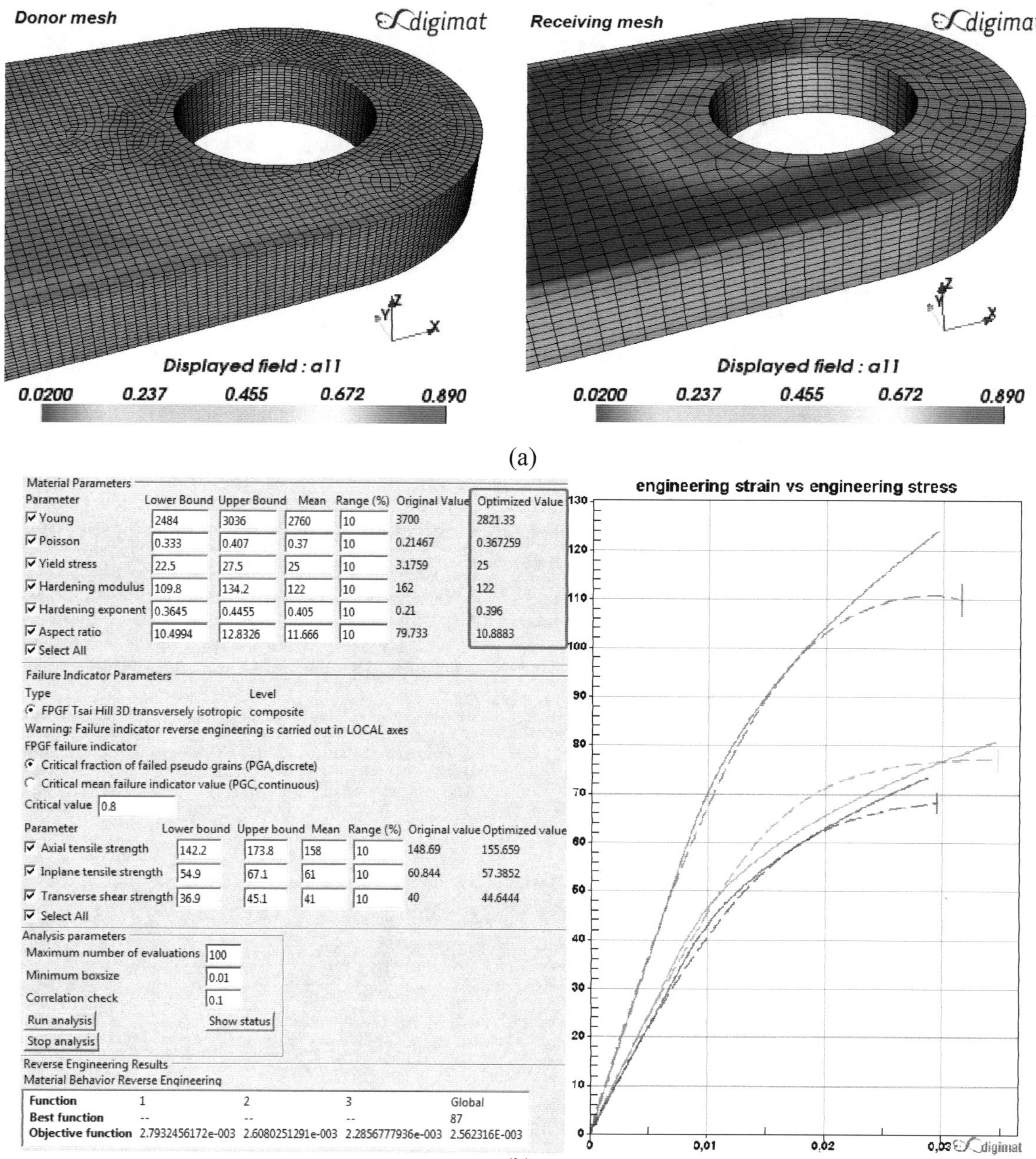

Figure 3. Predict anisotropic behavior using Digimat: (a) transferring orientation from molding (donor) to static structural (receiving) mesh, (b) Reverse Engineering of PA6 Aramid SV 30-1ETM material.

Static structural analysis of lug was carrying out in ANSYS Workbench, using Digimat CAE Wizard. Lug was fixed using Fixed Support of rear face and load by force, attached to the steel axis, connected to the lug during frictionless contact (figure 4). To exclude the lug and axis out-of-plane movement, the $\Delta Z = 0$ displacement condition is added on the axis sides and the lug edges. For failure prediction Tsai Hill 3D transversely isotropic model with FPGF (PGA) formulation was used [13]. As the ultimate load, we will consider the smallest force at which $\max(FPGF) = 1$. Total Deformation and Equivalent Stress (von Mises) at ultimate load are shown in figure 5.

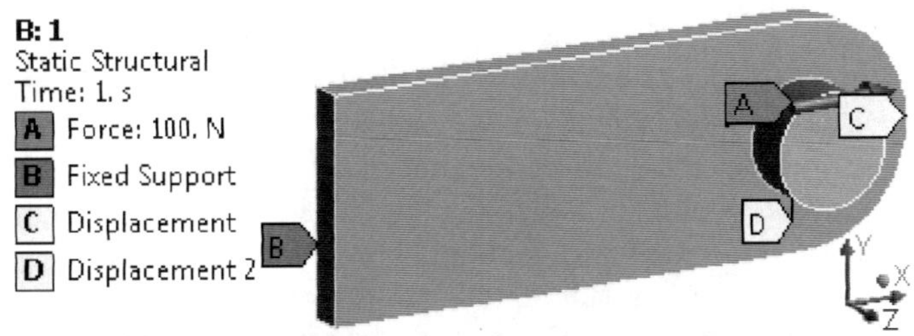

Figure 4. The lugs tension problem formulation.

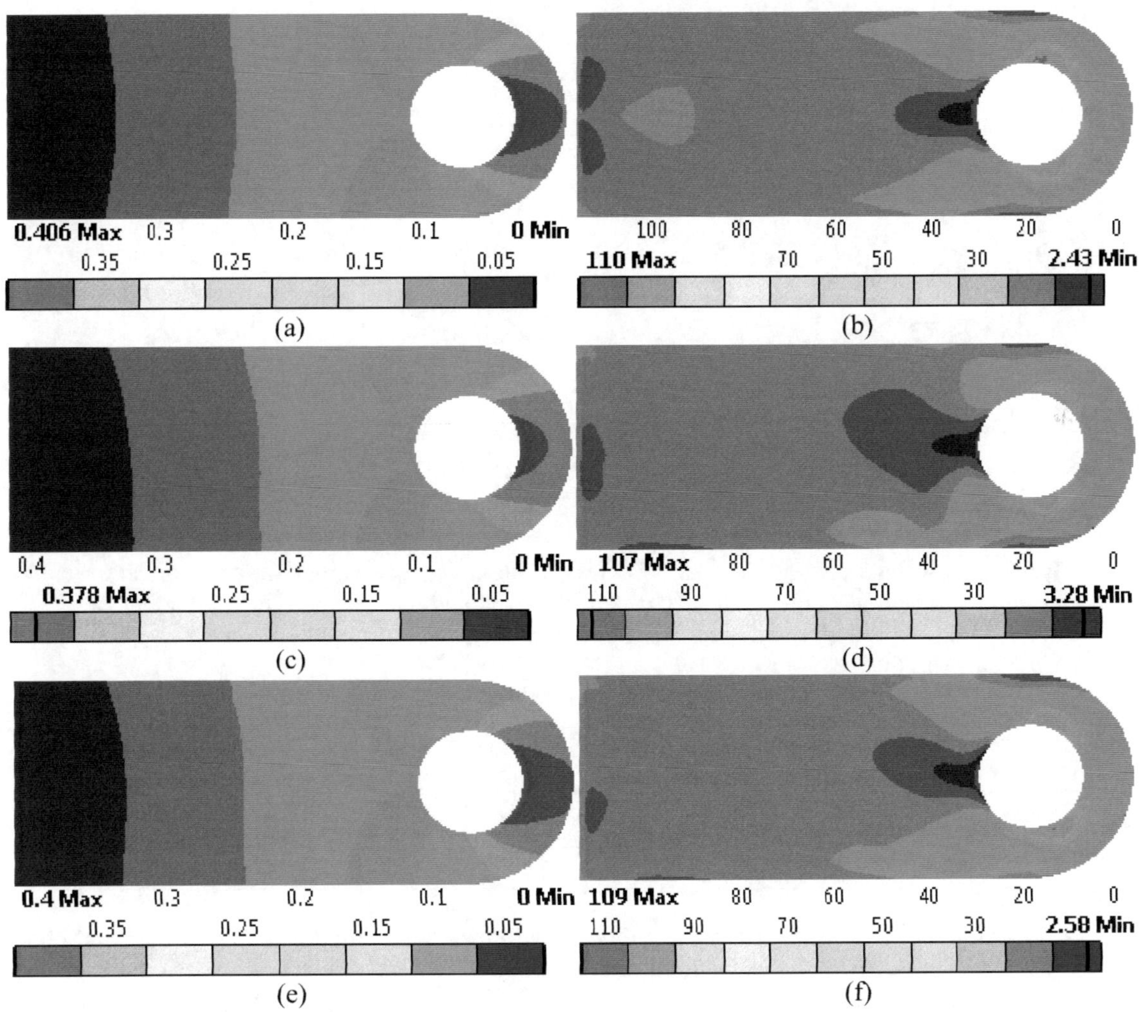

Figure 5. Total Deformation (a, c, e), mm and Equivalent Stress (b, d, f) (von Mises), MPa at ultimate load for melt entry point cases: (a, b) 1st; (c, d) 2nd; (e, f) 3rd.

The asymmetry of lug molding in 2nd and 3rd cases leads to the asymmetry of the fields of Total deformation and Equivalent stress.

To describe the tension process of lugs, we use the normal stress values $\sigma_N = \frac{F}{h \cdot d_0}$, where F – the applied load, h = 4 mm – the thickness of lug, d = 12 mm – the undeformed diameter of lug hole; and ovalization $\omega = \frac{d - d_0}{d_0} \cdot 100\%$, where d – the width of the deformed lug hole, mm, obtained by the difference in the deformations of the front and rear edges of lug (figure 6). The normal stress value does not depend on the thickness of lug and the ovalization of lug does not depend on the method of fixing the rear end of lug, therefore, the values σ_N and ω are well suited for comparison with other calculations and experimental data. The ultimate load is taken as the minimum load at which $\max FPGF = 1$. The values of ultimate loads for the three cases of molding of lugs are presented in table 1. The 1st case lug is most stiff and strength. The 2nd case lug shows the smallest stiffness and strength - its stiffness is 2.18% less and strength is 9.16% less than the stiffness and strength of 1st case lug. Mechanichal characteristics of 3rd case lug are in the range between 1st and 2nd case.

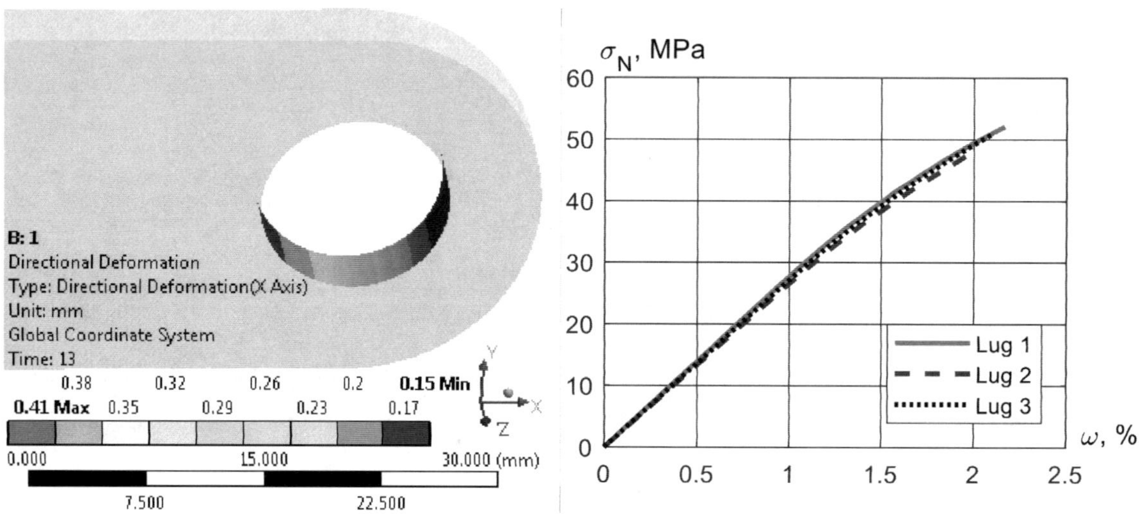

(a) (b)

Figure 6. Ovalization and normal stresses in lugs: (a) determination of ovalization through the difference in the deformations of the front and rear edges of lug, (b) a tension graph of lugs with three cases of location of the melt entry point.

Table 1. Ultimate stress values for three cases of lug molding.

N of melt entry case	Max Force (N)	Rear point Deform. (mm)	Front point Deform. (mm)	ω (%)	σ_N (MPa)
1	2500	0.406	0.146	2.166	52.08
2	2290	0.141	0.378	1.978	47.71
3	2450	0.148	0.401	2.107	51.04

Comparison of the fields of the strength criterion at ultimate load with the location of the weld line is shown in the figure 7. In the least strength 2nd case lug, the weld line is located closest to the location of the maximum values of the strength criterion.

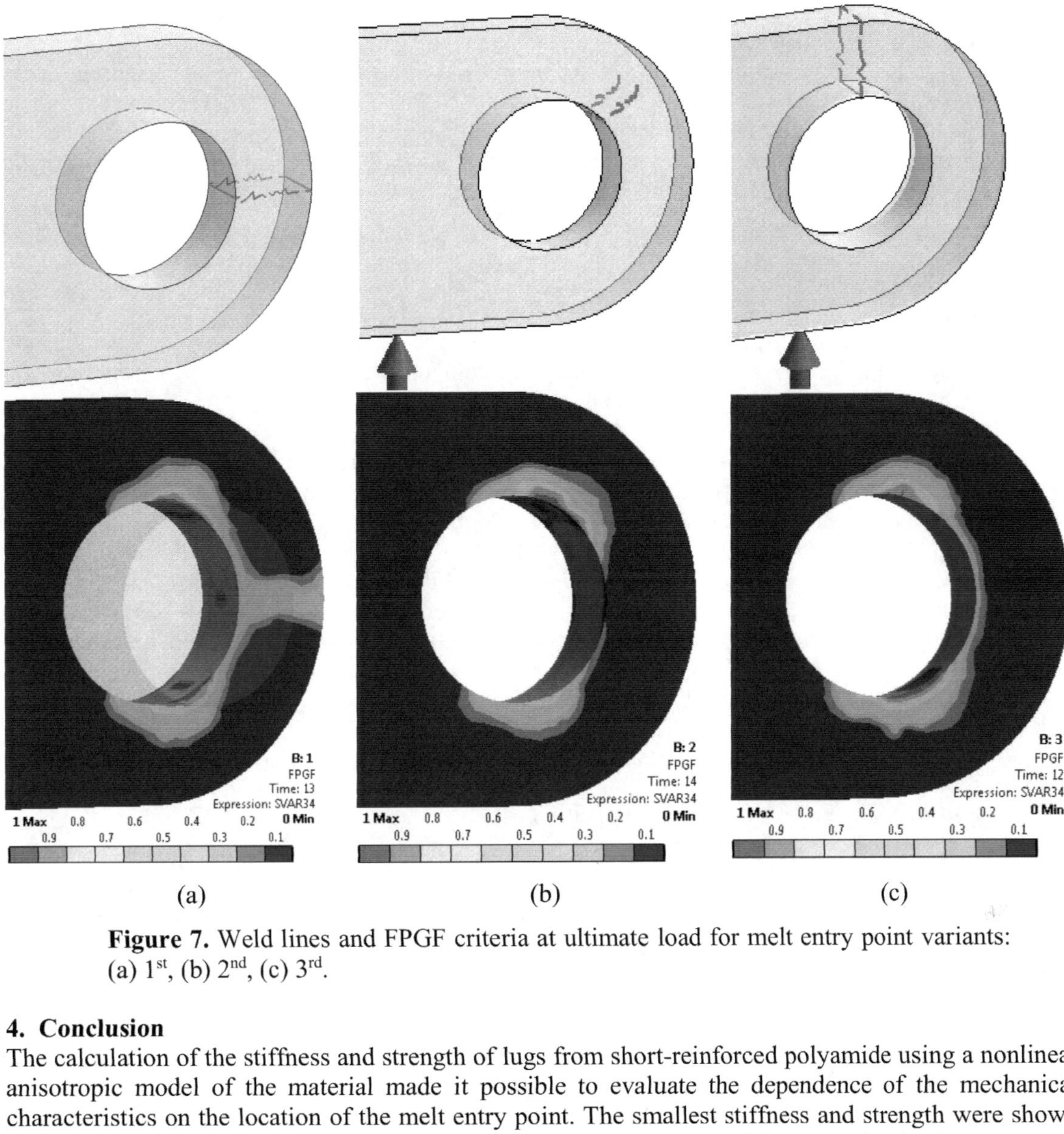

(a) (b) (c)

Figure 7. Weld lines and FPGF criteria at ultimate load for melt entry point variants: (a) 1st, (b) 2nd, (c) 3rd.

4. Conclusion

The calculation of the stiffness and strength of lugs from short-reinforced polyamide using a nonlinear anisotropic model of the material made it possible to evaluate the dependence of the mechanical characteristics on the location of the melt entry point. The smallest stiffness and strength were shown by lug in which the weld line located closest to the maximum values of the strength criterion.

Acknowledgments

The reported study was funded by Russian Science Foundation according to the research project N. 19-79-10205.

References

[1] Antoni N and Gaisne F 2011 Analytical modelling for static stress analysis of pin-loaded lugs with bush fitting. *Applied Mathematical Modelling* **35** 1 doi: 10.1016/j.apm.2010.05.002

[2] Murthy A V, Dattaguru B, Narayana H V L and Rao A K 1991 Stress and strength analysis of pin joints in laminated anisotropic plates. *Composite Structures* **19** 299 doi: 10.1016/0263-8223(91)90078-D

[3] Pierron F, Cerisier F and Grediac M 2000 A numerical and experimental study of woven

composite pin-joints. *Journal of Composite Materials* **34** 1028 doi: 10.1177/002199830003401204

[4] Hollmann K 1996 Failure analysis of bolted composite joints exhibiting in-plane failure modes. *Journal of Composite Materials* **30** 358 doi: 10.1177/002199839603000304

[5] Camanho P P and Matthews F L 1997 Stress analysis and strength prediction of mechanically fastened joints in FRP: a review. *Composites Part A: Applied Science and Manufacturing* **28** 529 doi: 10.1016/S1359-835X(97)00004-3

[6] *Ready for take-off: Doors on the A350 with a primary structural component made from Victrex polymer*, Victrex, available at: https://www.victrex.com/en/news/2015/09/airbus-bracket

[7] Scorza D, Carpinteri A and Vantadori S 2019 Tension failure assessment at lug hole edges. *International Journal of Fatigue* **121** 293 doi: 10.1016/j.ijfatigue.2018.12.010

[8] Komarov V A, Kishov E A, Charkviani R V and Pavlov A A 2015 Numerical and experimental study of the strength of fabric carbon-epoxy composite structures. *Vestnik of the Samara state aerospace university* **14**(2) 106 doi: 10.18287/2412-7329-2015-14-2-106-112

[9] Matthews F L and Rawlings R D 1999 *Composite Materials: Engineering and Science* (Cambridge: Woodhead Publishing) p 480 doi: 10.1016/C2013-0-17714-8

[10] Quinn K R, Duffy J A and Ellis C S 1997 Reinforced thermoplastics: understanding weld line integrity. *Journal of Injection Molding Technology* **1** 123

[11] Dzulkipli A A and Azuddin M 2017 Study of the effects of injection molding parameter on weld line formation. *Procedia Engineering* **184** 663 doi: 10.1016/j.proeng.2017.04.135

[12] Ozcelik B, Kuram E and Topal M M 2012 Investigation the effects of obstacle geometries and injection molding parameters on weld line strength using experimental and finite element methods in plastic injection molding. *International Communications in Heat and Mass Transfer* **39** 275 doi: 10.1016/j.icheatmasstransfer.2011.11.012

[13] Jain A, Lomov S V, Abdin Y, Verpoest I and Paepegem W V 2013 Pseudo-grain discretization and full Mori Tanaka formulation for random heterogeneous media: Predictive abilities for stresses in individual inclusions and the matrix. *Composites Science and Technology* **87** 86 doi: 10.1016/j.compscitech.2013.08.009

IOP Conf. Series: Materials Science and Engineering **868** (2020) 012029 doi:10.1088/1757-899X/868/1/012029

Pilot behavior in preview tracking task

A V Efremov, M S Tiaglik and I Kh Irgaleev*

Flight Dynamics and Control Department, *Moscow Aviation Institute* (National Research University), 4 Volokolamskoe shosse, Moscow 125993, Russia

*E-mail: pvl@mai.ru

Abstract. The influence of the input signal bandwidth and controlled element dynamics on the potential of the preview display for improving accuracy is considered. The pilot in the loop experiments in which the human operator tracked a program trajectory characterized by bandwidths of 0.2, 0.5, 1 rad/s were carried out with a number of simplified controlled element dynamics and with the predictive display/aircraft system dynamics. For the latter dynamics the experiments were performed for different preview times (from 0 up to 6 s) which allowed to define its optimal value.

1. Introduction

The majority of pilot-aircraft system investigations were carried out for the compensatory display. In that case only the error signal $e(t)$ is the stimulus that the pilot transforms in his actions (figure 1).

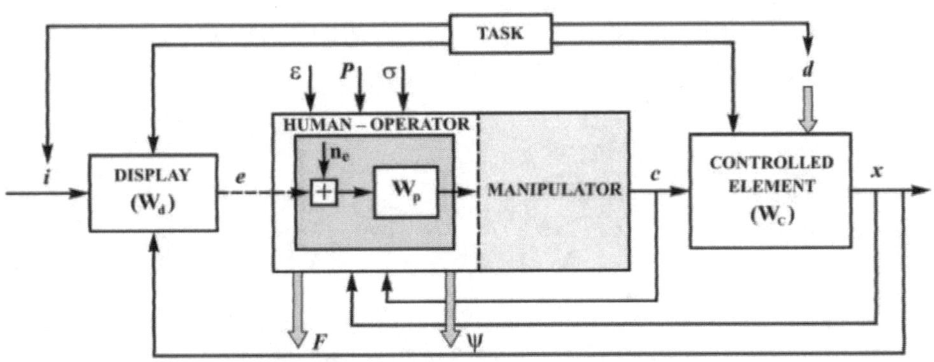

Figure 1. Pilot-aircraft system.

The fundamental regularities of human operator behavior were defined for such kind of system. They were used widely for solving different engineering problems (flight control system design, flying qualities criteria development, etc.). The compensatory tracking corresponds to a number of piloting tasks but not all of them. For example, in case of tracking a target flying against the background of clouds or the terrain and in the process of refueling, aside from the error signal, the pilot also perceives the input signal $i(t)$. In that case, the pilot-aircraft system corresponds to the so-called pursuit system. In case of flying through a mountain gorge or while driving a car, the pilot sees the planned trajectory $i(t+T_{prev})$ where T_{prev} is the preview time.

Initial studies [1, 2] of pursuit and preview tracking tasks carried out in the second half of the last century with simplified controlled element dynamics demonstrated that the perception of the planned trajectory leads to a considerable decrease in the variance of error for a preview time close to 0.5÷1 s. Increasing this time does not cause a decrease in the tracking error.

Broad investigations in this area were carried out at Delft University [3-10] for the gain coefficient and simple/double integrator controlled element dynamics, for a rectangular power spectrum input with different bandwidths ($\omega_i \geq 1.5$ rad/s). The maximum preview time T_{prev} was equal to 1 s in virtually all of these studies (except [10] where the preview time was equal to 2 s). The main accent in these studies was made on the definition of two pilot frequency response characteristics whose outputs (so-called "near/far view responses") were calculated after the experiments. The parameters of these characteristics were obtained from the preliminary measurements of two pilot describing functions, one describing the pilot response to the input signal $i(t+T_{prev})$ and the other describing the pilot response to the error signal $e(t)$. The methodology of these measurements and calculations is given in [6, 7]. All these studies exposed the effect of preview on pilot behavior characteristics and its potential for improving task performance, and the influence of the input signal bandwidth and controlled element dynamics on all these regularities. In spite of the importance of these results, all of them were performed for a high bandwidth of input signal which is not typical for the aircraft flight path motion. The planned aircraft trajectory is characterized by the spectrum density with lower bandwidths ($\omega_i = 0.2 \div 0.5$ rad/s) [11]. The interval of preview time accepted in these studies ($T_{prev} = 0 \div 2$ s) did not allow to obtain the optimum value of preview time. In addition, the dynamics of modern aircraft equipped with the display is defined by the dynamics of the aircraft-display system which is more complicated than the simple configurations investigated in [11]. As a matter of fact, the current state of technology has allowed to realize a display with a 3D presentation of the planned trajectory on the screen. This technology has caused a number of studies in defining the best way of presenting the information [2-6, 12]. Finally, the so-called "tunnel in the sky" display was proposed [12]. This display allows pilots to evaluate the current aircraft position in space and the future planned trajectory.

The following modification of this display is the so-called predictive display [13] where, aside from the tunnel, the surface moving inside of it and the aircraft velocity are also displayed (figure 2). The aircraft velocity vector is projected on this surface.

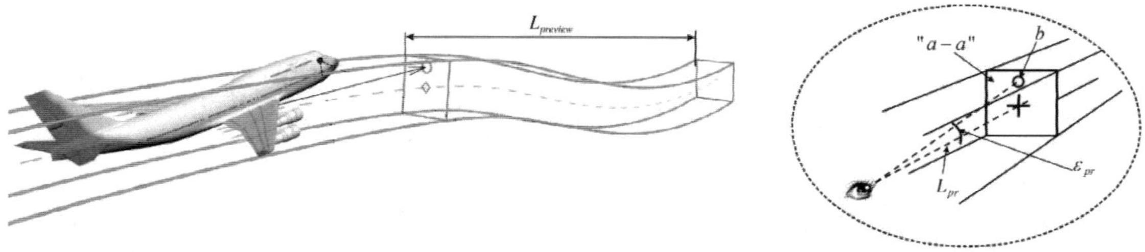

Figure 2. Predictive display with preview.

The selection of the distance between the pilot's eyes and the surface L_{pr} (or predictive time $T_{pr} = \dfrac{L_{pr}}{V}$) was carried out in [11] with the use of the traditional feedback control theory, aiming to provide the expected controlled element dynamics whose output signal is the predictive angle $\varepsilon_{pr} = \dfrac{h}{L_{pr}} + \gamma + \dot{\gamma}\dfrac{T_{pr}}{2}$, where γ is the path angle and h is the aircraft height displacement. It is shown in [11] that the controlled element dynamics of aircraft- display system in that case has the following describing function:

IOP Conf. Series: Materials Science and Engineering **868** (2020) 012029 doi:10.1088/1757-899X/868/1/012029

$$W_c(s) = \frac{\varepsilon_{pr}(s)}{X_e(s)} = \frac{K\left(T_{pr}s^2 + 2s + \frac{2}{T_{pr}}\right)}{s^2\left(s^2 + 2\zeta\omega s + \omega^2\right)}.$$

According to the technique proposed in [11], the selection of the predictive time is carried out by minimization of the variance of the current height error σ_Δ^2 where $\Delta = h - i(t)$. The results of this investigation demonstrated that the use of the predictive display allows to considerably improve tracking accuracy. In the current paper the integration of the predicted aircraft path with the preview of the planned trajectory on the same display and definition of the preview time providing the minimum task performance is considered. In addition, the influence of the planned trajectory bandwidth $\omega_i = 0.2 \div 1$ rad/s, which is more typical for path control tasks, on preview manual control behavior is studied too.

2. Experiment design

Two groups of experiments were performed with the goal of:

- exposing the effects of the planned trajectory with the bandwidth $\omega_i \le 1$ rad/s on characteristics of the pilot-aircraft system in pursuit and preview tracking;
- defining the optimum preview time providing the best task performance in case of using the predictive display in preview tracking and the influence of the input signal bandwidth on its value.

The first group of experiments was performed with the controlled element dynamics $W_c = \frac{K_c}{s}, W_c = \frac{K_c}{s^2}$ with different bandwidths of the input signal. The second group of experiments was performed with the dynamics $W_c(s) = \frac{\varepsilon_{pr}(s)}{X_e(s)}$ with the same input signal bandwidths. All experiments were performed on one of the fixed-based simulators in the Pilot-Vehicle Lab of Moscow Aviation Institute (figure 3).

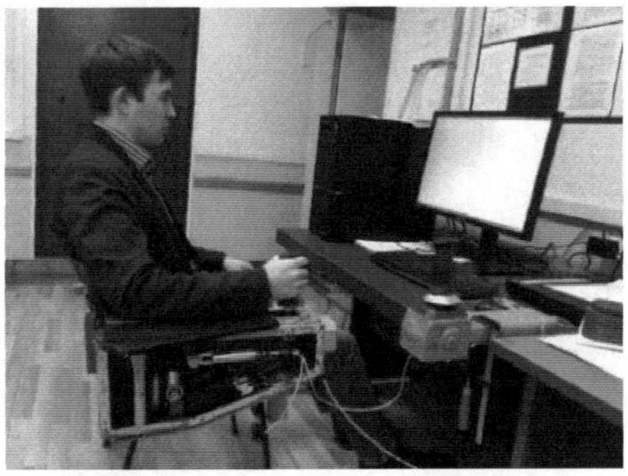

Figure 3. Fixed-based simulator.

The human operators were seated in front of the display and applied the control input with a side stick. The images corresponding to compensatory, pursuit, or preview tracking task (figure 4) were generated on the screen of this display.

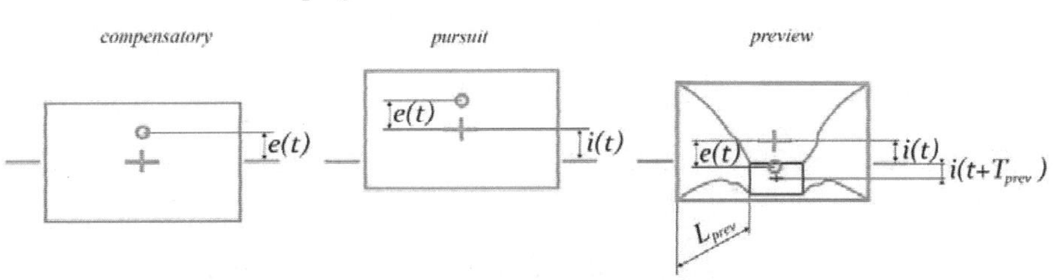

Figure 4. Different displays used in experiments.

The specific peculiarity of the preview display used in the current study is the image of the planned trajectory as a 3D tunnel with the dimensions of height and width of 3 and 4 meters respectively. Its centerline and the whole tunnel change vertical position according to the input signal $i(t+t')$. The error $e(t)$ and input $i(t)$ signals were displayed on a green surface presented on the screen. The distance between the green and grey surfaces $L_{prev} = T_{prev} * V$ is defined by the preview time. Along this distance the pilot perceives the future input signal $i(t+T_{prev})$. The input signal was presented by the polyharmonic signal $i(t) = \sum_{k=1}^{l} A_k \cos(\omega_k t)$ consisting of an orthogonal set of harmonics $\omega_k = k\omega_0$ where $\omega_0 = \frac{2\pi}{T}$ (T is the duration of the run equal to 144 sec).The selection of amplitudes and frequencies was carried out to provide the agreement of the power spectrum distribution with the spectral density of a random signal characterized by the spectral density $S_{ii} = \frac{K^2}{(\omega^2 + \omega_i^2)^2}$ according to the technique given in [14]. The experiments were performed for bandwidths $\omega_i = \{0.2, 0.5, 1\}$ rad/s for each controlled element dynamics. The variance of the input signal was equal to 4 cm^2 in all experiments.

For each combination of ω_i and W_c at least three runs were performed and the results were averaged. Three human operators participated in the experiments. The piloting task was to minimize the error signal.

The following set of pilot-aircraft system characteristics was calculated after the experiments: the frequency response characteristics of the pilot ($W_P = \frac{c(j\omega)}{e(j\omega)}$), closed-loop system ($\Phi = \frac{y(j\omega)}{i(j\omega)}$), variances of tracking error σ_e^2, control output σ_c^2 and their components correlated with the input signal ($\sigma_{e_i}^2$ and $\sigma_{c_i}^2$) and caused by the pilot remnant $n_e(j\omega)$ ($\sigma_{e_n}^2$ and $\sigma_{c_n}^2$). The calculation of Fourier transforms of all signals and components of the variances was carried out using the Fourier coefficients technique [14, 15] (figure 5).

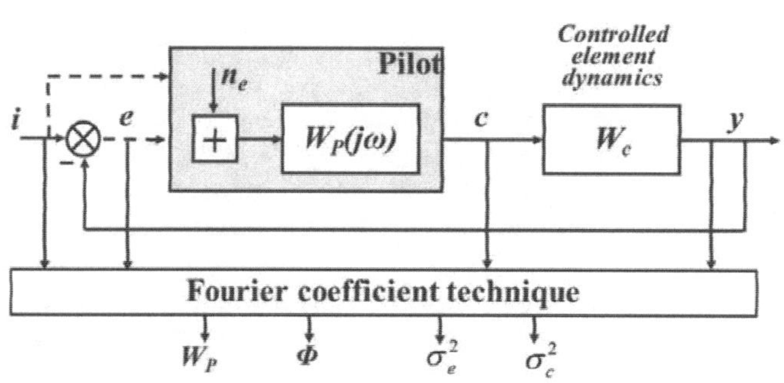

Figure 5. Characteristics measured in the experiments.

3. Results and discussions

3.1. The influence of the planned trajectory bandwidth on pursuit and preview manual control

The preview time in experiments with the preview display was equal to $T_{preview}$=1.5 s. The results of the experiments performed in cases of the controlled element dynamics $W_c = \frac{K_c}{s}$, $W_c = \frac{K_c}{s^2}$ for the compensatory, pursuit, and preview tracking tasks are shown in figures 6-7. All these experiments were performed with the bandwidth of the input signal spectrum $\omega_i = 0.5$ rad/s.

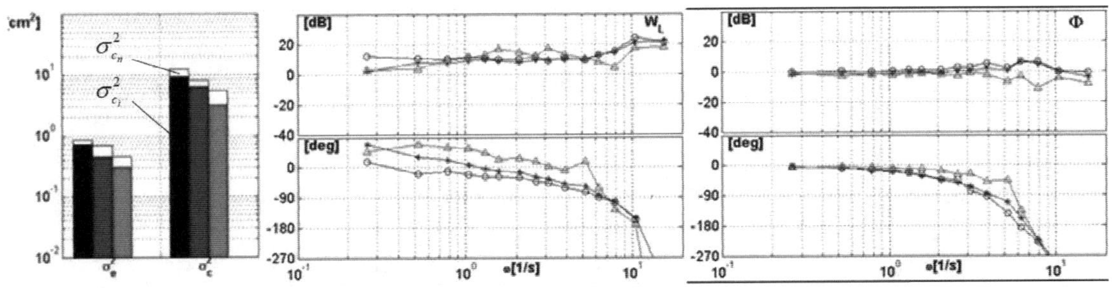

Figure 6. Influence of the piloting task, $W_c = \dfrac{K_c}{s}$.

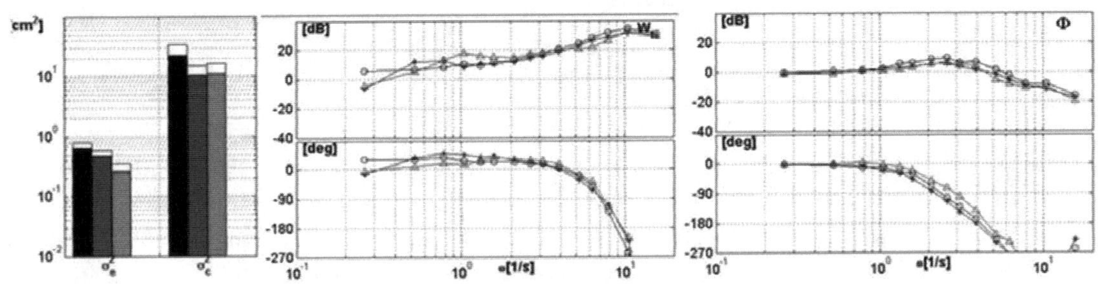

Figure 7. Influence of the piloting task, $W_c = \dfrac{K_c}{s^2}$.

The results demonstrated that in experiments with the controlled element dynamics $W_c = \dfrac{K_c}{s}$, the variance of error σ_e^2 stays approximately the same for all these displays. However, for the double integrator $W_c = \dfrac{K_c}{s^2}$ in the case of a pursuit task the variance of error is considerably lower in comparison to the experiments with the compensatory display, and in the case of the preview display it is more than twice as low.

It is also remarkable that for both dynamics the variance of the inceptor displacement σ_c^2 decreases, approximately by half in the case of the preview display in comparison with the compensatory display. The preview display allows the pilot to generate lead compensation in the crossover and low frequency ranges (see $W_L(j\omega)$ in figures 6, 7) to decrease the resonant peak by 3.2 dB and to increase the bandwidth ω_{BW} of the closed-loop system by 0.6 rad/s (see $\Phi(j\omega)$ in figures 6, 7).

For the small bandwidth of the input signal spectrum $\omega_i = 0.2$ rad/s, there is practically no difference in the frequency response and integral characteristics in compensatory, pursuit, and preview tracking tasks. In the case of the high bandwidth of the input signal spectrum $\omega_i = 1.0$ rad/s the difference between the characteristics is significant. For example, in that case for the controlled element dynamics $W_c = \dfrac{K_c}{s}$ (figure 8) the pursuit and especially the preview displays provide a decrease in the variance

IOP Conf. Series: Materials Science and Engineering **868** (2020) 012029 doi:10.1088/1757-899X/868/1/012029

of error σ_e^2 and the variance of the inceptor deflection σ_c^2 by 2 and by 2.4 respectively in comparison to the experiments with the compensatory display.

The variance of the inceptor deflection σ_c^2 decreases from 13 cm^2 for the compensatory display up to 5.5 cm^2 for the preview display for this input signal, and the pilot lead compensation in the crossover and low frequency range and the closed-loop system bandwidth increase as well.

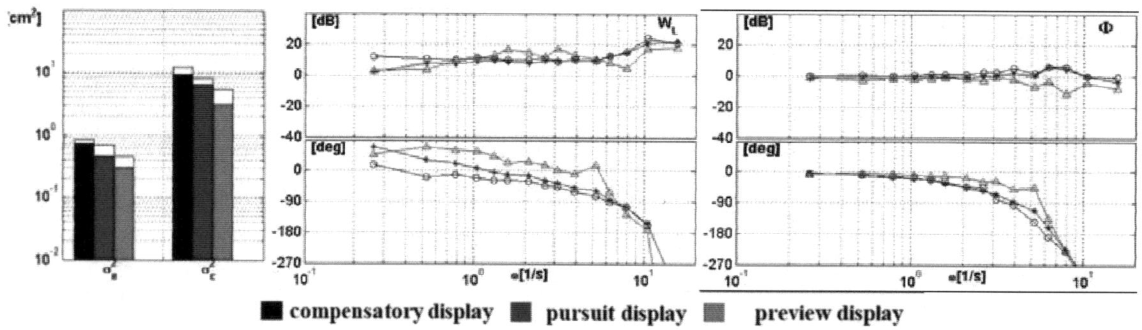

Figure 8. Influence of the piloting task, $\omega_i = 1.0$ 1/s.

3.2. *Definition of the optimal $T_{preview}$*

A set of experiments examining the influence of the preview time for a different bandwidth of the input signal was also carried out. As an example, the results for $\omega_i = 0.5$ rad/s are given in figure 9.

The analysis of these results demonstrates that an increase in the preview time from 0.5 s up to 2 s leads to a considerable improvement in tracking accuracy. In that case the pilot generates additional lead compensation in the crossover and low frequency ranges. It also causes an increase in the closed-loop system bandwidth by 1.2 rad/s and a decrease in the resonant peak by 3.5 dB. In the case of the following increase in the preview time up to 3 s, the variance of error became higher in comparison to the results of the experiments with $T_{preview}$=2 s.

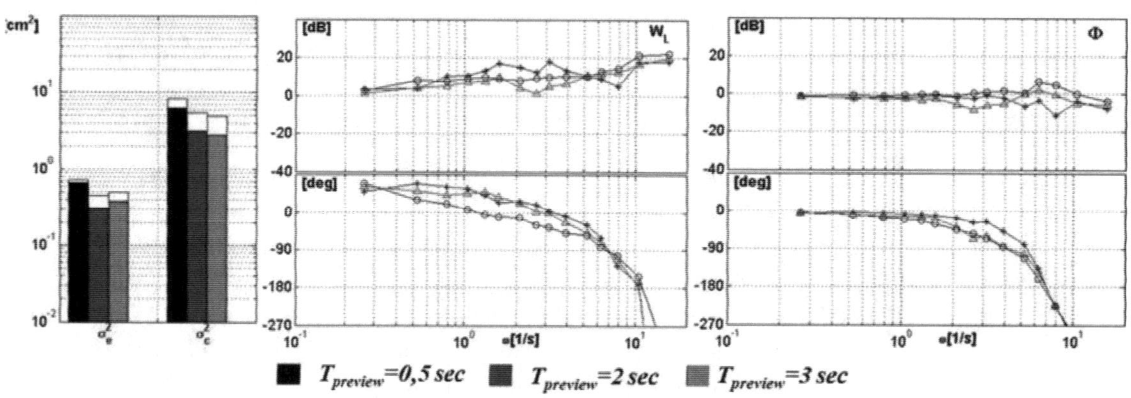

Figure 9. Influence of the preview time.

A separate series of experiments with the preview display were performed for different $T_{preview}$ from 0 up to 6 s. The experiments demonstrate the existence of an optimal value of $T_{preview}$ providing minimal variance of error whose value depends on the input bandwidth. This value is close to 2.4 s for $\omega_i = 0.5$ rad/s (figure 10). It was shown that for $\omega_i = 0.2$ rad/s – $T_{preview}$ = 3 s and for $\omega_i = 1.0$ rad/s – $T_{preview}$ = 2 s.

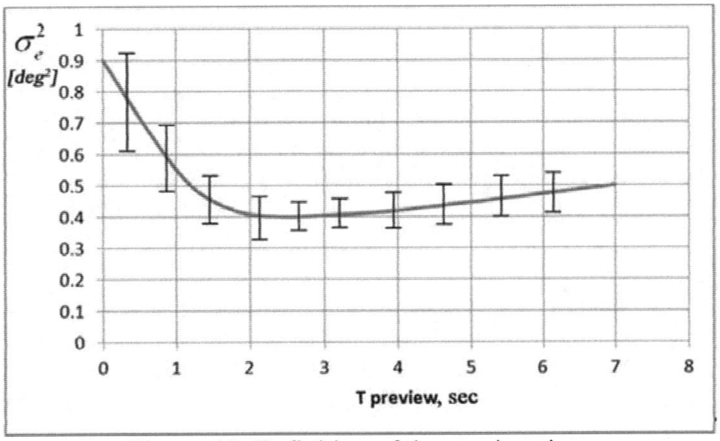

Figure 10. Definition of the preview time.

4. Conclusion

The results of the experiments performed for the planned trajectory bandwidth $\omega_i \leq 1$ rad/s demonstrated that the differences in all major characteristics in pursuit and preview tracking tasks in comparison with a compensatory task depend on the value of bandwidth and controlled element dynamics. It is practically unnoticeable in case of $\omega_i = 0.2$ rad/s and increases for $\omega_i = 0.5$ rad/s and $\omega_i = 1$ rad/s especially in a preview tracking task and in double integrator control. In these cases, the variances of error and inceptor deflection decrease by 2.3 times and 3 times respectively and the pilot generates considerable lead compensation. In a preview tracking task the close-loop system characteristics also improve (a decrease in the resonant peak by 3.2 dB and an increase in the bandwidth ω_{BW} of the closed-loop system by 1.2 rad/s). The preview time is characterized by the optimal value. It is equal to 2.4 s for $\omega_i = 0.5$ rad/s and it decreases from $T_{preview}$=3s to $T_{preview}$=2s when the bandwidth ω_i increases from 0.2 rad/s to 1 rad/s.

Acknowledgments
The results included in the paper are the results of the investigation that was sponsored by the Ministry of Science and Higher Education (contract ID: RFMEFI61619X0117).

References

[1] Reid L D, Drewell N H 1972 A pilot model for tracking with preview in process. *Proceedings of the 8th Annual Conference on Manual Control* (Michigan: University of Michigan Ann Arbor) pp 191-204
[2] Wilckens V, Schattenmann W 1968 Test Results with New Analog Displays for All Weather Landing. *AGARD Conference Proceedings "Problems of the Cockpit Environment"* (Amsterdam, Netherlands) pp 10.1–10.33
[3] Mulder M 1999 *Cybernetics of tunnel-in-the-sky displays*, PhD thesis, Delft University of Technology
[4] Mulder M and Mulder J A 2005 Cybernetic analysis of perspective flight-path display dimensions. *Journal of Guidance, Control, and Dynamics* **28(3)** 398 doi: 10.2514/1.6646
[5] Borst C, Mulder M, Van Paassen M M, Mulder J A 2006 Path-oriented control/display augmentation for perspective flight-path displays. *Journal of Guidance, Control, and Dynamics* **29(4)** 780 doi: 10.2514/1.16469
[6] Van der El K, Pool D M, Van Paassen M M, Mulder M 2018 Efects of Linear Perspective on Human Use of Preview in Manual Control. *IEEE Transactions on Human-Machine Systems* **48(5)** 496 doi: 10.1109/THMS.2017.2736882

[7] Van der El K, Pool D M, Damveld H J, Van Paassen M M, Mulder M 2016 An Empirical Human Controller Model for Preview Tracking Tasks. *IEEE Transactions on Cybernetics* **46(11)** 2609 doi: 10.1109/TCYB.2015.2482984

[8] Van der El K, Pool D M, Van Paassen M M, Mulder M 2018 Effects of Preview on Human Control Behavior in Tracking Tasks with Various Controlled Elements. *IEEE Transactions on Cybernetics* **48(4)** 1242 doi: 10.1109/TCYB.2017.2686335

[9] Van der El K, Padmos S, Pool D M, Van Paassen M M, Mulder M 2018 Effects of Preview Time in Manual Tracking Tasks. *IEEE Transactions on Human-Machine Systems* **48(5)** 486 doi: 10.1109/THMS.2018.2834871.

[10] Van der El K, Pool D M, Van Paassen M M, Mulder M 2020 Effects of Target Trajectory Bandwidth on Manual Control Behavior in Pursuit and Preview Tracking. *IEEE Transactions on Human-Machine Systems* **50(1)** 68 doi: 10.1109/THMS.2019.2947577

[11] Efremov A V, Tyaglik M S 2011 The development of perspective displays for highly precise tracking tasks. *Advances in Aerospace Guidance, Navigation and Control* (Germany: Springer) pp 163-174

[12] Grunwald A J 1985 Predictor Laws for Pictorial Flight Displays. *Journal of Guidance Control and Dynamics* **8(5)** 545 doi: 10.2514/3.20021

[13] Sachs G 2000 Perspective predictor flight – path display and minimum pilot compensation. *Journal of Guidance Control and Dynamics* **23(3)** 420 doi: 10.2514/2.4576

[14] Efremov A V, Rodchenko V V, Boris S 1996 Investigation of pilot induced oscillation tendency and prediction criteria development. *Final report WL-TR-96-3109,* Air Force Institute of Technology Wright-Patterson AFB (Dayton, Ohio) pp 1-138

[15] Efremov A V, Ogloblin A V, Predtechensky A N, Rodchenko V V 1992 *Pilot as a dynamic system* (Moscow: Mashinostroenie) p 332 (in Russian)

Head-up display architecture development for perspective civil aircraft

E S Neretin, S A Dyachenko*, V A Chufirin and D M Ilyashenko

Department 703 "Aircraft systems design", *Moscow Aviation Institute* (National Research University), 4 Volokolamskoe shosse, Moscow 125993, Russian Federation

*E-mail: dyachenkosergey33@yandex.ru

Abstract. Nowadays the civil aircraft head-up displays available on the market could not be easily integrated onboard due to architectural aspects. Particularly, these systems usually contain a dedicated computer unit performing the data processing to generate the appropriate image. This leads to increases in the overall avionics weight and size and limits the scope of head-up displays application in general. However, the mentioned characteristics could be improved by the usage of the integrated modular avionics concept for architecture design. According to that approach, the common hardware platform is used by different functional consumers to share computation resources. Thus, the required data processing function could be performed by the equipment already available onboard (without the need in additional hardware). The paper proposes the head-up display architecture for civil aircraft based on integrated modular avionics which significantly reduces the system's weight and size compared to characteristics of existing analogous. Also, this way allows reducing the time expenses for aircraft development, testing and certification, as well as the vehicle operational cost.

1. Introduction

Today, there are trends in aircraft avionics functionality expansion with a decrease in weight and size. This goal can be reached by using modern ways in the avionics design. For example, with the aid of the integrated module avionics (IMA) concept application.

IMA approach implies a high-integrity, partitioned environment that hosts multiple functions on a shared computing base. This provides advantages in weight and power since computing resources can be used more efficiently compared to previous avionics design concepts – distributed and federated principles [1-3].

Also, the application of the IMA concept allows to easily install new systems as a part of existing onboard equipment complexes. One of such systems is the head-up display (HUD) which allows pilots to read primary flight and navigation information without eyes refocusing from the external environment to the multifunctional displays on the instrument panel. It projects information on the special transparent screen (combiner) placed between the pilot's eyes and the cockpit windshield. Thus, the HUD usage reduces the crew workload and decision-making time in critical situations. The mentioned advantage is especially important during the takeoff and landing [4, 5].

Currently, HUDs are being developed under the federated approach i.e. the system represents by a number of self-contained units including separate computer. The purpose of this paper is the development of the HUD architecture designed following the IMA concept that is the novelty of the

IOP Conf. Series: Materials Science and Engineering **868** (2020) 012030 doi:10.1088/1757-899X/868/1/012030

work. The proposed solution allows to significantly reduce the system's weight and size, time expenses for the aircraft development, testing, and certification as well as operational cost.

2. Avionics architecture types

There are the following avionics design concepts at the current stage of aviation development:

- distributed architecture;
- federated architecture;
- IMA.

2.1. Distributed architecture

The distributed architecture was widespread at the beginning of avionics development. By this approach, the onboard equipment complex represents by a set of "self-sufficient" systems and instruments performing local functions. Such system usually consists of:

- sensors which collect the data about the external environment, aircraft position, etc.;
- converters transducing data into a format suitable for further processing;
- computers processing the information received;
- indicators displaying in the cockpit the measured values.

Each onboard system is aimed to measure and display several flight parameters. Signal processing is performed through elementary operations. Communication between systems is normally absent (the same data is measured independently by different equipment).

The advantages of this concept include ease of implementation and the no impact of one system failure on the other equipment performance. The main disadvantages are the duplication of data in individual systems and a large number of indicators cluttering the cockpit dashboard and leading to a crew workload increasing.

Nowadays the distributed architecture is recognized as obsolete and is not used practically.

2.2. Federated architecture

The growth in the number and complexity of the flight functions required an increase in the level of onboard equipment integration compared to the principle described earlier. The combination of autonomous systems performing similar functions has reduced the number of computers to one for each merged system. It was put in the federated architecture basis.

By this approach, onboard electronic equipment consists of:

- concentrator providing data obtaining from sensors;
- central computers that process signals from the concentrators and provide information to consumer equipment;
- aircraft systems using data from computers to perform the corresponding flight functions.

The systems exchange data via common interfaces, which allows the use of the same information by different consumers at the same time.

The advantages of the federated approach are the ease of certification, no need to collect information by each system separately, their high reliability, small overall dimensions of avionics relative to the distributed concept. However, the federated principle does not provide the flexibility for increasing and modification of the functionality since the integration of new systems negatively affects the avionics weight and size.

Aircraft developed following this concept are Boeing B737, Airbus A320, Tupolev Tu-204, etc.

2.3. IMA

The IMA is deprived of the federated approach disadvantages. It involves the usage of open network architecture and a common hardware platform for computing resources, which determines its flexibility in terms of modification. The functions of aircraft equipment are performed by software applications running a real-time operating system on a computing platform [1].

IOP Conf. Series: Materials Science and Engineering **868** (2020) 012030 doi:10.1088/1757-899X/868/1/012030

The IMA concept is used during modern and perspective aircraft design (in particular, Boeing B787, Irkut MC-21, CRAIC CR929, etc.).

The IMA concept has the following advantages compared to the federative architecture:

- the possibility of avionics functionality increase while maintaining the same weight and size characteristics;
- no need to recertify all software components when introducing changes to only one of the applications;
- reduction of time spent on the development, verification, and certification;
- overall aircraft operational cost reduction due to the equipment unification, etc.

The disadvantage of IMA is the complexity in managing the individual avionics functions design since it is impossible to provide independent development of software applications located on a single hardware.

3. Cockpit display systems (CDS)

The advantages of the IMA concept make it possible to integrate new aircraft systems into the avionics complex, which also include modern CDS.

At present, the main displays in the cockpit used in civil aviation are:

- traditional multi-functional displays located on the dashboard;
- HUD.

HUD projects graphic information on a special almost transparent screen (combiner) which is located between the pilot's eyes and the cockpit windshield. HUD usage can increase the situational awareness of the crew [6]. An example of a HUD image is shown in figure 1.

Figure 1. Example of a HUD image.

Currently, there are several HUD types differ in the optical system as refracting, catadioptric, and waveguide.

Refracting and catadioptric HUDs consist of an optical system, a high-brightness projector, and a combiner. The system is integrated into a single housing located above the pilot's head. The indicated HUD types have a similar principle of operation: the image is projected onto the lens which forming a collimated light beam and then projected onto the combiner. Combiner glass reflects the incident light in the direction of the pilot. The main difference between these HUD types is the combiner design: the refracting HUD has a flat shape, and the catadioptric one is curved.

Figures 2, 3 show the operation scheme of refracting and catadioptric HUDs (1 – electron-beam device, 2 – mirror, 3 – lens, 4 – combiner, 5 – point of view.

The perspective waveguide HUD is principally different from the types above mentioned. It does not have a projector; the overhead part is implemented as a compact monoblock that significantly reduces the overall system dimensions. The image generator is located at the top of the monoblock. The lens refracts the light at a certain angle. The combiner is two parallel transparent waveguides. The light arrives at the first waveguide at the full reflection angle. At the intersection of the beams and the alignment plate, the incident light is reflected towards the front plane of the waveguide. Thus, the first waveguide rotates the image horizontally, and the second – vertically and project it in the pilot's view direction [7].

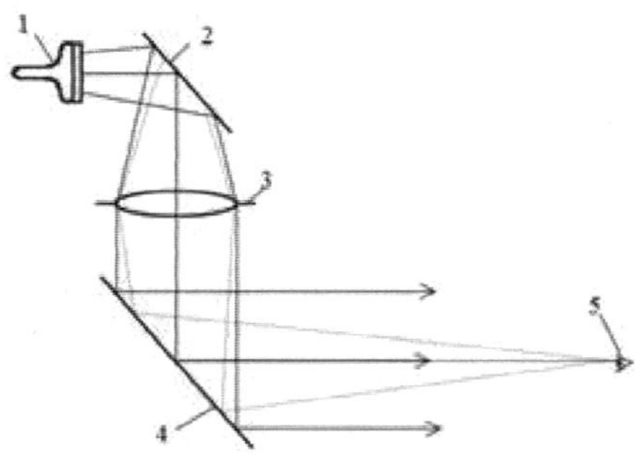

Figure 2. Refracting HUD operation scheme.

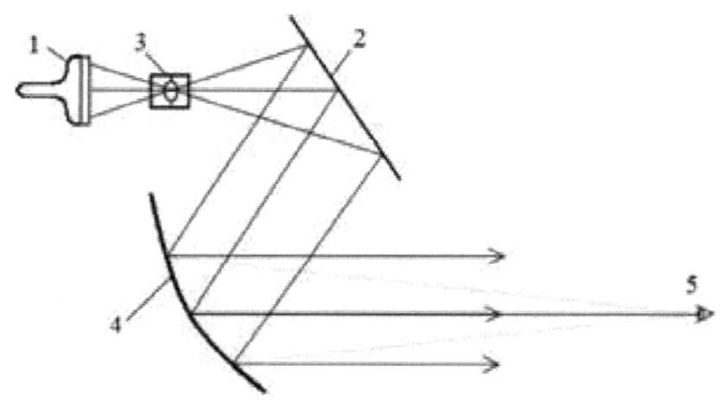

Figure 3. Catadioptric HUD operation scheme.

Figure 4 shows the operation scheme of the waveguide HUD (1 – image generator, 2 – lens, 3 – alignment plate, 4 – first waveguide, 5 – second waveguide).

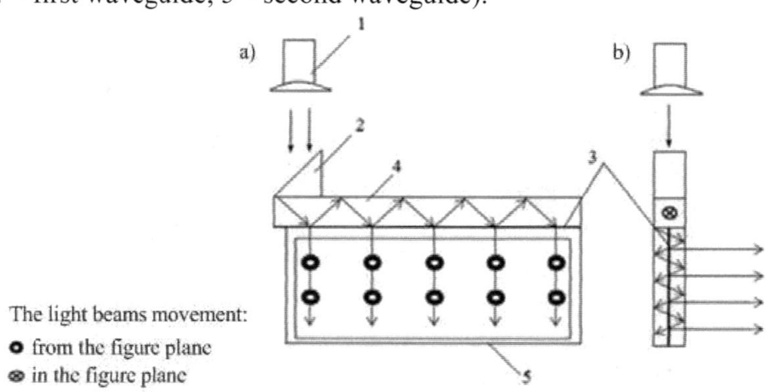

Figure 4. Waveguide HUD operation scheme (a – main view, b – side view).

Coupled with the mentioned means of information indication in the cockpit, technical vision systems (TVS) are used, which make it possible to increase the crew situational awareness regarding the external environment in hard weather conditions.

The following types of TVS are used in aviation [8]:

- enhanced vision systems (EVS) displaying infrared images according to the onboard cameras data;
- synthetic vision systems (SVS) that allow reconstructing an underlying earth surface three-dimensional model taking into account obstacles and airport infrastructure objects;
- combined vision systems representing the symbiosis of the two mentioned TVS types and combining of sensory and synthesized information;

The HUD and TVS are optional, but they are being installed on modern civil aircraft frequently since it allows reducing the crew workload and decision-making time in critical situations.

The HUD and TVS combination also called enhanced flight vision system allows a continuing approach in low visibility conditions. In other words, it provides operational credit in terms of decision height (DH) reduction.

According to the relevant flight operations standards for an instrumental approach, the DH is 200 feet (category I) [9]. Besides, the crew should evaluate the flight and navigation parameters and the external environment during the landing. In the traditional cockpit (without HUD and TVS) it is carried out by reading data from displays on the dashboard and transferring a view at the environment to search visually for relevant landmarks on the ground. If there are HUD and TVS on board, the use of them eliminates the pilot's eyes accommodation. Thus, the approach can be made even if the actual flight visibility is less than the instrumental approach required by reducing the DH to 100 feet [10].

4. Developed HUD architecture based on IMA concept

Now the HUD is being designed under the federated principle. Besides, a catadioptric HUDs are being chosen for installation onboard which is not optimal according to the weight and size criterion.

In this paper, the waveguide HUD architecture using the IMA concept for modern and perspective civil aircraft is proposed.

The developed architecture based on the IMA concept includes (figure 5):

- waveguide HUD;
- avionics central computer unit issuing data from onboard systems necessary for the displaying formats generation;
- CDS computer unit which receives and processes information from the central computer and also generates the resulting image as a combination of layers of symbolic information and TVS images;
- ambient light sensor that determines the current ambient light level;
- EVS;
- SVS;
- onboard data server containing SVS digital databases of the relief, obstacles and airports infrastructure;
- control panel that allows adjusting the HUD image parameters (in particular, brightness, contrast).

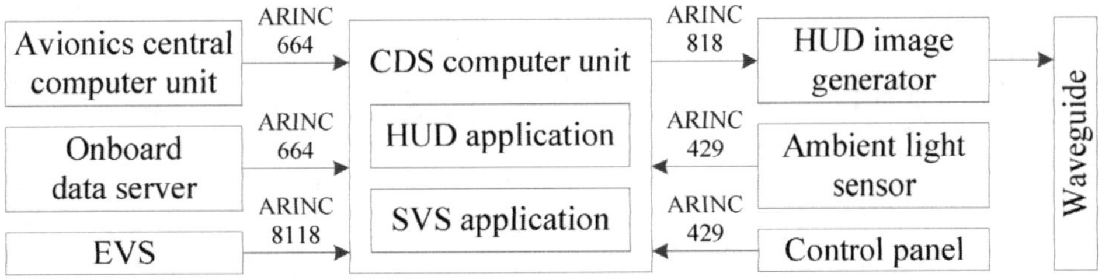

Figure 5. The developed HUD architecture based on the IMA concept.

The main components of the developed architecture are HUD and SVS software applications hosting in the CDS computer unit. The SVS application generates an external environment three-dimensional model based on information from the onboard data server. The HUD application generates a layer of symbolic information, combines it with TVS images and outputs the resulting format to the image generator.

Data from the avionics central computer unit and onboard data server are sent to the CDS computer via the ARINC 664 interface. The received information is stored in the HUD and SVS applications. The CDS computer and image generator interact via the ARINC 818 interface. The resulting image is converted into a light stream and projected onto the waveguide.

The ARINC 664 interface was chosen because of the high bandwidth required for transferring large data. In addition, it allows transmitting information with speed up to 100 Mbps, which is a key aspect for ensuring the requirements of the regulations in part of updating the image displayed.

The ARINC 818 interface is required for transmitting video to displays. The fiber-optic data channel speed is up to 28 Gbps.

Commands for adjusting the HUD image parameters are issued by the control panel via ARINC 429 interface. The ambient light sensor that makes it possible to change format brightness automatically depending on the ambient light level, also interact with the CDS computer via this interface. The choice of ARINC 429 is due to the lack of the need to transfer a large amount of data.

The proposed architecture meets the requirements of guidelines applied in civil aviation: DO-315B [10], DO-297 [11]. Compliance with DO-315B is achieved through HUD interaction with TVS. The matching with DO-297 requirements consists of the following aspects implied by the IMA concept:

- platform (CDS computer) resources are shared by multiple applications (particularly, by HUD and SVS);
- robust partitioning of shared resources;
- application may be designed independently of other applications (e.g. HUD and SVS);
- applications are independently modifiable and reusable etc.

5. Conclusion

The paper presents the analysis of current aviation CDS and TVS. Based on the requirements of the standards and the analysis of avionics architecture types, the HUD architecture based on the IMA approach is proposed. The developed architecture allows to significantly reduce the HUD weight and size compared to the characteristics of existing analogous.

References

[1] Watkins C B and Walter R 2007 Transitioning from federated avionics architectures to Integrated Modular Avionics *IEEE/AIAA 26th Digital Avionics Systems Conference* (USA: Dallas) pp. 2.A.1-1–2.A.1-10 doi: 10.1109/DASC.2007.4391842

[2] Gaska T, Watkins C B and Chen Y 2015 Integrated Modular Avionics – Past, present, and future. *IEEE Aerospace and Electronic Systems Magazine* **30(9)** 12 doi: 10.1109/MAES.2015.150014

[3] Butz H 2007 The Airbus approach to open Integrated Modular Avionics (IMA): technology, methods, processes and future road map. *Proc. AST 2007 Workshop on aircraft system technologies* (Germany, Hamburg), available at: https://pdfs.semanticscholar.org/08e3/4e1b048421631cc328fbbe874c9b58f31ec1.pdf

[4] Spitzer C R, Ferrell U and Ferrell T 2015 Digital avionics handbook (3rd edition) *Taylor & Francis* (United Kingdom: Oxfordshire) 848 p

[5] Burch D P and Braasch M S 2002 Enhanced Head-Up Display for General Aviation aircraft. *Proc. 21st Digital Avionics Systems Conference* (The USA: Irvine) pp. 11C6-11C6 doi: 10.1109/DASC.2002.1052982

[6] Moir I, Seabridge A and Jukes M 2013 Civil avionics systems (2nd edition) *John Wiley & Sons* (USA: Hoboken) 551 p

[7] Bigler C M, Blanche P-A and Sarma K 2018 Applied optics holographic waveguide heads-up display for longitudinal image magnification and pupil expansion. *Applied Optics* **57(9)** 2007

[8] Neretin E S and Dyachenko S A 2019 Architecture, algorithmic support and software development of aviation synthetic vision system for perspective transport civil aircraft. *J. Phys. Conf. Ser.* **1353(1)** 6 doi:10.1088/1742-6596/1353/1/012048

[9] FAR/AIM 2019 Federal Aviation Regulations / Aeronautical Information Manual *Aviation Supplies and Academics* (USA: Newcastle) 1182 p

[10] DO-315B 2011 Minimum Aviation System Performance Standards (MASPS) for Enhanced Vision Systems, Synthetic Vision Systems, Combined Vision Systems and Enhanced Flight Vision Systems *RTCA* (USA: Washington) 142 p

[11] DO-297 2005 Integrated Modular Avionics (IMA) Development. Guidance and Certification Considerations *RTCA* (USA: Washington) 137 p

IOP Conf. Series: Materials Science and Engineering **868** (2020) 012031 doi:10.1088/1757-899X/868/1/012031

The influence of network structure on quality of satellite corrections for precise point positioning in GNSS

A Podkorytov[1,2]

[1]Institute for radio-electronics, infocommunications and information security, *Moscow Aviation Institute* (National Research University), 4 Volokolamskoe shosse, Moscow 125993, Russian Federation
[2]*Joint Stock Company "Russian Space Systems"*, 53 Aviamotornaya Street, Moscow 111250, Russian Federation

E-mail: thepompous@gmail.com

Abstract. Network solution and user solution for precise point positioning in global navigation satellite systems are considered in the study. GPS ionosphere-free observation model with decoupled clocks is used for applying the procedure of ambiguity resolution of phase measurements. Four different variants of network solution based on European stations from a network of International GNSS service are considered. The other two variants of network solution are based on stations from a network of Russian satellite based augmentation system. Six investigated variants of network solution are varied in number of stations, total area, receiver types, configuration and measurements quality. Network products computed in different variants of network solution are then used for the user solution with applying the procedure of ambiguity resolution of phase measurements. The influence of network structure on accuracy and convergence time of precise point positioning with ambiguity resolution of phase measurements is analyzed. The properties of network structure are listed and described. Obtained precise point positioning results are averaged and discussed.

1. Introduction

Recently precise point positioning methods (PPP) [1] have been actively being developed in global navigation satellite systems (GNSS). The key features of PPP methods are using of precise ephemerides, compensation of multiple geophysical effects and applying of ambiguous phase measurements. Nowadays traditional ionosphere-free observation model with phase ambiguities estimated as real values is well developed (Float PPP methods, [2]). Main disadvantage of these methods is long convergence time. More complex methods which save integer nature of phase ambiguities and support the procedure of their ambiguity resolution are in progress currently (Integer PPP or PPP-AR methods [3]). Convergence time for PPP-AR methods is much shorter. The main problem of these methods is rank deficiency of corresponding system of linearized equations. There is a rank issue because of equipment biases in mathematical models of measurement and necessity to estimate phase ambiguities as integers. There are a few published PPP-AR approaches [4-6]. Author uses approach suggested in Natural Resources Canada which is based on Decoupled clock model [6].

Two types of solution can be distinguished for PPP in GNSS: network solution and user solution. Hereinafter "user solution" is determination of the user receiver coordinates in the PPP-AR methods with available precise satellite corrections. "Network solution" stands for determination of mentioned satellite corrections by common processing of code and carrier phase measurements formed by a network of ground stations with known coordinates. Precise satellites corrections (network products) are transmitted to user by means of communication lines. Approaches [4-6] differ in that part of network products which consists of satellite clocks and satellite equipment biases. Precise satellite coordinates are necessary for any approach. It is shown in [7] that in user solution network products corresponding to different approaches [4-6] can be strictly converted from one to another. In practice, however, there are some errors of such conversion because of specific aspects of network solutions for each approach.

In this study different variants of GPS network solution based on European stations from a network of International GNSS service (IGS) and stations from a network of Russian satellite based augmentation system (SBAS) are described. Investigated variants of network solution vary in number of stations, total area, receiver types, configuration and measurements quality. PPP user solution quality (accuracy and convergence time) with use of different network products is analyzed.

2. Methods and materials

2.1. Decoupled clock GPS observation model

Following ionosphere-free GPS observation model with decoupled clocks is used in the work [6]:

$$\begin{cases} P_3^j = R^j + m^j \Delta D_W + dT_{P3} - dt_{P3}^j + \varepsilon_{P3}^j \\ L_3^j = R^j + m^j \Delta D_W + dT_{L3} - dt_{L3}^j - \lambda_3 \left(17N_1^j + 60N_4^j\right) + \varepsilon_{L3}^j, \; j = \overline{1, J}, \\ A_4^j = \qquad b_{A4} - b_{A4}^j - \lambda_4 N_4^j + \varepsilon_{A4}^j \end{cases} \tag{1}$$

where P_3^j, L_3^j – code and phase ionosphere-free measurements of the j-th GPS satellite (m); A_4^j – ionosphere-free Melbourne–Wübbena linear combination of code and phase measurements of the j-th satellite (m); R^j – geometrical distance to j-th satellite (m); ΔD_W – uncompensated wet troposphere delay (m); m^j – mapping function for j-th satellite; dT_{P3}, dT_{L3} – code and phase ionosphere-free receiver clock offsets against GPS time scale including receiver equipment biases (m) (м); dt_{P3}^j, dt_{L3}^j – code and phase satellite clock offsets against GPS time scale including satellite equipment biases (m); b_{A4}, b_{A4}^j – equipment biases for receiver and *j*-th satellite in Melbourne–Wübbena combination (m); λ_3, λ_4 – wavelengths for GPS ionosphere-free combinations L_3^j and A_4^j (m); N_1^j and N_2^j – integer ambiguities for *j*-th GPS satellite (cycles); $N_4^j = N_1^j - N_2^j$ – widelane integer ambiguities (cycles); $\varepsilon_{P3}^j, \varepsilon_{L3}^j, \varepsilon_{A4}^j$ - measurements errors for combinations P_3^j, L_3^j, A_4^j ; J - the total number of available GPS satellites.

In user solution the model (1) contains $7 + 2J$ estimated parameters. Because of the presence of decoupled clocks in (1) the corresponding system of equations is singular and the rank deficiency equals two. In network solution the model (1) contains $3J + 4M + 2JM$ estimated parameters (M - the number of stations in network solution). Because of the presence of decoupled clocks in (1) the corresponding system of equations is singular and the rank deficiency equals $3 + 2(J + M - 1)$. According to the basic property of singular GNSS linear systems described in [3, 8], part of initial estimated parameters can be estimated as linear combinations. Due to this property inconsistent and underdetermined system of equations (1) can be transformed to inconsistent and overdetermined system of equations with unique least-square solutions.

In Float PPP methods traditional ionosphere-free observation model [1] is applied and the same receiver and satellite clock offsets are used in mathematical models of ionosphere-free code and phase measurements (unlike the model (1) with different clock offsets for P_3^j, L_3^j, A_4^j combinations). In this case ambiguity of phase measurement absorbs equipment biases and becomes real value estimate, rank deficiency disappears. As a result, Float PPP methods are simple in practice implementation (both network and user solutions), but they converge to 1 cm positioning accuracy longer PPP-AR methods [4-6].

2.2. Aspects determining accuracy and convergence time of precise point positioning

Let us consider the aspects which influence accuracy and convergence time of precise point positioning using ionosphere-free observation model with decoupled clocks (1) (meaning the aspects connected with quality of network products):

- **Statistical characteristics of network products** (standard deviation values of the satellite corrections $dt_{P3}^j, dt_{L3}^j, b_{A4}^j, j=\overline{1,J}$). During the first 1-2 hours of network solution the satellite corrections $dt_{P3}^j, dt_{L3}^j, b_{A4}^j, j=\overline{1,J}$ still have convergence period. The user needs to use some threshold values for standard deviations of $dt_{P3}^j, dt_{L3}^j, b_{A4}^j, j=\overline{1,J}$, which can indicate that convergence period is over and satellite corrections can be applied to precise point positioning. The issue here is that mentioned satellite corrections may have different convergence period depending on network configuration, quality of measurements and elevation angles of observed satellites.
- **Network configuration of stations** (number of stations, ranges between them and relative positions). Given a small number of stations with relative ranges 500-1000 km (local network solution), the quantity of network products for high elevation satellites can be not enough for PPP with ambiguity resolution of phase measurements. Increasing the number of stations and extending the network may lead to the lack of uniform (homogeneous) station receivers and their ineffective (irregular) relative distributing. Ineffective relative distributing of the stations inside the network also decreases the quality of the network products because from some areas of the network some satellites can be observed just with low elevation angles. Regular distributing of stations for network solution can not always be guaranteed. For instance, density of IGS stations in Russia is extremely low compared to Europe. In Russia to the north of Yekaterinburg (56 degrees north latitude) there are just five IGS stations, and three of them are located closely to the continental edges, i.e. near the edges of basic servicing area in Russia (stations in Tiksi, Saint Petersburg and Magadan).
- **Uniformity of receivers at network stations**. Uniformity (homogeneity) of receivers is based on one the same manufacturer, receiver model, list of measurement type and the same firmware version. It's difficult to fulfill this requirement with using open access IGS stations measurements. For this reason commercial network products providers choose to set up their own network consisting of uniform receivers. In Float PPP network solution it's admissible to process measurements from stations with different receivers because traditional ionosphere-free observation model is used for Float PPP methods. It's not the case for PPP-AR methods.
- **Ineffective location of one/several network stations.** In city area a lot of other powerful receivers and transmitters can operate not far from station receiver (communication systems, airport antenna systems, relay stations and others). As a result, multipath and noise errors can be high, and certain sector of visibility can be shadowed at all. Out of city area GNSS-stations are usually disposed at meteorological and geodetic buildings with limited space and probable shadow region. There are not so many stations with perfect location in terms of all the requirements. For instance, there are 2-3 IGS stations of high quality measurements and low multipath errors which are typically used as reference for network solution with traditional ionosphere-free observation model by IGS Analysis Centers. If station is located in not applicable place, it results in increased noise level and probable systematic biases in measurements which contribute an additional error to the network products.
- **User position inside the network and distance to stations.** Remoteness of the user from the network stations means that observed by the user satellites are visible with low elevation angles from the network stations, so precise satellite corrections can be noisy. PPP quality (accuracy, convergence time, reliability) for the user located at the edge of the network is usually worse compared to user at the center of the network. Most of observed satellites for such near-border users are observed from stations

with low elevation alngles. It reduces the quality of precise satellites corrections and user PPP quality for the locations near the edge of network.

- **Quality of measurements from network stations.** The quality of measurements used for the network solution is determined by their availability, continuality, low noise error and absence of systematic biases, multipath errors and cycle slips.

When the user applies the procedure of ambiguity resolution (fixed solution), in addition to requirements to accuracy and convergence time it appears the requirement to fixed solution reliability. The reliability of a fixed solution is connected with the value of probability of correct ambiguity resolution of phase measurements during the processing. It's determined by the quality of network products, i.e. by the all aspects described above.

3. Results and discussion

3.1. Comparative analysis of user positioning using precise satellite corrections obtained from different network configurations

Ultimate (best) results of user PPP accuracy and convergence time were estimated in [3] for the case of network solution with permanent satellite constellation (the simplest case). Three-dimensional error of user positioning not exceeding 1-3 cm is achieved in this case after 3-5 minutes of processing of measurements with 30 sec interval. Strict consideration of changing satellite constellation is quite complicated because of bulkiness of S-transformations theory [9, 10] applied for PPP-AR methods. For this reason obtained PPP-AR results for changing satellite constellation in terms of accuracy and convergence time are worse that mentioned above best results in current conditions (network solution is not optimal).

Four variants of network solution with European IGS stations and two variants of network solutions with Russian SBAS system stations were implemented. Network configuration and specific features for these variants are described in further:

1. Network solution with 10 IGS stations ("Network solution 1" in further) where the receivers are from the same manufacturer. Despite of a high density of IGS stations in Europe, it was impossible to choose 10 stations with good ranges and relative positions for any manufacturer. Network configuration for this variant is shown on figure 1 where 10 used stations are marked with red circles.
2. Network solution with 8 IGS stations ("Network solution 2" in further). In this variant of network solution two the most distant stations from the others shown on figure 1 were excluded. On figure 1 two excluded stations are marked as 1 and 2. Remaining 8 stations were used for processing.
3. Network solution with 8 IGS stations ("Network solution 3" in further). In this variant of network solution two stations with the most number of cycle slips and gaps in measurements (the worst stations in terms of quality of measurements) were excluded from the stations shown on figure 1. On figure 1 two excluded stations are marked as 3 and 4. Remaining 8 stations were used for processing.
4. Network solution with 6 IGS stations ("Network solution 4" in further). In this variant of network solution two stations with the most number of cycle slips and gaps in measurements and two the most distant stations from the others were excluded from the stations shown on figure 1. On figure 1 four excluded stations are marked as 1, 2, 3 and 4. Remaining 6 stations were used for processing.
5. Network solution with 10 stations of Russian SBAS system ("Network solution 5" in further) where the receivers are from the same manufacturer. Network configuration for this variant is shown on Figure 2 where 10 used stations are marked with red circles.
6. Network solution with 7 stations of Russian SBAS system ("Network solution 6" in further). In this variant of network solution three stations which are not uniform with others and which have the most noisy measurements were excluded from the stations shown on figure 2. On figure 2 three excluded stations are marked as 1, 2 and 3. Remaining 7 stations were used for processing.

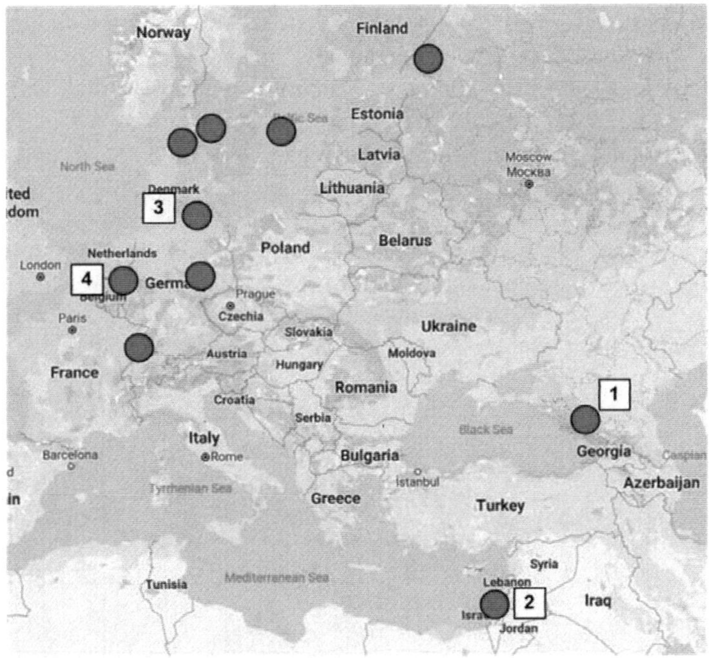

Figure 1. IGS stations used in four variants of network solution.

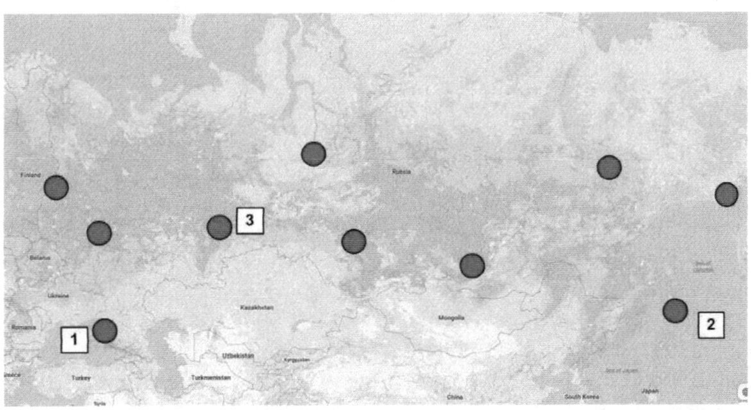

Figure 2. Stations of Russian SBAS system used in two variants of network solution.

Currently the quality of computed decoupled satellite corrections is not stable (PPP convergence time can significantly vary for different receivers and time periods). For this reason to illustrate in some way the current quality of PPP based on network products from considered variants of network solution averaging is applied. Figure 3 shows the averaged three-dimensional errors of PPP-AR fixed and float user solutions corresponding to four variants of network solution with IGS stations. Measurements from five IGS stations not included to corresponding network solutions were used as user measurements. The averaging among the described five stations and among five 50-minutes intervals was applied. On Figure 3 the curve “1” corresponds to the user fixed solution for Network solution 1; the curve “2” corresponds to the user float solution for Network solution 2; the curve “3” corresponds to the user float solution for Network solution 1; the curve “4” corresponds to the user float solution for Network solution 3; the curve “5” corresponds to the user float solution for Network solution 4 (on figure 3 the most specific curves are shown). The following features can be noted:

- The quality of network products computed in Network solution 1 is the worst (fixed user solution is considerably worse than float user solution in terms of accuracy and convergence time);
- The quality of fixed and float user solution is comparable for the network products from Network solution 2 and Network solution 3;
- Among the considered four variants of the network solution the best quality of network products is provided by Network solution 4 because corresponding fixed and float user solutions have the best accuracy and convergence time. In spite of the fact that stations for Network solution 2 look denser on figure 1 compared to stations for Network solution 4, in case of Network solution 2 two the worst stations (in terms of quality of measurements) degraded obtained PPP results.

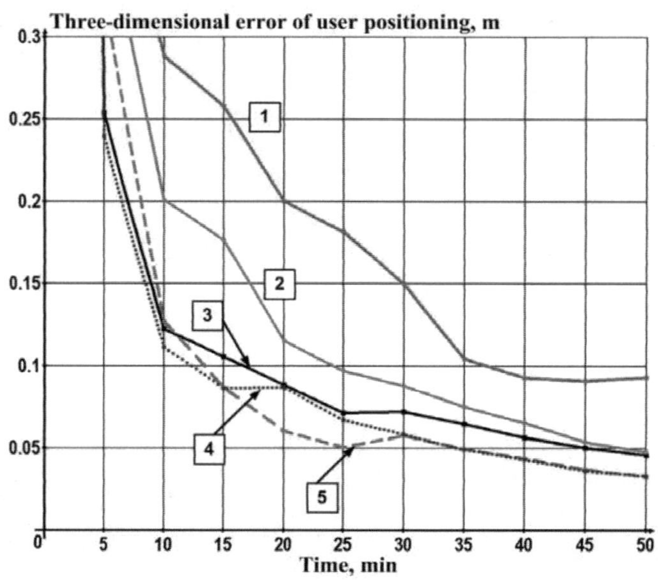

Figure 3. Float and fixed solution of averaged. Three-dimensional error of user positioning obtained with network products from network solution with IGS stations.

Figure 4 shows averaged Three-dimensional error of user positioning similar to one on Figure 3 but for network solution based on Russian SBAS system. Averaging among the five stations and among the five 50-minutes intervals was applied. On Figure 4 the curve “1” corresponds to the user fixed solution for Network solution 5; the curve “2” corresponds to the user fixed solution for Network solution 6; the curve “3” corresponds to the user float solution for Network solution 5; the curve “4” corresponds to the user float solution for Network solution 6; Following features can be noted:

- After 30 min of processing averaged Three-dimensional error of user positioning for all four solutions shown on figure 4 are almost concured;
- The shortest convergence time during the first 30 min of processing is provided by float user solutions with network products from Network solution 5 and Network solution 6;
- During the first 30 min of processing fixed user solution with using network products from Network solution 6 outstrips fixed user solution with using network products from Network solution 5;

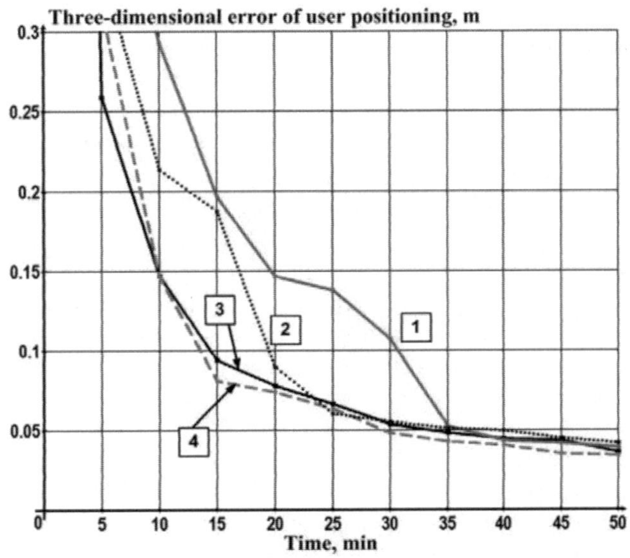

Figure 4. Float and fixed solution of averaged. Three-dimensional error of user positioning obtained withnetwork products from network solution with Russian SBAS stations.

3.2. Reliability of fixed solution

Positioning quality for PPP-AR methods is determined not only by stable, quick and correct ambiguity resolution but also by the difference between fixed (with use of ambiguity resolution procedure) and float (without ambiguity resolution procedure) user solutions. On other words, it is of great interest what is contribution of ambiguity resolution procedure to accuracy and convergence time of the user solution. Figures 5 and 6 show three-dimensional error of PPP-AR user positioning (fixed and float solutions). It's clear on figure 5 that fixed solution is reliable since 5 min of measurements processing, it's provide significant improving of convergence time compared with float solution. Figure 6 shows another case for comparison: convergence time for fixed and float solutions is comparable. It means that the procedure of ambiguity resolution hasn't brought any contribution in terms of accuracy and convergence time of user solution.

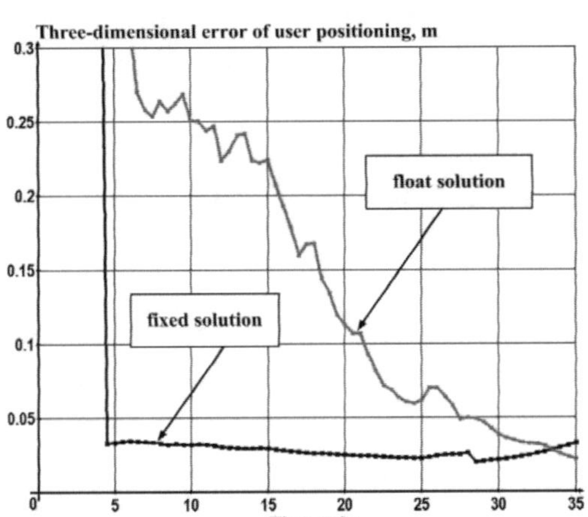

Figure 5. User 3D positioning error, example of reliable fixed solution.

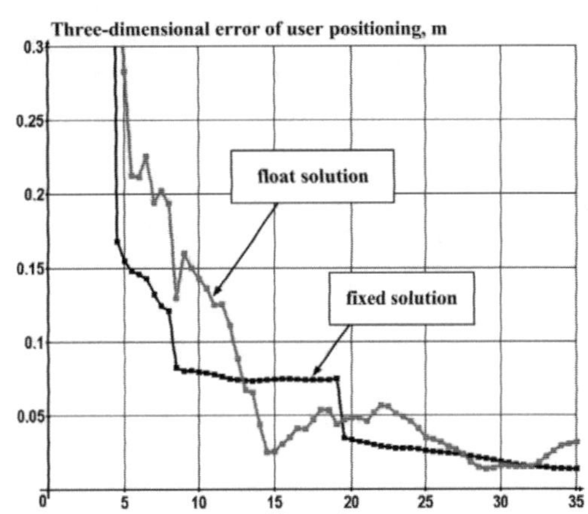

Figure 6. User 3D positioning error, example of unreliable fixed solution.

4. Conclusion

With help of six variants of network solution based on the measurements from IGS network and Russian SBAS network stations the influence of network structure on quality of PPP solution with ionosphere-free decoupled clock observation model was analyzed. It has been figured out that the quality of PPP results is implicitly depends on the network configuration and noisy measurement of certain stations.

For the structure of network involved in the satellite decoupled corrections computing the list of aspects which determine the accuracy and convergence time of user positioning was identified.

It was shown that for considered IGS and Russian SBAS system stations there were no way to use uniform receivers in network solution. It leads to reducing of decoupled satellite corrections accuracy and user positioning accuracy. Currently positioning accuracy inside the network for near-border users is significantly influenced by unreliability of ambiguity resolution procedure. In practice, it's not always possible to check PPP quality with receivers which were not included to the network solution (because of extremely low density of IGS stations in Russia).

References

[1] Kouba J 2009 Guide to Using International GNSS Service (IGS) Products, Jet Propulsion Lab., Pasadena, available at: http://igscb.jpl.nasa.gov/components/usage.html

[2] Gao Y, Shen X 2002 A new method for carrier phase based precise point positioning. *Navigation* 49(2) 109 doi: 10.1002/j.2161-4296.2002.tb00260.x

[3] Podkorytov A N 2014 *Precise point positioning in global navigation satellite systems due to ambiguity resolution of phase measurements*, PhD thesis, Moscow Aviation Institute (National Research University)

[4] Li P, Zhang X, Ren X, Zuo X, Li Y, Pan Y 2016 Generating GPS satellite fractional cycle bias for ambiguity-fixed precise point positioning. *GPS Solutions* **20(4)** 771 doi: 10.1007/s10291-015-0483-z

[5] Laurichesse D, Mercier F, Berthias J-P, Broca P, Cerri L 2009 Integer Ambiguity Resolution on Undifferenced GPS Phase Measurements and Its Application to PPP and Satellite Precise Orbit Determination. *Navigation* **56(2)** 135 doi: 10.1002/j.2161-4296.2009.tb01750.x

[6] Collins P, Lahaye F, Héroux P, Bisnath S Precise Point Positioning with Ambiguity Resolution using the Decoupled Clock Model. *Proc. of ION-GNSS-2008* (Savannah, Georgia, 16-19 September 2008) pp 1315-1322

[7] Seepersad G, Bisnath S 2017 An assessment of the interoperability of PPP-AR network products. *The Journal of Global Positioning Systems* **15(4)** doi:10.1186/s41445-017-0009-9

[8] Povalyaev A A, Podkorytov A N 2015 Precise Point Positioning in Global Navigation Satellite Systems with Ambiguity Resolution of Phase Measurements. *J. Commun. Technol. Electron.* **60(860)** doi: 10.1134/S1064226915070141

[9] Teunissen P J G 1985 Zero *Order Design: Generalized Inverses, Adjustment, the Datum Problem and S-Transformations. Optimization and Design of Geodetic Networks* (Berlin: Springer-Verlag) pp 11-55

[10] De Jonge P J 1998 A processing strategy for the application of the GPS in networks. *(*Netherlands Geodetic Commission, Delft, Netherlands)

IOP Conf. Series: Materials Science and Engineering **868** (2020) 012032 doi:10.1088/1757-899X/868/1/012032

A suboptimal algorithm for synthesis of a movable object control

A K Diyachuk*, B D Orkin and S D Orkin

Department of Aircraft Systems Design, *Moscow Aviation Institute* (National Research University), 4 Volokolamskoe shosse, Moscow 125993, Russian Federation

*E-mail: annakd16@mail.ru

Abstract. In this paper a suboptimal algorithm for synthesis of a movable object control, based on the known approach, according to which a control is formed depending on a control error – the difference between current and required values of a regulated variables vector is considered. The distinctive peculiarity of the algorithm was based on the vector of regulated parameters estimation approach. This vector is proposed to be estimated by solving a mobile left endpoint problem derived from the original one. It is shown by examples that the proposed approach allows to simplify the control synthesis problem solution and obtain effective control algorithms, providing a suboptimal control and, in some cases, presented in this paper, an optimal control. The suggested algorithm can be used to determine continuous and pulse controls of aircrafts (including spacecrafts) in problems of their approaching (guidance) and spacecraft docking, carrying out of correcting maneuvers, descent and soft landing of spacecrafts at al.

1. Introduction

The problem of effective movable objects control algorithms development is actual. Its solution, as a rule, is formed as follows:

1) firstly a required (basis) object trajectory is determined by a solution of a control (optimum control) problem with the fixed initial conditions [1, 2]. At the same time the required trajectory may be rebuilt [3], if the current object state is selected as initial conditions;
2) secondly a control, provides for the movement along the required path with using the control error – the difference between required (which corresponds to required trajectory) and current values of a regulated variables vector is generated.

The main problem is to solve the first task in the above list. This paper deals with the mentioned above approach, according to which a control is formed depending on a control error – the difference between current and required values of a regulated variables vector.

The main peculiarity and the novelty of the considered algorithm related with the required value of regulated variables vector determining method (see the task no 1 in the list above). We suggest to estimate this vector by solving of a mobile left endpoint problem instead of the fixed left endpoint problem [3], as it is generally accepted.

This approach allows to: 1) substantially simplify the optimal control problem solution; 2) synthesize the control which either coincides with the optimal one or it is close to the optimal control and this fact is confirmed by examples shown.

Moreover, at approximate solution of the optimal control problem with the mobile left endpoint one can use the widely applicable parametric representation of a desirable trajectory in the form of a given function of one of co-ordinates and a set of selected parameters even in cases when it cannot be made in the original optimum control problem with the fixed left endpoint.

When described problems solving, the widely known approaches – the *Pontryagin's maximum principle*, direct variational methods and others can be applied [4, 5].

In this paper we'll show that the proposed approach makes it possible to simplify the control synthesis problem solution and obtain effective control algorithms, providing a suboptimal control and, in some cases, presented in this paper, an optimal control.

2. Materials and research methods

The optimum control of a movable object problem, which we should call the "*original*", is considered:

$$\min_u I \tag{1}$$

with connections (motion equations)

$$\dot{x} = f(\bar{x}, \bar{u}, t), \tag{2}$$

under constraints

$$\bar{x} \in X, \ \bar{u} \in U, \tag{3}$$

initial

$$\bar{x}(0) = \bar{x}_0 \tag{4}$$

and terminal conditions

$$\bar{x}(T) \in M_T . \tag{5}$$

In formulas (1)–(5) the following designations are accepted: I – criterion; $\bar{x}$ – n-dimensional state vector; $\bar{u}$ – control vector; t – current time; X, U, M_T – given sets; $\bar{x}_0$ – given vector of initial conditions; T – time of movement.

For the approximate solution of "*original*" task let's use the approach, according to which the control $\bar{u}$ is set depending on the control error vector $\bar{E}$, estimating the deviation of m-dimensional ($m < n$) vector of adjustable parameters $\bar{x}_m$ from its given (required) value $\bar{x}_{m\mathrm{R}}$. Thus, a vector $\bar{x}_m$ during the movement should be modified so that to ensure $\bar{E} \to 0$ or, that is the same, $\bar{x}_m \to \bar{x}_{m\mathrm{R}}$. This approach is known as well as control algorithms, depending on an error $\bar{E}$ ($\bar{u} = g(\bar{E})$, g – an operator), denoted by A A Krasovskiy [4] and widely used in practice [5, 6]. Thereby, the implementation of a considered approach and the control problem solution optimization is related with the estimation of $\bar{x}_{m\mathrm{R}}$.

A vector $\bar{x}_m$ is a function of a state vector $\bar{x}$.

When forming a state vector $\bar{x}$ in appropriate way, one can achieve, that $\bar{x}_m$ includes last m components of the state vector $\bar{x}$. Then

$$\bar{x} = (\bar{x}_{n-m}, \bar{x}_m), \tag{6}$$

where $\bar{x}_{n-m}$ – the vector which includes first $n-m$ components of $\bar{x}$.

The required value of $\bar{x}_{m\mathrm{R}}$ we suggest to determine by solving the optimal control problem which we should call "*auxiliary*". It differs from the "*original*" one in that the condition (4) is substituted to

$$\bar{x}_{n-m}(0) = \bar{x}_{n-m,0}, \ \ \bar{x}_m(0) \in X_m . \tag{7}$$

where $\bar{x}_{n-m,0}$ – the first $n-m$ components of $\bar{x}_0$; X_m – the given set.

Consequently, the "*original*" problem with the fixed left endpoint transforms into the "*auxiliary*" problem with the mobile left endpoint.

The vector $\bar{x}_m$ and the set X_m are chosen from practical reasons according to control object's properties and received information. Often the value of $\bar{x}_m(0)$ is not restricted by the constraint $\bar{x}_m(0) \in X_m$.

The value of $\overline{x}_m(0)$, obtained when solving the "*auxiliary*" problem, also is the required $\overline{x}_{mR}$.

To obtain the approximate solution of the "*auxiliary*" problem and, therefore, to estimate $\overline{x}_{mR}$ we can apply different parametric methods (such as direct variational methods and others).

It should be noted that, that constraints imposed on state vectors and controls (3) and also motion equations (2) in the "*auxiliary*" problem may be more simple than in the "*original*" problem, because the exact accounting of the "*original*" problem constraints may be provided by forming the control which depends on the error $\overline{E}$.

Let's now consider solutions of some control problems with usage of the described approach.

3. Results and discussion

Example 1. The "*original*" optimal control problem:

$$\min_u \int_0^T dt, \tag{8}$$

$$\dot{x}_1 = x_2,\ \dot{x}_2 = u,\ |u| \le u_m, \tag{9}$$

$$x_1(0) = x_{10}, \tag{10}$$

$$x_2(0) = x_{20}, \tag{11}$$

$$x_1(T) = 0,\ x_2(T) = 0, \tag{12}$$

where $\overline{x} = (x_1, x_2)$, x_1, x_2 – position and velocity; u – control; u_m, $\overline{x}_0 = (x_{10}, x_{20})$ – given constraints on control and initial conditions.

The decision of this problem is known. Below it is written in the form of synthesis:

$$u = \begin{cases} -u_m \cdot sign(P), if\ P \ne 0; \\ u_m \cdot sign(x_1),\ if\ P = 0; \end{cases} \tag{13}$$

where

$$P = 2 \cdot u_m \cdot x_1 + x_2 \cdot |x_2|. \tag{14}$$

Let's take the velocity x_2 as the adjustable parameter and the error of control will be estimated by formula:

$$\overline{E} = x_{2R} - x_2, \tag{15}$$

where x_{2R} – is the required value of the adjustable parameter x_2.

When finding x_{2R} we shall not restricted $x_2(0)$. Then, according to the described approach, for to estimate x_{2R} the "*auxiliary*" optimum control problem which differs from the "original" problem (8)–(12) in that, the initial condition is absent, is to be solved. As the result, we should get (in the synthesis form):

$$x_{2R} = -\sqrt{2 \cdot u_m \cdot |x_1|} \cdot sign(x_1). \tag{16}$$

The control by the error $\overline{E}$ (15) we may construct, having used an obvious relation:

$$u = \begin{cases} u_m \cdot sign(\overline{E}),\ if\ \overline{E} \ne 0; \\ u_m \cdot sign(x_1),\ if\ \overline{E} = 0. \end{cases} \tag{17}$$

Let's note that expressions (13), (16) are easy to obtain when using the maximum principle of L S Pontryagin.

Controls evaluated by (13) and (17) coincide. Thus, control (17) is optimal for the "*original*" problem (8)–(13). Besides, to solve an "*auxiliary*" problem is easier, then an "original", because an "original" problem may have a switch of control, but control does not vary in an "auxiliary" problem.

IOP Conf. Series: Materials Science and Engineering **868** (2020) 012032 doi:10.1088/1757-899X/868/1/012032

In the use of parametric methods, when a trajectory or a control vector are determined by a set of unknown parameters, with reference to an "*auxiliary*" problem a trajectory $x_1(t)$ in the first case one may describe by the second order polynomial of time *t*, and control $u(t)$ – in the second case – by the zero order polynomial, because it is necessary to satisfy for three boundary conditions (10), (12). At the same time it is not difficult to get formula (16) in both cases by minimizing the motion time *T*, and then to synthesize the optimum control for an "original" problem by formula (17).

Therefore, with reference to an "*original*" problem polynomials orders, which describe a trajectory $x_1(t)$ or control $u(t)$, are to be increased, because it is necessary to fulfill not to three, but to four boundary conditions (10)–(12), so the obtained control in both cases should not be optimal by the criterion (8).

Example 2. The "*original*" optimal control problem: criterion (8), movement equations, constraints, initial and terminal conditions are

$$\dot{x}_1 = V \cdot \cos(x_3),\ \dot{x}_2 = V \cdot \sin(x_3),\ \dot{x}_3 = u,\ |u| \le u_m, \tag{18}$$

$$x_1(0) = x_{10},\ x_2(0) = x_{20}, \tag{19}$$

$$x_3(0) = x_{30}, \tag{20}$$

$$x_1(T) = x_{1T},\ x_2(T) = x_{2T},\ x_3(T) = x_{3T}, \tag{21}$$

where $\overline{x} = (x_1,\ x_2, x_3)$, $x_1,\ x_2$ – coordinates, x_3 – trajectory inclination; V – velocity; u – control; u_m, $\overline{x}_0 = (x_{10},\ x_{20},\ x_{20})$, $\overline{x}_T = (x_{1T},\ x_{2T},\ x_{2T})$ – given constraints on control, initial and terminal conditions.

The solution of the given problem is known and the optimal programmed control is a piecewise constant function of time, which takes at time intervals one of following values: $u = u_m$, $u = -u_m$, $u = 0$. Appropriate optimal paths are: "turn – straight line – turn", "turn – turn – turn".

Let's take the path inclination x_3 as the adjustable parameter and the error of control $\overline{E}$ we should estimate by the equation

$$\overline{E} = x_{3R} - x_3, \tag{22}$$

where x_{3R} – the required value of the adjustable parameter x_3.

We should not restrict $x_3(0)$ during the search of x_{3R}. Then, to evaluate the value of x_{3R} it is necessary to solve the "*auxiliary*" optimal control problem, which differs from the "*original*" problem (8), (18)–(21) in that, the initial condition (20) is absent. As a result we get (in a synthesis form):

$$x_{3R} = \begin{cases} \alpha_1 + \pi - \alpha_2,\ if\ \rho \ge R; \\ \beta_1 + \beta_2,\ if\ \rho < R; \end{cases} \tag{23}$$

where

$$R = \frac{V}{u_m};\quad \rho = \sqrt{x_1^2 + (x_2 - R)^2}; \tag{24}$$

$$\alpha_2 = \arcsin\left(\frac{R}{\rho}\right);\quad \alpha_1 = \begin{cases} \alpha_{10}, & if\ x_1 \ge 0; \\ \pi - \alpha_{10}, if\ x_1 < 0; \end{cases}\quad \alpha_{10} = \arcsin\left(\frac{x_2 - R}{\rho}\right); \tag{25}$$

$$\beta_1 = \beta_{12} - \beta_{11};\quad \beta_{12} = \arcsin\left(\frac{x_{1c}}{\rho_1}\right);\quad \beta_{11} = \arcsin\left(\frac{x_1}{\rho_1}\right); \tag{26}$$

$$x_{1c} = \sqrt{R^2 + (x_{2c} - R)^2};\quad x_{2c} = \frac{\rho_1^2 - R^2}{4R};\quad \rho_1 = \sqrt{x_1^2 + (x_2 + R)^2}; \tag{27}$$

$$\beta_2 = 2 \cdot arcsin\left(\frac{x_{2c}}{\rho_2}\right);\quad \rho_2 = \sqrt{x_{1c}^2 + x_{2c}^2}. \tag{28}$$

Relations (23)–(28) are obtained under conditions: $x_2 \geq 0,\ x_{1T} = 0,\ x_{2T} = 0,\ x_{3T} = 0$. These conditions are always satisfied at the expense of the transition to the new coordinate system, which origin is to be set in a point with coordinates x_{1T}, x_{2T} and its axes are oriented in appropriate way.

Optimal trajectories of the "*auxiliary*" task are: “straight line – turn”, if $\rho \geq R$ and “turn – turn” otherwise $(if\ \rho < R)$.

To solve an "*auxiliary*" problem is easier, because the amount of control switches in it is less than in the “*original*”.

Let's note that in situations of practical interest (the trajectory of the "*original*" problem is “turn – straight line – turn”) then, as a rule, the optimal controls of the "*original*" problem coincide with controls created by means of the above discussed approach.

Solutions of "*original*" and "*auxiliary*" problems, considered in the example 2, are easy to obtain by using of the maximum principle of L S Pontryagin.

Example 3. The "*original*" optimal control problem: (8), (18)–(21), where [7]:

$$u = \frac{g}{V} \cdot \left(n_y - \cos(x_3)\right), \tag{29}$$

$$x_{10} = 0\,\text{km},\ x_{20} = 10\,\text{km},\ x_{30} = 0 \text{ degrees}, \tag{30}$$

$$x_{1T} = 37 \text{ km},\ x_{2T} = 0 \text{ km},\ x_{3T} = -80 \text{ degrees},\ V = 300 \text{ m/s}. \tag{31}$$

In above formulas: x_1 – horizontal distance; x_2 – height of flight; n_y – normal load factor; g – acceleration of gravity.

In [7] the decision of this problem with the use of the parametric method is depicted, where the desirable trajectory was set in the form of a polynomial

$$x_2(x_1) = A_0 + A_1 \cdot x_1 + A_2 \cdot x_1^2 + A_3 \cdot x_1^3\ , \tag{32}$$

where unknown coefficients A_0, A_1, A_2, A_3 are determined by boundary conditions

$$A_0 = x_{20}\,;\ A_1 = \text{tg}(x_{30})\,; \tag{33}$$

$$A_2 = \frac{3(x_{2T} - x_{20})}{(x_{1T} - x_{10})^2} - \frac{\text{tg}(x_{3T})}{x_{1T} - x_{10}} - \frac{2\,\text{tg}(x_{30})}{x_{1T} - x_{10}}; \tag{34}$$

$$A_3 = -\frac{2(x_{2T} - x_{20})}{(x_{1T} - x_{10})^3} + \frac{\text{tg}(x_{30})}{(x_{1T} - x_{10})^2} + \frac{\text{tg}(x_{3T})}{(x_{1T} - x_{10})^2} \tag{35}$$

While moving along the trajectory (32) the control u and the load factor n_y were determined by relations

$$u = \left(2 \cdot A_2 + 6 \cdot A_3 (x_1 - x_{10})\right) \cdot V \cdot \cos^3(x_3), \tag{36}$$

$$n_y = \frac{V}{g} u + \cos(x_3). \tag{37}$$

Estimated results are depicted in the table 1 below, where the following assumptions had been taken: the velocity V does not vary; at the desired trajectory (32) the constraint on the control u (and, consequently, on the normal load factor n_y) was not imposed; the control had been formed according the above considered approach (see example 2) and constraints on u_m were obtained by the formula

$$u_m = \frac{g}{V} a, \qquad a = \begin{cases} 2, \text{ when estimation of } R \text{ is used } (see(24)); \\ 3, \text{ when control is obtained depending on } \overline{E}\,. \end{cases} \tag{38}$$

Table 1. Results obtained in example 3.

Method	$T\,(s)$	n_{ymax}	W (s)
Inverse dynamics problems concept	242	3.4	152
Concept described in this article	133	1.4	119

According to table 1:

$$n_{ymax} = \max_{0 \le t \le T} n_y(t); \quad W = \int_0^T n_y^2\,dt \tag{39}$$

Note that, as a rule, constraints are imposed on n_y instead of u. In order to ensure demands to n_y according to considered methodic, we have to vary u_m in appropriate way (see the relation (29)) during x_{3R} estimation and the control error "elimination".

Example 4. The "*original*" optimal control problem: (8), (18)–(21), (29)–(31) and additional constraint imposed on x_3:

$$x_3 \le 0. \tag{40}$$

Let's take the path inclination x_3 as the adjustable parameter and the control error we should estimate by the formula (22).

In the "*auxiliary* problem we should not set constraints on $(x_3(0), u)$ and eliminate the condition (40).

While the "*auxiliary* problem (to find x_{3R}) approximate solving let's use the inverse dynamics problems concept (see example 3) and set the desirable trajectory in the polynomial form (32) of:

– second order (not third order, as it is shown in the example 3);
– third order, thus, let's demand, that the condition $n_y(T) = 0$ is to be satisfied.

As the result, we should get (in the synthesis form):

– for the second order polynomial

$$x_{3R} = \mathrm{arctg}\left(\frac{2(x_{2T} - x_2)}{x_{1T} - x_1} - \mathrm{tg}(x_{3T})\right); \tag{41}$$

– for the third order polynomial

$$x_{3R} = \mathrm{arctg}\left(-\frac{0.5 \cdot g \cdot (x_{1T} - x_1)}{V^2 \cdot \cos^2(x_{3T})} + \frac{3(x_{2T} - x_2)}{x_{1T} - x_1} - 2 \cdot \mathrm{tg}(x_{3T})\right). \tag{42}$$

It is easy to observe, that (41) appears from conditions $A_3 = 0$ и $n_y(T) = 0$ (see (32), (35)–(37)), but parameters $x_{10}, x_{20}, x_{30}, x_1$ there should be replaced with x_1, x_2, x_{3R}, x_{1T} respectively.

When using the control depending on $\bar{E}$, the constraint on u_m is determined by (38) and condition (40) is considered while the control error "elimination" by the correction of x_{3R} as follows: if $x_{3R} > 0$, then it is set $x_{3R} = 0$. Results are shown in the table 2 below.

Table 2. Results of control estimation obtained in example 4.

The order of the polynomial	$T\,(s)$	n_{ymax}	W (s)
2	154	2.3	195
3	150	1.8	152

Let's note, that the representation of a desirable trajectory by polynomials in the form (32) in the "*original*" problem is possible, if $|x_{30}| < \pi / 2$, whereas in the "*auxiliary*" problem this demand is removed, because of $|x_{3R}| < \pi / 2$.

The analyses of the direct methods applicability to design optimal short-term spatial maneuvers for an unmanned vehicle in a faster than real-time scale is performed in [8].

The task of the aerial vehicles adaptive control synthesis is considered in [9, 10]. Based on an advanced nonsingular fast terminal sliding mode control scheme and adaptive control, an adaptive nonsingular fast terminal sliding mode guidance law is there proposed in the presence of the target acceleration as an unknown bounded external disturbance.

4. Conclusion

The above considered examples demonstrate that:

– the proposed approach allows to simplify a solution of a control problem and to obtain effective control algorithms;
– at approximate solution of the optimal control problem with the mobile left endpoint one can use the widely applicable parametric representation of a desirable trajectory in the form of a given function of one of coordinates and a set of selected parameters even in cases when it cannot be made in the original optimum control problem with the fixed left endpoint (see examples 3 and 4 – the representation of a desirable trajectory by polynomials as (32) in the original problem is impossible at $|x_{30}| \geq \pi / 2$, but it is possible in the auxiliary problem because the value of x_{30} is selective).

In this paper we have shown that the proposed approach allows to simplify the control synthesis problem solution and obtain effective control algorithms, providing a suboptimal control and, in some cases, an optimal control.

The suggested algorithm can be used to determine continuous and pulse controls of aircrafts [10, 11] (including spacecrafts) in problems of their approaching (guidance) and spacecraft docking, carrying out of correcting maneuvers, descent and soft landing of spacecrafts at al. [12].

References

[1] Enkin A A 2017 Methodology for aircraft rational control law forming while performing maneuvers in horizontal plane. *Electronic journal "Trudy MAI"* **97**
[2] Lam T T Solving boundary value problem of trajectory planning for aircraft executing spatial maneuvers. 2014 *Electronic journal "Trudy MAI"* **78**
[3] Krasil'shchikov M N and Sypalo K I 2011 Self-guiding of a high velocity pilotless aircraft on the terminal flight leg in atmosphere. *Journal of Computer and Systems Sciences International.* 50(6) 983 https://doi.org/10.1134/S106423071106013X
[4] Shopin S A and Lovchakov V I 2017 Design of quasi-optimal regulators for quadratically-cubic control objects using generalized work criterion. *21st International Conference on System Theory, Control and Computing, ICSTCC.* vol 21 (Sinaia, Romania) pp 809-815 DOI: 10.1109/ICSTCC.2017.8107136
[5] Velischanskii M A 2013 Synthesis of a quasi-optimal trajectory of an unmanned aerial vehicle. *Science and education* **12** DOI: 10.7463/1213.0646471
[6] Kanatnikov A N 2013 Design of aircraft trajectories with non-monotonic change in energy. *Science and education* **4** DOI: 10.7463/0413.0554666
[7] Krasilshikov M N, Sebryakov G G 2005 *Control and guidance of unmanned maneuvering flying machines on the basis of modern information technologies* (Moscow: Phizmatlit) p 280
[8] Basset G, Xu Y and Yakimenko O A 2010 Computing short-time aircraft maneuvers using direct methods. *Journal of Computer and Systems Sciences International.* **49** 481 DOI:10.1134/S1064230710030159
[9] Song J, Song S and Zhou H 2016 Adaptive nonsingular fast terminal sliding mode guidance law

with impact angle constraints. *International Journal of Control, Automation and Systems* **14** 99 DOI: 10.1007/s12555-014-0155-8

[10] Prabhakar N, Painter A, Prazenica R, Balas M 2018 Trajectory-Driven Adaptive Control of Autonomous Unmanned Aerial Vehicles with Disturbance Accommodation. *Journal of Guidance, Control and Dynamics*. **41(9)** 1976 DOI: 10.2514/1.G003341

[11] Hodgson J A and Lee D W 2003 Terminal Guidence using Doppler Beam Sharpening Radar. *AIAA Guidance, Navigation, and Control Conference and Exhibit* (Austin, Texas) DOI: 10.2514/6.2003-5447

[12] Lu K F, Xia Y Q, Yu C M and Liu H L 2016 Finite-Time Tracking Control of Rigid Spacecraft Under Actuator Saturations and Faults. *IEEE Transactions on Automation Science and Engineering* **13(1)** 368 DOI: 10.1109/TASE.2014.2379615

IOP Conf. Series: Materials Science and Engineering **868** (2020) 012033 doi:10.1088/1757-899X/868/1/012033

Numerical calculation of gas exchange in two-stroke engine for aircraft

I Borisenko[1*], Yu Grishin[1] and V Bakulin[2,3]

[1]Department of Piston Engines, *Bauman State Technical University*, 5, 2 Baumanskaya Street, Moscow 107005, Russia
[2]Department of Aircraft Design and Sertification, *Moscow Aviation Institute* (National Research University), 4 Volokolamskoye Avenue, Moscow 125993, Russia
[3]Laboratory of Non-Classical Models of Mechanics of Composite Materials and Structures, *Institute of Applied Mechanics*, *Russian Academy of Sciences*, 7 Leningradskaya Avenue, Moscow 125040, Russia

[*]E-mail: iw-borisenko@yandex.ru

Abstract. General aviation light aircrafts and unmanned aerial vehicles are equipped mainly with internal combustion piston four-stroke and two-stroke engines. Despite of the significant amount of research and development works, the ways of the further performance improving for such engines are nearly exhausted. However, there are fundamentally new developments, which combine the advantages of four-stroke valvetrains with the two-stroke cycle, which provides significant increasing of the engine performance. This article presents the results of computational studies of a *Rotax* 914 four-stroke aircraft engine, modified to the two-stroke cycle. The calculations were carried out with the software named *Diesel-RK*, which made it possible to select the optimal engine settings, mostly timing, and software *AVL FIRE* for 3D numerical simulation of cylinder scavenging through valves to confirm the residual gas amount in cylinder with such type of gas exchange. The calculations confirmed the possibility of increasing performance of the engine almost in 1.5 - 2 times for the entire operating speed speed, which is confirmed by experimental results published in the known open articles.

1. Introduction

On the general aviation light aircrafts and unmanned aerial vehicles usually are installed internal combustion piston engines. They are distinguished by design simplicity, reliability, lower fuel consumption and lower cost in comparison to the gas turbine powertrains. Piston engines usually have two-stroke and four-stroke cycles. Many scientific papers have been devoted to improving designs and improving the characteristics of both two-stroke [1] and four-stroke [2,3] engines used in aviation. Two-stroke engines of the simplest design with a crank-case scavenging and a loop scavenging of the cylinder through intake ports and exhaust ports have a power up to 80 kW. They are characterized by the best ratio of power and weight, and are used on ultralight class aircrafts. Four-stroke engines have a more efficient valve gas exchange system; to improve technical and economic parameters, they can be equipped with turbochargers [4-6], inertial superchargers [7-9], have a power up to 400 kW and are widely used on the light-engines for larger dimensioned aircrafts. The effective power N_e of piston engines is expressed by the formula:

$$N_e = \frac{i \cdot V_h \cdot P_e \cdot n}{k \cdot \tau}$$

IOP Conf. Series: Materials Science and Engineering **868** (2020) 012033 doi:10.1088/1757-899X/868/1/012033

where i – the number of cylinders, V_h – engine displacement, P_e – the effective pressure of the operation cycle, n – the crankshaft speed, τ – number of strokes in cycle (2 for two-stroke, 4 for four-stroke), k – the proportionality coefficient (depends on the units of measurement).

This formula shows power advantage up to two times for two-stroke engines compared to four-stroke ones with the same values of V_h and n. However, in reality this advantage is in general not so big. This happens due to the insufficient scavenging of the cylinders with the loop scavenging through the ports, symmetric timing of gas exchange and, as a consequence, low effective pressure P_e. In addition, two-stroke engines are characterized by high toxicity of the exhaust and, most importantly, increased fuel consumption due to the high amount of unburned fuel – up to 25-40% [10, 11].

However, the fundamental possibilities of increasing of power performance and other characteristics by account of the two-stroke cycle, force the engines' developer to find a new ways of the further improvement. The most important is the combination of the usage of the general valvetrain mechanisms with a two-stroke cycle. Some results of such developments have already been patented in USA, Australia, Japan, and France [12-14]. For example - prototype engines *Orbital ES*, *GM CDS2*, *IAPAC* with a crank-case cylinder scavenging. Here the fresh charge is pathed not by the inlet ports, but through special valves in the cylinder head, fuel (gasoline) is fed to manifold before the valves with the special nozzles. These nozzles are using the pneumatic spray made by compressed in the crankcase gas. In other engines, for example, *Subaru*, *Toyota S 23L*, a direct injection system of gasoline into the cylinder is implemented, the engines are mechanically supercharged. It is important to note that *Toyota* engine has neither inlet nor outlet ports, whole gas exchange is carried out with the typical valves like in four-valve cylinder heads.

Tests results showed that for such engines the main disadvantages of traditional two-stroke engines are completely eliminated: high fuel consumption, lean combustion (especially at low loads). It is noted [12-14], that for such engines the fuel consumption is reduced two times compared to traditional two-stroke engines, in the load ranges up to 50%, fuel consumption is even 20% lower than in high-performance four-stroke engines.

The most important factors for aviation application is the weight to power ratio. For such engines it is 30 to 40% less than for similar in power range four-stroke engines. The volumetric power for such engines almost two times higher than the typical values for general four stroke engines and reaches power values of 80 - 90 kW per litter.

It is also important to note the possibility of modifying already set in serial production and well accepted by market designs of four-stroke engines to a two-stroke cycle after the appropriate refinement of the control systems and valvetrain.

However, a number of problems still require further research. Among them are the optimization of the timing, fuel injection and ignition, the adjustment of the boost systems, and most important, it is necessary to verify the feasibility of the residual gas scavenging from cylinder through the valves in a short period of time, typical for the two-stroke engine. Without these researches, a two-stroke engine with a valve gas distribution system is simply unable for operation.

The purpose of this work was the numerical calculation study, feasibility study of modification for the real four-stroke engine to the two-stroke cycle with an assessment of the main technical and economic characteristics.

2. Materials and research methods

The four-stroke four-cylinder *Rotax* 914 [15] horizontally opposed (boxer) aviation spark ignited engine was chosen as an object for study. Fuel - *RON* 98 gasoline, compression ratio $\varepsilon = 9{:}1$, $V_h = 1211.2$ cc, bore 79.5 cm, stroke 61 cm, boost pressure - up to 2 bar. The numerical calculation studies were carried out in two stages.

2.1. Work process in Diesel-RK

At the first stage, using the thermodynamic simulation software *Diesel-RK* [16], the numerical calculation and optimization of the working process for *Rotax* 914 engine were executed. This engine was modified to two-stroke cycle, with assumption that existing two intake and two exhaust valves provide enough scavenging and admission. After simulation work with this software it was possible to compare the engine parameters both for the basic four-strike cycle and for the two-stroke cycle after the engine parameters modification and optimization for such two- stroke cycle.

2.2. Gas exchange simulation in AVL FIRE
At the second stage, in order to confirm the possibility of the required scavenging of the cylinder and admission by the charge through two exhaust and two intake valves, it was necessary to use 3D numerical simulation of the flow in the following simulation system - “inlet valve port - cylinder - exhaust valve port”.

There are many published papers with description of different gas dynamics numerical methods, that could be potentially applied for such task [17, 18]. However, it appears that the most suitable software here is *AVL FIRE* ™ [19]. This software was developed by the well-known company *AVL List* specifically for such numerical 3D simulation of gas exchange and in-cylinder processes for piston engines. This software allows to simulate 3D fields for gas velocity and other parameters, as well as the final parameters of the engine cylinder scavenging and admission in the real time conditions for intake and exhaust valves with the optimized for combustion of the piston engines turbulence simulation model.

3. Results and discussion
Figure 1 and figure 2 represent plots of calculated characteristics comparison for the *Rotax* 914 engine for ground test conditions with the basic four-stroke and two-stroke modes. Figure 1 shows a significant advantage in N_e of the two-stroke operating cycle to the four-stroke operating cycle on the entire speed range. Particularly, at speed $n = 3000$ rpm, an increase N_e by 88% was obtained, for $n = 6000$ by 71%.

The dependence of the torque M_e from the engine's speed in the two-stroke cycle also significantly exceeds the torque values obtained for the engine with the original four-stroke cycle. The increase of M_e at the maximum speed points is 89%.

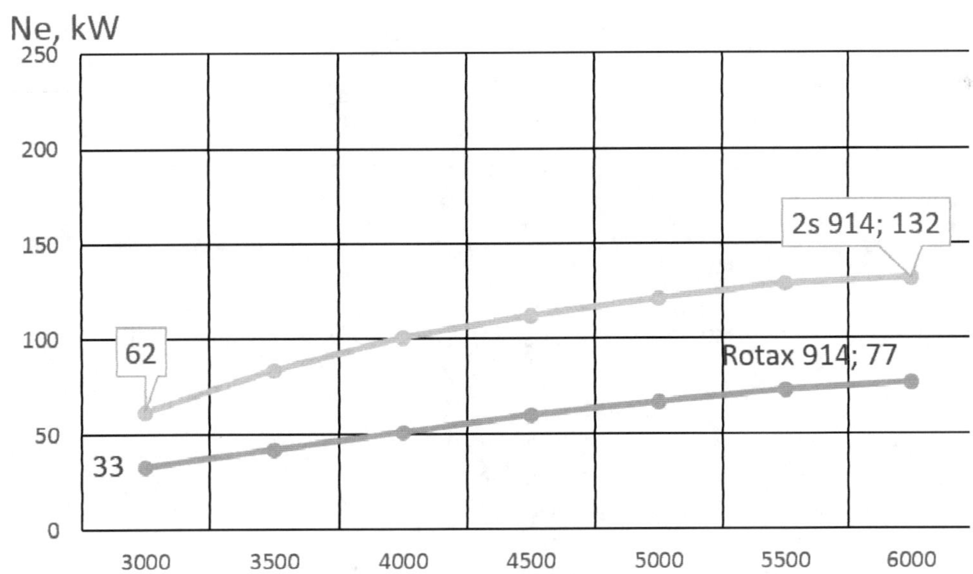

Figure 1. 2-stroke engine`s power in comparison with common engines (kW at speed).

IOP Conf. Series: Materials Science and Engineering **868** (2020) 012033 doi:10.1088/1757-899X/868/1/012033

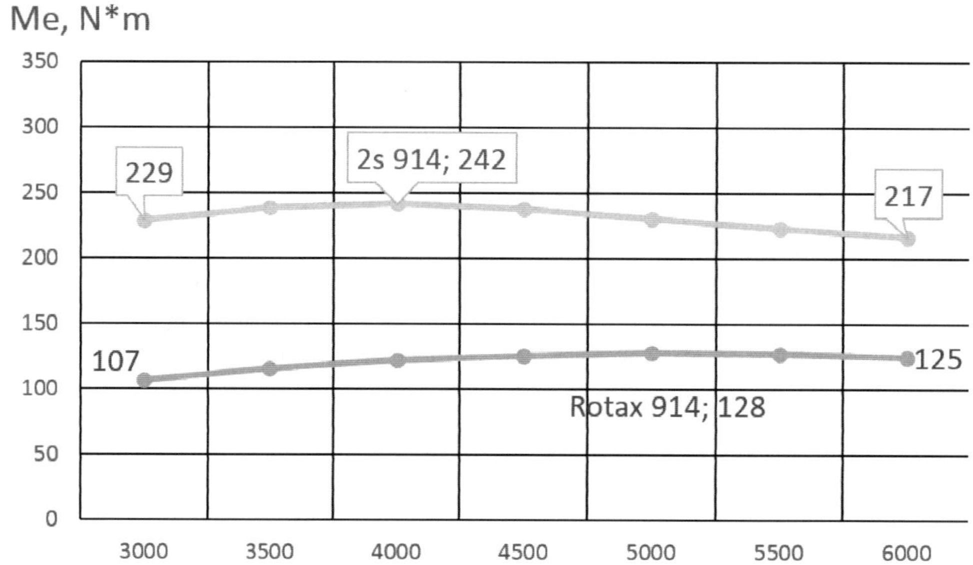

Figure 2. 2-stroke engine`s torque in comparison with common engines (Nm at speed).

In order to perform the numerical 3D calculations of the engine's scavenging-admission in two-stroke cycle, at first was created a 3D numerical mesh (regular grid) consisting of cells with edges not exceeding 1 mm size. This mesh is necessary for finite-different gas dynamic calculations, it fits into the complex geometry of the computational domain of the system "inlet valve ports - cylinder - outlet valve ports". The mesh is adaptive and provides calculations in the conditions of movable valves and piston.

The resulting mesh is presented on figure 3.

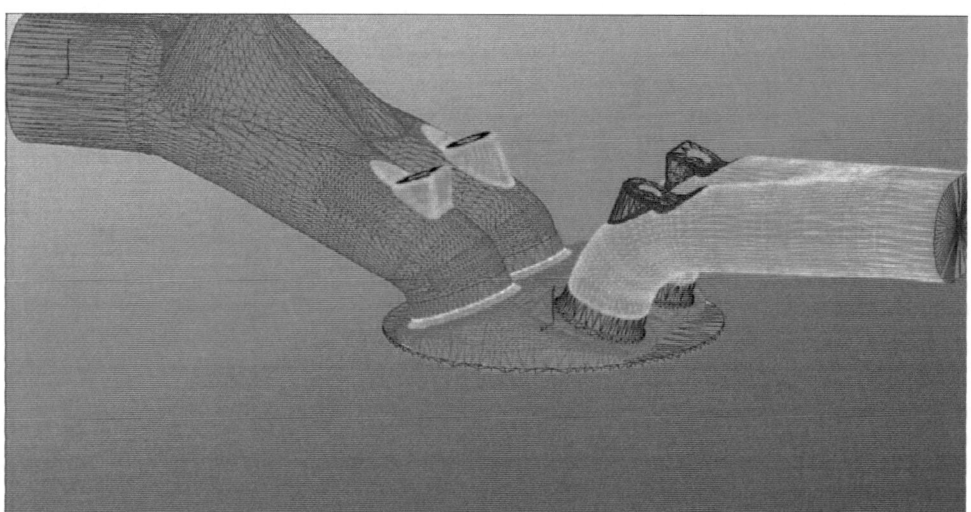

Figure 3. 2-stroke engine's mesh boundary conditions.

Numerical calculations were performed with the initial conditions and boundary conditions, previously obtained at the first calculation stage in the *Diesel-RK* software. Among them were the changing temperature distribution over the hot surfaces inside the cylinder; variable pressure and mass air flow at the inlet and outlet, consistent with the characteristics from the compressor map and turbine map; variable valve timing, timing of ignition and fuel injection. The whole process was carried out in stages, based on the set of the main influencing factor - the valve timing parameters combination. The change in the corresponding angles was carried out in increments of 1 degree separately for each step of changing valve timing (for opening and closing, intake and exhaust - all separate). It was obtained a table of results, and the best combination of parameters was chosen after this. The crankshaft rotation angle at the time of opening the exhaust valve is 66° before bottom dead center (BDC), the inlet opening angle is 44° before BDC, the closing angles coincide and are 25° after BDC. The ignition timing and fuel injection angles vary widely and are selected using the optimization module integrated in the *Diesel-RK* software. In real engine, this task is performed by electronic control system.

According to numerical calculation, as a result of an integrated assessment of the scavenging and admission of the cylinders for each of the modes, the value of gas residual coefficient γ was obtained. It varies depending on the crankshaft speed in the range from 11% to 18%. These values are higher than in an engine operating on a four-stroke cycle, but are sufficient for the stable operation of a two-stroke engine. It should be noted here that an increased amount of exhausted gases provides the effect of internal recirculating (the EGR system's effect without system itself), which significantly improves engine emissions.

The calculation results are also presented on figure 4 and figure 5 in a form of the streamlines at the time of the most intensified scavenging. It can be seen that there are no zones of local stagnation in the entire volume of the cylinder, air displaces exhaust gases from all zones, and reliable scavenging is provided.

Figure 4. 2-stroke engine's gas exchange (small streamlines).

4. Conclusion

This paper represents the results of a numerical calculation study of the possibility of modifying the four-stroke piston aircraft engine *Rotax* 914 to a two-stroke operation cycle. It is shown that, with a relevant setting of the valvetrain and setting of the well-optimised phases of the intake and exhaust valves, ignition and fuel injection timing, it is possible to receive a significant increase of the engine's power and torque.

The calculations showed that with a dry engine's weight of 64 kg at a nominal mode matching to a speed n = 6000 rpm, weight to power ratio of the engine decreases from 0.83 to 0.48 kg / kW, i.e. by 42%, the volume power N_l increases from 63.6 to 109.0 kW per litter, by 71%. These results are fully coincide to the data given in [12-14], which presents the results of practical work with similar piston engines.

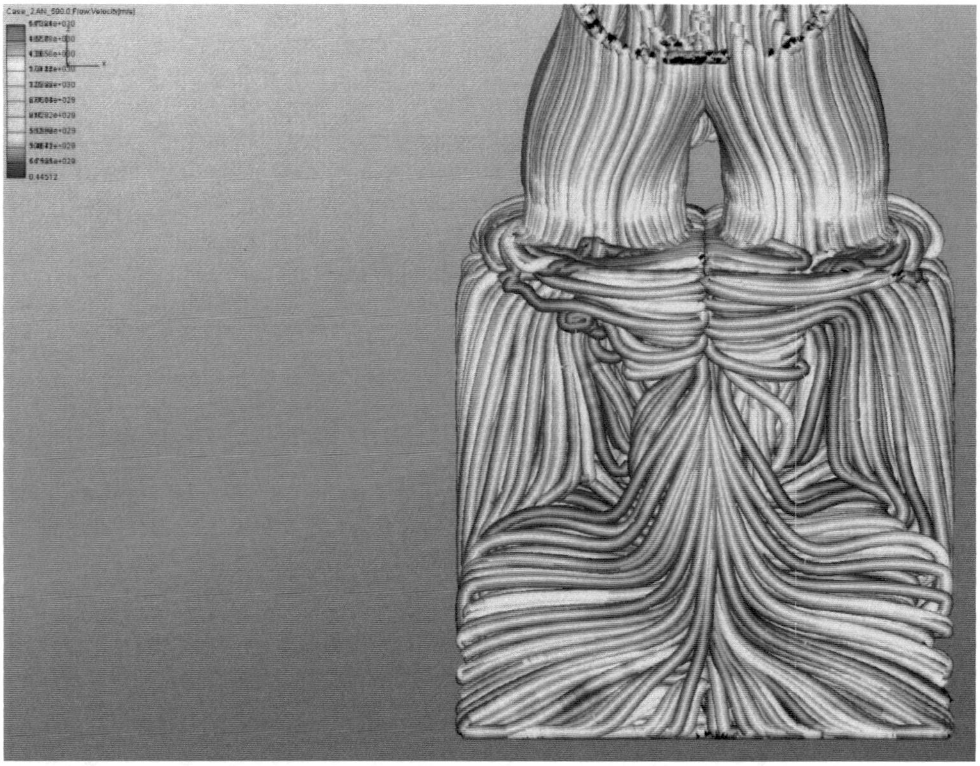

Figure 5. 2-stroke engine`s gas exchange (large streamlines).

It should also be noted that the part of exhausted gases in the cylinder by the time of ignition is 11-18%, which ensures normal mixing of the fuel and air as well as the combustion for such two-stroke engine, it creates the effect of internal exhaust gas recirculation (the EGR system's effect without system itself). The experiments showed [12-14] that this effect, in addition to the fact that the valve gas exchange in a two-stroke engine prevents the emission of 20–40% of unburned hydrocarbons and reduces nitrogen oxides, the exhaust toxicity of such engines is reduced to 80%.

It is obvious that the continuation of numerical calculations and practical researches on the modify of four-stroke engines to a two-stroke cycle with the usual valvetrain system seems to be very actual.

References

[1] Szlachetka M, Paszko M and Baranski G 2018 Modeling Loads Applied to Main and Crank Bearings in the Compression Ignition Two-Stroke Engine. *International Journal of Aerospace and Mechanical Engineering* **12(4)**

[2] Czarnigowski J, Skiba K and Dubieński K 2019 Investigations of the temperature distribution in the exhaust system of an aircraft piston engine. *Combustion Engines.* **177(2)** 12 DOI: 10.19206/CE-2019-203

[3] Czarnigowski J and Jakliński P 2019 The influence of the ignition control on the performance of an aircraft radial piston engine. *Combustion Engines.* **177(2)** 60 DOI: 10.19206/CE-2019-211

[4] Grishin Yu A 2013 Pressure-charging device. *Internal Combustion engines, Mechanical Engineering, Encyclopedia* (Moscow: Machine Building) vol IV-14 pp 362-382 [in Russian]

[5] Zinner K 1975 *Aufladung von Verbrennungs-Motoren* (Berlin Heidelberg New York: Springer Verlag) p 264

[6] Weaving J H 1990 *Internal Combustion Engineering: Science & Technology* (London and New York: Elsevier Science Publishers LTD) p 873

[7] Grishin Yu A and Sonkin V I 2013 Intake Systems. *Internal Combustion engines, Mechanical Engineering, Encyclopedia* ed. A A Aleksandrov and N A Ivashchenko (Moscow: Machine Building) vol IV-14 pp 461-474 [in Russian]

[8] Stehlig J, Taylor J, Dingelstadt R and Gurney D 2012 Variable-length air intake module for turbocharged engines *MTZ Worldw* **73** 34 https://doi.org/10.1365/s38313-012-0130-9

[9] Weber O, Wild S and Kiener T 2002 Highly-variable intake systems. *MTZ Worldw* **63** 7 https://doi.org/10.1007/BF03227532

[10] Mattarelli E, Cantore G and Rinaldini C 2013 Advances in The Design of Two-Stroke, High Speed, Compression Ignition Engines (London, IntechOpen Limited) DOI: 10.5772/54204

[11] Gossler H and Deutschmann O 2015 Numerical optimization and reaction flow analysis of syngas production via partial oxidation of natural gas in internal combustion engines. *International Journal of Hydrogen Energy* **40(34)** 11046 DOI: 10.1016/j.ijhydene.2015.06.125

[12] Kenny R G 1992 Developments in Two-Stroke Cycle Engine Exhaust Emissions. *Proceedings of the Institution of Mechanical Engineers, Part D: Journal of Automobile Engineering* **206(2)** 93 https://doi.org/10.1243/PIME_PROC_1992_206_165_02

[13] A New Generation of Two-Stroke Engines 1994 *IFP Advanced Technologies* (Paris: Institut Francais du Petrole) p 6

[14] Grishin Yu A 2000 *Analysis and development prospects of piston aircraft engines* (Moscow: Central Aerohydrodynamic Institute) p 52 [in Russian]

[15] Rotax website, avaliable at: https://www.rotax.com/en/products/rotax-powertrains/details/rotax-914-f-ul.html

[16] Kuleshov A 2006 Use of Multi-Zone DI Diesel Spray Combustion Model for Simulation and optimization of Performance and Emissions of Engines with Multiple Injection. *SAE Paper.* No 2006–01–1385 pp 1–17

[17] Fletcher C A J 1989 *Computational Techniques for Fluid Dynamics 2. Specific Techniques for Different Flow Categories* (Berlin–Heidelberg- New York-London-Paris-Tokio: Springer Verlag) p 552

[18] Toro E F 1997 *Riemann Solvers and Numerical Methods for Fluid Dynamics. A Practical Introduction* (Berlin: Springer) p 724

[19] FIRE. Users Manusal Version 2014 / AVL List GmbH. Graz (Austria), 2009. (License Agreement for Use of the Simulation Software AVL FIRE between Bauman Moscow State Technical University and AVL List GmbH, 2015)

IOP Conf. Series: Materials Science and Engineering **868** (2020) 012034 doi:10.1088/1757-899X/868/1/012034

Research of the influence of rectangular cutout dimensions on the stress-strain state of shells using an effective finite element model

V N Bakulin[1,2]

[1]Department of Aircraft Design and Certification, *Moscow Aviation Institute* (National Research University), 4 Volokolamskoe shosse, Moscow 4125993, Russia
[2]Laboratory of Non-Classical Models of Mechanics of Composite Materials and Structures, *Institute of Applied Mechanics, Russian Academy of Sciences*, 7 Leningradsky Avenue, Moscow 125040, Russia

E-mail: vbak@yandex.ru

Abstract. The effect of the dimensions of a rectangular cut on the stress-strain state of a cylindrical shell using an effective finite element model is investigated. A cylindrical finite element rectangular in the plan is used for thin moment shells built on the Kirchhoff – Love hypotheses.The considered finite element is based on functions of the form received at approximation of the generalized deformations with the subsequent satisfaction to the equations of compatibility of deformations. The functions of the form accurately consider displacements as rigid bodies. A comparison of convergence of results obtained with the help of the considered finite element and other known finite elements of foreign authors is given.The example of a cylindrical shell with notches shows a good correspondence of the solution radiated by the considered finite element with the method of grids, as well as the efficiency of the applied finite element having 20 degrees of freedom as compared to the common finite element having 24 degrees of freedom. The influence of the cutout dimensions on the maximum values of deflections and longitudinal forces for a cylindrical shell is shown.

1. Introduction

When creating aviation and space technology, it is often necessary to break the continuity of structural elements, including shells, of various kinds of cutouts (having structural, technological or other purposes), which lead to a weakening of the carrying capacity of structures. Insufficient knowledge of the true picture of the stress-strain state can lead to destruction or an increase in the weight of the structure.The study of the stress-strain state of shells weakened by cuts is a current scientific and technical problem, despite many works devoted to solving this problem. This explains the continuous improvement of methods for calculating stress-strain state (SSS) of structural elements, especially shells with cutouts.

A number of reviews are related to the analysis and development of methods for the calculation of shells with notches and shells with variable parameters, for example [1, 2]. Thus, paper [1] gives an overview of publications on analytical, numerical and analytical computational approaches, as well as experimental methods for the study of homogeneous and heterogeneous, including orthotropic shells with cutouts of various shapes and sizes.

The review [2] deals with the studies of static and dynamic deformation of isotropic and anisotropic elastic shells with variable parameters using classical and refined theories. To solve two-dimensional boundary value problems, an unconventional discrete-continuum approach is used which is based on a spline approximation of unknown equation functions in partial derivatives with variable coefficients. It allows to reduce an initial problem to system of the one-dimensional problems solved by a method of discrete orthogonalization. The analysis of numerical results was carried out. Emphasis is placed on the accuracy of the results.

The tasks of shell statics with holes were considered in [3, 4]. A numerical and experimental study of a cylindrical shell with a neckline is presented in [5]. The influence of circular holes, their size and location on mechanical behavior and strength of cylindrical shells was studied in [4]. The influence of circular holes on mechanical behavior of a three-layer carbon fiber cylinder with a lattice filler was considered in [6].

The analysis of the given and many other works shows that in the overwhelming majority of them the strained - deformed condition of shells with curvilinear holes and mainly circular holes was studied.

The most common in practice are rectangular cutouts in aspect of shape. If for the shells weakened by curvilinear cutouts it is still possible to try to apply analytical methods to calculate the stress-strain state, then for the shells with rectangular cutouts the use of analytical methods faces great, and often impossible to overcome, mathematical difficulties [1]. Therefore, to solve this scientific and technical problem, experimental methods are widely used to study the stress-strain state of shells with cutouts, for example [5].But the most common and appropriate are numerical calculation methodsand especially the finite element method.

The finite element under consideration in this paper is based on form functions obtained by the approximation of generalized deformations with subsequent satisfaction of deformation compatibility equations. The shape functions take exact account of movements as rigid bodies. Efficiency of the presented approach (high speed of convergence of numerical results) that allows to reduce necessary for calculation number of FE and consequently to reduce an order of systems of equilibrium, i.e. to lower dimension of solved problems in comparison with traditionally used approximations is shown. This is especially relevant in the problems of calculation of shells with notches, because of the observed concentration of stresses near the notches it is necessary to thicken the grid of breaks in FE, i.e. increase the number of FE, reducing their size.This leads to cumbersome finite element models (FEM), which significantly complicates and limits the use of such models, have to work with large amounts of information, which greatly complicates the processing of the numerical results obtained.At the large dimension of the tasks the calculation errors, the necessary computer resources and the time of calculation increase.This makes it difficult to research such structures with proper accuracy.In this connection, the problem of decreasing the dimensions of the FEM for calculating envelope elements of structures with cut-outs is important.

The purpose of the research in this article is:

- consideration and description of the approach that allows reducing the dimensionality of the tasks to be solved in comparison with the traditionally used approximations;
- the study of the convergence of the results obtained with the finite element in question by comparison with other known data;
- study of the effect of cutout sizes on the largest values of deflections and longitudinal forces for a cylindrical shell.

2. Description of the final element in question

In this paper, we study the effect of the dimensions of a rectangular cutout in aspect of shape on the stress-strain state of the cylindrical shells of rotation by means of an effective finite element model. A two-dimensional rectangular cylindrical finite element (FE) is considered for calculating thin moment shells whose theory is based on the hypotheses of Kirchhoff-Lava.

The form functions of the considered FE, as well as in [7], are based on approximation of generalized deformations $(\varepsilon_1, \varepsilon_2, \gamma, æ_1, æ_2, \chi)$ with subsequent satisfaction of the deformation compatibility equations and include movements as rigid bodies. In contrast to work [7], in which the three deformation compatibility equations are not justified because of mathematical difficulties, the three deformation compatibility equations are replaced by four equations in obtaining functions that approximate generalized deformations.

Such approach, when functions of the finite element form are based on approximation of generalized deformations with the subsequent satisfaction of the equations of deformation compatibility and include displacements as rigid bodies, allows to reduce the number of finite elements necessary for calculation, that is to increase the rate of convergence of numerical results.Such finite elements with a high rate of convergence of numerical results will be called effective FEs.

Efficiency of the considered finite element of shells of rotation of the cylindrical form is explained by efficiency of finite element approximations of functions of the generalized movings (further we will name them as approximating functions of movings).

At first, the approximating functions of the displacements are obtained, which are responsible for the displacements of the considered finite element as a rigid body. These approximating functions of movements as a rigid body are defined by integration of the relations connecting components of a vector of deformations with movements, at zero values of the generalized deformations.

Then we obtain the approximating functions of displacements responsible for deformation of the considered finite element. These finite element approximation functions of displacements responsible for the deformation of cylindrical shells are based on the initial approximation of generalized deformations of these shells. At the same time, functions which approximate deformations and satisfy the equations of deformation compatibility are constructed.

By integrating relations connecting components of a strain vector with displacements, at the received expressions of approximating functions of the generalized deformations, functions, approximating the displacements responsible for deformation of a considered final element are defined.

Adding to approximating functions of displacements as a rigid body the received expressions of approximating functions of the displacements responsible for deformation of the considered final element, we define finite element approximations of functions of displacements.

The forces and moments are calculated through the components of the deformation vector using physical relations (Hook's Law).Further solution of the problem is carried out using FEM procedures in the variant of the displacement method.

Reliability, convergence, accuracy of the results obtained with the help of the considered model and efficiency of the constructed finite element of the shell of cylindrical form rotation were investigated on a number of well-known and widely spread for shell finite element tests by comparison with analytical solutions and numerical results obtained with the help of known finite elements of foreign authors.One such test was an example of the solution of S.P. Timoshenko's problem [8] for a free circular cylindrical shell loaded with two symmetrical diametrically opposed concentrated radial forces.

3. Results and discussion

Reliability and accuracy of the results obtained with the help of the considered model, and the efficiency of the constructed finite element for the calculation of the tensely deformed state of the shells of rotation of cylindrical shape with rectangular in plan notches was checked by the following example.

For a circular cylindrical shell with four symmetrically located on a circle rectangular in plan cutouts at action of the axial compressive force uniformly distributed on faces of a shell, good conformity of the decision is shown, obtained with the help of the considered finite element of cylindrical shells (diagrams marked by dots), with the method of grids (diagrams marked by triangles), as well as the efficiency of the considered FE, which has 20 degrees of freedom, in comparison with the widely known finite element of circular cylindrical shell G Cantin, R W Glagh [9], having 24 degrees of freedom (graphs shown by lines without points) (figure 1).

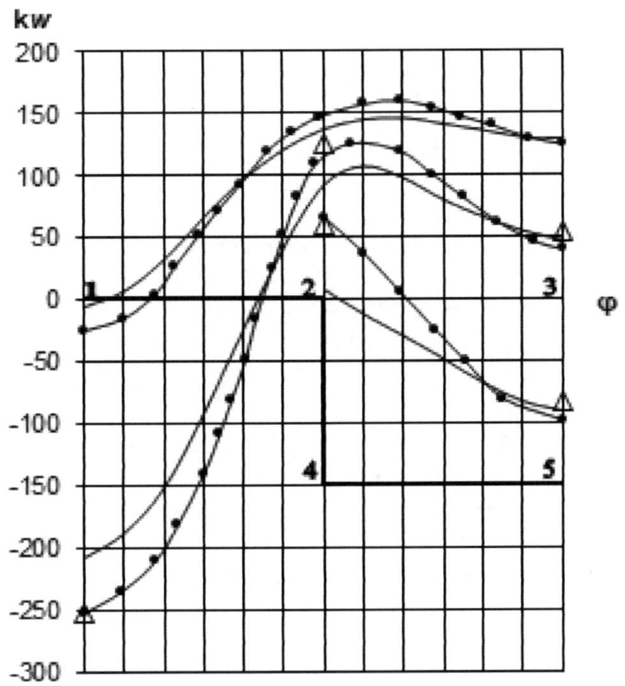

Figure 1. Comparison of numerical results obtained using the considered finite element of the circular cylindrical shell and the widely spread finite element G Cantin, R W Glagh [9], and also comparison with a method of grids on an example of calculation of the stress-strain state of a shell of rotation of the cylindrical form with four symmetrically located on a circle rectangular in the plan cutouts.

Parameters of the investigated shell of rotation of the cylindrical form with rectangular in shape cutouts and notches are as follows:

$$L/R = 2.24; \; h/R = 0.225 \cdot 10^{-2}; \; \ell_1/L = 3/7; \; \ell_2 = \pi/4 \cdot R;$$

where L – shell length, R – shell radius, h – shell thickness, ℓ_1 and ℓ_2 – the cut dimensions in the axial and circumferential directions respectively.

Due to symmetry, 1/16 of the shell was considered, which was divided into finite elements as follows: in the circumferential direction into 12 elements, in the axial direction into 7 elements. The total number of unknowns is n = 366, the half-width of the tape of the system of equations is m = 46.

Figure 1 shows the graphical dependencies of the coefficient $kw = w - Eh / pR$ for the points of the considered shell rotation of cylindrical form with rectangular in plan notches for three cross sections perpendicular to the axis of the shell. Points 1-3 of the investigated circular cylindrical shell with rectangular cutouts are located in the cross section perpendicular to the axis of the shell and passing on the curvilinear edge of the cutout, and points 4, 5 are located in the cross section of the shell perpendicular to the axis of the shell and passing through the middle of the rectangular edge of the cutout. The upper two graphs refer to points located in the cross section of the shell, perpendicular to the axis of the shell and passing at an equal distance from the edge of the shell and curved edge of the cutout.

The comparison of numerical results obtained by the mesh method, as well as by means of the considered finite element of the circular cylindrical shell and the widely spread finite element G Cantin, R W Glagh was carried out [9], has shown reliability and accuracy of the model used for construction of a finite element, and also efficiency (higher convergence speed) of applied FE that has allowed to reduce number of finite elements necessary for calculation in comparison with widespread finite elements on an example of calculation of tensely - deformed state of a shell of rotation of the cylindrical form with four rectangular in the plan of cutouts.

At present, the most accurate models for calculating inhomogeneous shells is an models based on layer-by-layer analysis of the stress-strain state. Due to its effectiveness, the considered approach to obtaining the functions of the FE form will be relevant for the construction of models of layer-by-layer analysis of the stress-strain state of non-homogeneous cylindrical irregular shells, especially weakened by cut-outs.

The problem of investigating the influence of rectangular cut sizes on the tensely deformed state was solved for a hinged cylindrical shell rotation with a rectangular cutout under the action of an axial compression load uniformly distributed at the ends of the shell.

The parameters of the investigated shell are as follows: L/R = 2.24; h/R = 0.0025; $E=19.6\cdot10^4$MPa; $\mu = 0.3$, where L is shell length, R is shell radius, h is shell thickness, Young modulus and Poisson's coefficient respectively.

As an example, we considered a cut with an axial dimension l_c of 20% of the shell length, and in the circumferential direction, the angle of the cutout was varied within $36^\circ \leq \theta \leq 96^\circ$ with a step of 12°.

Due to the symmetry, 1/4 of the shell was considered, which was divided into the same finite elements as follows: in the axial direction of 10 elements, in the circumferential direction - into 30 elements.The kinematic boundary conditions at the ends of the shell correspond to an articulated support and look:

$$v=0 \; w=0 \; \vartheta_x=0,$$

where v and w are moves in radial and tangential directions; ϑ_x – rotation angle relative to the axis of the shell.

The influence of the cutout size on the maximum values of deflections and longitudinal forces is shown in figure 2. It shows the graphical dependences $k_w = w \cdot Eh / pR$ at points located in the middle of the curved (maximum values are observed here) and the rectilinear edges of the cutout (graphs 1 and 2, respectively) , as well as $k_N=N_1/p$ (where p-axial compressive forces are p-uniformly distributed over the shell ends) in the vicinity of the cut-out corner point (graph 3) from the cut-out solution angle θ.

The value of deflections and longitudinal forces increases significantly as the size of the cut increases, and the dependencies of k_w and k_N on the angle θ are nonlinear. The effect of changing the size of the cutout on k_w and k_N in the circumferential direction is stronger than in the axial direction.

The received graphic dependences can be used at designing of designs in the form of a shell of rotation of the cylindrical form with rectangular in the plan cut-outs.

4. Conclusion

The approach and the algorithm of construction of effective finite element model for calculation of the stress-strain state of a shell of rotation of the cylindrical form with rectangular cutouts in the plan is considered. The reliability and accuracy of the results obtained with the help of the considered model is shown, as well as the efficiency of the constructed finite element of the circular cylindrical shell with four symmetrically arranged rectangular notches on the circumference under the action of the axial compression force uniformly distributed at the ends of the shell, by comparison with the method of grids and with the widespread finite element of the circular cylindrical shell G Cantin, R W Glagh. The influence of rectangular cut solution angle on the stress-strain state of circular cylindrical shell is studied. Graphic dependencies of the cutout size influence on the largest values of deflections and longitudinal forces for the rotation shell of cylindrical shape with rectangular cutout are given.

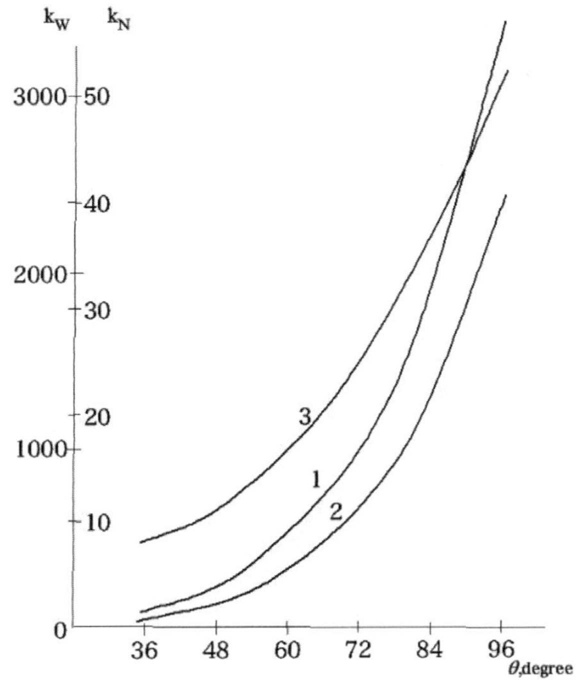

Figure 2. The influence of the angle of the cutout solution on the maximum values of deflections observed in the area of the middle of the curvilinear and straight edge of the cutout (Graphs 1 and 2, respectively) and longitudinal forces in the vicinity of the corner point of the cutout (Graph 3).

Acknowledgments
I would like to thank V V Repinskii for help in carrying out the calculations. The work was carried out within the framework of state task No. AAAA-A19-119012290177-0.

References

[1] Bakulin V N 2019 Layer-by-Layer Analysis of the Stress-Strain State of Three-Layer Shells with Cutouts *Mech. Sol.* **54(3)** 448 doi: 10.3103/S 0025654419050054
[2] Grigorenko Ya M, Grigorenko A Ya 2013 Static and Dynamic Problems for Anisotropic Inhomogeneous Shells with Variable Parameters and Their Numerical Solution. (Review) *Int. Appl. Mech.* **49(2)** 123 doi: 10.1007/s10778-013-0558-x
[3] Lee S E, Sahin S, Rigo P, Park M and Paik J K 2017 Ultimate strength of cylindrical shells with cutouts. *Ships and Offshore Structures* 12(sup1) 153 doi: 10.1080/17445302.2016.1271592
[4] Celebi M, Gürdal Z, Tatting B, Blom-Schieber A, Rassaian M and Wanthal S P 2017 Effects of Size and Location of a Circular Cutout on Buckling and Failure of a Cylindrical Shell in Bending. *In 58th AIAA/ASCE/AHS/ASC Structures, Structural Dynamics, and Materials Conf.* (Grapevine, Texas) 0433 https://arc.aiaa.org/doi/abs/10.2514/6.2017-0433
[5] Hatami H, Shariati M 2019 Numerical and Experimental Investigation of SS304L Cylindrical Shell with Cutout Under Uniaxial Cyclic Loading. *Iran J Sci Technol Trans Mech Eng.* **43** 139 https://doi.org/10.1007/s40997-017-0120-2
[6] Sun F, Wang P, Li W, Fan H, Fang D 2017 Effects of circular cutouts on mechanical behaviors of carbon fiber reinforced lattice-core sandwich cylinder. Composites. *Part A: Appl. Sci. and Man.* **100** 313 https://doi.org/10.1016/j.compositesa.2017.05.029
[7] Ashwell D G, Sabir A B 1972 A new cylindrical shell finite element, based on simple independent

strain functions. *Int. J. of Mech. Sci.* **14(3)** 171 doi: 10.1016/0020-7403(72)90074-4

[8] Timoshenko S, Woinovsky-Krieger S 1969 *Theory of Plates and Shells* Second Edition (New York:McGraw-Hill Book Company INC) p 636

[9] Cantin G, Glagh R W 1968 A curved cylindrical shell finite element. *AIAA Journal* **6(6)** 1057 https://doi.org/10.2514/3.4673

IOP Conf. Series: Materials Science and Engineering **868** (2020) 012035 doi:10.1088/1757-899X/868/1/012035

Time series forecasting by global optimization method

V Balyk[1], E Balyk[2] and V Vakulchuk[1*]

[1]Department of design of aerohydrocosmic systems, *Moscow Aviation Institute* (National Research University), 4 Volokolamsk highway, Moscow 125993, Russia
[2]*Federal Institute of Industrial Property*, 30 Berezhkovskaya nab., building 1, Moscow 125993, Russia

[*]E-mail: vakulvi@yandex.ru

Abstract. The actual problem of time series forecasting associated with the choice of the type of forecasting functions is considered. The aim of the work is to develop a new forecasting method, in which, along with the construction of the predictive function, the values of the predicted characteristics are directly optimized at the entire depth of the forecast. The developed method belongs to the class of probabilistic statistical forecast methods and partially to the class of feedback methods. Typical representatives of the methods of these classes are correlation and regression analysis, factor and variance analysis, statistical modeling. In the new method the predictive function is constructed in the basis of trigonometric polynomials, which leads to multi extremality of the forecasting problem. In this regard, to solve the problem of forecasting, the method of global optimization, hereinafter called the criterion shifts method, is used. The developed forecasting method allows solving a wide range of practical forecasting problems. Since the method of global solution search is used, it guarantees the exact finding of the predicted characteristics obtained as a result of their optimization by the criterion of regularity. The results obtained in this article showed high accuracy.

1. Introduction

The problem of forecasting time series, the solution of which is based on the theoretical provisions presented in [1, 2], is considered. The predictive function is constructed in the basis of trigonometric polynomials, and it, along with high approximating qualities, is multi-extreme. A minimum of regularity corresponds to an accurate prediction, and we consider the method of searching the global minimum based on the construction of process of extremum search in the extended space of variables, the dimension of which is determined by the dimensionality of the coefficients of the prediction function and the number of predictable characteristics. The peculiarity of the method is that the ordinates of the predicted process for the entire depth of the forecast for the entire sampling interval are taken as variables.

The relevance of the problem is due to numerous practical applications. So, in [3], an algorithm is considered based on innovative models of the state space that underlie the methods of exponential smoothing. It also provides a step-by-step forecasting algorithm using ARIMA models. In [4], the ecological problem of developing a model for forecasting based on time series is considered. In [5], approaches related to forecast aggregation, the so-called 'coherence forecast' methods, are presented. Clustering of time series data, used in various scientific fields to identify patterns, is given in [6, 7]. A similar problem of clustering, but for granular time series, is discussed in [8].

Forecasting of time series is shown on the problem of fuel cost forecasting according to statistics taken in the period from 2002 to 2008. The predictive function is constructed in the class of

IOP Conf. Series: Materials Science and Engineering **868** (2020) 012035 doi:10.1088/1757-899X/868/1/012035

trigonometric polynomials, which describe quite well the various processes occurring in time, but the problem of choosing a prediction function remains. Optimization of forecast points, as is done in this method, allows you to implement a parallel forecast, which gives additional justification when choosing a forecast function.

The novelty in this work is that here, in addition to the coefficients of the predictive trigonometric polynomial, the actual predicted parameters vary, the optimal values of which are taken as the final forecast. This is the fundamental difference between this work and similar works [3-9].

The aim of the work is to develop a new method for forecasting time series and compare the quality of the forecast from [9] with the forecast obtained by this method. All comparisons are made with the training of 20, 40 and 60 points respectively.

2. Time series forecasting by the criterion of regularity

The initial data for prediction is the implementation of the investigated random process specified with a certain sampling interval:

$$X(t_k), \qquad k=\overline{0,n}, \qquad t_{k+1}-t_k=\Delta,$$

in this case, the sampling interval should be less than the expected characteristic interval. The statistical sample size of N must be greater than the decay period of its autocorrelation function. It should be noted that the choice of too small sampling interval significantly increases the volume of initial data and complicates their processing-estimation of statistical characteristics and forecasting.

This method assumes the formalization of the problem of forecasting a random process by the criterion of regularity (minimum of the mean square of the error of the predictive function), which is the most common in statistical optimization problems. In General, the optimal operator is such an operator that provides an extreme value of the criterion for the quality of the functioning of the system, the input of which is a known random process $X(t)$ [1, 2].

In the statement of the problem, it is assumed that at the system output it is required to obtain a given output signal $Y_h(t)$, where h is the observation interval, when the system generates an output signal $Y(t)$. In the process prediction problem $X(t)$ for the interval γ, the required output signal is determined by the expression:

$$Y_h(t)=X(t+\gamma).$$

The forecast error is defined as:

$$E(t)=Y(t)-Y_h(t),$$

which in the case where $Y(t)$ and $Y_h(t)$ are random processes is also a random process.

The initial statistical sample in the forecasting problem has the form presented in table 1.

Table 1. Statistical sampling in the forecasting problem.

t	$X(t)$; $Y_h(t)$
t_1	$X(t_1)$
t_2	$X(t_2)$
...	...
t_{N_A}	$X(t_{N_A})$
t_{N_A+1}	$Y_h(t_{N_A+1})$
t_{N_A+2}	$Y_h(t_{N_A+2})$
...	...
t_N	$Y_h(t_N)$

First part of the statistical sample (N_A) is the training part. Second part is the predicted part of the statistical sample. A predictive function is constructed on a segment $\left[t_1, t_{N_A}\right]$ of long *h*. The forecast $\left(Y_h\left(t_{N_A+1}\right), Y_h\left(t_{N_A+2}\right), \ldots, Y_h\left(t_N\right)\right)$ is based on the segment $\left[t_{N_A+1}, t_N\right]$.

The predicted values $Y_h\left(t_{N_A+1}\right), \ldots, Y_h\left(t_N\right)$ are also variable parameters $x_1, x_2, \ldots, x_{N-N_a}$. Construction of the predictive function is carried out on a statistical sample from table 2.

Table 2. Statistical sampling of predictive function construction.

t	$X(t)$
t_1	$X(t_1)$
t_2	$X(t_2)$
...	...
t_{N_A}	$X(t_{N_A})$
t_{N_A+1}	x_1
t_{N_A+2}	x_2
...	...
t_N	x_{N-N_A}

A trigonometric polynomial was adopted as a predictive function:

$$Y_h(t) = \frac{a_0}{2} + \sum_{i=1}^{m}\left[a_i \cos(\omega_i t) + b_i \sin(\omega_i t)\right]$$

where a_0 - the average value of the predicted process, m - the Harmonic number, a_i, b_i - Fourier coefficients, ω_i - the frequency of the process.

Parameters a_i, b_i, ω_i, $i = \overline{1, m}$ as well as parameters x_i, $i = \overline{1, N - N_A}$ are determined from the minimum condition of the regularity criterion:

$$J^{(1)} = \min_{\{x_i, i=\overline{1,N-N_A}\}} \frac{\sum_{i=N_A+1}^{N-N_A} [Y_h(t_i) - x_{i-N_A}]^2}{\sum_{i=1}^{N-N_A} x_i^2} \quad (1)$$

$$J^{(2)} = \min_{\{a_i, b_i, \omega_i, i=\overline{1,m}\}} \frac{\sum_{i=1}^{N_A}[X(t_i) - Y_h(t_i)]^2 + \sum_{i=N_A+1}^{N} [Y_h(t_i) - x_{i-N_A}]^2}{\sum_{i=1}^{N_A}[X(t_i)]^2 + \sum_{i=1}^{N-N_A} x_i^2} \quad (2)$$

For the purpose of generality, a criterion is introduced *J*. Under the criterion *J*, we will understand either a criterion $J^{(1)}$ or a criterion $J^{(2)}$ depending on which variables are optimized. If the criterion *J* is optimized for forecast points x_i, $i = \overline{1, N - N_A}$, then $J = J^{(1)}$. If the criterion *J* is optimized by Fourier coefficients and frequencies of the trigonometric polynomial a_i, b_i, ω_i, $i = \overline{1, m}$, then $J = J^{(2)}$.

The minimization tasks (1), (2) are multi-extreme and in practical applications - of large dimension. To solve such problems, a method of criterion shifts has been developed, which is given below.

3. Research methods

According to the considered approach, solution of the time series forecasting problem is reduced to the problem of minimizing the regularity criterion of types (1) and (2) by choosing variable parameters $x_1, x_2, \ldots, x_n$, where x_1, x_2, x_3 - Fourier coefficients and the frequencies of the trigonometric polynomial, and $x_4, x_5, \ldots, x_n$ - variable predicted parameters. The global extremum search method from [9] was chosen as the minimization method.

Let J_0 be the initial value of the criterion and let $J(x_1, x_2, \ldots, x_n)$ be the criterion function. Since the criterion function must be minimized, a shift is made:

$$J_1 = J_0 - \Delta J$$

and the condition is checked:

$$J(x_1, x_2, \ldots, x_n) \le J_1 .$$

This check is carried out by solving the following problem:

$$\inf h(x), \tag{3}$$

where $h(x) = |J(x_1, x_2, \ldots, x_n) - J_1|$.

Let's replace the variables:

$$\begin{cases} x_1 = a_1 \cos(\omega_1 J) + b_1 \sin(\omega_1 J); \\ x_2 = a_2 \cos(\omega_2 J) + b_2 \sin(\omega_2 J); \\ \quad \ldots \\ x_n = a_n \cos(\omega_n J) + b_n \sin(\omega_n J). \end{cases}$$

Then task (3) will be rewritten as:

$$\inf_{\{a_i, b_i, \omega_i, i=\overline{1,n}\}} h(x). \tag{4}$$

If in the process of solving problem (4) it turns out that

$$J(x_1, x_2, \ldots, x_n) \le J_1,$$

is done shift:

$$J_2 = J_1 - \Delta J$$

and the described process continues.

If in the process of solving problem (4) it turns out that

$$J(x_1, x_2, \ldots, x_n) > J_1,$$

the shift is made in the opposite direction, i.e.

$$J_2 = J_1 + \Delta J$$

and further similar checks are carried out.

The search step of the method is based on the rule:

$$J_{\lim}^{(k+1)} = \begin{cases} J_{\lim}^{(k)} - \delta, \ \delta = \dfrac{\left|J_{\lim}^{(1)}\right|}{2k} \quad if\ H\left[f, J_{\lim}^{(k)}\right] < \varepsilon; \\ J_{\lim}^{(k)} + \delta, \ \delta = \dfrac{\left|J_{\lim}^{(1)}\right|}{2k} \quad if\ H\left[f, J_{\lim}^{(k)}\right] > \varepsilon, \end{cases} \tag{5}$$

$$k = 1, 2, \ldots$$

Note that the process of minimizing the incompatibility measure of equation (4) stops if the inequality is established at some step $s(k)$: $f(x_1, x_2, \ldots, x_n) \le J_{\lim}^{s(k)}$.

As a result of the iterative process (5), a sequence of criteria is formed:

$$J_{\lim}^{(1)}, J_{\lim}^{(2)}, \ldots, J_{\lim}^{(k)},$$

from which a decreasing subsequence can be distinguished:

$$J_{\lim}^{(1)} > J_{\lim}^{(3)} > J_{\lim}^{(5)} > \ldots > J_{\lim}^{(2k+1)}$$

and increasing subsequence:

$$J_{\lim}^{(2)} < J_{\lim}^{(4)} < J_{\lim}^{(6)} < \ldots < J_{\lim}^{(2k)},$$

the limits of which at $k \to \infty$ are equal to each other, and this limit corresponds to the global extremum.

The iterative process is terminated if

$$\left| J_{\lim}^{(2k+1)} - J_{\lim}^{(2k)} \right| \le \varepsilon .$$

The method is described. We prove the convergence of the method by value.

Theorem:

A bounded, continuous function is considered $J = J(x)$.

$$J(x): X \to \mathbb{R}^n, \ where\ x \in X = \mathbb{R}^n,$$

where X is the compact set.

Let the global minimum be defined as:

$$J_* = \underset{x \in X}{\text{globmin}}\, J(x) = J(x_*)$$

where x_* is any point of the global minimum equal to J_*.

Let the sequence of records $\{J_k\}$ be constructed by the method (1) – (4). Then

$$\lim_{k \to \infty} J_k = J_* .$$

Evidence:

Let's replace the variables:

$$\begin{cases} x_1 = a_1 \cos(\omega_1 J) + b_1 \sin(\omega_1 J); \\ x_2 = a_2 \cos(\omega_2 J) + b_2 \sin(\omega_2 J); \\ \ldots \\ x_n = a_n \cos(\omega_n J) + b_n \sin(\omega_n J). \end{cases}$$

$a_1, b_1, \omega_1, \ldots, a_n, b_n, \omega_n$ - selectable parameters that can be interpreted as Fourier coefficients and frequency, J - a new argument, it is also the criterion of optimality.

Suppose that a point $J^{(k)}$ is already known at any $k \ge 0$. Calculate the value $J^{(k+1)}$:

$$\begin{aligned} J^{(k+1)} &= J^{(k)} - \frac{\left|J^{(1)}\right|}{2k} \quad if\ f(x_1, x_2, \ldots, x_n) \le J^{(k+1)}; \\ J^{(k+1)} &= J^{(k)} + \frac{\left|J^{(1)}\right|}{2k} \quad if\ f(x_1, x_2, \ldots, x_n) > J^{(k+1)}. \end{aligned} \tag{6}$$

Rule (6) forms an iterative process whereby the optimality criterion decreases as long as the criterion function is less than or equal to the value of the sequentially decreasing argument $J^{(k)}$. If there is no such solution for some k, then $k = k+1$ the criterion $J^{(k)}$, according to rule (6), increases sequentially. The stepwise increase of the criterion stops when the inequality is fulfilled $f(x_1, x_2, \ldots, x_n) \le J^{(k+1)}$, is accepted $k = k+1$, and the points obtained by rule (6) will be called 'search points', as it is done in [2]. Each search point has a double designation: if the point appeared for the first time and it happened on the s step, it receives the number s and is denoted by J_s. At each subsequent step with a number $k \ge s$, the same point J_s, depending on the location on the segment $[J_1, J_2]$, receives one of the numbers i $(0 \le i \le k)$ and is denoted by J_i. Here the number $i = i(k)$ as well as the number $s = s(k)$ depend on the step number k.

We show that the point J_{k+1} satisfies inequalities $J_{s-1} < J_{k+1} < J_s$.

The search for the minimum of the function $f(x)$ starts from any valid point $x_1 \in X$ at which it is calculated $J_2 = f(x_1)$.

Suppose *k* iterations are made. The point at which the inequality $f(x_1, x_2, \ldots, x_n) \le J^{(k+1)}$ is changed to inequality $f(x_1, x_2, \ldots, x_n) > J^{(k+1)}$ is taken as the left boundary $J^{(k)}$ of the segment $[J_{s-1}, J_s]$. The next iteration $k = k+1$ starts from a point J_{s-1}, but with a step twice smaller than in the previous step. This ensures that the new point J_s of change of inequality $f(x_1, x_2, \ldots, x_n) > J^{(k+1)}$ to inequality $f(x_1, x_2, \ldots, x_n) \le J^{(k+1)}$ does not exceed the point $J^{(k+1)}$, i.e. $J_s \le J^{(k+1)}$. Similarly, the point J_{s-1} obtained by changing inequality $f(x_1, x_2, \ldots, x_n) \le J^{(k+1)}$ to inequality $f(x_1, x_2, \ldots, x_n) > J^{(k+1)}$ is not less than the point $J^{(k)}$, i.e. $J_s > J^{(k)}$. It follows that

$$J_{s-1} < J_{k+1} < J_s .$$

Thus, following the new search point J_{k+1} on $k+1$ the step, two new segments $[J_{s-1}, J_{k+1}]$ and $[J_{k+1}, J_s]$ are obtained. As a result of iterative process (6) the system of nested segments is formed:

$$\begin{gathered} [J_2, J_1], \\ [J_4, J_3], \\ \ldots \\ [J_{2k}, J_{2k-1}]. \end{gathered} \qquad (7)$$

Let J_* be the limit point of the sequence $\{J_k\}$, then in the system of segments (7) each segment contains a point J_* and finite set of search points other than J_*.

We introduce a set consisting $N(J_*)$ of all those numbers $k \ge 1$ for which either $J_{m(k)-1} < J_{m(k+1)-1}$, or $u_{m(k+1)} < u_{m(k)}$. This means that at $k \in N(J_*)$, the search point J_{k+1} falls inside the segment, $\left[J_{m(k)-1}, J_{m(k)}\right]$ forming one of the ends of the next segment $\left[J_{m(k+1)-1}, J_{m(k+1)}\right]$.

Since each of the segments $\left[J_{m(k)-1}, J_{m(k)}\right]$ contains infinitely many search points, the set $N(J_*)$ also contains infinitely many numbers.

The length of the segment $\left[J_{m(k)} - J_{m(k)-1}\right]$ is estimated by analogy with the method of dividing the segment in half. The formation of points in the considered method and in the method of dividing the segment in half is organized by a single mechanism: if the sign of inequality in system (5), at the new point, is reversed, then $k = k+1$. In the method of dividing a segment in half, if at the *k-th* point $J(x_{2k-1}) \le J(x_{2k})$, then the left border of the segment $[a_k, b_k]$ does not change, if $J(x_{2k-1}) > J(x_{2k})$, then the right border of the segment $[a_k, b_k]$ does not change. The length of the *k-th* segment in the method of dividing the segment in half is equal to

$$b_k - a_k = \frac{b - a - \delta}{2^k} + \delta .$$

Here δ is a method parameter, $0 < \delta < b - a$, and it determines the accuracy of the calculations.

When estimating the length of a segment $\left[J_{m(k)} - J_{m(k)-1}\right]$, the value δ is neglected, then

$$J_{m(k)} - J_{m(k)-1} = \frac{J_0 - J_{\min}}{2^k},$$

where $J_{\min}$ is the lower estimate of the investigated function, J_0 is the initial value.

Hence, and from the countable set, $N(J_*)$ it follows that

$$\lim_{k\to\infty}\left(J_{m(k)} - J_{m(k)-1}\right) = 0$$

and

$$\lim_{k\to\infty} J_{m(k)-1} = \lim_{k\to\infty} J_{m(k)} = J_* .$$

4. Results and discussion

All of the following results were obtained on the basis of retrospective prediction, that is, the test part of the sample $\left[Y_h(t)\right]$ is considered unknown and its elements are considered as variable parameters.

Studies of the dependence of the forecast quality on the volume of the training sample were modeled by the volume of the testing part of the sample. Here the depth of the forecast was consistently taken equal $\gamma = 12,\ 32,\ 52$.

The initial data were taken from state statistics on fuel prices [10] during the period from 2002 to 2008. The forecast was obtained for the initial sample. Then it was compared with the data from the international report [9], the results of which were obtained on the basis of the same initial data.

Figure 1 shows the initial process of changing costs over time $c(t)$. Dates are deferred on the abscissa axis. The sample size of 72 points.

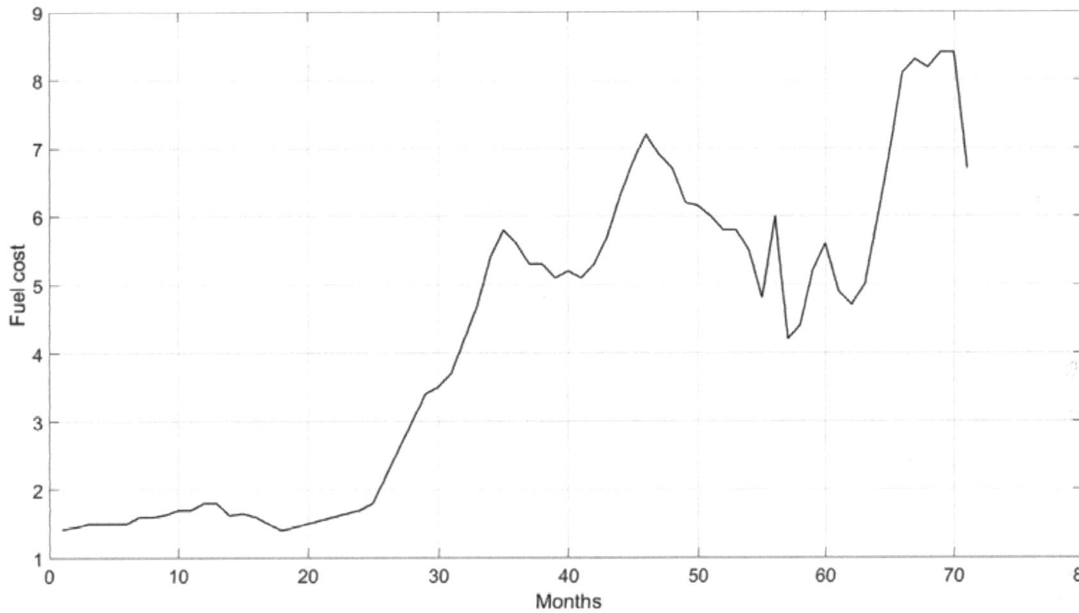

Figure 1. Initial statistical dependence.

The following three graphs (figures 2-4) show the results of forecasting the cost of fuel taken from [9]. Figure 2 shows the forecast made for the training period of 20 points, obtained on the basis of data from [10]. As can be seen from figure 2, the forecast and the tabular curve diverge significantly from each other.

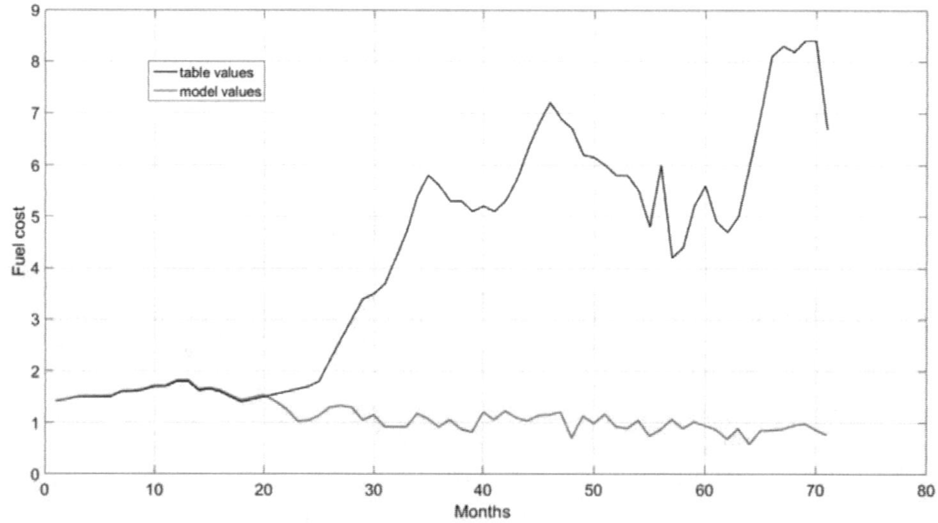

Figure 2. Forecast, based on the algorithm from [9], made by the period of training 20 points.

The same large divergence is observed in the forecast for 40 points (figure 3) and 60 points (figure 4).

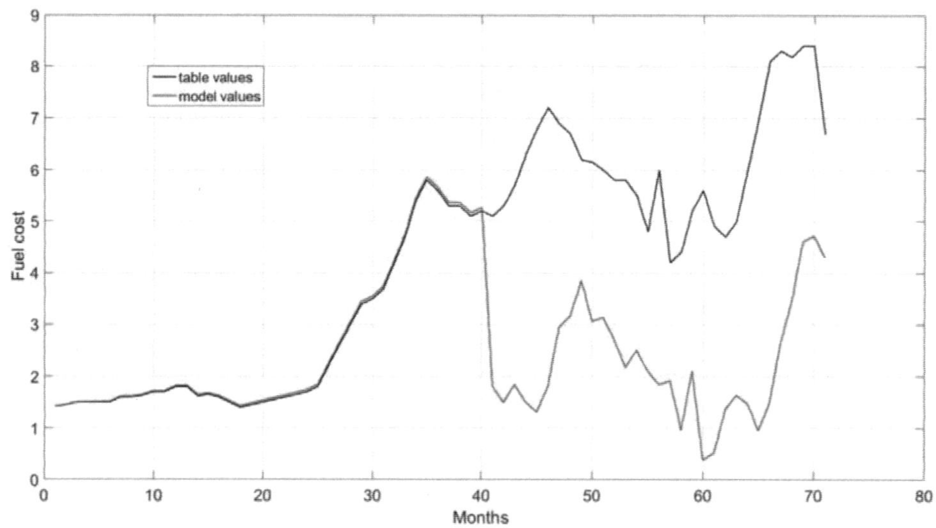

Figure 3. Forecast, based on the algorithm from [9], made by the period of training 40 points.

As can be seen from figures 2-4, the forecast obtained by the algorithm from [9] differs significantly from the forecast made by the new method, figures 5-7. The calculation of the mean absolute percentage error (MAPE) was made for each forecast. Among all the forecasts obtained by the algorithm from [9], for most accurate the MAPE value is 62.0397%. This forecast was made on the basis of the training sample containing 40 values for fuel cost (figure 3). Whereas the most accurate of the forecasts obtained by the new method corresponds to the MAPE value of 2.6917%. This forecast was obtained on the basis of the training sample of 60 points (figure 5). Thus, the new method of time series forecasting using global optimization demonstrates the most accurate results from the reviewed. Figure 6 shows the prediction obtained by the new method when training on 40 points. Figure 7 shows the prediction obtained by the new method when training on 20 points.

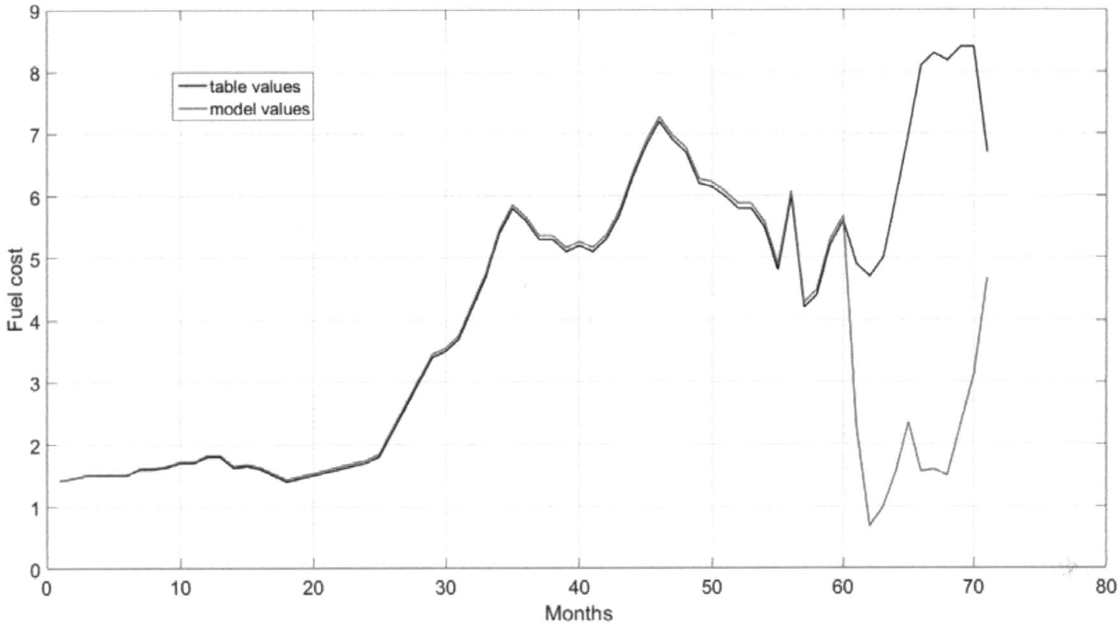

Figure 4. Forecast, based on the algorithm from [9], made by the period of training 60 points.

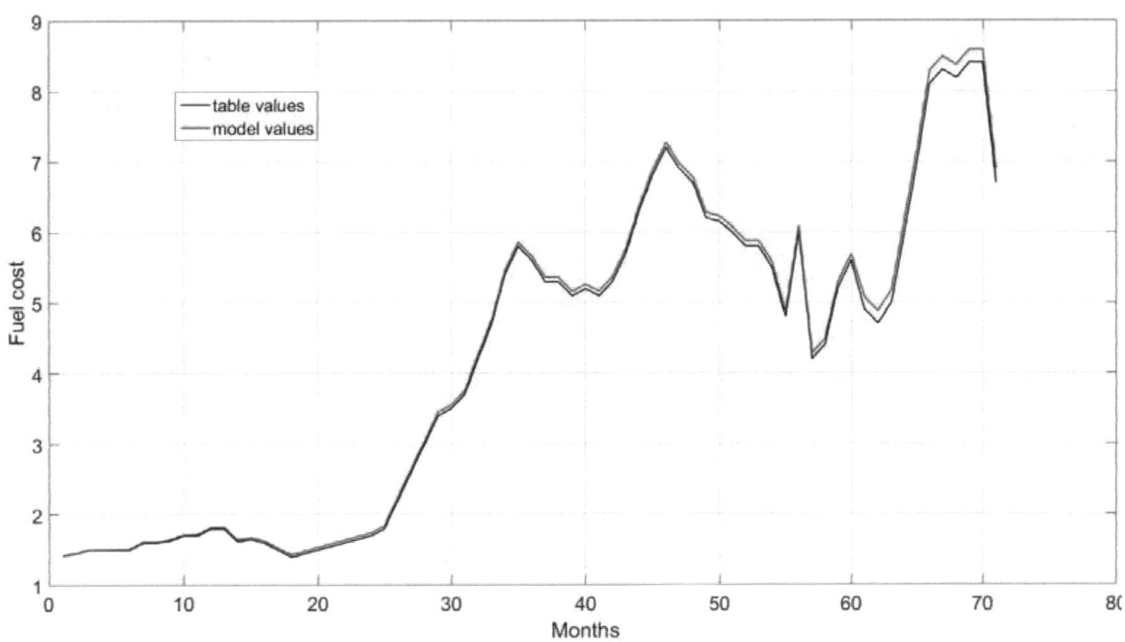

Figure 5. Forecast, based on the new method, made by the period of training 60 points.

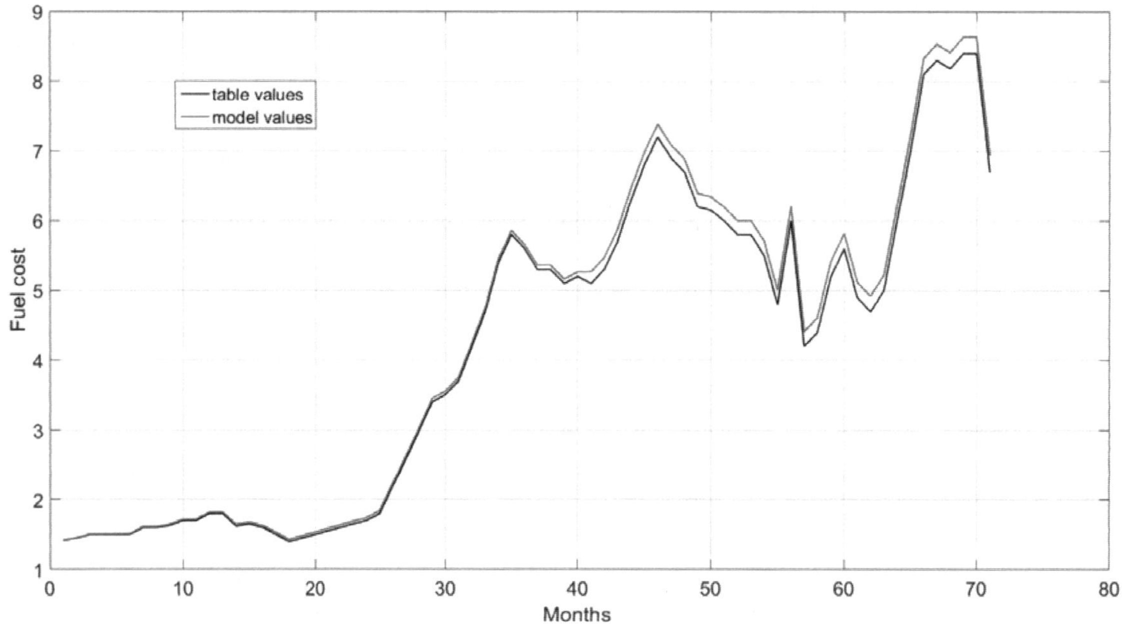

Figure 6. Forecast, based on the new method, made by the period of training 40 points.

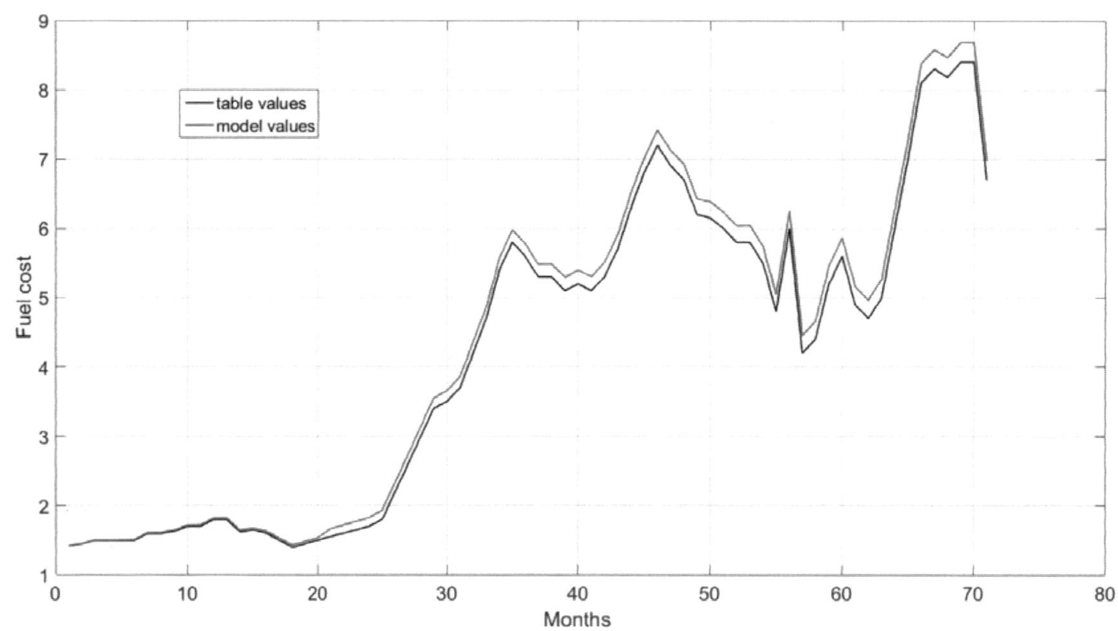

Figure 7. Forecast, based on the new method, made by the period of training 20 points.

5. Conclusion

As a result of numerical experiments the following results were obtained:

1. Figures 2-4 show a significant divergence of the predicted process from the real one, which is due to the unsuccessful choice of the predictive function in the form of a trigonometric polynomial.

2. Application of the new method to the problem solution shows its efficiency. As can be seen from figures 5-7, the difference between the actual process and its forecast is not more than 7.5%.
3. The accuracy of the forecast is related to the size of the training part of the sample. The more elements are considered in the process of training the model, the higher the accuracy of the forecast.

References

[1] Box G E P, Jenkins G M, Reinsel G C, Ljung G M 2015 *Time Series Analysis: Forecasting and Control* (New York: John Wiley & Sons. Inc)
[2] Anderson T W 1971 *The Statistical Analysis of Time Series* (New York: John Wiley & Sons. Inc.)
[3] Hyndman R J, Khandakar Y 2008 Automatic time series forecasting: The forecast package for R. *Jour. of Statistical Software* **27(1)** 1 doi: 10.18637/jss.v027.i03
[4] Leigh C, Kandanaarachchi S, McGree J M, Hyndman R J, Alsibai O, Mengersen K, Peterson E E 2019 Predicting sediment and nutrient concentrations in rivers using high-frequency water quality surrogates, available at: https://arxiv.org/abs/1810.12499
[5] Wickramasuriya S L, Athanasopoulos G, Hyndman R J 2019 Optimal forecast reconciliation for hierarchical and grouped time series through trace minimization. *J. American Statistical Association* **114(526)** 804 doi: 10.1080/01621459.2018.1448825
[6] Aghabozorgi S, Shirkhorshidi A S, Wah T Y 2015 Time-series clustering – A decade review. *Information Systems* **53** 16 doi: 10.1016/j.is.2015.04.007
[7] Liao T W 2005 Clustering of time series data – a survey. *Pattern Recognition* 38(11) 1857 doi: 10.1016/j.patcog.2005.01.025
[8] Duan L, Fusheng Yu, Pedrycz W, Xiao W, Xiyang Y 2018 Time-series clustering based on linear fuzzy information granules. *Applied Soft Computing* 73 1053 doi: 10.1016/j.asoc.2018.09.032
[9] Balyk V M, Balyk E V, Thuong N Q, Osadchuk M A 2019 Inverse Forecasting Problems of Time Series in the Harmonic Basis. *Int. Jour. of Applied Mathematics and Statistics* 58(4) 86
[10] *Central Statistical Database under Federal Service of State Statistics*, available at: https://www.fedstat.ru/indicator/31448.

IOP Conf. Series: Materials Science and Engineering **868** (2020) 012036 doi:10.1088/1757-899X/868/1/012036

Acceptable range parameters of asymmetry of spacecraft descending in the Martian atmosphere

E V Kurkina

Department of Further Mathematics, *Samara National Research University,* 34 Moskovskoe shosse, Samara, 443086, Russian Federation

E-mail: ekaterina.kurkina@mail.ru

Abstract. The article discusses the technique to determine the region of admissible deviations of the asymmetry parameters of the spacecraft descending in the atmosphere from their nominal values. The effect of the principal resonance is taken into account. The uncontrolled descent of a spacecraft in the Martian atmosphere is considered for small asymmetry and a small angle of attack. A numerical simulation was carried out using this technique to determine the range of admissible restrictions of the parameters of the displacement of the mass centre of the spacecraft, the moments of inertia and aerodynamic moments due to small shape distortions.

1. Introduction

As a rule, when developing uncontrolled spacecraft descending, there are strive to provide dynamic symmetry and give them an external axially symmetrical shape. In practice, a small asymmetry usually arises [1]. Spacecraft asymmetry leads to the fact that the oscillatory motion of the body symmetry axis with respect to the approach flow and the rotational motion of the body around the symmetry axis become interdependent and resonance phenomena arise [2]. Resonances that persist for a sufficiently long time period can lead to significant perturbations of the parameters of the descent trajectory in the atmosphere: an increase in the amplitude of oscillations of the angle of attack, an increase in overload, the spinning of the spacecraft around its longitudinal axis and other undesirable consequences [3]. When descending in the atmosphere of the planet, it is necessary to observe the limitations of values of the angle of attack and angular velocity [4]. To describe the angular motion of the spacecraft, a low-frequency system of equations can be used [5]. To calculate the probability of capture descending spacecraft in resonance the method proposed by Neustadt [6, 7] can be used [8]. The atmosphere of Mars is characterized by a significant difference in characteristics during the year [9, 10], which makes the task of evaluating resonant phenomena even more relevant. The aim of the work is to solve the problem of finding the maximum values of the asymmetry parameters at which the angular velocity does not reach its resonance values in the case of uncontrolled descent of the spacecraft in the Martian atmosphere, provided that the asymmetry is small. It is assumed that the spacecraft makes a movement at small angles of attack.

2. Statement of the problem

A nonlinear low-frequency system of equations of motion for a spacecraft with small mass and aerodynamic asymmetry, which describes the motion of the spacecraft relative to the center of mass (figure 1), has the form [3, 8]:

$$\bar{I}_x \frac{d\omega_x}{dt} = -\bar{m}_x^A \omega^2 \sin(\theta+\theta_2), \quad \frac{F_a}{4\omega_a^2}\frac{d\alpha}{dt} = -\frac{tg\alpha}{2\omega_a^2}\frac{\omega^2}{2q}\frac{dq}{dt} - \mathrm{sign}(\omega_x)\frac{\bar{m}^A\omega^2}{2\omega_a}\cos(\theta+\theta_1), \quad \frac{d\theta}{dt} = \omega_x - \omega_{1,2}, \quad (1)$$

where ω_x - spacecraft angular velocity relative to its longitudinal axis OX; α - spatial angle of attack, $\theta = \varphi - \pi/2$, φ - aerodynamic roll angle, relative parameters are marked with bar: $\bar{m}_x^A = m_x^A \cdot \omega^{-2}$, $\bar{m}^A = m^A \cdot \omega^{-2}$; $\bar{I}_x = I_x / I$, m^A, m_x^A - parameters characterizing the magnitude of mass, inertial and aerodynamic asymmetries; I_x and $I_y = I_z = I$ spacecraft inertia moments in spacecraft related coordinate system, θ_1, θ_2 - determines the relative position of mass-inertial and aerodynamic asymmetries: $\cos\theta_1 = -\frac{m_2^A}{m^A}$, $\cos\theta_2 = \frac{m_{x2}^A}{m_x^A}$, $\sin\theta_1 = \frac{m_1^A}{m^A}$, $\sin\theta_2 = -\frac{m_{x1}^A}{m_x^A}$;

$F_\alpha(\omega_x, \alpha, q) = -\frac{M_{zn}}{I} + \frac{\omega_{1,2}^2}{\cos^2\alpha} + (\bar{I}_x\omega_x - \omega_{1,2})(\bar{I}_x\omega_x - 2\omega_{1,2})$, M_{zn} - aerodynamic moment, $\omega_{1,2} = \frac{1}{2}\bar{I}_x\omega_x + \mathrm{sign}(\omega_x)\cdot\omega_a$, $\omega_a = \sqrt{\frac{1}{4}\bar{I}_x^2\omega_x^2 + \omega^2}$, $\omega = \sqrt{-\frac{1}{I}m_{zn}qSL\cot\alpha}$ - frequency of angular velocity ω_x precession, $m_{zn} = \frac{M_{zn}}{qSL}$ - coefficient of aerodynamic moment, $q = \frac{1}{2}\rho V^2$ - dynamic pressure, V- center of mass velocity, $\rho = \rho_0 \exp(-\rho_{coeff} H)$ - density, H - altitude.

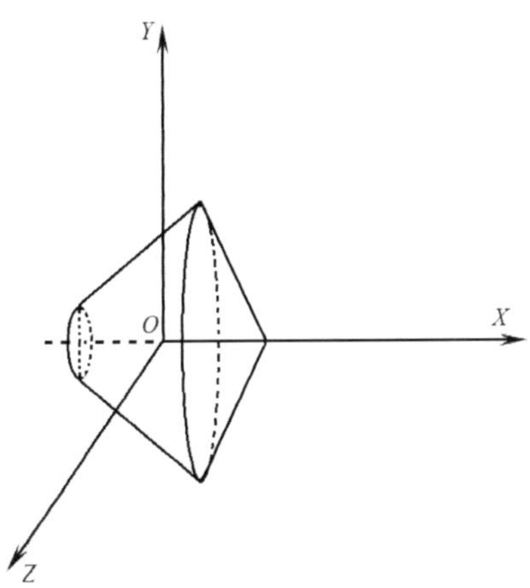

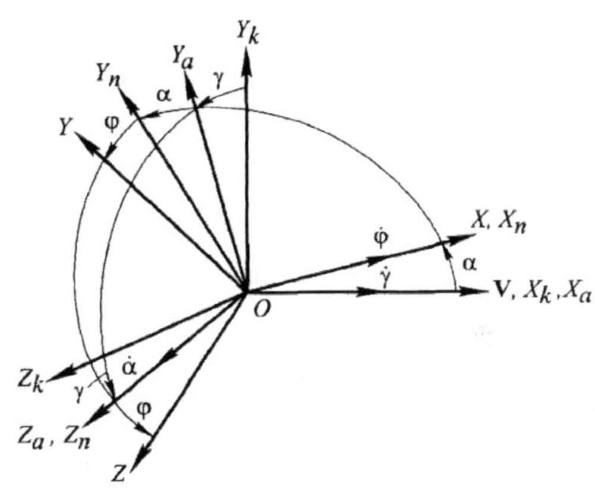

Figure 1. Coordinate systems: (a) Spacecraft related, (b) Angles of transition from Spacecraft related $OXYZ$ to Aerodynamic $OX_aY_aZ_a$ and Trajectory $OX_kY_kZ_k$ coordinate systems.

The system of spacecraft motion equations (1) is solved together with three equations characterizing the motion of the center of mass of an axisymmetric spacecraft:

$$\frac{dV}{dt} = -g\sin\vartheta - C_{xa}q\frac{S}{m}, \qquad \frac{d\vartheta}{dt} = -\frac{\cos\vartheta}{V}\left(g - \frac{V^2}{R}\right), \qquad \dot{H} = V\sin\vartheta, \qquad (2)$$

where $C_{xa}(\alpha)$ - drag coefficient, ϑ- trajectory angle, g - gravity acceleration , m - spacecraft mass, R – planet radius.

For a quasilinear system of differential equations describing the rotational motion of the spacecraft around the mass center of function that characterize the aerodynamic and mass asymmetries are of the form [11]:

$$m^A = \sqrt{\left(m_1^A\right)^2 + \left(m_2^A\right)^2}\ ,\ m_x^A = \sqrt{\left(m_{x1}^A\right)^2 + \left(m_{x2}^A\right)^2} \tag{3}$$

$$m_1^A = -\frac{\omega^2}{m_{z1}} m_{yo}^f + \frac{\omega^2}{m_{z1}} C_{x1}\Delta\overline{z} - \overline{I}_{xz}\omega_x^2,\quad m_2^A = -\frac{\omega^2}{m_{z1}} m_{zo}^f - \frac{\omega^2}{m_{z1}} C_{x1}\Delta\overline{y} + \overline{I}_{xy}\omega_x^2, \tag{4}$$

$$m_{x1}^A = -\frac{\omega^2}{m_{z1}} C_{y1}\Delta\overline{y} - \overline{I}_{xy}\omega_{1,2}^2,\quad m_{x2}^A = -\frac{\omega^2}{m_{z1}} C_{y1}\Delta\overline{z} - \overline{I}_{xz}\omega_{1,2}^2, \tag{5}$$

where $C_{y1}, C_{x1}, m_{z1}, m_{yo}^f, m_{zo}^f$ - coefficients characterizing the aerodynamic characteristics of the spacecraft; $\Delta y, \Delta z$ - displacement of the mass center of the spacecraft relative to the longitudinal axis; I_{xy}, I_{xz} - the moments of inertia of the spacecraft.

Approximate expressions for estimating the probability of capture in the principal resonance at an angle of attack $\alpha \in [0, \pi/2]$ with a positive low torque $a > 0$ [3]:

$$P = \frac{Q}{2\pi|a| + Q/2},\ if\ 0 < Q < 4\pi|a|;\quad P = 1,\ if\ Q \geq 4\pi|a(\sigma)|;\quad P = 0,\ if\ Q < 0, \tag{6}$$

where
$$Q = \frac{16\left(\overline{I}_x - 1\right)\omega^2\sqrt{\overline{m}_x^A\,\overline{m}^A}\cos\left(\theta_1 - \theta_2\right)}{\left(2\,\alpha\,\overline{I}_x\right)^{1/2}\left(2 - \overline{I}_x\right)^{3/2}} + 16\left(2\frac{1 - \overline{I}_x}{2 - \overline{I}_x}\cdot\frac{\alpha\,\overline{m}_x^A}{\overline{I}_x}\right)^{\frac{1}{2}}\left(1 - \frac{1 - \overline{I}_x}{\left(2 - \overline{I}_x\right)^2}\right)\frac{d\omega}{dt},$$

$$|a| = \frac{2\left(1 - \overline{I}_x\right)^{1/2}}{2 - \overline{I}_x}\left|\frac{d\omega}{dt}\right|.$$

The task is to find the maximum values of the asymmetry parameters at which the angular velocity ω_x does not reach its resonance values ω_x^r. The restrictions on the angular velocity ω_x are:

$$\min\omega_x \leq \omega_x \leq \max\omega_x = \omega_x^r \tag{7}$$

For small angles of attack α the derivative of the angular velocity ω_x is proportional to the generalized asymmetry parameter [3]:

$$\Omega = \sqrt{\overline{m}_x^A\,\overline{m}^A}\cos(\theta_1 - \theta_2), \tag{8}$$

where $\cos(\theta_1 - \theta_2) = \cos\theta_1\cos\theta_2 + \sin\theta_1\sin\theta_2$.
In the resonance case:

$$\omega_x^r = \frac{\omega}{\sqrt{1 - \overline{I}_x}},\quad \omega_x = \omega_{1,2}.$$

3. Acceptable range parameters of asymmetry finding

Condition (7) can be ensured by the following inequality:

$$|\Omega| = \left|\sqrt{\overline{m}_x^A\,\overline{m}^A}\cos(\theta_1 - \theta_2)\right| \leq |\Omega_r|, \tag{9}$$

where Ω_r – maximum allowed parameter value Ω, under which condition (1) is still satisfied. To find the value Ω_r it is necessary to choose the calculated case according to the initial conditions of motion and asymmetry parameters. Substituting expressions (3) - (5) in equation (8) and simplify, we obtain the function Ω:

$$\Omega = \frac{-k_1\overline{I}_{xz}\Delta\overline{y} + k_1\overline{I}_{xy}\Delta\overline{z} + k_2 m_{yo}^f\Delta\overline{y} + k_2 m_{zo}^f\Delta\overline{z} + k_3 m_{zo}^f\overline{I}_{xz} + k_3 m_{yo}^f\overline{I}_{xy}}{\left(k_4\left(\Delta\overline{y}\right)^2 + k_4\left(\Delta\overline{z}\right)^2 + k_5\overline{I}_{xy}\Delta\overline{y} + k_5\overline{I}_{xz}\Delta\overline{z} + k_6\left(\overline{I}_{xy}\right)^2 + k_6\left(\overline{I}_{xz}\right)^2\right)^{1/4}}, \tag{10}$$

where

$$k_1 = \frac{(C_{x1}+C_{y1})}{m_{z1}(1-\overline{I}_x)},\ k_2 = -\frac{C_{y1}}{m_{z1}^2},\ k_3 = -\frac{1}{m_{z1}(1-\overline{I}_x)},\ k_4 = \frac{C_{y1}^2}{m_{z1}^2},\ k_5 = \frac{2C_{y1}}{m_{z1}(1-\overline{I}_x)},\ k_6 = \frac{1}{(1-\overline{I}_x)^2}. \quad (11)$$

Inequality (9) has six unknowns; therefore, it has an infinite set of solutions. To unambiguously determine the accessible regions of change in the parameters $\Delta\overline{y}$, $\Delta\overline{z}, \overline{I}_{xy}, \overline{I}_{xz}$, m_{yo}^f, m_{zo}^f we set the weight relations between these values. Let the equalities be satisfied:

$$\Delta\overline{y}\frac{-p_1C_{y1}}{m_{z1}} = \Delta\overline{z}\frac{p_2C_{y1}}{m_{z1}} = \overline{I}_{xy}\frac{p_3}{1-\overline{I}_x} = \overline{I}_{xz}\frac{p_4}{1-\overline{I}_x} = m_{yo}^f\frac{p_5}{m_{z1}} = m_{zo}^f\frac{-p_6}{m_{z1}} = u,$$

where u - parameter that determines the choice of the solution surface section, p_i – positive real numbers (weights) for which the identity is true $\sum_{i=1}^{6} p_i \equiv 6$. Therefore:

$$\Delta\overline{y} = \frac{-m_{z1}}{p_1C_{y1}}u,\ \Delta\overline{z} = \frac{m_{z1}}{p_2C_{y1}}u,\ \overline{I}_{xy} = \frac{(1-\overline{I}_x)}{p_3}u,\ \overline{I}_{xz} = \frac{(1-\overline{I}_x)}{p_4}u,\ m_{yo}^f = \frac{m_{z1}}{p_5}u,\ m_{zo}^f = -\frac{m_{z1}}{p_6}u. \quad (12)$$

The signs in expression (12) are selected with the condition of ensuring the maximum value of the parameter Ω

Use equation (12) to inequality (9) allows us to find the desired region of admissible values. However, at this stage, the question of the nature of the change in the function Ω remains unattended, which does not allow us to unambiguously determine the values of the asymmetry parameters at which this function is maximum. Therefore, we examine the extremum function Ω.

From the necessary conditions for the existence of an extremum, we obtain stationary points defined in six-dimensional space by the plane:

$$\Delta\overline{y} = \frac{k_3}{k_1}m_{zo}^f,\ \Delta\overline{z} = -\frac{k_3}{k_1}m_{yo}^f,\ \overline{I}_{xy} = -\frac{k_2}{k_1}m_{zo}^f,\ \overline{I}_{xz} = \frac{k_2}{k_1}m_{yo}^f,\ m_{yo}^f = m_{yo}^f,\ m_{zo}^f = m_{zo}^f. \quad (13)$$

Let us verify the fulfillment of the sufficient condition for the existence of an extremum at points defined by the equation (13) (Sylvester's criterion [12]). We compose hessians from second-order partial derivatives:

$$H = \begin{bmatrix} 0 & 0 & 0 & -\frac{k_1}{c} & \frac{k_2}{c} & 0 \\ 0 & 0 & \frac{k_1}{c} & 0 & 0 & \frac{k_2}{c} \\ 0 & \frac{k_1}{c} & 0 & 0 & \frac{k_3}{c} & 0 \\ -\frac{k_1}{c} & 0 & 0 & 0 & 0 & \frac{k_3}{c} \\ \frac{k_2}{c} & 0 & \frac{k_3}{c} & 0 & 0 & 0 \\ 0 & \frac{k_2}{c} & 0 & \frac{k_3}{c} & 0 & 0 \end{bmatrix}, \quad (14)$$

where $c = \left(\frac{k_4J_6^2k_3^2}{k_1^2} - \frac{k_5J_6^2k_2k_3}{k_1^2} + \frac{k_4J_5^2k_3^2}{k_1^2} - \frac{k_5J_5^2k_2k_3}{k_1^2} + \frac{k_6J_6^2k_2^2}{k_1^2} + \frac{k_6J_5^2k_2^2}{k_1^2}\right)^{1/4}$.

We consider the principal minor determinant of hessian: $\Delta_1 = 0$, $\Delta_2 = 0$, $\Delta_3 = 0$, $\Delta_4 = k_1^4$, $\Delta_5 = 0$, $\Delta_6 = 0$.

Since in equation (14) $\det H = 0$, an additional study of the matrix of second-order partial derivatives is necessary. Given that the fourth-order angular minor is positive, it is possible to prove the sign-definiteness of the Hessian matrix. We calculate all the principal minors of this matrix. The calculations showed that there are sign-negative principal minors of the fourth order. Therefore, the Hessian matrix is alternating-sign and there are no extrema at points defined by equation (13).

Therefore, the function Ω reaches its maximum value at the boundary of region defined by inequality (9). Use relations for weights (12) to equation (10), we obtain

$$\Omega = \frac{\dfrac{(C_{x1}+C_{y1})u^2}{p_4 C_{y1}} + \dfrac{(C_{x1}+C_{y1})u^2}{p_3 C_{y1}} + \dfrac{u^2}{p_1 p_5} + \dfrac{u^2}{p_2 p_6} + \dfrac{u^2}{p_4 p_6}}{\left(\dfrac{u^2}{p_1^2} + \dfrac{u^2}{p_2^2} - \dfrac{2u^2}{p_1 p_3} + \dfrac{2u^2}{p_2 p_4} + \dfrac{u^2}{4p_3^2} + \dfrac{u^2}{4p_4^2}\right)^{1/4}}. \quad (15)$$

We write equation (15) in a more compact form: $\Omega = B\sqrt{u^3}$,

$$B = \frac{A}{\sqrt[4]{C}},\ A = \frac{(C_{x1}+C_{y1})}{p_4 C_{y1}} + \frac{(C_{x1}+C_{y1})}{p_3 C_{y1}} + \frac{1}{p_1 p_5} + \frac{1}{p_2 p_6} + \frac{1}{p_4 p_6}, \quad (16)$$

$$C = \frac{1}{p_1^2} + \frac{1}{p_2^2} - \frac{2}{p_1 p_3} + \frac{2}{p_2 p_4} + \frac{1}{4p_3^2} + \frac{1}{4p_4^2}.$$

Given inequality (9) $|\Omega| \le |\Omega_r|$ and solving it with allowance for equation (15) we find the desired range of admissible values in the form:

$$0 \le \Delta\bar{y} \le -\sqrt[3]{\left(\frac{\Omega_r}{B}\right)^2 \frac{m_{z1}}{p_1 C_{y1}}}, \qquad 0 \le \Delta\bar{z} \le -\sqrt[3]{\left(\frac{\Omega_r}{B}\right)^2 \frac{m_{z1}}{p_2 C_{y1}}},$$

$$0 \le \bar{I}_{xy} \le \sqrt[3]{\left(\frac{\Omega_r}{B}\right)^2 \frac{(1-\bar{I}_x)}{p_3}}, \qquad 0 \le \bar{I}_{xz} \le \sqrt[3]{\left(\frac{\Omega_r}{B}\right)^2 \frac{(1-\bar{I}_x)}{p_4}}, \quad (17)$$

$$0 \le m_{yo}^f \le -\sqrt[3]{\left(\frac{\Omega_r}{B}\right)^2 \frac{m_{z1}}{p_5}}, \qquad 0 \le m_{zo}^f \le -\sqrt[3]{\left(\frac{\Omega_r}{B}\right)^2 \frac{m_{z1}}{p_6}}.$$

The following scheme is proposed that defines the region of admissible values of the asymmetry parameters:

1. Value calculation Ω_r.
2. Setting weights p_i.
3. Defining the asymmetry parameter area given by equation (14)
4. Condition check $|\Omega| \le |\Omega_r|$.
5. Verification calculation using initial nonlinear equations.

4. Discussion

If the asymmetry parameters are inside this region defined by inequalities (17), then the angular velocity ω_x, changing from small values, does not reach the values at which the principal resonance is realized.

We apply this technique to determine the region of admissible deviations of the asymmetry parameters of the spacecraft descending in the Martian atmosphere “Mars Polar Lander”. The calculated spacecraft is close to axially symmetrical and has the following characteristics: the largest radius of the cone base r = 1.25 m, the cone height l = 2 m, the mass m = 576 kg, the axial moments of inertia I_x = 270 kg·m^2, I_y = 443 kg·m^2, I_z = 443 kg·m^2. When choosing the simulation case, the following initial conditions for the motion of the spacecraft were adopted: velocity of the center of mass $V(0) = 3400$ m·s^{-1}, the flight altitude $H(0) = 120$ km, the local flight-path angle $\vartheta(0) = -0.017$ rad. Weight functions were defined as follows: $p_1 = p_2 = 1.5$, $p_3 = p_4 = 1$, $p_5 = p_6 = 0.5$. For these initial data, the following parameter value was obtained by calculation, using which, using expressions (14), we can determine the desired region of admissible restrictions on the asymmetry parameters of the spacecraft. As a result, we obtain for these initial data, the following parameter value $\Omega_r = 0.1$ was obtained by calculation, using equation (17), we can determine the desired region of admissible deviations on the asymmetry parameters of the spacecraft. As a result, we obtain:

$$\Delta\overline{y},\ \Delta\overline{z} \in [0,\ 0.0037];\quad \overline{I}_{xy},\ \overline{I}_{xz} \in [0,\ 0.0439];\quad m_{yo}^{f},\ m_{zo}^{f} \in [0,\ 0.0133]. \qquad (18)$$

Numerical simulation of the solution of system (1) and (2) for "Mars Polar Lander" case presented in the figure 2 shows that the values represented by upper range of conditions (18) correspond absence of principal resonance on the upward dynamic pressure curve, whereas the excess of these values is 1.5 times corresponds to the beginning of principal resonance near dynamic pressure maximum time.

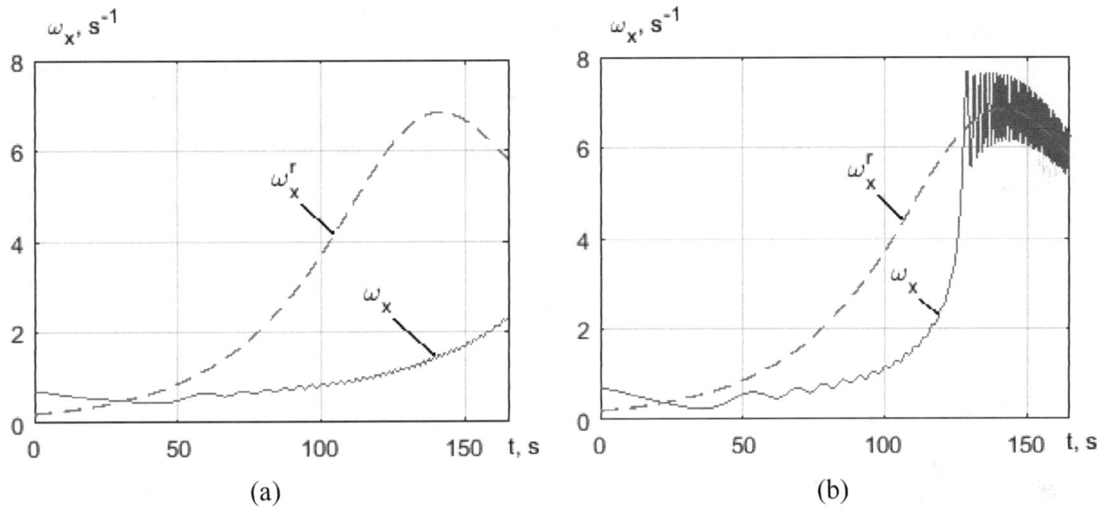

Figure 2. Angular velocity ω_x and resonance values ω_x^r for "Mars Polar Lander" case on the upward dynamic pressure curve: (a) $\Delta\overline{y} = \Delta\overline{z} = 0.0037$, $\overline{I}_{xy} = \overline{I}_{xz} = 0.0439$, $m_{yo}^{f} = m_{zo}^{f} = 0.0133$ (b) $\Delta\overline{y} = \Delta\overline{z} = 0.00555$, $\overline{I}_{xy} = \overline{I}_{xz} = 0.06585$, $m_{yo}^{f} = m_{zo}^{f} = 0.01995$.

5. Conclusion

A technique is proposed for determining the region of admissible values during the motion of a spacecraft in the atmosphere with a small initial angular velocity of rotation relative to the longitudinal axis, taking into account the effect of the principal resonance phenomenon on the rotational motion of the spacecraft. This technique can be used to determine region of admissible values at the design of specific spacecraft with small shape asymmetry. The performed numerical simulation confirms the possibility of practical use of the proposed technique at the descent spacecraft design stage - asymmetry range estimates determined using this technique correspond to a non-resonant case, while exceeding asymmetry estimates leads to principal resonance on the upward dynamic pressure curve.

References

[1] Koryanov V V and Kazakovtsev V P 2018 Dynamics of angular motion of landing vehicle in Martian atmosphere with allowance for small asymmetries. *International Journal of*

Mechanical Engineering and Robotics Research **7** 385 doi: 10.18178/ijmerr.7.4.385-391

[2] Lyubimov V V 2015 Dynamics and control of angular acceleration of a re-entry spacecraft with a small asymmetry in the atmosphere in the presence of the secondary resonance effects. *2015 International Siberian Conference on Control and Communications (SIBCON)* (Omsk: IEEE) p 4 doi: 10.1109/SIBCON.2015.7147134

[3] Lyubimov V V and Kurkina E V 2017 Probability of a resonance capture for an asymmetric capsule during its controlled descent in the Martian atmosphere. *Mekhatronika, Avtomatizatsiya, Upravlenie* **18** 564 doi: 10.17587/mau.18.564-571

[4] Lorenz R D 2010 Attitude and angular rates of planetary probes during atmospheric descent: Implications for imaging. *Planetary and Space Science* **58** 838 doi: 10.1016/j.pss.2010.01.003

[5] Belokonov V M, Belokonov I V and Zabolotnov Yu M 1983 Accelerated calculation of the descent trajectory in the atmosphere of unguided spacecraft taking into account their motion relative to the center of mass. *Cosmic Research* **21** 415

[6] Neishtadt A 1990 Averaging, capture into resonances, and chaos in nonlinear systems. *Soviet-American Conference on Chaos* (Woods Hole: American Institute of Physics) p 496

[7] Neishtadt A I 2014 Averaging, passage through resonances, and capture into resonance in two-frequency systems. *Russ. Math. Surv.* **69** 771 doi: 10.1070/RM2014v069n05ABEH004917

[8] Zabolotnov Yu M and Lyubimov V V 2003 Application of the method of integral manifolds for construction of resonant curves for the problem of spacecraft entry into the atmosphere. *Cosmic Research* **41** 453 doi: 10.1023/A:1026046232640

[9] Gierasch P J and Goody R M 1973 A model of Martian great dust storm. *Journal of the atmospheric science* **30** 169 doi: 10.1175/1520-0469(1973)030<0169:AMOAMG>2.0.CO;2

[10] Kliore A J, Cain D L, Fjeldbo G, Seidel B L and Rasool S I 1972 Mariner 9 S-Band Martian occultation experiment: initial results on the atmosphere and topography of Mars. *Science* **175** 313 doi: 10.1126/science.175.4019.313

[11] Belokonov V M and Zabolotnov M Y 2002 Estimation of the probability of capture into a resonance mode of motion for a spacecraft during its descent in the atmosphere. *Cosmic Research* **40** 467 doi: 10.1023/A:1020647017797

[12] Gilbert G T 2018 Positive definite matrices and Sylvester's criterion. *The American Mathematical Monthly* **98** 44 doi: 10.1080/00029890.1991.11995702

IOP Conf. Series: Materials Science and Engineering **868** (2020) 012037 doi:10.1088/1757-899X/868/1/012037

Comparative analysis of calibration variants for inertial measurement unit based on microelectromechanical system

A A Krylov* **and K K Veremeenko**

Department 305 "Automated complexes of orientation and navigation systems", *Moscow Aviation Institute* (National Research University), 4 Volokolamskoe shosse, Moscow 125993, Russian Federation

*E-mail: akril91@rambler.ru

Abstract. This paper is about comparison of two methods of calibration – calibration method for inertial measurement unit with gyroscope and accelerometer direct information outputs and calibration method for strapdown inertial navigation system with coordinate, velocity and orientation outputs. Two mathematical methods - Least Squares and Kalman Filter methods are applied respectively. Both methods assume the same mathematical error model. This model includes zero drift, nonlinearity coefficients, nonorthogonal projections and also noises of these parameters The research compare these methods and make attempt to give approximate numerical results. The results reveal features of each case, e.g. optimal calibration time for Kalman Filter and opportunity to find unknown noise parameters in inertial measurement unit variant. These features can help a researcher to choose a suitable variant, all else being equal it is recommended to use the inertial measurement unit variant.

1. Introduction

IMU is an inertial measurement unit which includes three mutually orthogonal gyroscopes and accelerometers. IMU calibration is a test process consisting in finding calibration coefficients of the IMU error model. Depending on the type of sensors, calibration may include the determination of these errors of gyroscopes and accelerometers as part of the IMU, such as zero drift and non-linearity of the scale factor, as well as the non-orthogonality of the sensor installation in relation to the instrumental axes. Calibration is important to eliminate systematic errors with the known instability (noise) of the calibrated parameters.

For blocks based on microelectromechanical system (MEMS) sensors, both pre-operational and operational preliminary calibrations are important and allow to take into account the systematic component that changes during long-term storage. Based on this, it is important not only to evaluate errors accurately, but also to do it by the quickest way.

When calibrating blocks based on MEMS gyroscopes, the most significant errors are zero drift [1] and non-linearity of the scale factor [2]. The significant non-orthogonality caused by the assembly process of blocks [3] should also be taken into account. To find gyroscope drift, the method of several fixed positions is usually used, which provides full observability of zero drifts [4-5].

Two of the most common variants for information outputs of sensor blocks are: IMU outputs – three mutually orthogonal measurements of angular velocities and linear accelerations relative to the instrumental coordinate system (that is, relative to the axes of the unit itself) [6-8] and strapdown inertial navigation system (SINS) outputs – three orientation angles relative to the geographical coordinate system, as well as three linear velocities and three coordinates relative to them [3, 9]. Depending on the outputs, calibration approaches can vary significantly. In the literature [10] about the estimation of calibration parameters, only one method is usually given, often even without its explicit declaring, however, questions regarding a qualitative comparison of calibration options are not posed, as well as a comparison of the estimated parameter accuracy and the calibration time.

There are some works that compare different mathematical methods to the same data but there are so not many papers that compare the device with the same sensors and different information output and explore the features of each variant. So it is an idea to compare calibration accuracy and other features for IMU and SINS outputs.

2. Properties of calibrated IMU

The investigated IMU (its characteristics are in table 1) includes three MEMS gyroscopes and MEMS accelerometers (their characteristics are in table 2) of the linear-linear (LL) type (one gyroscope and one accelerometer in one MEMS case), as well as a microcontroller that allows to correct sensor readings taking into account the calculated systematic errors.

Table 1. Mass and size characteristics of the IMU block.

Characteristic	Value
IMU Mass	140g
IMU sizes	80x43x34

Table 2. The declared characteristics of the sensors.

Characteristic	Value
Frequency of data output	1000 Hz
Dynamic Range of gyroscope	±500°/s
Dynamic Range of accelerometer	±100g
Zero bias stability of gyroscope	0.02°/s
Zero bias stability of accelerometer	0.02g
In-run zero drift (for 5 minutes) of gyroscope	0.02°/s
In-run zero drift (for 5 minutes) Of accelerometer	0.02g
Zero signal noise of gyroscope	0.006 °/s/$\sqrt{Hz}$
Zero signal noise of accelerometer	0.002 g/$\sqrt{Hz}$

3. Variants of calibration

IMU calibration is a test process consisting in determining calibration coefficients of the IMU error model. Calibration includes the determination of these errors of gyroscopes and accelerometers as part of the instrumentation, such as zero drift and non-linearity of the scale factor, as well as the non-orthogonality of the installation of sensors relative to the instrument axes. Calibration is important to eliminate systematic errors and take into account known instability (noise) of the calibrated parameters if they exist.

In case of the IMU, at the output of the device there are three angular velocities ($\omega_x, \omega_y, \omega_z$) and three accelerations (a_x, a_y, a_z):

$$Z_{IMU} = \left[\omega_x, \omega_y, \omega_z, a_x, a_y, a_z\right] \tag{1}$$

A "direct" calibration can be carried out, the values of the errors of the sensors and the angle of non-orthogonality can be determined for each sensor independently.

In the case of SINS, at the output of the device there are three coordinates (R_x, R_y, R_z), three projections of linear velocity (V_x, V_y, V_z) and three orientation angles ($\Theta_x, \Theta_y, \Theta_z$):

$$Z_{SINS} = \left[R_x, R_y, R_z, V_x, V_y, V_z, \Theta_x, \Theta_y, \Theta_z \right] \tag{2}$$

The errors of gyroscopes and accelerometers can be estimated in an "indirect" way, for example, using an Kalman Filter, provided the system is observable.

It poses a question how differs the accuracy and other calibration features using these two methods with the same sensors (MEMS sensors of the LL type).

4. IMU calibration variant

To simplify the general mathematical model of sensors, the dependence of zero drift and other parameters on temperature changes is not taken into account. Each sensor can be calibrated independently of the others, therefore the same type of error equation is used for each gyroscope and for each accelerometer.

Gyroscope error model for any considered axis $L(0,1,2$ for $X, Y, Z)$:

$$\omega_{mL} = Sfg_L(\omega_{aL} + \delta\omega_L), \tag{3}$$

where ω_m – the measured value by the gyroscope, ω_a – the adjusted by the rotary bench angular velocity value, $\delta\omega$ – the total error of the gyroscope.

$$\delta\omega_L = \upsilon g_{xL}\omega_{ax} + \upsilon g_{yL}\omega_{ay} + \upsilon g_{zL}\omega_{az} + \omega_a^{\ 2} K\omega_{nlL} + \Delta\omega_{drL} + \varepsilon\omega_L \tag{4}$$

Accelerometer error model for any considered axis L:

$$a_m = Sfa_L(a_a + \delta a_L), \tag{5}$$

where a_m – the measured value by the accelerometer, a_a – the linear acceleration value set by the bench, δa – the total error of the accelerometer.

$$\delta a_L = \upsilon_x a_{ax} + \upsilon_y a_{ay} + \upsilon_z a_{az} + a_a^{\ 2} Ka_{nl} + \Delta a_{dr} + \varepsilon a_L \tag{6}$$

Example for a gyroscope X:

$$\upsilon_x = \cos(\alpha_L) \tag{7}$$

– projection of deviation angle of the sensor along the axis of rotation relative to the same instrument axis (figure 1);

$$\upsilon_y = \cos(\beta_L) \tag{8}$$

– projection of deviation angle of the sensor along the axis of rotation relative to another instrument axis (figure 1);

$$\upsilon_z = \cos(\gamma_L) \tag{9}$$

– projection of deviation angle of the sensor along the axis of rotation relative to another instrument axis (figure 1); L – axis of the calibrated sensor.

Similarly, for the *Y* and *Z* axes, the projection will be the cosine of the angle of deviation from the corresponding axis, and the projections on the other axes will be the sines of the angles of deviation to them.

$$K\omega_{nli} = \widehat{K\omega_{nl}} + \varepsilon Sf\omega_i \tag{10}$$

– scale factor nonlinearity, $\varepsilon Sf\omega_j$ – instability of the scale factor for the i-th measurement;

$$\Delta\omega_{drj} = \widehat{\Delta\omega_{dr}} + \varepsilon\upsilon\omega d_j \tag{11}$$

– zero drift; $\varepsilon\upsilon\omega d_j$ – instability of zero drift for the j-th measurement; $\varepsilon\omega$ – zero signal noise of gyroscope.

IOP Conf. Series: Materials Science and Engineering **868** (2020) 012037 doi:10.1088/1757-899X/868/1/012037

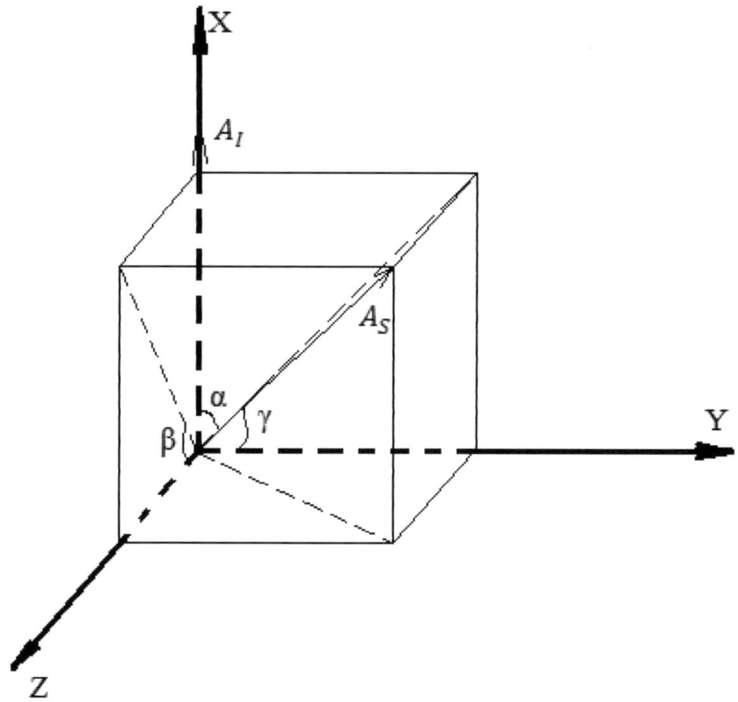

Figure 1. An example of non-orthogonal angles along the X axis.α is the deviation angle of the X axis of the sensor $X(A_S)$from the instrumental axis (A_I). β и γ – are the deviation angles of the axis of the sensor $X(A_S)$ to the instrumental axes of other sensors (Z and Y respectively). The angle is greatly increased for visibility.

A priori information on the noise parameters $\varepsilon Sf\omega$, $\varepsilon \upsilon\omega d$, $\varepsilon\omega$ is absent or precisely accurate, therefore, in the model they are considered a random variable with zero arithmetic mean (white noise). The same method can be applied for Y and Z gyroscopes and for accelerometers. This calibration method does not take into account such typical elements of the SINS model as the Schuler pendulum and the Earth's rotation speed due to their smallness in comparison with the parameters of the sensors indicated in table 1.

To determine the systematic component of each error, a sufficient number of test repetitions is necessary. In the case of the IMU, tests are carried out 5 times. To obtain sufficient statistics within one record with a known noise $\varepsilon\omega$ under adjusted angular velocity, at least 20,000 measurements will be required. In a static position zero drift of these sensors changes about 5 minutes, hence we take this interval for zero drift coefficient finding.

Calibration plan:

1) Finding of zero drift of gyroscopes and accelerometers.

Fixed position, orientation y axis up. For one repetition in one dimension:

$$\Delta\omega_{drj} = \omega_a + \varepsilon\upsilon\omega d_j + \varepsilon\omega \quad (12)$$

$$\omega_a = 0 \quad (13)$$

For gyroscopes to eliminate earth angular velocity we should put sensor to the east or west direction.

For accelerometer drift finding we should use 6-position method as in [7]. Implied that we do 5 runs for each of 6 positions from table 3.

Table 3. Attitude configuration for accelerometer test

Test No.	Direction of axes			Name of measured value		
	X	Y	Z	X	Y	Z
1	Up	East	North	a_{drxxn}	a_{drxyn}	a_{drxzn}
2	Down	West	North	a_{drxxp}	a_{drxyp}	a_{drxzp}
3	West	Up	North	a_{dryxn}	a_{dryyn}	a_{dryzn}
4	East	Down	North	a_{dryxp}	a_{dryyp}	a_{dryzp}
5	East	North	Up	a_{drzxn}	a_{drzyn}	a_{drzzn}
6	West	North	Down	a_{drzxp}	a_{drzyp}	a_{drzzp}

$$\Delta a_{drxj} = \frac{\dfrac{a_{drxxp} * a_{dryxn} - a_{drxxn} * a_{dryxp}}{a_{drxxp} + a_{dryxn} - a_{drxxn} - a_{dryxp}} + \dfrac{a_{drxxp} * a_{drzxn} - a_{drxxn} * a_{drzxp}}{a_{drxxp} + a_{drzxn} - a_{drxxn} - a_{drzxp}}}{2} \tag{14}$$

$$\Delta a_{drxj} = \frac{\dfrac{a_{dryyp} * a_{drxyn} - a_{dryyn} * a_{drxyp}}{a_{dryyp} + a_{drxyn} - a_{dryyn} - a_{drxyp}} + \dfrac{a_{dryyp} * a_{drzyn} - a_{dryyn} * a_{drzyp}}{a_{dryyp} + a_{drzyn} - a_{dryyn} - a_{drzyp}}}{2} \tag{15}$$

$$\Delta a_{drzj} = \frac{\dfrac{a_{drzzp} * a_{drxzn} - a_{drzzn} * a_{drxzp}}{a_{drzzp} + a_{drxzn} - a_{drzzn} - a_{drxzp}} + \dfrac{a_{drzzp} * a_{dryzn} - a_{drzzn} * a_{dryzp}}{a_{drzzp} + a_{dryzn} - a_{drzzn} - a_{dryzp}}}{2} \tag{16}$$

For one run on n measurements:

$$\widehat{\Delta\omega_{aj}} = \frac{\sum_{i=1}^{n} \omega_{mi}}{n} + \varepsilon\upsilon\omega d_j \tag{17}$$

For 5 runs:

$$\widehat{\Delta\omega_{dr}} = \frac{\sum_{i=1}^{5} \widehat{\Delta\omega_{drj}}}{5} \tag{18}$$

Similarly, accelerometer drift coefficients $\widehat{\Delta a_{dr}}$ are calculated from the same records.

2) Finding of non-orthogonality angles and scale factors of gyroscopes.

When rotating at a constant angular velocity for three installations relative to each of the instrument axes, three projections are measured along one axis ng^{I}_{Lmean} (for X_I), ng^{II}_{Lmean} (for Y_I), ng^{III}_{Lmean} (for Z_I) for each set:

$$\omega_m = ng^{I}_{Lmean} = \omega_a(Sfg_L + \varepsilon Sf\,\omega_j)\cos(\alpha_L) + \varepsilon\omega \tag{19}$$

While rotating at a constant angular velocity for three installations relative to each of the instrument axes, three projections are measured along one axis:

$$\omega_m = ng^{II}_{Lmean} = \omega_a(Sfg_L + \varepsilon Sf\,\omega_j)\cos(\beta_L) + \varepsilon\omega \tag{20}$$

For any another axis:

$$\omega_m = ng^{III}_{Lmean} = \omega_a(Sfg_L + \varepsilon Sf\,\omega_j)\cos(\gamma_L) + \varepsilon\omega \tag{21}$$

Herewith:

$$\alpha_L = arc\cos((1-\cos(\beta_L)^2 - \cos(\gamma_L)^2))^{1/2} \tag{22}$$

Scale factors and nonorthogonality angles finding:

$$\widehat{mk_{Lj}} = \frac{(ng^{I\ 2}_{Lmean} + ng^{II\ 2}_{Lmean} + ng^{III\ 2}_{Lmean})^{1/2} - \Delta\omega_{dr}}{\omega_a} + \varepsilon Sf\,\omega_j \tag{23}$$

$L-$ axis (sensor); ng^{I}_{Lmean}, ng^{II}_{Lmean}, ng^{III}_{Lmean} – measurements are carried out while the bench rotating with angular velocity ω_a for three mutual orthogonal sets; $\varepsilon Sf\,\omega_j$ – scale factor instability.

For 5 runs:

$$\widehat{Sfg_L} = \frac{\sum_{j=1}^{5} \widehat{Sfg_{Lj}}}{5} \tag{24}$$

Three nonorthogonalities for L axis are finding from three projections:

$$\widehat{v_{xLj}} = \cos(\widehat{\alpha_L}) = \frac{\sum_{j=1}^{5} ng^{I}_{Lmeanj} - \Delta\omega_{dr}}{\omega_a * \widehat{Sfg_L}} \tag{25}$$

$$\widehat{v_{yLj}} = \cos(\widehat{\beta_L}) = \frac{\sum_{j=1}^{5} ng^{II}_{Lmeanj} - \Delta\omega_{dr}}{\omega_a * \widehat{Sfg_L}} \tag{26}$$

$$\widehat{v_{zLj}} = \cos(\widehat{\gamma_L}) = \frac{\sum_{j=1}^{5} ng^{III}_{Lmeanj} - \Delta\omega_{dr}}{\omega_a * \widehat{Sfg_L}} \tag{27}$$

3) Finding of the scale factor nonlinearity for gyroscopes.

While rotating with angular acceleration U_ω from 0 to ω_a.

$$\omega_m = \omega_a^2 (Sfg_L + nlg + \varepsilon Sf\,\omega) \tag{28}$$

Fn – frequency of instrument data output.

The common case is maximum angular velocity $\max(\omega_a)$=500°/s and U_ω=10 °/ s^2.

4) Finding of the scale factors of the accelerometers.

It is carried out in a centrifuge similar to the determination of the scale factors of the gyroscopes, taking with non-orthogonality coefficients applied (the misalignment angle between the gyroscope and the accelerometer in the MEMS housing is small).

$$a_m = na^{I}_{Lmean} = a_a (Sfa_L + \varepsilon Sfa_j) * \cos(\alpha_L) + \varepsilon a \tag{29}$$

$$\cos(\alpha_L) \sim 1 \tag{30}$$

5) Finding of the scale factor nonlinearity for accelerometers.

While rotating with angular acceleration A_a from 0 to a_a.

$$a_m = a_a^2 (Sfa_L + nla + \varepsilon Sfa) \tag{31}$$

Fn – frequency of instrument data output.

The common case is maximum angular velocity $\max(\omega_a)$ = 500 °/ s and U_ω = 10 °/ s^2.

The obtained equations are quite simple and may be solved by the simplified least squares (LS) method.

5. SINS calibration variant

An alternative option for calibrating gyroscopes and accelerometers is a method based on the use of the output signals of a working inertial navigation system (INS), which they include. One of the main reasons for switching to such a calibration method is the lack of direct data from inertial sensors at the INS output.

When calibrating gyroscopes and accelerometers as part of INSs, the following mathematical approaches are most often used to evaluate their errors:

1) Least Squares Method
2) Bayesian methods of statistical estimation
3) Asymptotic filters (Luenberger filters)
4) Kalman filter

The following important points arising from the imposed restrictions and properties of the algorithms used can be attributed to the features of the application of these methods.

Asymptotic filters have a significant advantage due to the adjustable rate of convergence of estimates: it depends on the magnitude of the real part of the roots of the characteristic equation of the evaluator, and this is determined by the selected gain. However, a negative feature of such filters is the complete absence of accounting for noise in measurements and in the system, which leads to significant estimation errors, especially at high amplification factors and, accordingly, short estimation times.

The best estimate for linear systems for unbiased, white, Gaussian noise of system and measurements is theoretically given by the KF. However, this filter requires a very significant amount of a priori information about noise, which is often very unreliable, which leads to disturbances in the operation of the films and a decrease in the accuracy of estimation, up to the divergence of the estimates. Nevertheless, KF have become one of the most widely used estimation algorithms in technical systems, and it is proposed to take it as a basis when considering the calibration of inertial sensors in the INS.

It is also worth noting that the the strapdown version of INS (SINS), is the most used nowadays in a number of practical problems.

Calibration is carried out by a series of indirect measurements of errors using the optimal Kalman filter. For maximum compliance of gyroscopes and accelerometers mathematical models, their description is the same as in the formulas (3)-(6).

The methodology for calibrating inertial sensors based on INS signals during its operation mode differs significantly from the methodology for calibrating the IMU. This is primarily due to the fact that in this case there are no direct measurements of the angular velocity of the equipment rotation and its apparent acceleration, because this is not provided by the the INS output. In this case, only measurements with respect to coordinates, velocity projections and orientation angles remain accessible measurements. For simplicity, we will assume that the coordinates determined by the INS correspond to its mass center, which, when the system is mounted on the turntable, coincides with its center of rotation. This assumption is not fundamental because in case of mismatch between the center of mass of the system under test and the center of rotation, you can always recalculate the readings taking into account the mismatch arm, which is measured before calibration. Taking into account the reservation made, the following measurements can be formed during calibration of inertial sensors of operating INS:

$$Z=\begin{bmatrix} R_{SINSx}-R_{Tx} \\ R_{SINSy}-R_{Ty} \\ R_{SINSz}-R_{Tz} \\ V_{SINSx}-V_{Tx} \\ V_{SINSy}-V_{Ty} \\ V_{SINSz}-V_{Tz} \\ \Theta_{SINSx}-\Theta_{Tx} \\ \Theta_{SINSy}-\Theta_{Ty} \\ \Theta_{SINSz}-\Theta_{Tz} \end{bmatrix}=\begin{bmatrix} \delta R_{SINSx}-\delta R_{Tx} \\ \delta R_{SINSy}-\delta R_{Ty} \\ \delta R_{SINSz}-\delta R_{Tz} \\ \delta V_{SINSx}-\delta V_{Tx} \\ \delta V_{SINSy}-\delta V_{Ty} \\ \delta V_{SINSz}-\delta V_{Tz} \\ \delta\Theta_{SINSx}-\delta\Theta_{Tx} \\ \delta\Theta_{SINSy}-\delta\Theta_{Ty} \\ \delta\Theta_{SINSz}-\delta\Theta_{Tz} \end{bmatrix} \quad (32)$$

These differences can be represented through components of INS errors and errors of its inertial sensors. In this case, the measurements can be reduced to:

$$Z = HX + V, \quad (33)$$

where
$$H = [H_{11}, 0_{9\times 30}]; \tag{34}$$

$$H_{11} = \begin{bmatrix} \frac{1}{R+h} & 0 & 0 & 0 & 0 & 0 & 0 & 0 & 0 \\ 0 & \frac{1}{(R+h)\cos\varphi} & 0 & 0 & 0 & 0 & 0 & 0 & 0 \\ 0 & 0 & 1 & 0 & 0 & 0 & 0 & 0 & 0 \\ \frac{\dot{H}}{R+h} + \Omega_x tg\varphi & -\omega_{az} & \omega_{ay} & 1 & 0 & 0 & 0 & 0 & 0 \\ 0 & \frac{\dot{H}}{R+h} & -\omega_{ax} & 0 & 1 & 0 & 0 & 0 & 0 \\ 0 & 0 & 0 & 0 & 0 & 1 & 0 & 0 & 0 \\ 0 & -\frac{1}{R+h} & 0 & 0 & 0 & 0 & 1 & 0 & 0 \\ \frac{1}{R+h} & 0 & 0 & 0 & 0 & 0 & 0 & 1 & 0 \\ \frac{1}{R+h} tg\varphi & 0 & 0 & 0 & 0 & 0 & 0 & 0 & 1 \end{bmatrix} \tag{35}$$

where φ - latitude of the location of the test-bench with installed INS.

Above written nine measurements and their relations with INS errors are the basis for development of calibration algorithms for inertial sensors of INS in its operation mode. This technique uses the output of location, velocity, and angles that come from the output of the INS. An optimal Kalman filter is used to realize this technique. Let's write the equations of the system in state space to be able to apply the KF. The complete linear model contains the system state equation and the measurement equation:

$$\begin{cases} \dot{\overline{x}} = F\overline{x} + G\overline{w} \\ \overline{z} = H\overline{x} + \overline{v} \end{cases}, \tag{36}$$

where $\overline{x}$ - state vector; $\overline{w}$ - system noise vector; $\overline{v}$ - measurement noise vector; F - system dynamic matrix; G - system noise matrix; H - measurement matrix; $\overline{z}$ - measurement vector formed as differences between INS output and reference values from the test-bench.

Let's set the error model of INS, as well as the error models of gyroscopes and accelerometers, represented in the formulas (3-6).

The errors of INS gyroscopes and accelerometers are reduced to the navigation coordinate system by the following formulas:

$$\delta\omega_x = C_{11}\delta\omega_1 + C_{12}\delta\omega_2 + C_{13}\delta\omega_3, \tag{37}$$

$$\delta\omega_y = C_{21}\delta\omega_1 + C_{22}\delta\omega_2 + C_{23}\delta\omega_3, \tag{38}$$

$$\delta\omega_z = C_{31}\delta\omega_1 + C_{32}\delta\omega_2 + C_{33}\delta\omega_3, \tag{39}$$

где $\delta\omega_1$, $\delta\omega_2$, $\delta\omega_3$ - gyroscope errors, finding by formula 4 for different L.

$$\delta a_x = C_{11}\delta a_1 + C_{12}\delta a_2 + C_{13}\delta a_3 \tag{40}$$

$$\delta a_y = C_{21}\delta a_1 + C_{22}\delta a_2 + C_{23}\delta a_3 \tag{41}$$

$$\delta a_z = C_{31}\delta a_1 + C_{32}\delta a_2 + C_{33}\delta a_3 \tag{42}$$

where δa_1, δa_2, δa_3 - accelerometer errors, finding by formula 4 for different L.

here $C_{11}...C_{33}$ - elements of the cosine guide matrix between the body coordinate system and the selected rotating coordinate system, calculated by INS orientation algorithms.

The system state vector has a dimension of 39 and can be written as follows:

$$\begin{aligned}\bar{X} = [&R_x, R_y, R_z, V_x, V_y, V_z, \Theta_x, \Theta_y, \Theta_z, \upsilon a_{xx}, \upsilon a_{xy}, \upsilon a_{xz}, \upsilon a_{yx}, \upsilon a_{yy}, \upsilon a_{yz}, \upsilon a_{zx}, \upsilon a_{zy}, \upsilon a_{zz},\\ &Ka_{nlx}, Ka_{nly}, Ka_{nlz}, \Delta a_{drx}, \Delta a_{dry}, \Delta a_{drz}, \upsilon g_{xx}, \upsilon g_{xy}, \upsilon g_{xz}, \upsilon g_{yx}, \upsilon g_{yy}, \upsilon g_{yz}, \upsilon g_{zx}, \upsilon g_{zy}, \upsilon g_{zz},\\ &K\omega_{nlx}, K\omega_{nly}, K\omega_{nlz}, \Delta\omega_{drx}, \Delta\omega_{dry}, \Delta\omega_{drz}]^T\end{aligned} \quad (43)$$

System noise vector $\bar{W}$ has an appearance:

$$\bar{W} = [0,0,0,\varepsilon a_x, \varepsilon a_y, \varepsilon a_z, \varepsilon\omega_x, \varepsilon\omega_y, \varepsilon\omega_z, 0_{1\times 30}]^T \quad (44)$$

System noise intensiveness matrix Q_1 has an appearance:

$$Q_1 = \begin{bmatrix} \sigma^2_{SINS11} & 0 & 0 & 0 & 0 & 0 \\ 0 & \sigma^2_{SINS22} & 0 & 0 & 0 & 0 \\ 0 & 0 & \sigma^2_{SINS33} & 0 & 0 & 0 \\ 0 & 0 & 0 & \sigma^2_{SINS44} & 0 & 0 \\ 0 & 0 & 0 & 0 & \sigma^2_{SINS55} & 0 \\ 0 & 0 & 0 & 0 & 0 & \sigma^2_{SINS66} \end{bmatrix} \quad (45)$$

where $\sigma^2_{SINS11}, \sigma^2_{SINS22}, \sigma^2_{SINS33}$ - accelerometer noise RMS; $\sigma^2_{SINS44}, \sigma^2_{SINS55}, \sigma^2_{SINS66}$ - gyros noise RMS.

Dynamic matrix F in this case has an appearance:

$$F = \begin{bmatrix} F_{11} & F_{12} & F_{13} \\ 0_{30\times 9} & 0_{30\times 15} & 0_{30\times 15} \end{bmatrix} \quad (46)$$

where components of this matrix are as follows:

$$F_{11} = \begin{bmatrix} 0 & 0 & 0 & 1 & 0 & 0 & 0 & 0 & 0 \\ 0 & 0 & 0 & 0 & 1 & 0 & 0 & 0 & 0 \\ 0 & 0 & 0 & 0 & 0 & 1 & 0 & 0 & 0 \\ \omega_y^2 + \omega_z^2 - \omega_0^2 & -\omega_x\omega_y + \dot{\omega}_z & -\omega_x\omega_z - \dot{\omega}_y & 0 & 2\omega_z & -2\omega_y & 0 & -a_z & a_y \\ -\omega_x\omega_y - \dot{\omega}_z & \omega_x^2 + \omega_z^2 - \omega_0^2 & -\omega_y\omega_z + \dot{\omega}_x & -2\omega_z & 0 & 2\omega_x & a_z & 0 & -a_x \\ -\omega_x\omega_z + \dot{\omega}_y & -\omega_y\omega_z - \dot{\omega}_x & \omega_y^2 + \omega_z^2 + 2\omega_0^2 & 2\omega_y & 2\omega_x & 0 & -a_y & a_x & 0 \\ 0 & 0 & 0 & 0 & 0 & 0 & 0 & \omega_z & \omega_y \\ 0 & 0 & 0 & 0 & 0 & 0 & -\omega_z & 0 & \omega_x \\ 0 & 0 & 0 & 0 & 0 & 0 & \omega_y & -\omega_x & 0 \end{bmatrix} \quad (47)$$

$$F_{12} = \begin{bmatrix} 0 & 0 & 0 & 0 & 0 & 0 & 0 & 0 & 0 & 0 & 0 & 0 & 0 & 0 & 0 \\ 0 & 0 & 0 & 0 & 0 & 0 & 0 & 0 & 0 & 0 & 0 & 0 & 0 & 0 & 0 \\ 0 & 0 & 0 & 0 & 0 & 0 & 0 & 0 & 0 & 0 & 0 & 0 & 0 & 0 & 0 \\ C_{11} & C_{12} & C_{13} & C_{11} & C_{12} & C_{13} & C_{11} & C_{12} & C_{13} & C_{11}a_x & C_{12}a_y & C_{13}a_z & C_{11} & C_{12} & C_{13} \\ C_{21} & C_{22} & C_{23} & C_{21} & C_{22} & C_{23} & C_{21} & C_{22} & C_{23} & C_{21}a_x & C_{22}a_y & C_{23}a_z & C_{21} & C_{22} & C_{23} \\ C_{31} & C_{32} & C_{33} & C_{31} & C_{32} & C_{33} & C_{31} & C_{32} & C_{33} & C_{31}a_x & C_{32}a_y & C_{33}a_z & C_{31} & C_{32} & C_{33} \\ 0 & 0 & 0 & 0 & 0 & 0 & 0 & 0 & 0 & 0 & 0 & 0 & 0 & 0 & 0 \\ 0 & 0 & 0 & 0 & 0 & 0 & 0 & 0 & 0 & 0 & 0 & 0 & 0 & 0 & 0 \\ 0 & 0 & 0 & 0 & 0 & 0 & 0 & 0 & 0 & 0 & 0 & 0 & 0 & 0 & 0 \end{bmatrix} \quad (48)$$

$$F_{13}=\begin{bmatrix} C_{11} & C_{12} & C_{13} & C_{11} & C_{12} & C_{13} & C_{11} & C_{12} & C_{13} & C_{11}\omega_x & C_{12}\omega_y & C_{13}\omega_z & C_{11} & C_{12} & C_{13} \\ C_{21} & C_{22} & C_{23} & C_{21} & C_{22} & C_{23} & C_{21} & C_{22} & C_{23} & C_{21}\omega_x & C_{22}\omega_y & C_{23}\omega_z & C_{21} & C_{22} & C_{23} \\ C_{31} & C_{32} & C_{33} & C_{31} & C_{32} & C_{33} & C_{31} & C_{32} & C_{33} & C_{31}\omega_x & C_{32}\omega_y & C_{33}\omega_z & C_{31} & C_{32} & C_{33} \\ 0 & 0 & 0 & 0 & 0 & 0 & 0 & 0 & 0 & 0 & 0 & 0 & 0 & 0 & 0 \\ 0 & 0 & 0 & 0 & 0 & 0 & 0 & 0 & 0 & 0 & 0 & 0 & 0 & 0 & 0 \\ 0 & 0 & 0 & 0 & 0 & 0 & 0 & 0 & 0 & 0 & 0 & 0 & 0 & 0 & 0 \\ 0 & 0 & 0 & 0 & 0 & 0 & 0 & 0 & 0 & 0 & 0 & 0 & 0 & 0 & 0 \\ 0 & 0 & 0 & 0 & 0 & 0 & 0 & 0 & 0 & 0 & 0 & 0 & 0 & 0 & 0 \\ 0 & 0 & 0 & 0 & 0 & 0 & 0 & 0 & 0 & 0 & 0 & 0 & 0 & 0 & 0 \end{bmatrix} \quad (49)$$

System noise matrix G has an appearance:

$$G=\begin{bmatrix} G_{11} & 0_{9\times 33} \\ 0_{30\times 6} & 0_{30\times 33} \end{bmatrix} \quad (50)$$

where matrix G_{11} can be written in the form:

$$G_{11}=\begin{bmatrix} 0 & 0 & 0 & 0 & 0 & 0 \\ 0 & 0 & 0 & 0 & 0 & 0 \\ 0 & 0 & 0 & 0 & 0 & 0 \\ C_{11} & C_{12} & C_{13} & 0 & 0 & 0 \\ C_{21} & C_{22} & C_{23} & 0 & 0 & 0 \\ C_{31} & C_{32} & C_{33} & C_{31} & 0 & 0 \\ 0 & 0 & 0 & C_{11} & C_{12} & C_{13} \\ 0 & 0 & 0 & C_{21} & C_{22} & C_{23} \\ 0 & 0 & 0 & C_{31} & C_{32} & C_{33} \end{bmatrix} \quad (51)$$

The measurement noise vector will be written in the following form:

$$\bar{V}=\begin{bmatrix} \delta R_{Tx} & \delta R_{Ty} & \delta R_{Tz} & \delta V_{Tx} & \delta V_{Ty} & \delta V_{Tz} & \delta\Theta_{Tx} & \delta\Theta_{Ty} & \delta\Theta_{Tz} \end{bmatrix}^T, \quad (52)$$

where δR_{Tx} – latitude reference error obtained as a result of test-bench location measurement; δR_{Ty} – longitude reference error obtained as a result of test-bench location measurement; δR_{Tz} – altitude reference error obtained as a result of test-bench location measurement; δV_{Tx} – northern velocity error of the test-bench; δV_{Ty} – eastern velocity error of the test-bench; δV_{Tz} – vertical velocity error of the test-bench; $\delta\Theta_{Tx}, \delta\Theta_{Ty}, \delta\Theta_{Tz}$ - angular error of the test-bench.

Measurements noise intensiveness matrix can be written as:

$$R=\begin{bmatrix} \sigma_{T11}^2 & 0 & 0 & 0 & 0 & 0 & 0 & 0 & 0 \\ 0 & \sigma_{T22}^2 & 0 & 0 & 0 & 0 & 0 & 0 & 0 \\ 0 & 0 & \sigma_{T33}^2 & 0 & 0 & 0 & 0 & 0 & 0 \\ 0 & 0 & 0 & \sigma_{T44}^2 & 0 & 0 & 0 & 0 & 0 \\ 0 & 0 & 0 & 0 & \sigma_{T55}^2 & 0 & 0 & 0 & 0 \\ 0 & 0 & 0 & 0 & 0 & \sigma_{T66}^2 & 0 & 0 & 0 \\ 0 & 0 & 0 & 0 & 0 & 0 & \sigma_{T77}^2 & 0 & 0 \\ 0 & 0 & 0 & 0 & 0 & 0 & 0 & \sigma_{T88}^2 & 0 \\ 0 & 0 & 0 & 0 & 0 & 0 & 0 & 0 & \sigma_{T99}^2 \end{bmatrix} \quad (53)$$

где σ_{T11}^2, σ_{T22}^2, σ_{T33}^2, σ_{T44}^2, σ_{T55}^2, σ_{T66}^2, σ_{T77}^2, σ_{T88}^2, σ_{T99}^2 — latitude, longitude, altitude, velocity and angle noise measurement RMS of bench data.
The relations between matrices are:

$$\Phi = E + FT \tag{54}$$

$$\Gamma = GT \tag{55}$$

Full equations of discret Kalman Filter algorithm in Joseph form are:

$$\widehat{X_k} = \Phi \widehat{x_{k-1}} + K_k (Z_k - H\Phi \widehat{X_{k-1}}) \tag{56}$$

$$S_k = \Phi P_{k-1} \Phi^T + \Gamma Q_1 \Gamma^T \tag{57}$$

$$K_k = S_k H^T (H S_k H^T + R_1)^{-1} \tag{58}$$

$$P_k = [E - K_k H] S_k [E - K_k H]^T + K_k R_1 K_k^T \tag{59}$$

where X_k – system state vector; Z_k – measurement vector; Φ – transition matrix of the system; Γ – random noise matrix; S_k – a priori covariance matrix on the k-th step; P_k – a posteriori covariance matrix on the k-th step; Q_1 – symmetric non-negative defined matrix of system noises; R_1 – symmetric positive defined density matrix of measurement noises;

There are some ways to organize calibration motions (stages) for providing consecutive error observability. It can be like stages in section 4.

6. Results and discussion

Some results of semi-natural simulation of errors in two calibration variants are given below. Differences in calibration methods are highlighted.

The example of gyroscope zero drift finding according to the IMU methodology is shown on figure 2. During the stage 1 in section 4 we find zero drift value in one repetition by formula 12 and for five repetitions by formula 18. If we take parameters from acquiring data and apply them to gyro model with the same characteristics we have residual error about 0.001°/s. It means that initial error reduced by 90%.

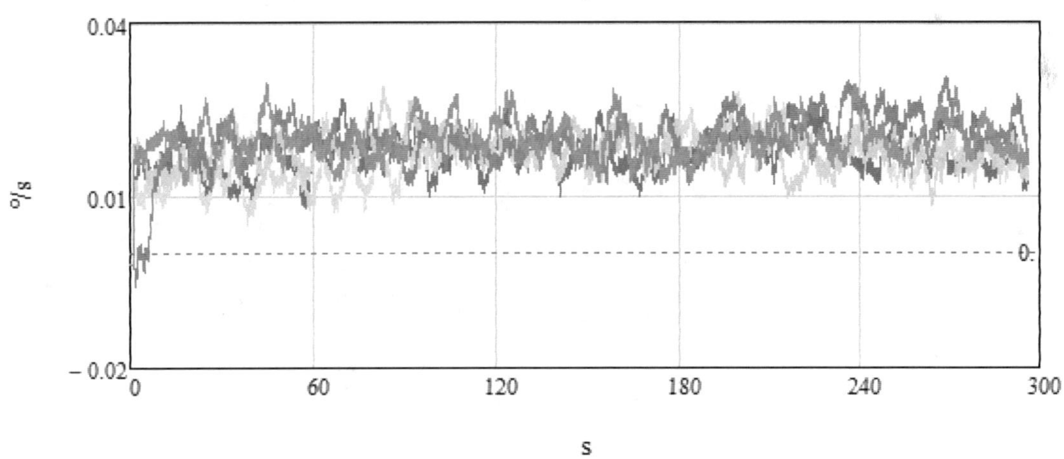

Figure 2. 5 runs of gyroscope zero drift with the applied calibration coefficients (IMU variant).

The instability of the gyroscope scale factor obtained by the IMU variant is shown in figure 3. Stage 2 from section 4 is used. Applying this method by semi-natural experiment (finding coefficients from real data and applying for new model data) we have residual error about 0.01% of in-run instability and 0.1% of run-to-run instability.

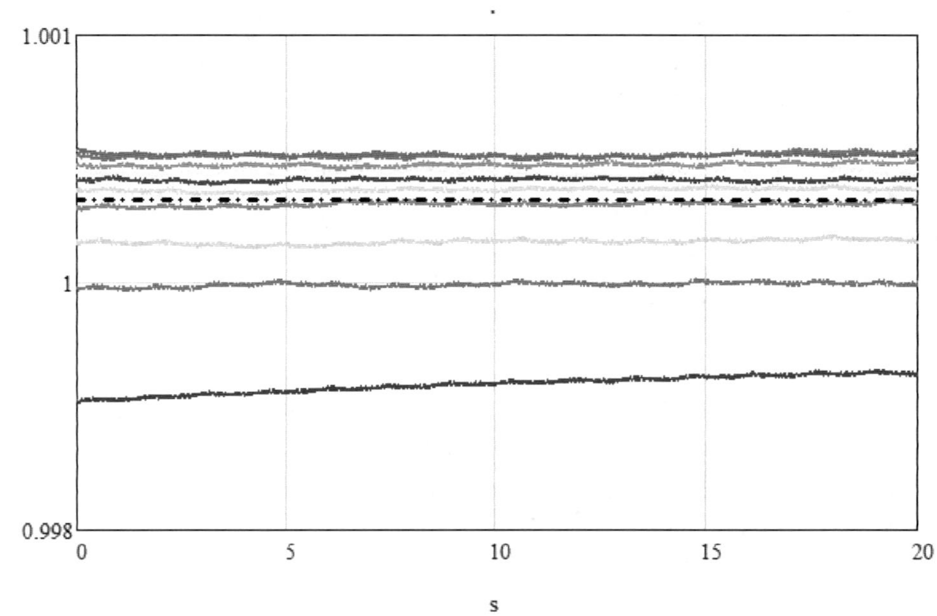

Figure 3. Modeling scale factors for IMU variant at a speed of 500°/s for 10 runs.

In relation to accelerometer parameters we have similar percentage values of residual errors to gyroscope.

For comparison graphic results of semi-natural simulation for gyroscope zero drift are shown on figure 4. On this figure curves of estimated angle and angular velocity errors for 2 measurements are shown. Residual error is about 0.0012 °/s (12%). Time of estimation is about 90 seconds and it's less than set time in the IMU variant.

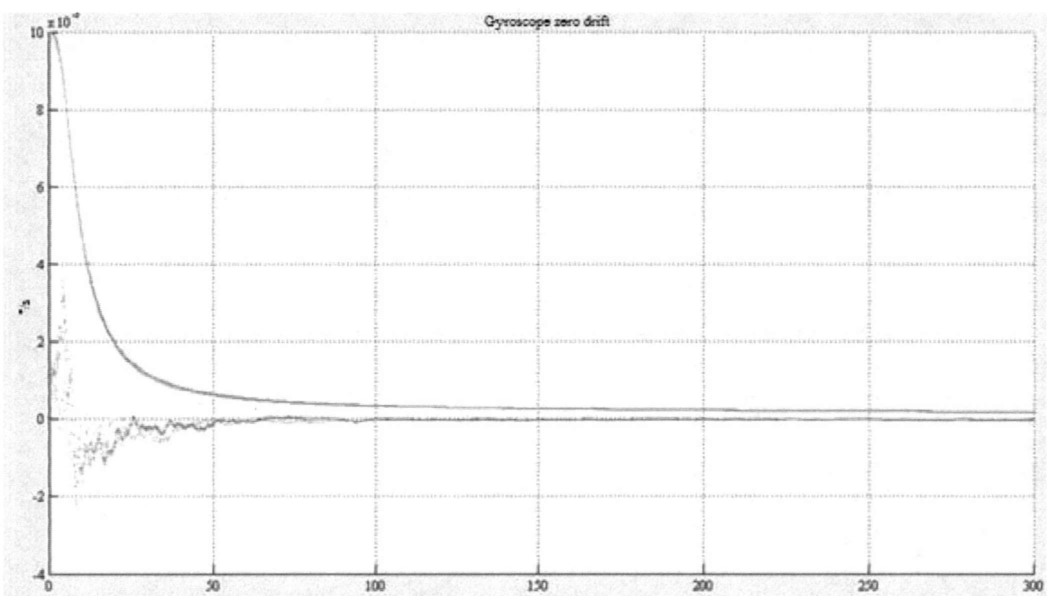

Figure 4. Gyroscope drift estimation for SINS variant with KF method.

Calibration time for SINS is optimal in case of good preliminary estimated noises but in theory IMU calibration at the same stage can be shorter because of direct outputs. The results of the zero drift calibration of gyroscopes and accelerometers showed are similar (table 4), IMU variant is better on insignificant 2-3 percent. It should be noted that the drift values approximately correspond to the declared certification characteristics.

Table 4. Comparison of calibrated errors in run-to-run test by semi-natural simulation.

Characteristic	IMU calibration method (°/s)	SINS calibration method (°/s)
In-run gyroscope zero drift	0.001	0.0012
In-run accelerometer zero drift	0.0011	0.0013

Most international works describe the one method that is more suitable or habitual for researches. For IMU variant we can find both mathematical approaches – Least Squares [7] and Kalman Filter [3]. For SINS variant we can find works with only Kalman Filter approach as in [1]. Some works compare approaches – in [11] LS and KF are compared and in [12] Constraint Kalman Filter and the Sliding Window Filter. The lack of these works for our purpose is limiting itself by simple matrix description. The work [8] describes full diagram which strictly documents all stages of IMU calibration by LS method. In [13] we find full diagram for IMU calibrated by KF method. Unlike these works we are trying to detail the comparison of these methods with full matrices and stages description.

7. Conclusion

This work compares two different devices based on the same sensors with direct and indirect information output. Two different mathematical methods are applied – Least Squares and Kalman Filter respectively. Main points about the comparative analysis of these methods applied to information outputs are:

1) IMU calibration is suitable when we have lack of instability data or bad data in device passport.

2) SINS calibration shows optimal time for measuring in case of correct noise values and IMU calibration time is initially set.

3) Parameter observability depends on motion stages of calibration but no on information output.

4) IMU calibration allows to find in-run instabilities (noises).

5) Both variants allow us to find run-to-run instabilities by a number of repetitions.

6) If we use Extend Kalman Filter we can also find instabilities but it was outside our task.

7) Calibration SINS (KF) in case of unsuccessful determination of system noise may be delayed or dispersed.

8) Results of semi-natural modeling of error parameters are almost identical with insignificant advance of the IMU variant.

These facts should be taken into account if a researcher had both variants and variable conditions for calibration conducting. All else being equal it is recommended to prefer a variant with IMU outputs. The study will be continued, since the comparison can be carried out more correctly with more correct mathematical error model and more close calibration stages.

Acknowledgments

The reported study was funded by Russian Foundation for Basic Research (RFBR) according to the research project No. 19-08-01223.

References

[1] Du Sh, Sun W and Yang 2016 MEMS IMU Error Mitigation Using Rotation Modulation Technique. *Sensors* **16(12)** 20 doi: https://doi.org/10.3390/s16122017

[2] Tang Q, Wang X and Yang Q 2017 Scale factor model analysis of MEMS gyroscopes. *Microsystem Technologies* **23** 1215 doi: https://doi.org/10.1007/s00542-016-2825-3

[3] Jafari M, Sahebjameyan M, Moshiri B and Najafabadi T A 2015 Skew redundant MEMS IMU calibration using a Kalman filter. *Measurement Science and Technology* **26(10)** 15 doi: 10.1088/0957-0233/26/10/105002

[4] Feng Yi, Li Xi and Zhang Xi 2015 An Adaptive Compensation Algorithm for Temperature Drift of Micro-Electro-Mechanical Systems Gyroscopes Using a Strong Tracking Kalman Filter. *Sensors* **15(5)** 11222 doi: 10.3390/s150511222

[5] Zhang H, Wu Yu, Wu M, Hu Xi and Zha Ya 2008 A Multi-Position Calibration Algorithm for Inertial Measurement Units. *AIAA Guidance, Navigation and Control Conference and Exhibit* (Honolulu, Hawaii) 15 doi: 10.2514/6.2008-7437

[6] Syed Z F, Aggarwal P, Goodall P C, Niu X and El-Sheimy N 2007 A new multi-position calibration method for MEMS inertial navigation systems. *Measurement Science and Technology* **18(7)** 1897 doi: 10.1088/0957-0233/18/7/016

[7] Lee D, Lee S, Park S and Ko S 2011 Test and Error Parameter Estimation for MEMS-Based Low Cost IMU Calibration. *International Journal of Precision Engineering and Manufacturing.* **12** 597 doi: 10.1007/s12541-011-0077-9

[8] Zhang B, Chua H, Suna T, Guo L 2017 Thermal calibration of a tri-axial MEMS gyroscope based on Parameter-Interpolation method. *Sensors and Actuators A: Physical* **261** 103 doi: 10.1016/j.sna.2017.04.013

[9] Duong D Q, Sun J, Nguyen T P, Luo L 2016 Attitude Estimation by Using MEMS IMU with Fuzzy Tuned Complementary Filter. *IEEE Int. Conf. on Electronic Information and Communication Technol.* **372** doi: 10.1109/iceict.2016.7879720

[10] Veremeenko K K and Savel'ev V M 2013, In-Flight Alignment of a Strapdown Inertial Navigation System. *Journal of Computer and Systems Sciences International* **52** 106 doi: 10.1134/S1064230712060147

[11] Jørgensen M J, Paccagnan D, Poulsen N K, Larsen M B 2014 IMU calibration and validation in a factory, remote on land and at sea. *IEEE/ION Position, Location and Navigation Symposium - PLANS 2014* (Monterey, CA, USA) doi: 10.1109/PLANS.2014.6851514

[12] Clement L E, Peretroukhin V, Lambert J, Kelly J 2015 The Battle for Filter Supremacy: A Comparative Study of the Multi-State Constraint Kalman Filter and the Sliding Window Filter. *12th Conference on Computer and Robot Vision* (Halifax, NS, Canada) doi: 10.1109/CRV.2015.11

[13] Yang H, Zhou B, Wang L, Xing H and Zhang R 2018 A Novel Tri-Axial MEMS Gyroscope Calibration Method over a Full Temperature Range. *Sensors* **18(9)** 15 doi: 10.3390/s18093004

IOP Conf. Series: Materials Science and Engineering **868** (2020) 012038 doi:10.1088/1757-899X/868/1/012038

Consideration of aeroservoelasticity requirements in the development of highly maneuverable unmanned aerial vehicle

S Parafes'* and I Turkin

Chair of Design and Strength of Aviation, Rocket and Space Products, *Moscow Aviation Institute (National Research University)*, 4 Volokolamskoe Shosse, 125993 Moscow Russian Federation

*E-mail: s.parafes@mail.ru

Abstract. An approach to the joint design construction and stabilization system of the highly maneuverable unmanned aerial vehicle account with requirements of aeroservoelacticity, the most important of which is to prevent self-oscillations in the unmanned aerial vehicle stabilization loop. The approach includes three stages. The first stage is the preliminary choice of structure and basic parameters of the unmanned aerial vehicle stabilization loop account with aeroservoelacticity requirements. The second stage is the separate design of subsystems (airframe, stabilization system) by traditional methods in compliance with the requirements formulated at the first stage. The goals to be achieved at this stage are to obtain a rational (from the standpoint of mass and technological perfection) design of the unmanned aerial vehicle airframe, and a stabilization system that provides the required dynamics and the accuracy of execution of commands, as well as the minimum cost, dimensions and weight. The third stage is the coordination of the selected structure and design parameters of the airframe and stabilization system, aimed at meeting the aeroservoelasticity requirements (the authors use the model of the unmanned aerial vehicle stabilization loop stability study, reflecting the specifics of the classes of highly maneuverable aircraft). An example of the process of developing the highly maneuverable unmanned aerial vehicle account with aeroservoelacticity requirements is given.

1. Introduction

The most important requirement of aeroservoelasticity, which must be provided in the development of highly maneuverable unmanned aerial vehicle (UAV), is to prevent self-oscillations in the UAV stabilization loop. Self-oscillations lead to "clogging" of automatic control system (ACS) channels with high-frequency components of signals, which negatively affects the controllability of the UAV, disrupting the normal operation of on-Board equipment, and leads to a significant decrease in the quality of its functioning up to failure.

The problems associated with ensuring the stability of the UAV stabilization loop should be solved starting from the preliminary design stage. Detection of the presence of oscillations in the loop only at the final stages of design (during ground and flight tests) leads to significant additional costs for carrying out measures to eliminate them. The best results should be expected in the joint design of the UAV and stabilization system. Analysis of the state of the problem of ensuring aeroservoelasticity requirements of in the design of UAVS shows that to date, well-developed methods for studying the aeroelastic stability of UAVS with ACS, when the parameters of the latter are considered known [1-6]. However, the proposed approaches in the field of joint design of UAV and stabilization system [7-10] are still far from solving real full-size problems.

IOP Conf. Series: Materials Science and Engineering **868** (2020) 012038 doi:10.1088/1757-899X/868/1/012038

2. Methodology

2.1. Statement and General scheme of the problem solution

The problem of joint design of the airframe structure and the stabilization system of highly maneuverable UAV, when as a limitation imposed by the joint functioning of subsystems (airframe design and stabilization system), is the requirement of stability stabilization loop of the elastic UAV has the following formulation. To be known: aeroballistic and mass-geometric parameters of the UAV; its internal layout; the stabilization system structural scheme; parameters of the calculated flight modes; requirements to the airframe design (strength, stability, etc.) and stabilization system (accuracy, speed, etc.); aeroservoelasticity requirements. It is required to obtain an agreed decision on the structure and parameters of the airframe design and UAV stabilization system, taking into account the complex requirements.

To solve the problem of joint design of the airframe and stabilization system, taking into account aeroservoelasticity requirements, an iterative approach was developed, including three stages (figure 1):

- choice of structure and main parameters of UAV stabilization loop;
- separate design of the airframe structure and the stabilization system;
- coordination of the selected parameters of the airframe and the stabilization system, aimed at fulfilling of the aeroservoelasticity requirements.

2.2. Choice of structure and main parameters of UAV stabilization loop

Let's consider an axisymmetric UAV of the cross-wing or wingless aerodynamic scheme with the aerodynamic control. Since the UAV is axisymmetric, it has identical pitch and course channels. We next consider the UAV motion on the pitch.

The initial equations of the axisymmetric UAV with the most common aerodynamic method of creating control forces and moments are the equation of forces and moments. The equations can be written by applying dynamic coefficients that are used to describe the aircraft trajectory in flight dynamics: the damping coefficient a_1, the static stability coefficient a_2, the efficiency coefficient of the aerodynamic rudders a_3, the coefficient of lift force created by the attack angle a_4 and the coefficient of lift force created by the aerodynamic rudders deflection a_6 ($a_6 = 0$).

The transfer functions of the rigid UAV as a control object in the pitch channel for angular velocity ω and linear acceleration W (in the installation places of sensors) have the form:

$$W_\omega(p)\Big|_{x_{\text{AVS}}} = \frac{\omega}{\delta}\Big|_{x_{\text{AVS}}} = \mp k_r \frac{1 + T_{1c}p}{1 + 2\xi_r T_r p + T_r^2 p^2}; \quad (1)$$

$$W_W(p)\Big|_{x_{\text{LAS}}} = \frac{W}{\delta}\Big|_{x_{\text{LAS}}} = \mp V k_r \frac{1}{1 + 2\xi_r T_r p + T_r^2 p^2}; \quad (2)$$

$$k_r = \frac{a_3 a_4}{a_2 + a_1 a_4}; \quad T_{1c} = \frac{1}{a_4}; \quad \xi_r = \frac{a_1 + a_4}{2\sqrt{a_2 + a_1 a_4}}; \quad T_r = \frac{1}{\sqrt{a_2 + a_1 a_4}};$$

$$a_1 = -\frac{m_z^{\omega_z} q S L^2}{V I_z}; \quad a_2 = -\frac{c_y^\alpha (\bar{x}_m - \bar{x}_p) q S L}{I_z}; \quad a_3 = \mp \frac{c_y^\delta (\bar{x}_m - \bar{x}_r) q S L}{I_z}; \quad a_4 = \frac{c_y^\alpha q S}{mV} + \frac{P}{mV};$$

V, q – the speed and the dynamic pressure of the UAV; $m_z^{\omega_z}$ – the derivative of coefficient of the moment of aerodynamic damping; c_y^α, c_y^δ – the derivative of lift coefficient of the UAV by angle of attack and rudder deflection; P – the engine thrust; $\bar{x}_m, \bar{x}_p, \bar{x}_r$ – coordinates of the center of mass, the center of pressure, and the axis of rotation of rudder, related to the length of the UAV L; m, I_z, S – mass, moment of inertia and characteristic area of the UAV. In the expressions "minus" sign corresponds to the normal aerodynamic layout, "plus" sign – tail-first layout.

IOP Conf. Series: Materials Science and Engineering **868** (2020) 012038 doi:10.1088/1757-899X/868/1/012038

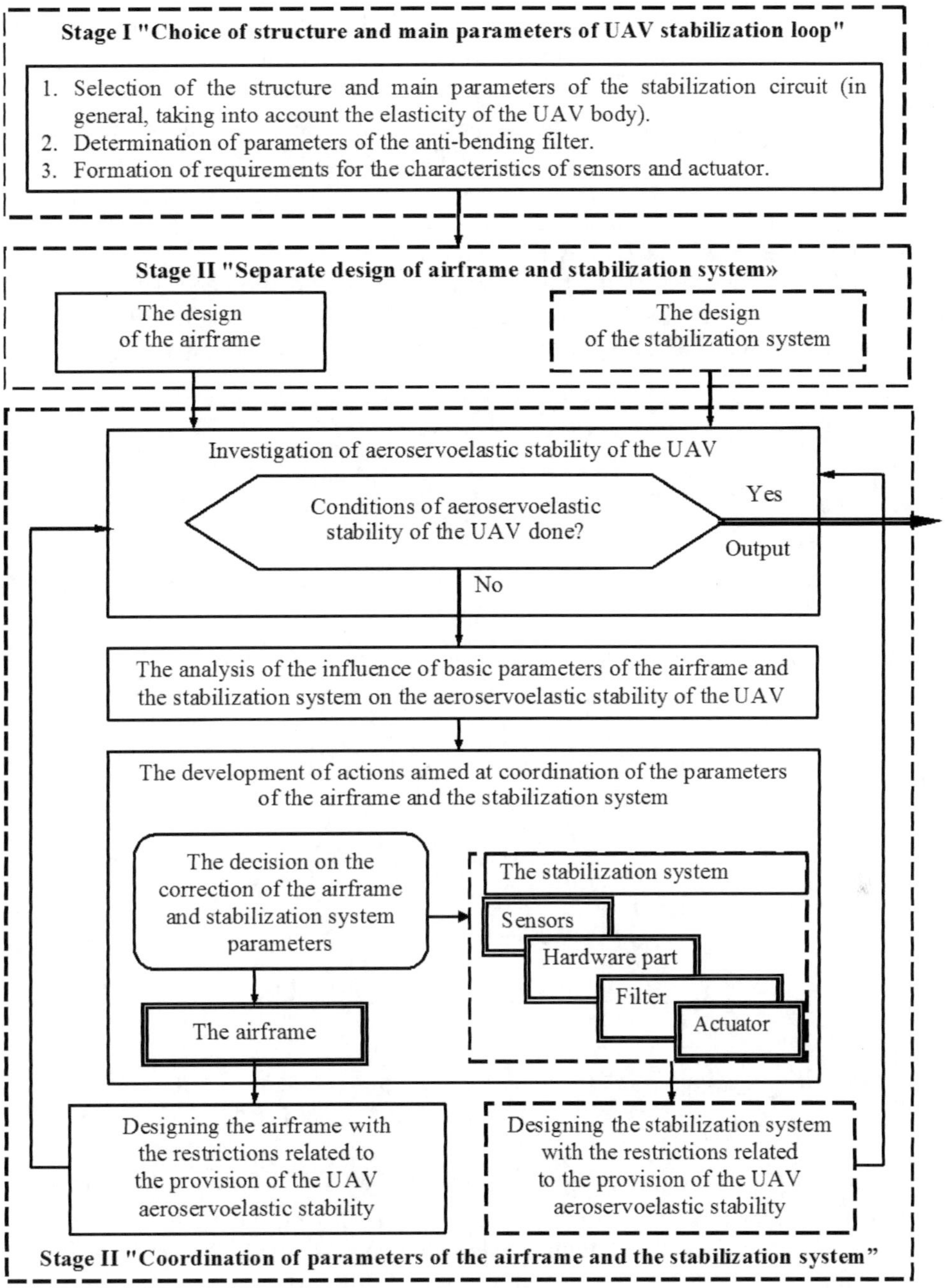

Figure 1. Enlarged scheme of solving the problem of joint design.

Transfer functions defined by the expressions (1), (2) is obtained under the conditions of neglecting the rudder lift force (because of its smallness in comparison with the lift force of the UAV) and the without Coriolis acceleration generated by the rotation of the UAV from the effects outflowing jet of the jet engine. Full UAV transfer functions include transfer functions of rigid and elastic UAV (the latter, as a rule, are approximate and take into account only the elastic vibrations of the body on the 1st and 2nd tone).

On the highly manoeuvrable UAV, the most widely used stabilization system for the pitch and course channels with two feedbacks: angular velocity ω (formed by the angular velocity sensor (AVS) with gain k_{AVS}) and linear acceleration W (formed by the linear acceleration sensor (LAS) with gain k_{LAS}). Feedbacks form two closed circuit: first circuit, with the angular velocity feedback, called the damping circuit, the second - with the linear acceleration, called the stabilization circuit.

Assuming the transfer functions of the drive and sensors of the ACS equal to one (which is acceptable at frequencies that characterize the trajectory of the UAV), we get an expression for the transfer function of a closed-loop of the stabilization system with two feedbacks.

At the frequencies that characterize the trajectory of the UAVs motion and determine its dynamic properties as the link of the control loop and the accuracy of targeting, the transfer function of the closed-loop of the stabilization system can be represented as an oscillatory link:

$$W_{st}(p) = \frac{W}{\lambda} = \frac{k_{st}}{1 + 2\xi_{st} T_{st} p + T_{st}^2 p^2};$$

$$k_{st} = \frac{V a_3 a_4 k_{LAS}}{V a_3 a_4 k_{LAS} + a_2};\ \omega_{st} = \frac{1}{T_{st}} = \sqrt{V a_3 a_4 k_{LAS} + a_2};\ \xi_{st} = \frac{a_3 k_{AVS}}{2\omega_{st}},$$

where k_{st} is the gain factor; ξ_{st} – the damping rate; T_{st} – the time constant; ω_{st} – the natural frequency. The parameters ω_{st} and ξ_{st} of the closed-loop stabilization system clearly define its response time τ and the amount of emission Δ_1 on the intermittent control command (figure 2). The parameters of the transition process τ, Δ_1 are specified in the technical specification for development of the UAV.

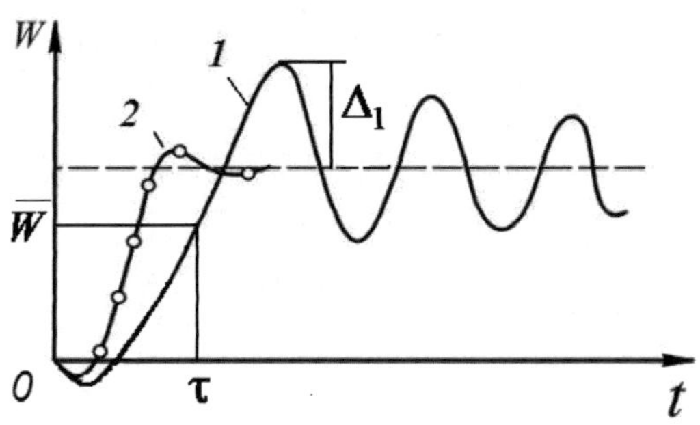

Figure 2. The reaction of the UAV by a step-input command: *1* – the statically sustainable unstabilized UAV; *2* – the UAV with on-board stabilization system.

The coefficients k_{AVS} and k_{LAS} affect the parameters of the transfer function $W_{st}(p)$, the transition process and thus on the dynamic properties of the UAV, improving them (figure 2). Providing the required dynamic properties of the UAV with the stabilization system and requirements of its stability and determine, ultimately, the choice of the required values of the coefficients k_{AVS} and k_{LAS}.

To ensure of the stabilization system stability at the frequencies of elastic oscillations of the UAV, a filter with a transfer function of the form is introduced into its composition

$$W_f(p) = (1 + 2\xi_1 T_1 p + T_1^2 p^2)/(1 + 2\xi_2 T_2 p + T_2^2 p^2). \quad (3)$$

The filter is tuned to the frequency of the suppression of the 1st and optionally the 2nd tone of the bending vibrations of the UAVs body. The ratio of the time constants T_2/T_1 in equation (3) is typically in the range of from 0.5 to 2, the damping coefficients ξ_1 is from 0 to 0.2, and ξ_2 – 0.3 to 1.

IOP Conf. Series: Materials Science and Engineering **868** (2020) 012038 doi:10.1088/1757-899X/868/1/012038

2.3. Separate design of the airframe structure and the stabilization system

The next stage of the considered problem (figure 1, stage II) is separate design of subsystems: airframe design and stabilization system. The purpose of this stage is to obtain a rational design and technological solutions of UAV from the standpoint of functional and technological perfections, as well as a stabilization system with the required characteristics of accuracy, speed and reliability with minimum size and weight. At this stage, the design of individual UAV subsystems is carried out by traditional methods, the most acceptable for each subsystem, in compliance with the requirements formulated at the previous stage of solving the problem under consideration – the problem of choosing the parameters of the stabilization loop of the elastic UAV.

2.4. Coordination of the selected parameters of the airframe and the stabilization system

The solution of this problem (figure 1, stage III) is aimed at ensuring the requirements of aeroservoelasticity. The mathematical model of UAV stabilization loop stability study is based on the computational model of aeroelastic oscillations of UAV. The construction of the computational model is associated with the choice of the most important forms of motion. For the considered highly maneuverable UAV, these are: bending and torsional oscillations of the rudders (generalized coordinates q_1, q_2) and bending oscillations of the UAV body on the 1st and 2nd tones (generalized coordinates q_3, q_4). Oscillations of the wing lower tones are taken into account through the bending oscillations of the UAV body. The system of equations, compiled by the method of generalized coordinates and recorded in the operator form, has the form:

$$\sum_{j=1}^{4} f_{ij}(p) = 0;\ f_{ij}(p) = m_{ij}p^2 + (h_{ij} + d_{ij}V)p + g_{ij} + b_{ij}V^2;\ (i, j) = 1,2,3,4\,, \tag{4}$$

where m_{ij}, h_{ij}, g_{ij} – the coefficients of inertia, damping and stiffness; h_{ij}, b_{ij} – aerodynamic coefficients (damping and stiffness).

As an example, we give expressions for the nonzero coefficients of equation (4) for a wingless UAV of the normal aerodynamic layout and "+" scheme equipped with aerodynamic rudders with a straight axis of rotation in a supersonic flow:

- coefficients of inertia:

$$m_{11} = 2J_{xx};\ m_{12} = m_{21} = -2J_{xz};\ m_{22} = 2J_{zz};\ m_{13} = m_{31} = 2S_x\psi_1(x_0) + 2J_{xz}\dot{\psi}_1(x_0);$$
$$m_{14} = m_{41} = 2S_x\psi_2(x_0) + 2J_{xz}\dot{\psi}_2(x_0);\ m_{23} = m_{32} = -2S_z\psi_1(x_0) - 2J_{zz}\dot{\psi}_1(x_0);$$
$$m_{24} = m_{42} = -2S_z\psi_2(x_0) - 2J_{zz}\dot{\psi}_2(x_0);\ m_{33} = \int_0^{l_b} m_b(x)\psi_1^2(x)dx;\ m_{44} = \int_0^{l_b} m_b(x)\psi_2^2(x)dx,$$

where J_{xx}, J_{zz}, J_{xz} and S_x, S_z – the moments of inertia and the static moments of the rudder around the axes Ox and Oz (the Oz-axis coincides with the axis of rotation); $\psi_1(x), \psi_2(x)$ – the forms of the deflections of the UAV body axis during its deformation by the 1st and 2nd tones, respectively (normalized so that $\dot{\psi}_1(x_0) = \dot{\psi}_2(x_0) = 1$); l_b, $m_b(x)$ – the length and the mass per unit of the UAV body; x_0 – the coordinate of the axis of rotation of the rudder;

- coefficients of stiffness and structural damping:

$$g_{ii} = m_{ii}(2\pi f_i)^2;\ h_{ii} = 2\nu_i f_i m_{ii};\ i = 1, 2, 3, 4,$$

where f_1, f_2 – frequencies and logarithmic decrements of respective natural oscillations of the UAV;

- aerodynamic coefficients:

$$b_{12}=\overline{b}_{12};\ b_{13}=-\overline{b}_{12};\ b_{14}=-\overline{b}_{12};\ b_{22}=\overline{b}_{22};\ b_{23}=-\overline{b}_{22};\ b_{24}=-\overline{b}_{22};$$
$$b_{32}=-\overline{b}_{22}-\overline{b}_{33}\psi_1(x_0);\ b_{33}=\overline{b}_{22}+\overline{b}_{33}\psi_1(x_0);\ b_{34}=\overline{b}_{22}+\overline{b}_{33}\psi_1(x_0);$$
$$b_{42}=-\overline{b}_{22}-\overline{b}_{33}\psi_2(x_0);\ b_{43}=\overline{b}_{22}+\overline{b}_{33}\psi_2(x_0);\ b_{44}=\overline{b}_{22}+\overline{b}_{33}\psi_2(x_0);$$
$$d_{11}=\overline{d}_{11};\ d_{12}=\overline{d}_{12};\ d_{13}=-\overline{d}_{12}-\overline{b}_{12}\psi_1(x_0);\ d_{14}=-\overline{d}_{12}-\overline{b}_{12}\psi_2(x_0);$$
$$d_{21}=\overline{d}_{21};\ d_{22}=\overline{d}_{22};\ d_{23}=-\overline{d}_{22}-\overline{b}_{22}\psi_1(x_0);\ d_{24}=-\overline{d}_{22}-\overline{b}_{22}\psi_2(x_0);$$
$$d_{31}=-\overline{b}_{12}\psi_1(x_0)-\overline{d}_{21};\ d_{32}=-\overline{d}_{22}-\overline{b}_{22}\psi_1(x_0);\ d_{33}=\overline{b}_{22}\psi_1(x_0)+\overline{b}_{33}\psi_1^2(x_0)+\overline{d}_{22}+\overline{b}_{22}\psi_1(x_0);$$
$$d_{34}=\overline{b}_{22}\psi_1(x_0)+\overline{b}_{33}\psi_1(x_0)\psi_2(x_0)+\overline{d}_{22}+\overline{b}_{22}\psi_2(x_0)_4;\ d_{41}=-\overline{b}_{12}\psi_2(x_0)-\overline{d}_{21};\ d_{42}=-\overline{d}_{22}-\overline{b}_{22}\psi_2(x_0);$$
$$d_{43}=\overline{b}_{22}\psi_2(x_0)+\overline{b}_{33}\psi_1(x_0)\psi_2(x_0)+\overline{d}_{22}+\overline{b}_{22}\psi_1(x_0);\ d_{44}=\overline{b}_{22}\psi_2(x_0)+\overline{b}_{33}\psi_2^2(x_0)+\overline{d}_{22}+\overline{b}_{22}\psi_2(x_0).$$

The coefficients $\overline{b}_{12}$, $\overline{b}_{22}$, $\overline{b}_{33}$, $\overline{d}_{11}$, $\overline{d}_{12}$, $\overline{d}_{21}$, $\overline{d}_{22}$ are defined for two rudders at the motionless of the UAV body:

$$\overline{b}_{12}=-\rho c_y^\delta\int_{z_0}^{l+z_0} bz\,dz;\ \overline{b}_{22}=-\rho c_y^\delta\int_{z_0}^{l+z_0} b^2(\overline{x}_0-\overline{x}_F)dz;\ \overline{b}_{33}=\rho c_y^\delta\int_{z_0}^{l+z_0} b\,dz;$$

$$\overline{d}_{11}=\rho c_y^\delta\int_{z_0}^{l+z_0} bz^2dz;\ \overline{d}_{12}=\overline{d}_{21}=\rho c_y^\delta\int_{z_0}^{l+z_0} b^2(\overline{x}_0-\overline{x}_F)z\,dz;\ \overline{d}_{22}=\rho c_y^\delta\int_{z_0}^{l+z_0} b^3\left[(\overline{x}_0-\overline{x}_F)^2+\frac{1}{12}\right]dz,$$

where ρ – the density of air; c_y^δ – the derivative of lift coefficient by angle of rudder deflection; *b*, *l* – the chord and the span of the rudder; z_0 – the distance from the place of fastening of the rudder shaft to the side chord; *z* – the distance to the considered chord section of the rudder; $\overline{x}_0=x_0/b$, $\overline{x}_F=x_F/b$ – the distance from the rudder nose to the axis of rotation and the aerodynamic focus, respectively, divided by the rudder chord.

From equation (4) can determine transfer functions $W_1(p)$ and $W_2(p)$ from the generalized coordinate q_2, representing the deviation of the rudder, δ to coordinates q_3 and q_4, characterizing the UAV body motion. Considering the deflection forms of the UAV body axis in the AVS installation place ($\psi_1'(x_{\text{AVS}})$, $\psi_2'(x_{\text{AVS}})$) and derived forms in the LAS installation place ($\psi_1(x_{\text{LAS}})$, $\psi_2(x_{\text{LAS}})$ when the flexural deformations of the UAV body on the 1st and 2nd tone, respectively, we write the transfer function of the elastic UAV in the channel pitch by angular velocity (in the place of installation AVS) and linear acceleration (in place of LAS):

$$W_\omega^{\text{e}}(p)\Big|_{x_{\text{AVS}}}=\frac{\omega}{\delta}\Big|_{x_{AVS}}=\psi_1'(x_{\text{AVS}})pW_1(p)+\psi_2'(x_{\text{AVS}})pW_2(p);\qquad(5)$$

$$W_\omega^{\text{e}}(p)\Big|_{x_{\text{AVS}}}=\frac{\omega}{\delta}\Big|_{x_{AVS}}=\psi_1'(x_{\text{AVS}})pW_1(p)+\psi_2'(x_{\text{AVS}})pW_2(p),\qquad(6)$$

The complete transfer functions of the UAV $W_{\text{UAV}}^{\omega}(p)$ and $W_{\text{UAV}}^{W}(p)$ can be obtained by summing the corresponding transfer functions (1), (5) and (2), (6).

One possible way of analyzing the UAV stabilization loop stability is the comparison of the amplitude and phase frequency characteristics of the transfer function $W_{\text{UAV}}(p)$ and feedback transfer function $W_{\text{ACS}}^{-1}(p)$. According to the frequency the Nyquist criterion the product of these transfer functions on the boundary of stability is equal to one. The transfer function $W_{\text{UAV}}(p)$ includes a transfer function of the hard and the flexible UAV, and the transfer function $W_{\text{ACS}}(p)$ – the transfer functions of the sensors, the amplifier-converter tract of the stabilization system and the control surface actuator.

IOP Conf. Series: Materials Science and Engineering **868** (2020) 012038 doi:10.1088/1757-899X/868/1/012038

Depending on the studied circuit (damping or stabilization), $W^{\omega}_{\text{UAV}}(p)$ or $W^{W}_{\text{UAV}}(p)$ are considered as $W_{\text{UAV}}(p)$, and $W^{\omega}_{\text{ACS}}(p)$ or $W^{W}_{\text{ACS}}(p)$ as $W_{\text{ACS}}(p)$.

Using the proposed model, the study of the UAV stabilization loop stability (figure 1, stage III). In case of noncompliance of the requirements of aeroservoelasticity performed parametric analysis of the influence of the main parameters of the airframe and system stabilization on the stability of the loop, which produced measures to the parameters of the airframe and system stabilization. These measures can be associated with the correction of the characteristics of the airframe design (primarily, natural oscillation frequencies) or the parameters of the stabilization system (optimization of the location of sensors along the length of the UAV body, correction of parameters or installation of new filters (configured to suppress elastic vibrations of the UAV), changing the frequency characteristics of the steering actuators). In the first case, a new problem of designing the UAV design is solved, in the second-the stabilization system. The iterative design process ends when the requirements of aeroservoelasticity will be satisfied.

3. Results and discussion

Let's consider an example of application of the offered approach to joint design of the airframe structure and the stabilization system of the UAV taking into account requirements of aeroservoelasticity. We consider the hypothetical highly maneuverable UAV of the normal aerodynamic layout, having aerodynamic rudders and the stabilization system with two feedbacks. For the design mode is adopted, the mode with the maximum dynamic pressure ($q = q_{\max} = 1$ MPa), corresponding to the end of the active leg of the UAVs flight.

When forming the stabilization loop (figure 1, stage I), the requirements of aeroservoelasticity were provided, including due to:

- the rational placement of the AVS near the antinode of the elastic line of the 1st tone of elastic oscillations of the UAVs body (taking into account the possibility of UAV layout);
- the filter settings to suppress the elastic oscillations of the body on the 1st tone.

Next, the tasks of separate design of the airframe structure and stabilization system were solved (figure 1, stage II). The characteristics of the natural oscillations of the UAV body are shown in the figure 3. There are also two characteristic cross sections of the body: the section with AVS and LAS (section I-I) and the section with axis of rotation of the rudders (section II-II). The parameters of the rudder are presented in table 1. The logarithmic decrements of the rudder and UAV body oscillations were assumed to be 0.05.

Analysis of the UAV stabilization loop stability in the framework of the design parameters harmonization of the airframe and stabilization system (figure 1, stage III) showed a lack of stability in the damping circuit at the frequency of the 2nd tone of flexural oscillations of the of the UAV body, which is evidenced by the intersection of the frequency response of transfer functions $W^{\omega}_{\text{UAV}}(p)$ (curve *1*) and $(W^{\omega}_{\text{ACS}}(p))^{-1}$ (curve *2*) in figure 4. Since we will use only the amplitude methods to ensure stability, we will limit ourselves to considering only the amplitude frequency characteristics of the elastic UAV and the stabilization system.

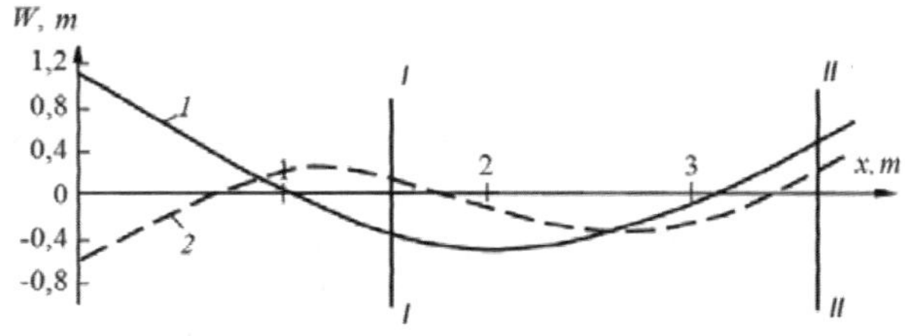

Figure 3. Forms of natural bending oscillations of the UAV body:

IOP Conf. Series: Materials Science and Engineering **868** (2020) 012038 doi:10.1088/1757-899X/868/1/012038

1 – 1st tone, f_3 = 40 Hz; 2nd tone, f_4 = 96 Hz.

Table 1. Rudder parameters.

Rudder parameter	Modified Rudder	Source Rudder
$J_{xx}(kg \cdot m^2)$	0.135	0.15
$J_{zz}(kg \cdot m^2)$	0.0085	0.01
$J_{xz}(kg \cdot m^2)$	-0.005	0.005
f_1(Hz)	97	107
f_2(Hz)	285	320

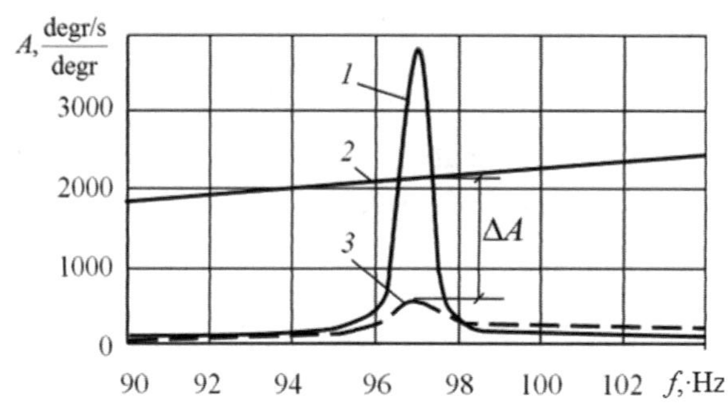

Figure 4. To the study of measures aimed at ensuring the damping circuit stability.

A detailed analysis of the presented dependencies allows us to conclude that a significant rise in the module $W^{\omega}_{\text{UAV}}(p)$ (figure 4, curve *1*) is caused by the proximity of the frequencies of the body's own bending oscillations along the 2nd tone and the bending oscillations of the rudder, which led to a resonant amplification of the amplitude at this frequency. Therefore, it is advisable as a measure to ensure the damping circuit stability to modify the design of the rudder in order to increase its frequency of bending oscillations (up to 107 Hz). As the result of solving a new design problem with an additional restriction on the frequency of natural bending oscillations of the rudder ($f_1 \geq 107Hz$), the rudder with new parameters is obtained (table 1). Figure 4 presents the results of the study of the stability of the damping circuit of the UAV with a modified rudder, from which it follows that now the circuit has sufficient reserves of stability modulo ΔA-modulo margin): stability is achieved by significantly reducing the modulus of the transfer function of the elastic UAV with the modified rudder $W^{\omega}_{\text{UAV}}(p)$ (figure 4, curve *2*).

4. Conclusion

The problem of joint design of the airframe structure and the stabilization system of the highly maneuverable UAV is formulated taking into account the requirements of aeroservoelasticity and the iterative approach to its solution is proposed, including:

- choice of structure and main parameters of UAV stabilization loop;
- separate design of the airframe structure and the stabilization system;
- coordination of the selected parameters of the airframe and the stabilization system, aimed at fulfilling of the aeroservoelasticity requirements.

IOP Conf. Series: Materials Science and Engineering **868** (2020) 012038 doi:10.1088/1757-899X/868/1/012038

The problems of selecting the main parameters of the UAV stabilization loop and matching the parameters of the airframe design and stabilization system in order to meet the requirement of aeroservoelasticity are considered in more detail.
An example of the matching problem of the parameters of the airframe design and stabilization system of the highly maneuverable UAV is solving.

The novelty of the manuscript lies primarily in the authors' original approach to the joint design of the construction and the stabilization system of the highly maneuverable unmanned aerial vehicles with requirements of aeroservoelasticity, as well as original models, reflecting the specific use of such UAVS.

References

[1] Idan M, Karpel M and Moulin B 1999 Aeroservoelastic Interaction Between Aircraft Structural and Control Design Schemes. *Journal of Guidance, Control, and Dynamics* **22(4)** 513

[2] Karpel M 2001 Procedures and Models for Aeroservoelastic Analysis and Design. *Journal of Applied Mathematics and Mechanics* (ZAMM) **81(9)** 579

[3] Balas G J, Moreno C and Seiler P J 2012 Robust Aeroservoelastic Control Utilizing Physics-Based Aerodynamic Sensing. *AIAA Guidance, Navigation, and Control Conference*, (Minneapolis, Minnesota) AIAA 2012-4897 11 p

[4] Luber W 2012 Aeroservoelastic Flight Control Design for a Military Combat Aircraft Weapon System. *28th International Congress of the Aeronautical Sciences (ICAS),* (Brisbane, Australia) 12 p

[5] Nalci M O, Kayran A 2014 Aeroservoelastic Modeling and Analysis of a Missile Control Surface with a Nonlinear Electromechanical Actuator. *AIAA Atmospheric Flight Mechanics Conference, AIAA AVIATION Forum*, (Atlanta, Georgia). AIAA 2014–2055 28 p. DOI: 10.2514/6.2014-2055

[6] Bykov A V, Kondrashev G V, Parafes' S G, Turkin I K 2016 Methods for investigating the unmanned aerial vehicle electric actuator performance in aeroelasticity tasks. *Russian Aeronautics* **59(3)** 331 [in Russian]

[7] Stanford B 2018 Aeroservoelastic Optimization under Stochastic Gust Constraints. *Applied Aerodynamics Conference, AIAA AVIATION Forum,* (Hyatt Regency, Atlanta), 17 p. DOI: 10.2514/6.2018-2837

[8] Stanford B K 2019 Gradient-based aeroservoelastic optimization with static output feedback. *Journal of Guidance, Control, and Dynamics* **42(10)** 2314 https://doi.org/10.2514/1.G004373

[9] Broughton-Venner J, Palacios R and Wynn A 2017 Aeroservoelastic optimisation of aerofoils with compliant flaps via reparameterization and variable selection. *AIAA Journal* **56(3)** 1· DOI: 10.2514/1.J056141

[10] Moulin B, Idan M and Karpel M 200) Aeroservoelastic structural and control optimization using robust design schemes. *Journal of Guidance, Control, and Dynamics* **25(1)** 152

AUTHOR INDEX